Microalgae for Carbon Sequestration and Wastewater Valorization:

Pathways to a Sustainable Bioeconomy

Imran Ahmad
https://orcid.org/0000-0001-9774-9111
Durban University of Technology, South Africa

Faizal Bux
Durban University of Technology, South Africa

Nilamjyoti Kalita
https://orcid.org/0009-0007-9115-2980
Murkong Selek College, Jonai, India

IGI Global Scientific Publishing
Publishing Tomorrow's Research Today

Vice President of Editorial	Melissa Wagner
Director of Acquisitions	Mikaela Felty
Director of Book Development	Jocelynn Hessler
Production Manager	Mike Brehm
Cover Design	Jose Rosado

Published in the United States of America by
IGI Global Scientific Publishing
701 East Chocolate Avenue
Hershey, PA, 17033, USA
Tel: 717-533-8845 | Fax: 717-533-7115
Website: https://www.igi-global.com E-mail: cust@igi-global.com

Library of Congress Cataloging-in-Publication Data

Names: Ahmad, Imran, 1982- editor | Bux, F. (Faizel) editor | Kalita, Nilamjyoti, 1996- editor
Title: Microalgae for carbon sequestration and wastewater valorization : pathways to a sustainable bioeconomy / edited by Imran Ahmad, Faizal Bux, Nilamjyoti Kalita.
Description: Hershey, PA : IGI Global Scientific Publishing, [2027] | Includes bibliographical references and index. | Summary: "This book aims to advance understanding of how microalgal systems can simultaneously address global carbon emissions"-- Provided by publisher.

Identifiers: LCCN 2026007420 (print) | LCCN 2026007421 (ebook) | ISBN 9798337376301 hardcover | ISBN 9798337376318 paperback | ISBN 9798337376325 ebook
Subjects: LCSH: Sewage--Purification--Biological treatment | Microalgae--Industrial applications | Microalgae--Biotechnology | Carbon sequestration
Classification: LCC TD755 .M47 2026 (print) | LCC TD755 (ebook)
LC record available at https://lccn.loc.gov/2026007420
LC ebook record available at https://lccn.loc.gov/2026007421

British Cataloguing in Publication Data
A Cataloguing in Publication record for this book is available from the British Library.

Table of Contents

Detailed Table of Contents

Microalgae are highly diverse photosynthetic microorganisms found in aquatic and terrestrial ecosystems. Their evolutionary diversity provides exceptional physiological flexibility, allowing adaptation to varying light, nutrient, and stress conditions. This chapter reviews major microalgal groups, emphasizing their structural, biochemical, and functional traits. Key growth processes including photosynthesis, nutrient assimilation, pigment composition, and stress responses are discussed in relation to ecological roles. Particular attention is given to carbon fixation via the Calvin–Benson–Bassham cycle, CO_2-concentrating mechanisms, and alternative carbon acquisition pathways that enhance carbon capture efficiency. Molecular regulation, metabolic plasticity, and biotechnological applications in carbon sequestration and bioresource production are also highlighted.

The study of Biological CO_2 Sequestration Mechanisms in Microalgal systems shows that the continuous rise of atmospheric carbon dioxide (CO_2) during the past several decades has become one of the most critical environmental problems which the world currently endeavors to solve. . The result has been rising global temperatures which now cause unpredictable weather patterns and ice cap melting and sea level rise and severe ecological damage. The existing carbon seizing and storage methods which include chemical absorption and geological storage and physical scrubbing systems face multiple operational restrictions. Microalgae represent one of the top biological systems which scientist study for their potential to store carbon over extended periods. Microalgae represent tiny photosynthetic organisms which scientists discover in both marine and freshwater ecosystems, which they use because of their exceptional ability to absorb CO_2 and turn it into organic matter at a rapid pace.

Salma Hameed, Independent Researcher, Pakistan
Muhammad Saqlain, University of Jhang, Pakistan
Nadia Tehseen, Independent Researcher, Pakistan
Qudrat Ullah, Independent Researcher, Thailand
Tauqeer Ahmed Qadri, Independent Researcher, Pakistan
Rashid Abbas Khan, University of South Africa, South Africa

Microalgae is a disruptive technology for Carbon Capture and Utilization in a circular bioeconomy, showing volumetric productivity and CO_2 fixation rates much higher than terrestrial plants. These organisms act as efficient biological carbon sinks due to specialized Carbon Concentrating Mechanisms that enhance inorganic carbon concentration at RuBisCO to maximize the Calvin–Benson cycle. Reported sequestration rates among microalgal strains range from 0.06 to 2.57 g L^{-1} d^{-1}, with Haematococcus sp. at the upper end. Genetic and metabolic engineering improved performance, Chlamydomonas reinhardtii manipulation increased carbon fixation by 35%, while transgenic Chlorella sorokiniana achieved 88.5% carbon capture. Coupling microalgal systems with wastewater treatment enables nutrient recovery, removing up to 93% nitrates and 83% phosphates, yielding 3.21 g L^{-1} biomass. Optimal growth occurs at ~2000 lux and up to 35 °C. Although 5 vol% CO_2 enhances biomass, 9 vol% can cause culture collapse. High PBR capital costs and energy-intensive harvesting remain major commercialization barriers.

Rahul Kumar, Department of Microbiology, Graphic Era (Deemed to be) University, Dehradun, India

Microalgae are promising biotechnological tools for carbon sequestration and wastewater valorization, with the dual benefit of greenhouse gas emission mitigation and nutrient recovery from waste streams. However, the natural metabolic efficiency of microalgae for carbon fixation and nutrient uptake is suboptimal for large-scale applications. Recent technological breakthroughs in genetic engineering and synthetic biology have enabled the selective manipulation of photosynthetic, metabolic, and regulatory networks to improve the efficiency of carbon and nutrient capture.

frameworks are explored, including their functionality in monitoring and control. Data-driven approaches for optimizing carbon fixation, biomass productivity, and metabolite synthesis are presented with case studies in CO_2 sequestration, wastewater treatment, and biofuel generation. AI enables pattern recognition, IoT provides real-time monitoring, and digital twins serve as virtual replicas for optimization. Critical challenges including data standardization, costs, and cybersecurity are addressed. Future prospects include autonomous biorefineries.

Global environmental challenges, necessitate innovative resource recovery strategies. This chapter explores the "stress-induced biorefinery" paradigm, which leverages the heterogeneous nature of wastewater as a tunable metabolic signal for microalgae. By manipulating environmental stressors such as nutrient limitation, light intensity, and salinity, microalgal metabolism can be "programmed" to accumulate high-value compounds, including triacylglycerols for biofuels, carotenoids for nutraceuticals, and carbohydrates for biopolymers. Integrated systems utilizing algal-bacterial consortia offer energy-efficient treatment by replacing mechanical aeration with photosynthetic oxygenation. Furthermore, the coupling of nutrient recovery with carbon sequestration provides a pathway for industrial carbon management. This chapter synthesizes biological mechanisms, engineering strategies for diverse effluent streams, and cascading valorization routes to transform wastewater treatment from a cost-intensive disposal process into a sustainable, value-generating manufacturing platform.

Anaerobic reactors and algal sludge conversion have become potential technologies of sustainable bioenergy generation and waste management. Anaerobic digestion can be defined as a biological process where the microorganisms decompose the organic matter to produce biogas with high levels of methane and carbon- dioxide. The algal sludge is formed as a result of wastewater treatment, as well as the harvesting of algal biomass, which is a useful substrate because it contains a large amount of organic and nutrient matters. The process is improved and stabilized by the next

to enhance carbon capture are discussed, highlighting their ecological and industrial significance.

Diatoms produce intricate silica-based cell walls (frustules) with nanoscale precision, representing millions of years of evolutionary optimization in biomineralization. These structures offer unprecedented opportunities for developing functional nanomaterials with applications spanning drug delivery, biosensors, photonics, catalysis, and environmental remediation. This chapter will explore the emerging field of algal nanotechnology, with particular focus on diatoms as natural nanofabrication platforms.

The reason is that sustainable remediation methods are necessary since the water pollution caused by the heavy metals, organic pollutants, and industrial effluent entails severe risks to both the environment and human health. Microalgae is a green platform to produce Nanoparticles (NPs) because it is fast growing, biocompatible, and

million tons of CO_2 and produce more than 3 million tons of biomass. Recent growth technologies comprise photobioreactors, hybrid systems, artificial carbon concentration systems, and genetic and metabolic engineering to enhance the rate of fixation. Carbon credit regimes, technological-economic analyses, and policy frameworks for enabling quantifiable and verifiable algal systems in carbon markets are also studied. Overall, the innovation of algal carbon sequestration blocks scientific, economic, and environmental policy progress to global net-zero carbon emissions and a circular bioeconomy.

Preface

Microalgae have emerged as one of the most promising biological resources for addressing some of the world's most pressing environmental challenges. Their remarkable ability to capture atmospheric carbon dioxide, recover nutrients from wastewater, and generate valuable bioproducts positions them at the forefront of sustainable technologies that support the transition toward a circular bioeconomy. As nations seek innovative strategies to mitigate climate change while enhancing resource efficiency, microalgal systems offer an integrated approach that combines environmental remediation with renewable energy production, industrial biotechnology, and sustainable metabolite production.

Microalgae for Carbon Sequestration and Wastewater Valorization: Pathways to a Sustainable Bioeconomy brings together contributions from international experts to provide a comprehensive examination of the biological, technological, and policy dimensions of microalgal research. The volume explores the diversity and physiology of microalgae, advanced carbon-fixation mechanisms, genetic engineering, multi-omics technologies, artificial intelligence, digital twins, wastewater valorization, bioenergy production, nanotechnology, sustainable construction materials, and carbon policy frameworks. Together, these topics illustrate the expanding role of microalgae as a platform for environmental remediation and sustainability.

Beyond presenting recent scientific advances, the book highlights the interdisciplinary nature of this rapidly evolving field. Integrating microbiology, biotechnology, environmental engineering, computational technologies, materials science, and public policy, the chapters emphasize how collaborative approaches can accelerate the development and commercialization of scalable microalgal systems. They also address the technical and economic challenges that remain, including cultivation efficiency, harvesting costs, process optimization, regulatory considerations, and market implementation.

Designed for researchers, environmental scientists, engineers, policymakers, graduate students, and industry professionals, this volume serves as both a reference and a roadmap for harnessing the full potential of microalgae. By connecting fundamental science with practical applications, the book demonstrates how microalgal technologies can contribute significantly to carbon neutrality, wastewater treatment, renewable energy production, and sustainable economic development.

CHAPTER OVERVIEW

Chapter 1: Introduces the diversity, physiology, and carbon fixation pathways of microalgae. It examines their evolutionary adaptations, photosynthetic mechanisms, nutrient assimilation processes, and carbon-concentrating mechanisms while highlighting their significance in carbon sequestration and bioresource production.

Chapter 2: Explores the biological mechanisms through which microalgae capture and sequester atmospheric carbon dioxide. The chapter discusses photosynthetic carbon fixation, compares biological approaches with conventional carbon capture technologies, and evaluates the environmental benefits of microalgal systems for long-term carbon storage.

Chapter 3: Examines advanced biological carbon sequestration in microalgal systems with an emphasis on carbon-concentrating mechanisms, genetic engineering, and wastewater integration. It also discusses the challenges encountered in commercialization and strategies to improve carbon capture efficiency and nutrient recovery.

Chapter 4: Focuses on genetic engineering and synthetic biology techniques for enhancing carbon fixation and nutrient uptake. The chapter reviews genome editing technologies, metabolic pathway optimization, synthetic promoter design, and systems biology approaches for developing high-performance microalgal strains.

Chapter 5: Presents multiomics-driven strategies for improving photosynthetic efficiency, carbon capture, and nutrient utilization. Integrating genomics, metabolomics, and fluxomics, the chapter demonstrates how systems biology can optimize strain development and reactor-scale deployment for circular bioeconomy applications.

Chapter 6: Explores the integration of artificial intelligence, the Internet of Things (IoT), and digital twins into microalgal bioprocesses. It highlights intelligent monitoring, predictive analytics, virtual modeling, and autonomous optimization to enhance carbon sequestration, wastewater treatment, and biomass production.

Chapter 7: Examines stress-induced biorefinery approaches that utilize environmental stressors to enhance microalgal productivity and wastewater valorization. The chapter discusses algal-bacterial consortia, nutrient recovery, carbon management, and cascading valorization pathways that transform wastewater into valuable bio-based products.

Chapter 8: Discusses anaerobic reactors and algal sludge conversion as sustainable technologies for bioenergy generation. It reviews anaerobic digestion processes, advanced reactor configurations, operational factors, and their contributions to greenhouse gas reduction and circular bioeconomy objectives.

Chapter 9: Investigates the integration of biohydrogen production with carbon capture technologies. The chapter examines recent advances in biological hydrogen generation, process intensification, reactor design, and the potential of combined carbon capture and renewable hydrogen production to support the low-carbon energy transition.

Chapter 10: Highlights the ecological importance of diatoms as major contributors to global carbon sequestration. It explores their physiological characteristics, silica biomineralization, photosynthetic efficiency, and emerging applications in biofuels, biosensors, nanotechnology, and carbon-neutral materials.

Chapter 11: Explores the emerging field of algal nanotechnology, focusing on diatom biosilica as a naturally engineered nanomaterial. The chapter discusses applications in drug delivery, biosensors, catalysis, photonics, and environmental remediation.

Chapter 12: Examines the synthesis and application of microalgal nanoparticles for wastewater treatment. It discusses green nanoparticle production, pollutant-removal mechanisms, scalability challenges, and the integration of artificial intelligence to optimize sustainable nanotechnology processes.

Chapter 13: Presents innovative approaches for integrating microalgal carbon sequestration with engineered cementitious materials. The chapter explores microbial biomineralization, self-healing concrete technologies, and carbon-negative construction materials that combine structural performance with environmental sustainability.

Chapter 14: Concludes the volume by examining carbon credit systems, policy frameworks, and economic perspectives supporting algal carbon sequestration. It evaluates technological advances, market opportunities, and regulatory strategies that can facilitate the large-scale adoption of microalgal technologies in achieving global net-zero emissions and circular bioeconomy goals.

The chapters presented in this volume collectively demonstrate that microalgae represent one of the most versatile and promising biological platforms for achieving sustainable environmental and industrial development. Through advances in carbon sequestration, wastewater treatment, bioenergy production, synthetic biology, artificial intelligence, nanotechnology, and sustainable materials, the book illustrates how microalgal systems can simultaneously address climate change mitigation, resource recovery, and economic growth.

While significant scientific and technological progress has been achieved, widespread implementation will require continued innovation in cultivation technologies, genetic improvement, process optimization, policy development, and economic scalability. Equally important will be interdisciplinary collaboration among researchers, engineers, industry stakeholders, and policymakers to translate laboratory discoveries into commercially viable and environmentally impactful solutions.

By presenting both foundational knowledge and emerging innovations, *Microalgae for Carbon Sequestration and Wastewater Valorization: Pathways to a Sustainable Bioeconomy* provides a comprehensive resource for advancing research, informing policy, and inspiring future developments. As global efforts intensify to achieve carbon neutrality and sustainable resource management, the technologies and strategies explored in this volume offer practical pathways to build a resilient, low-carbon, and circular bioeconomy.

Imran Ahmad
Durban University of Technology, South Africa

Faizal Bux
Durban University of Technology, South Africa

Nilamjyoti Kalita
Murkong Selek College, Jonai, India

Chapter 1
Diversity, Physiology, and Carbon Fixation Pathways of Microalgae

Muhammad Nauman Khan
http://orcid.org/0000-0003-3790-0902
University of Chakwal, Punjab, Pakistan

Hansa Gul
http://orcid.org/0009-0001-5554-617X
University of Mianwali, Punjab, Pakistan

Alevcan Kaplan
http://orcid.org/0000-0001-6738-7527
Dicle University, Diyarbakır, Turkey

ABSTRACT

Microalgae are highly diverse photosynthetic microorganisms found in aquatic and terrestrial ecosystems. Their evolutionary diversity provides exceptional physiological flexibility, allowing adaptation to varying light, nutrient, and stress conditions. This chapter reviews major microalgal groups, emphasizing their structural, biochemical, and functional traits. Key growth processes including photosynthesis, nutrient assimilation, pigment composition, and stress responses are discussed in relation to ecological roles. Particular attention is given to carbon fixation via the Calvin–Benson–Bassham cycle, CO_2-concentrating mechanisms, and alternative carbon acquisition pathways that enhance carbon capture efficiency. Molecular regulation, metabolic plasticity, and biotechnological applications in carbon sequestration and bioresource production are also highlighted.

DOI: 10.4018/979-8-3373-7630-1.ch001

1. INTRODUCTION

Microalgae are an enormous and critical group of unicellular photosynthetic organisms that play a key role in freshwater and marine ecosystems. Their high growth potential is based on the ability of *Chlorella* to take up CO_2 and necessary nutrients (i.e., phosphorus, nitrogen) directly from an aquatic environment in a highly efficient manner (Sukačová et al., 2020). Their contribution to the carbon cycle is enormous on a planetary level. Marine microalgae, particularly phytoplankton, contribute to roughly 50 gigatons (Gt) of CO_2 fixation annually and lead half of the world's biologic carbon assimilation. Just diatoms contribute to 20% of the total CO_2 sequestration worldwide, which is similar to that achieved by all the rainforests on earth combined (Hoque et al., 2025). However, while their innate performance is 10–50 folds greater than terrestrial plants, the transition from laboratory success to industrial-scale engineering remains a significant bottleneck due to high operational costs and mass transfer limitations (Hoque et al., 2025; Zhou et al., 2017). Beyond providing a descriptive overview of the physiological compromises and economic assessments to enable microalgal systems to become a cornerstone of the green economy, this chapter also critically evaluates them.

Their biotechnological potential, in addition to their ecological relevance, have placed highly interesting microalgae as a suitable and promising partner for the green economy. Microalga-based systems provide an efficient, sustainable alternative to the classic energy-intensive greenhouse gas-producing methods of wastewater treatment (Sun et al., 2025). This method, termed phycoremediation, cleanses water while also turning pollutant into a useful biomass source. The end use of this biomass as a biofertilizer or biofuel feed stock creates an example of circular economy by increasing the recycling of the valuable nutrients extracted from wastewaters into agricultural and energy system (Das et al., 2019; Khan et al., 2019). Some species e.g., Chlorella vulgaris, have shown sequestration rates of 1.6–2 tonnes CO_2 per ton biomass produced and in optimized photobioreactors some systems can reach a CO_2 capture efficiency up to 93.7% (Hoque et al., 2025). In this chapter, we will discuss the amazing physiological and metabolic adjustments that render these applications feasible, particularly emphasizing their unusual carbon assimilating mode.

2. ADVANCED PHYSIOLOGICAL ADAPTATIONS TO NUTRIENT STRESS

Microalgae are physiologically highly adaptable and can survive even in environments with very low levels of the basic nutrients. Among them, phosphorus (P) is a key element that controls cellular growth and metabolism, especially lipid synthesis. On the one hand it is a sophisticated mechanism of survival, but on the other hand it is a challenge from the biotechnological point of view because the energy for membrane preservation may be taken at the expense of the total biomass productivity. There are current gaps in research that need to be filled, such as elucidating how to activate these high-value lipid pathways without causing growth arrest as would nutrient starvation. Several studies have been shedding light on the close relationship between nutrient availability and lipid metabolism in microalgae (Yang et al., 2024). In P-limited conditions, metabolic changes occur in microalgae in order to maximize nutrient retention and cell tolerance levels.

Remodeling of membrane lipids should be one of the most interesting responses to Pi deficiency. When P is limiting, the cells of the microalgae switch their strategy from unnecessary investment in phospholipid waste products like phosphatidylcholine (PC), phosphatidylethanolamine (PE) and phosphatidylglycerol (PG) to focusing on other lipids in the membrane. Such replacements are mainly glycolipids, particularly sulfoquinovosyldiacylglycerol (SQDG) and betaine lipids e.g., diacylglyceroltrimethylhomoserine (DGTS). This remodeling of membranes is manifest as an efflux of phosphorus from those structures and a redirection of the atoms towards essential cellular activities such as nucleic acid synthesis, energy generation and regulation signalling.

The phosphorus-induced lipid remodeling is evolutionarily conserved across photosynthetic organisms, which could serve as a model to study low P adaptation in higher plants and offers an application view of P recovery from wastewater for sustainable agriculture (Sun et al., 2025). In addition, there is a phenomenon that occurs in the growth of microalgae, called 'Luxury Uptake,' where the microalgae take up and store more phosphorus than is required for its immediate metabolic rate, in the form of intracellular polyphosphate granules. Microalgae have a physiological characteristic that makes it possible for them to remove as much as 90% of phosphorus from secondary effluents of wastewater.The physiological property of the microalgae is that they can remove up to 90% of the P from the secondary effluents of wastewater, compared to many other conventional chemical precipitation methods.

3. CHALLENGES IN CULTIVATION

A major hurdle is the efficient supply of carbon for the cultivation of microalgae to obtain carbon capture and biomass production. There are two main ways of providing carbon to microalgal cultures: gaseous CO_2 and bicarbonate. The different methods each have their own advantages and drawbacks, which affect the efficiency and cost-effectiveness of microalgal cultivation systems. This analysis is based on technical performance, operational needs and implications of large-scale application of these two approaches.

3.1 Gaseous CO_2 Method

Although microalgal cultivation using gaseous CO_2 has several challenges to ensure minimal efficiency and scale, it is a common practice. One of the disadvantages of using CGY as a biodegradable polymer is its poor water solubility, which results in significant CO_2 emission during cultivation, reducing the carbon sequestration and increasing the operational cost (Yang et al., 2024). Furthermore, capture, compression and transportation of gaseous CO_2 is energy consuming and costly which makes it impractical for large-scale application. Unlike the microalgae, however, it cannot be taken up during the dark period and therefore, continuous supply of industrial CO_2 is not assimilated and efficiency is decreased. Although continuous CO_2 sparging can be helpful for pH control, this approach is not able to be scaled up in industry without adding an additional energy cost. Overall, these restrictions demonstrate the need for other approaches for carbon supplementation or engineered solutions for more effective CO_2-based microalgae growing.

3.2 Bicarbonate Method

Addition of bicarbonate is a much more effective alternative to gaseous CO_2 for microalgal cultivation. It has far higher CO_2 fixation efficiency with lower energy requirement for its process, where bicarbonate-based process showed much higher net energy ratio (NER ≈ 7.29) in comparison to conventional CO_2 bubbling (NER ≈ 0.85) (Nugroho et al., 2025). It is more economical to transport a solution or solid bicarbonate than to extract and compress CO_2 and to regenerate further bicarbonate from CO_2 in the bicarbonate solution.

Bicarbonate also serves as a carbon source and being a natural buffer, creates an even growth environment without loss in biomass and lipid production. Furthermore, it is reported that bicarbonate supplementation can enhance the growth rate of microalgae, and their lipid productivity, especially beneficial for biofuel production and other value added bioproducts.

To conclude, bicarbonate proves to be an economic, sustainable and practical approach to control the culture of large-scale microalgae in an indoor setting. Bicarbonate-based systems are in general more efficient, energy-saving, and environmentally friendly than gaseous CO_2, as they can minimize the demand for a continuous carbon supply and have potential for better integration into industrial flue gases (Nugroho et al., 2025) (Yang et al., 2024). Their efficacy varies among microalgal species and culture conditions; thus, strain selection is essential. For example, *Chlorella vulgaris* is a good producer of lipids while growing with both methods the alkaliphilic species such as *Spirulina platensis* shows better results when using bicarbonate owing to pH tolerance (Nugroho et al., 2025). Technical, operational and environmental comparison of CO_2 and bicarbonate-based strategies (Nugroho et al., 2025) Table 1.

Table 1. Comparative Evaluation of Carbon Supply Strategies for Microalgal Cultivation

	Aspect / Parameter	CO_2-Based System (Gas Bubbling)	Bicarbonate-Based System
Technical Performance	Carbon Fixation Efficiency	Typically very low, often <10% due to limited gas–liquid mass transfer.	Can approach 100% when bicarbonate concentration is adequately supplied.
	Operational Stability & pH Control	Carbon availability fluctuates; continuous bubbling leads to acidification, requiring strict pH regulation.	Offers stable and predictable carbon supply; bicarbonate acts as an inherent pH buffer.
	Biomass Quality & Density	Often achieves higher peak cell densities, facilitating downstream harvesting.	Higher biomass productivity is common, though peak cell density may be lower than optimized CO_2 systems.
	System Complexity	Simple designs are inexpensive, but efficiency improvement requires complex and costly reactor arrangements.	Cultivation system is simpler, but depends on an additional centralized CO_2 absorption unit.
Economic Viability	Net Energy Ratio (NER)	Low NER (≈0.85), indicating poor energy feasibility.	Very high NER (≈7.79), demonstrating strong energetic and economic performance.
	Specific Energy Consumption	High energy demand (~7.29 kWh kg^{-1} dry biomass) due to gas compression and delivery.	Very low energy requirement (~0.79 kWh kg^{-1} dry biomass) because liquid transport is more efficient.
	Investment Cost (CAPEX)	Low for basic systems, but rises sharply for high-efficiency configurations.	Higher initial cost due to CO_2 absorption and chemical handling infrastructure.
	Operational Cost (OPEX)	Elevated long-term costs driven by high energy use for gas handling.	Substantially reduced energy costs (up to 80–90% lower); alkaline chemical costs can be mitigated via recycling.
Environmental Impact	Fugitive Emissions	High loss potential; >90% of unabsorbed CO_2 may escape to the atmosphere.	Low emission risk; CO_2 release is confined to a single absorption unit, simplifying monitoring and control.
	Algal Species Suitability	Best suited for robust, CO_2-tolerant species (e.g., *Chlorella vulgaris, Dunaliella salina*).	Particularly suitable for alkaliphilic species (e.g., *Spirulina platensis*); also compatible with *Chlorella vulgaris*.
	Resource Consumption	No chemical inputs required unless buffering is needed to offset acidification.	Requires continuous supply of alkaline reagents (e.g., NaOH) for CO_2 capture

In addition, large-scale cultivation requires substantial water resources and continuous monitoring, which increases operational costs and energy consumption. Harvesting and downstream processing of microalgal biomass remain technically demanding and expensive due to the small cell size and dilute culture conditions. Furthermore, the scalability of laboratory-optimized cultivation systems to industrial levels is still a major obstacle, limiting the commercial feasibility of microalgae-based biofertilizers and bioproducts.

3.3 Hybrid Biofilm Systems

To overcome the harvesting bottleneck, recent advancements have introduced Hybrid Systems (HS) that combine suspended growth (HRAPs) with Biofilm Reactors (BR). In these systems, biomass grows attached to surfaces (such as polyester), allowing for manual or mechanical scraping. This method has been shown to increase biomass recovery efficiency from 22% in traditional ponds to over 61%, while reducing environmental impacts—such as global warming potential and fossil depletion—by up to 40% (Calijuri et al., 2025).

Hybrid biofilm technology combining the high productivity of floating systems with ease of harvesting attached biofilm has emerged as one of the most promising technologies for sustainable large-scale microalgal culture. Harvesting microalgae in standard open pond systems is very energy-intensive because algal suspensions are often dilute, requiring centrifugation, flocculation or filtration to extract algae. In contrast, there is a method of cultivating the biomass directly onto the supporting materials (via biofilm), which decreases water handling and energy requirements during downstream processing.

Different media like nylon fibres, stainless steel meshes, rotating drums and polymeric membranes have been tried to improve the formation of the biofilm and nutrient transfer. Research shows that algal biofilm reactors can yield biomass productivities over 20-30 g m-2 day-1, significantly higher than many raceway ponds. Moreover, the efficiency of light utilization is enhanced by hybrid systems since the efficiency of the attached cells is better and they do not suffer from self-shading effects. Greater efficiency of CO_2 transfer and nutrient uptake have also been reported, which renders these systems applicable to simultaneous wastewater treatment and carbon sequestration systems.

Another significant benefit of the hybrid biofilm systems is that they are stable during operational changes in the environment. Biofilms are more tolerant of contamination, grazing organisms, and hydraulic washout, than are suspended cultures. Moreover, these systems need a lower amount of chemical input in harvesting reducing operational costs and secondary pollution. In the recent life-cycle assessment studies, it has been proven that hybrid biofilm technologies can significantly lower the energy demand, freshwater usage, and greenhouse gas emissions, which align them well for the bioresource production in the circular bioeconomy (Assis et al., 2023).

4. ADVANCEMENTS IN OMICS AND GENETIC ENGINEERING

Recent work using "Omics" (genomics, transcriptomics, and metabolomics) has developed a blueprint that will provide the next generation of "super-strains" to address the performance gaps in wild-type microalgae. In the future, scientists are planning to use a gene editing technique known as CRISPR/Cas9 to enhance the efficiency of Rubisco by targeting the Carbon Concentrating Mechanism (CCM) (Kupriyanova et al., 2023). Recent efforts are targeting the *DGAT* (Diacylglycerol acyltransferase) and *TGD* (Trigalactosyldiacylglycerol) genes. By overexpressing these specific lipid-flux regulators, researchers have achieved a two-fold increase in triacylglycerol (TAG) accumulation without the traditional trade-off of reduced biomass growth (Cai et al., 2025). In addition, metabolic engineering is being employed to "rewire" lipid pathways so that cells can store a high level of oil at the stress-free active growth stage rather than when stressed. A key challenge to the conversion of microalgae from simple biological units into high performance catalysts in industry will be the development of these molecular tools. Beyond genetic engineering, 'Spectral Conversion' is emerging as a novel tool. The application of Quantum Dots (QDs) can convert unusable ultraviolet (UV) light into visible wavelengths that match the absorption spectra of microalgal pigments. This physical manipulation optimizes the light-harvesting complex (LHC) efficiency, potentially bypassing the photosynthetic saturation limits of wild-type strains (Udaypal et al., 2024).

4.1 Technological Innovations in Biomass Processing and Extraction

Suitable cell disruption and extraction methods are required to convert microalgae waste-stream biomass to high value biofertilizer. Microalgae possess a tough and robust cell wall, so mechanical or physical disruption is typically used to liberate bioactive compounds from the intracellular fluid (Nur et al., 2025).

Supercritical CO_2 extraction is one of the most promising emerging technologies. Bioactive compounds that are sensitive to heat (thermolabile) can be retained through the use of CO_2 as a solvent at specific temperatures (usually 50°C) and pressures (200-500 bars) because they would otherwise be destroyed by conventional heat drying (Ronga et al., 2019). In addition, enzymatic methods are used that involve the use of proteolytic enzymes to produce protein hydrolysates. The hydrolysates contain a high concentration of free amino acids and soluble peptides, directly involved with the synthesis of chlorophyll and defense mechanisms in plants. Microalgal biomass can be used in three main forms: (1) liquid extracts from microalgae for foliar application to deliver hormones instantly; (2) dried microalgal biomass as a slow-release soil amendment; and (3) thermochemical conversion of microalgal biomass to produce biochar, which is a carbon-rich product, enhances soil aeration and provides a long-term carbon sink(Wu, Tham, Chew, Munawaroh, Tan, Wan-Mohtar, et al., 2023).

5. THE ROLE OF MICROALGAE IN INTEGRATED NUTRIENT RECYCLING AND SUSTAINABLE AGRICULTURE

5.1 Nutrient Recovery and Wastewater Purification

Microalgae are very effective in assimilating nutrients and an effective means of bioremediation. They are used for nutrient-rich wastewater treatment, and they provide an environmentally friendly technology to the traditional methods. Types of algae such as Chlorella vulgaris can accumulate excessive amounts of phosphorus and nitrogen, thereby reducing the eutrophication caused by industrial and agricultural drainage. Production in waste streams, whereby pollutants are converted into concentrated biomass he allowed to purify water and the beginning of nutrient recycling (Schreiber et al., 2018).

Balancing nutrients with CO_2 injection is a common practice, however there has been a rise in the practice of 'Effluent Balancing' as a more sustainable option. For instance, the industrial wastewater (e.g. from paint booths, brewery) containing low nutrient content but high carbon content can be combined with domestic sewage to give ideal C/N/P ratio (e.g. 62:35:1), which can lead to better removal of pollutants and to a stimulation of the production of carotenoids without the use of synthetic gas (Calijuri et al., 2025).

5.2 Microalgae as Bio fertilizer and Soil Conditioner

Reprocessed microalgal biomass can be an excellent biofertilizer, thereby providing a means of linking supply and demand in terms of critical elements. Application of dried biomass improves soil organic carbon and nutrient availability, whereas processed to liquid extracts or hydrolysates can be utilized as foliar sprays and seed primers. Residual biomass, which is available after the extraction of biofuel, also works as a potential soil conditioner and contributes to waste valorization, crop yielding and sustainable agriculture (Zhang et al., 2023).

Algal biofertilizers act as slow-release nutrients because the phosphorus in polyphosphate molecules is released gradually by way of microbial degradation. This enhances nutrient utilization efficiency, minimizes leaching and provides macronutrients, micronutrients as well as bioactive ingredients like phytohormones and polysaccharides to promote plant grow the, roots formation and overall crop yield (Wang & Zhang, 2022).

5.3 Bio stimulant and Bio control Effects

Apart from providing nutrients, microalgae also produce bioactive compounds that can serve as plant biostimulant and natural biocontrol agent. Regulation of nutrient cycle, recruitment dormancy and carotenoid and chlorophyll formation is mediated by GFPs induced by HRPs.-Other three kinds of DMSO-soluble protein are HRP-induced proteins. Some allelopathic compounds from microalgae are natural herbicides or biopesticides, it makes microalgal as a sustainable pest management tool (Hoque et al., 2025).

In addition, microalgae produce phytohormones like auxins, cytokinins, gibberellins and abscisic acid, which promote root growth, germination of seeds, flowering and plant growth. The metabolites increase the efficiency of nutrient assimilation and increase the resistance of plants to environmental stresses such as drought, salinity and temperature changes. Soil microbial activity will also be improved by the presence of microalgal polysaccharides and amino acids which will enhance soil fertility and rhizosphere health.

In addition to the above, the phenolic compounds, fatty acids, peptides, and pigments (phycocyanin and carotenoids) produced by microalgae also possess antimicrobial and antifungal properties. They inhibit the growth of harmful pest, bacteria and pathogenic fungi, thereby decreasing the reliance on chemical pesticides. Certain cyanobacteria can also trigger a plant systemic resistance that further enhances plant resistance to infection. Furthermore, the use of extracts from the microalgae has been proven to improve crop production, leaf chlorophyll content and photosynthetic efficiency in different crops cultivation systems. Microalgal biostimulants and biocontrol agents are gaining recognition as potential tools for environmentally-friendly agriculture and integrated pest management due to their biodegradability, low toxicity and renewability (Asimakis et al., 2022; Cichoński & Chrzanowski, 2022; Hassan et al., 2022).

5.4 Carbon Fixation and Photosynthetic Efficiency

Microalgae play a crucial role in global carbon cycles, fixing atmospheric CO_2 into organic biomass and producing approximately half of the Earth's oxygen. Each kilogram of algal biomass sequesters ~1.3 kg of CO_2, giving them 10–50× higher carbon adsorption capacity than terrestrial plants without competing for arable land (Ashour et al., 2024). This is facilitated by the Carbon Concentrating Mechanism (CCM), centered on the pyrenoid organelle, which enriches CO_2 around Rubisco to maximize photosynthetic efficiency (Kupriyanova et al., 2023) (Figure 1).

Figure 1. Conceptual illustration of carbon fixation by phytoplankton at global and cellular scales. (A) Global scale representation showing atmospheric CO_2 exchange with oceanic systems and the contribution of phytoplankton (microalgae) to approximately 50% of Earth's total carbon fixation. (B) Cellular scale depiction of a microalgal cell highlighting CO_2 uptake and concentration within the pyrenoid, where the enzyme Rubisco facilitates carbon fixation inside the chloroplast. Key cellular components including the cell wall, cell membrane, nucleus, and pyrenoid are indicated.

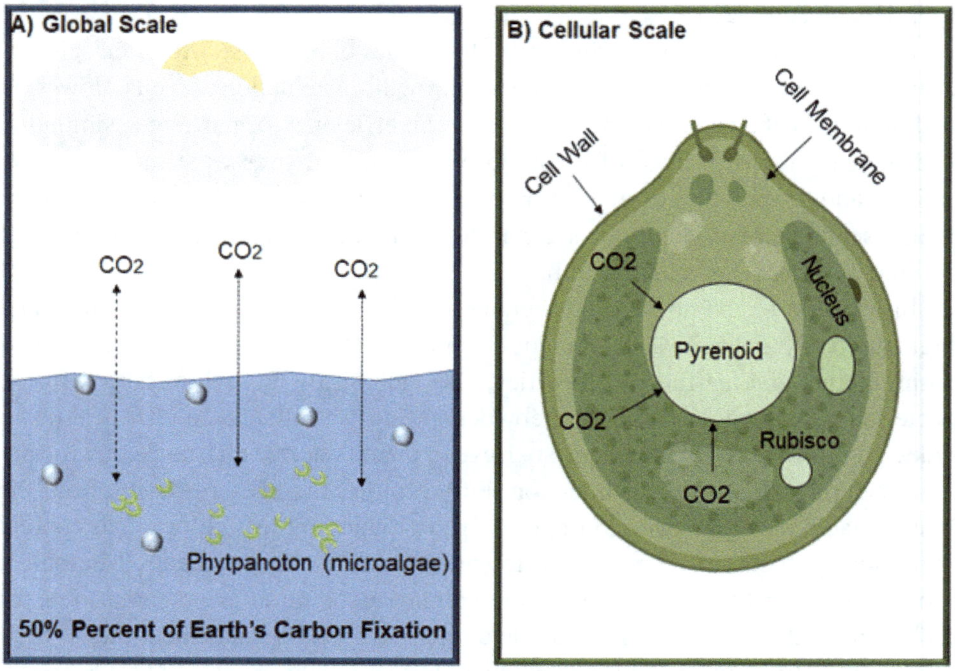

5.5 Sustainable Agriculture: Biofertilizers and Yield Metrics

Microalgal biomass has emerged as a promising and sustainable biofertilizer for modern agriculture due to its rich nutritional composition and environmentally friendly nature. The rising demand for sustainable farming has led to the adoption of biological solutions as alternatives to chemical fertilizers in the last few years. Of these options, the use of microalgae has received significant interest due to their possible addition to the soil as an amendment that contains essential nutrients, amino acids, vitamins, phytohormones, carbohydrates, proteins, and bioactive compounds which improve soil fertility and crop productivity. Use of microalgal biomass as soil conditioner enhances soil physical and biochemical properties, resulting in improved agricultural performance and sustainability (Panday et al., 2024; Penfold et al., 1995).

The ability of microalgae to effectively immobilize atmospheric nitrogen and to make phosphorus more available to plants plays an important role in the cycling of these nutrients. Species like Chlorella, Spirulina and Anabaena are preferred for biofertilizer formulation due to their plant growth promoting activity. The organic matter in the microalgal biomass helps to enhance soil structure, aeration, water holding capacity and microbial activity. These enhancements provide a positive and conducive environment for root growth and nutrient absorption. In addition, microalgal biomass can help to lower salinity in soils and help plants to withstand environmental stressors, thus improving their growth and development.

Microalgal biofertilizers have been found to have beneficial effects on seed germination rate, plant height, chlorophyll content, fruit quality, biomass accumulation, and yield of the crops. Research has shown that crops fed with microalgal preparations may show increased productivity compared with crops fertilized by conventional means. Furthermore, limited use of synthetic agrochemicals will decrease pollution in the environment and help to keep soil ecosystems healthy (Gonçalves et al., 2023).

The use of *Chlorella* and *Spirulina* as biofertilizers has been documented to increase rice yields by 20.9% and common bean yields by as much as 219–656 kg per hectare. Unlike synthetic urea, which can lose 40–60% of its nitrogen through leaching and volatilization, algal-based nutrients are "slow-release," locked in polyphosphate molecules that degrade over 30–60 days, matching the plant's growth cycle (Kabato et al., 2025).

5.5.1 Practical Applications in Crop Systems

Microalgae as fertilizers and biostimulants improve soil health and crop productivity. Notable examples include in table 2. By enhancing nutrient availability and microbial activity, these biological agents offer a sustainable alternative to synthetic chemical fertilizers, effectively promoting a circular agricultural economy.

Table 2. Types of different Microalgae as fertilizers

Microalgal Species / Treatment	Crop / System	Key Observed Benefits	Reference
Chlorella vulgaris, Spirulina platensis	Rice	Yield increase up to 20.9%	(Dineshkumar et al., 2018)
Mixed microalgae	Maize	Improved germination, early growth, yield traits	(Dineshkumar et al., 2019)
Monoraphidium sp.	Tomato	Shoot biomass ↑ 32%; chlorophyll-a ↑ 12%	(Jimenez et al., 2020)
Nostoc piscinale	Maize	Enhanced vegetative growth and grain yield	(Ördög et al., 2021)
Anabaena spp.	Wheat and cereals	Improved yield, N fixation, disease resistance	(Hakkoum et al., 2025)
Wastewater microalgae + compost	Wheat field	Microbial C ↑ 31.8–67%; N fertilizer ↓ 25%	(Renuka et al., 2016)
Chlorella vulgaris + PGPR	Common bean	Yield ↑ 219–656 kg ha^{-1}; profit ↑ 25.6%	(de Oliveira et al., 2025)
Nostoc, Anabaena	Cropping systems	N fixation 25–40 kg ha^{-1}; leaching ↓ to 7%	(Solomon et al., 2023)

5.6. Microalgae in Wastewater Treatment and Resource Recovery

Microalgae as an effective treatment agent for wastewater and resource recovery has become a central element with respect to circular integrated bioeconomy approaches. Genera such as *Chlorella* and *Scenedesmus* on the other hand, demonstrates high removal of inorganic nutrients (nitrate, phosphate and ammonium) from wastewater stream together with heavy metals (Cu, Zn, Pb, Cr, Cd), synthetic dyes and organic impurities at removal efficiency of 90–98% (Popa et al., 2025; Shabbirahmed et al., 2025). The pollutants are taken in by the algal biomass and thus water is treated as well as the valuable feed stock produced.

By the same token, microalgae can employ CO_2-enriched flue gases from industrial processes to promote growth, converting gaseous pollutants into useful biomass (Chen et al., 2022). As an instance, CO_2 assimilation rates of ~137 mg L^{-1} day^{-1}, lipid productivities close to 32% and biomass yields up to 1.3 g L^{-1} were reported for *Micractinium pusillum* when growing on flue gas streams (Singh Chauhan et al., 2022). This bivalent feature would allow co-treatment scenarios, which convert waste water nutrients and industrial gases into biomass, bolstering overall sustainability by namely valorization of resources (Hoque et al., 2025).

In the kind of system under consideration by SIMBRA, harvested biomass is converted to multiple biofuels and high-value products such as biodiesel, biogas, omega-3 fatty acids, proteins (produced in higher concentration than other currently popular photosynthetic autotrophic crops), phycobiliproteins, biofertilizer for different uses from nitrogen-rich arable land substrate to trace element decontaminants), bioplastics in a zero-waste context with principles for cascading valorization (Hoque et al., 2025). Some strains such as *Schizochytrium sp.* and *Nannochloropsis sp.*, which contain lipids that make up more than 45% of their DW and obtain a lipid accumulation of 49% alongside a fatty acid methyl ester (FAME) yield of up to 32% when grown on food-waste hydrolysates (Wu et al., 2025).

Commercial wastewater treatment with cyanobacteria and red microalgae, including *Nostoc sp., Arthrospira platensis* and *Porphyridium purpureum* has obtained removal efficiencies of up to 98% for COD, 94% for nitrogen, and 100% total phosphate removal as well as valuable pigments, such as phycocyanin at yields up to the high value proposition of 103 mg g^{-1}dry weight (Arashiro et al., 2020). Detritus, in particular residual algal biomass can be processed to biochar which supports higher soil fertility, improve water storage capacity and sequestrate carbon for longer period (Wu, Tham, Chew, Munawaroh, Tan, Wan, et al., 2023).

The concepts have been confirmed with pilot and demonstration projects such as the EU Horizon initiative WWTBP-by-Microalgae. holds potential for the production of phycocyanin, biogas and biochar from brewery effluents with ~90% nutrient and CO_2 recovery (valued at EUR 221 kg^{-1}) but also meeting desired fertilizer content reducing waste generation (Hoque et al., 2025). Pilot-scale units demonstrate >90% removal of nitrogen and phosphorus for municipal and industrial wastewater flows (Velásquez-Orta et al., 2024). Passive pond systems for warm regions: A hybrid system for low-energy, economic options to high-tech wastewater treatment (Sarma et al., 2024).

Cascading biorefinery schemes to extract lipids, proteins, pigments, and carbohydrates in series for resource recovery have also been proposed and would improve economic viability while minimizing waste (Malik et al., 2022; Wu et al., 2025). Although there are hurdles that need to be overcome in terms of harvesting, drying, extraction yields, regulatory framework and market infrastructure recent techno-economic studies show development. In the case of *Scenedesmus almeriensis* tubular photobioreactor in Almeria, Spain biomass production cost were reported as EUR 69 kg^{-1} DW, expected to fall to EUR 12–13 kg^{-1} at scale up of 200 t y^{-1} (Acién et al., 2012). Also in Italy, a 1-ha Green Wall Panel system growing *Tetraselmis suecica* had costs of EUR 12.4 kg−1, which were reduced to EUR 3.2 kg^{-1} in optimal conditions (Tredici et al., 2016). These figures highlight the necessity of Techno-Economic Analysis (TEA) and Life Cycle Assessment (LCA) in current research. While capital costs (CAPEX) remain high, the "environmental credit" from wastewater remediation and the cascading extraction of multiple products (lipids +

proteins + pigments) are the only pathways to making these systems competitive with petroleum-based alternatives (Figure 2).

Figure 2. Schematic overview of the multifunctional roles of microalgae in climate mitigation and sustainable development. The diagram illustrates key processes including microalgal physiological traits (rapid growth, high photosynthetic efficiency, carbon-concentrating mechanisms, and stress tolerance), carbon and nutrient uptake (CO_2 fixation, nutrient assimilation, heavy metal absorption), applications in cultivation and remediation systems (photobioreactors, wastewater treatment, brackish soil systems), and biomass generation (nutrient-rich biomass and bioactive compounds). The figure further highlights environmental and socio-economic benefits such as greenhouse gas reduction, air and water purification, soil improvement, land restoration, biofuels and bioenergy production, biorefineries, biodegradable materials, and contributions to a circular bioeconomy

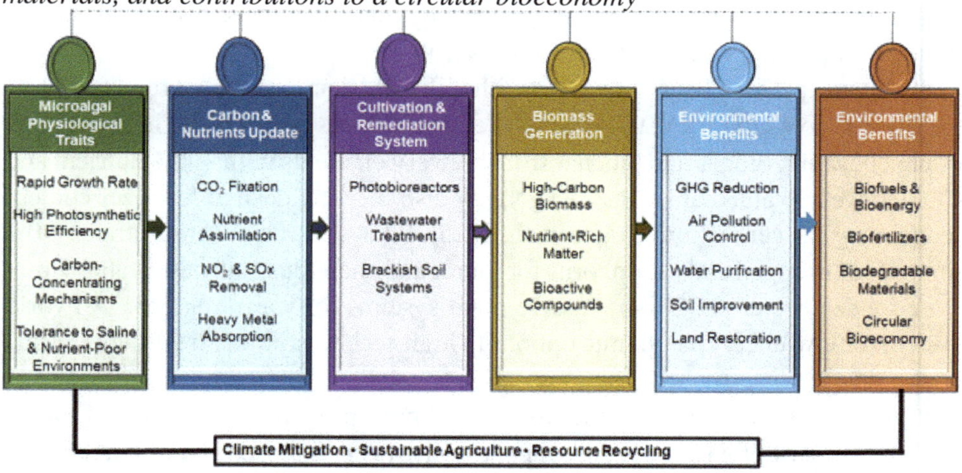

6. ROLE OF MICROALGAE IN THE CIRCULAR BIOECONOMY

In the emerging circular bioeconomy, for which microalgae as biological recycling units and CO_2 fixating organisms with high biomass production capacities are gaining interest, conversion of renewable resources and waste streams to valuable bioproducts is a key issue. Microalgal biotechnology can convert various forms of organic or inorganic solid, liquid and gaseous wastes to nutrient rich biomass which further can be utilized as feedstocks for extraction of high-value products systems, like bioplastics, pharmaceuticals and biofuels through a biorefinery concept enhancing economy of the process.

Chlorella vulgaris; developed for 100% textile wastewater, and can also be applied for removal of pollutants in other water when pollutants decrease to permissible limits and the extracted pigment can serve as an additive in cosmetic industry (Hoque et al., 2025). Economical-dominant genera such as *Chlorella* and *Spirulina* (belonging to Arthrospira) are competing in the area of nutraceuticals and venturing other industries like waste water management, biofuels and green technologies. Selecting appropriate species is important, because performance depends on environmental conditions and the types of pollutant concerned. Some of these species additionally contribute to the phytoremediation by removing heavy metals from petroleum-impacted soils, and rehabilitating the land for agriculture in the future (Abreu et al., 2023).

Although the scale-up has been problematic, the industrial prospects of microalgae are very promising. The global microalgae market is anticipated to increase from USD 782.6 million in 2024 to USD 1.38 billion by A1802032, at a CAGR of 7.29% according to Fortune Business Insights (Hoque et al., 2025). In combination, the microalgae-based wastewater treatment and biorefinery systems could integrate CO_2 management, environmental restoration processes into high-value product production and circular economics demonstrating transformational potential with scientific literature support.

7. CONCLUSION AND OUTLOOK

7.1 Implication in Practice and Future Industrial Application

The large-scale industrialization of algal Carbon Capture and Utilization (CCU) necessitates more rapid progress in biology and engineering, including the over-current technical issues are circumvented. Algal-CCU could make use of concentrated CO_2 from industrial point sources, transforming waste into high-value biomass. However, low efficiency in mass transfer and low CO_2 dissolution are still challenges to overcome and this necessitates the development of efficient bioreactor designs as well as the development of advancements in CO_2 delivery. On the other hand, soluble carbonate such as bicarbonate avoids mass transfer limitation and ensures pH stability, which could ideal for an economically viable industrial process like this.

Challenges of the coming years should be pointed out for both the microorganisms and the relevant cultivation systems. The development of genetically and metabolism engineered algal strains with robust tolerance to industrial conditions, high growth rate and customized production of value added bioproducts is of essential importance.

Meanwhile, optimization of these systems (e.g., increasing the efficiency of CO_2 delivery and capturing low-energy CO2 and utilizing the full biomass) will spawn a circular and economical practice from concept to industrially applied algae-based CO_2-CCU (Yang et al., 2024). The future of the field is the "Biorefinery" approach which means process a zero waste system where all parts of the cell are used. To achieve this, a shift in policy and the creation of standardized carbon-credit markets are as important as the biological breakthroughs themselves. By bridging the gap between molecular biology and industrial engineering, microalgae will transition from a promising alternative to a primary driver of global sustainability.

7.2 Economic Drivers, Market Trends, and Regulatory Challenges for Microalgae Bioproducts

The economic viability and market integration of the microalgae are the critical steps that needs to happen to shift the field from a promising laboratory study to a powerhouse of an industry. Although the global microalgae market is expected to grow to a massive scale from USD 782.6 million in 2024 to USD 1.38 billion by 2032 (Fortune Business Insights, Hoque et al., 2025), the market's growth is influenced by various factors such as market drivers, high value product streams, and ongoing regulatory challenges that outline the market's commercial path.

At present, the most promising areas for microalgae derivatives are in nutraceuticals, functional foods and cosmetics. Some species such as Spirulina (Arthrospira) and Chlorella have become well established as protein-rich dietary supplements for their complete nutrient profile including vitamins, minerals and antioxidants. But the fine chemistry and extraction of specialized high value compounds are increasingly becoming the reality of the economic potential. Astaxanthin is, mostly produced from Haematococcus pluvialis, a strong antioxidant that is marketed at high prices for health and wellness applications, with the global market for this alone estimated to be more than USD 1 billion by 2027. Likewise, omega-3 fatty acids (EPA and DHA) produced by microalgae, like Schizochytrium sp. and Nannochloropsis sp., are in a surge of demand. The algal-derived omega-3s are a sustainable alternative, aiding in meeting the increasing consumer and industry concerns about overfishing and marine ecosystem health. These specialised extracts typically involve a higher value-added, often through the use of sophisticated extraction methodologies such as supercritical CO_2 and this is crucial to the biorefinery concept.

In addition to human food, microalgae is playing an important role in the animal feed market, especially in aquaculture. Microalgae can be used to improve the growth and health of fish and shrimp larvae, while cutting down on the use of unsustainable fishmeal, as it is a sustainable, nutrient-rich feed ingredient. Within the cosmetic industry, microalgal extracts are highly valued for their anti-aging, moisturizing and anti-inflammatory properties, and pigments such as phycocyanin and various carotenoids, are used as natural, colorful cosmetic dyes in place of synthetic dyes.

While these are promising market opportunities, there are still important economic barriers to overcome. Advanced photobioreactors still require high capital costs (CAPEX), and operational costs (OPEX) remain high, due to the energy consumption for CO_2 compression, nutrient delivery and energy-intensive downstream processing (such as harvesting, dewatering, extraction etc.). Although projected biomass production costs (EUR 12-13 kg-1 DW at scale - Acién et al., 2012), are encouraging, they still need to be competitive with traditional agricultural feedstocks for bulk commodity markets, otherwise microalgae are unlikely to enter lower value markets without some co-product valorization.

One of the most important concepts in enhancing the economic viability is that of "environmental credit". The combination of microalgal cultivation with wastewater treatment and capture of flue gas CO_2 reduces considerably the operational costs by using the waste streams for the nutrient and carbon supply. This dual benefit cuts production cost of biomass and generates environmental services which can be monetized via carbon credits or a reduction in pollution fees. So it is essential that there are policy regimes that are supportive, such as strong carbon-credit markets and incentives for green biotechnologies, in order to fill the current economic gap.

Yet, there's a significant hurdle on the regulatory side to overcome, especially the use of wastewater-grown biomass for food and feed, among other uses. The market requires strict international safety standards for these new products in order to assure consumer confidence and gain market access. Defining universally accepted guidelines for test procedures, processing methods, and labeling will be key for the future industrial use and broader societal acceptance of microalgae as a key component of the circular bioeconomy. The solution of these interrelated economic and regulatory issues is essential for the realization of the potential of microalgae in sustainable development.

7.3 Challenges, Limitations, and Solutions

With a broad range of applications, the utilization of microalgae has its limitations in terms of technical challenges, high process costs and lack of environmental-friendly practices which hinder commercialization in biofuels, food industry, pharmaceuticals and wastewater treatment processes (Piyatilleke et al., 2025; Webster et al., 2024). Open raceway ponds are economical, but suffer from contamination and varying productivity (Novoveská et al., 2023), whereas photobioreactors are more stable but with high capital and energy costs (Yu et al., 2024). Heavy cultures can be limited for light and show photoinhibition, with a decreased photosynthetic performance (Lizárraga, 2016).

Energy-demanding lipids, pigments and proteins extraction is part of the downstream process leading to high cost and limited sustainability. Environmental responses to temperature, light, salinity, and pH also limit open-pond cultivation the result is that products of microalgae are still more expensive than established alternatives (even in co-product scenarios) (Hoque et al., 2025).

However, microalgae have the potential of offering considerable improvement to sustainable agriculture. High-nutrient species such as *Spirulina* and *Chlorella* improve the soil's fertility, microbial activity, and nutrient availability in order to reduce synthetic fertilizer application as well as minimize soil and water contamination. The incorporation of algal biomass into biochar results in long-term C sequestration and soil amendment, contributing to climate smart agriculture (Hoque et al., 2025).

- Industrial CO_2 capture can also benefit from the addition of microalgae, enabling emission reduction concomitant with high-value biomass production. The possibility to cultivate them on non-cultivable surfaces such as arid and wastewaters, together with the natural antimicrobials metabolites will strengths the resilience of crop and control pests without coupling with food production (Parmar et al., 2023), making them interesting organisms in circular and resilient agroecosystems (Stiles et al., 2018).
- Subsequent improvements will rely on genetic engineering and synthetic biology to generate stress-tolerant high productivity strains, together with low energy consumption cultivation, harvesting and processing technologies. It will be important for the economic sustainability that integrated algae based biorefineries which cater multiple value chain product categories, like biofertilizers, biofuels, nutraceuticals, bioplastics and feed stocks come up in future.
- A variety of other supportive policy and financial instruments also are needed for broad uptake, and environmental soundness. Microalgae could represent a cornerstone in the global sustainable bioeconomy and climate-resilient agriculture, thanks to the integrated science, technology, and policy efforts. The integrated science, technology and policy approaches can make microalgae a keystone of climate-resilient agriculture and sustainable global bioeconomy (Hoque et al., 2025). The circular bioeconomy is technically possible, but there is a considerable 'Regulatory Bottleneck'. Existing systems in many areas limit the application of wastewater-grown biomass for human food and animal feed. The creation of new international safety standards for algal products from waste to allow consumer confidence and market access is essential for future industrial success.

REFERENCES

Abreu, A. P., Martins, R., & Nunes, J. J. B. (2023). Emerging applications of Chlorella sp. and Spirulina (Arthrospira) sp. *10*(8), 955.

Acién, F., Fernández, J., Magán, J., & Molina, E. (2012). Production cost of a real microalgae production plant and strategies to reduce it. *Biotechnology Advances*, *30*(6), 1344–1353. DOI: 10.1016/j.biotechadv.2012.02.005 PMID: 22361647

Arashiro, L. T., Boto-Ordóñez, M., Van Hulle, S. W., Ferrer, I., Garfí, M., & Rousseau, D. P. (2020). Natural pigments from microalgae grown in industrial wastewater. *Bioresource Technology*, *303*, 122894. DOI: 10.1016/j.biortech.2020.122894 PMID: 32032937

Ashour, M., Mansour, A. T., Alkhamis, Y. A., Elshobary, M. J. F. i. B., & Biotechnology. (2024). Usage of Chlorella and diverse microalgae for CO2 capture-towards a bioenergy revolution. *12*, 1387519.

Asimakis, E., Shehata, A. A., Eisenreich, W., Acheuk, F., Lasram, S., Basiouni, S., Emekci, M., Ntougias, S., Taner, G., May-Simera, H., Yilmaz, M., & Tsiamis, G. (2022). Algae and their metabolites as potential bio-pesticides. *Microorganisms*, *10*(2), 307. DOI: 10.3390/microorganisms10020307 PMID: 35208762

Assis, L. R., Ferreira, J., Assemany, P. P., Teixeira, J. S., Castro, J. S., Pereira, H. A., & Calijuri, M. L. (2023). Environmental benefits of a hybrid system for algal biomass production, harvesting and nutrient recovery under a life-cycle assessment. *Algal Research*, *73*, 103163. 10.1016/j.algal.2023.103163. DOI: 10.1016/j.algal.2023.103163

Cai, W.-L., Yu, S.-Y., & Hu, Y.-H. (2025). Synergistic mechanisms of DGAT and PDAT in shaping triacylglycerol diversity: evolutionary insights and metabolic engineering strategies [Review]. *Frontiers in Plant Science, Volume 16 - 2025*. DOI: 10.3389/fpls.2025.1598815

Calijuri, M. L., Do Couto, E. A., Assemany, P. P., Ribeiro, V. J., Lorentz, J. F., Castro, J. S., de Assis, L. R., Oliveira, A. P. S., Pereira, A., Marangon, B. B., Magalhães, I. B., Silva, T. A., Ferreira, J., Braga, M. Q., da Gama, R. C. N., & Henriques, B. S. (2025). Microalgae-Based Wastewater Treatment and Biomass Valorization: Insights, Challenges, and Opportunities from 15 Years of Research. *ACS Omega*, *10*(42), 49273–49299. DOI: 10.1021/acsomega.5c04364 PMID: 41179135

Chen, J., Dai, L., Mataya, D., Cobb, K., Chen, P., & Ruan, R. (2022). Enhanced sustainable integration of CO2 utilization and wastewater treatment using microalgae in circular economy concept. *Bioresource Technology, 366*, 128188. DOI: 10.1016/j.biortech.2022.128188 PMID: 36309175

Cichoński, J., & Chrzanowski, G. (2022). Microalgae as a source of valuable phenolic compounds and carotenoids. *Molecules (Basel, Switzerland), 27*(24), 8852. DOI: 10.3390/molecules27248852 PMID: 36557985

Das, P., Quadir, M. A., Thaher, M., Alghasal, G., & Aljabri, H. (2019). Microalgal nutrients recycling from the primary effluent of municipal wastewater and use of the produced biomass as bio-fertilizer. *International Journal of Environmental Science and Technology, 16*(7), 3355–3364. DOI: 10.1007/s13762-018-1867-8

de Oliveira, K. S., Volsi, B., Telles, T. S., Mendes, A. D. R., Yunes, J. S., & Andrade, D. S. (2025). Co-inoculation with Rhizobium, Azospirillum, and microalgae increases common bean yield and profitability. *Agronomy Journal, 117*(1), e21719. DOI: 10.1002/agj2.21719

Dineshkumar, R., Kumaravel, R., Gopalsamy, J., Sikder, M. N. A., & Sampathkumar, P. (2018). Microalgae as bio-fertilizers for rice growth and seed yield productivity. *Waste and Biomass Valorization, 9*(5), 793–800. DOI: 10.1007/s12649-017-9873-5

Dineshkumar, R., Subramanian, J., Gopalsamy, J., Jayasingam, P., Arumugam, A., Kannadasan, S., & Sampathkumar, P. (2019). The impact of using microalgae as biofertilizer in maize (Zea mays L.). *Waste and Biomass Valorization, 10*(5), 1101–1110. DOI: 10.1007/s12649-017-0123-7

Gonçalves, J., Freitas, J., Fernandes, I., & Silva, P. (2023). Microalgae as biofertilizers: A sustainable way to improve soil fertility and plant growth. *Sustainability (Basel), 15*(16), 12413. DOI: 10.3390/su151612413

Hakkoum, Z., Minaoui, F., Chabili, A., Douma, M., Mouhri, K., & Loudiki, M. (2025). Biofertilizing effect of soil Cyanobacterium Anabaena cylindrica–based formulations on wheat growth, physiology, and soil fertility. *Agriculture, 15*(2), 189. DOI: 10.3390/agriculture15020189

Hassan, S., Meenatchi, R., Pachillu, K., Bansal, S., Brindangnanam, P., Arockiaraj, J., Kiran, G. S., & Selvin, J. (2022). Identification and characterization of the novel bioactive compounds from microalgae and cyanobacteria for pharmaceutical and nutraceutical applications. *Journal of Basic Microbiology, 62*(9), 999–1029. DOI: 10.1002/jobm.202100477 PMID: 35014044

Hoque, M. M., Iannelli, V., Padula, F., Radice, R. P., Saha, B. K., Martelli, G., Scopa, A., & Drosos, M. J. B. (2025). Microalgae: Green engines for achieving carbon sequestration, circular economy, and environmental sustainability—A review based on last ten years of research. *12*(9), 909.

Jimenez, R., Markou, G., Tayibi, S., Barakat, A., Chapsal, C., & Monlau, F. (2020). Production of microalgal slow-release fertilizer by valorizing liquid agricultural digestate: Growth experiments with tomatoes. *Applied Sciences (Basel, Switzerland)*, *10*(11), 3890. DOI: 10.3390/app10113890

Kabato, W. S., Hailegnaw, N., Ermias, T., Samuel, A., De Silva, G., & Molnár, Z. (2025). Microalgae-Based Strategies for Soil Health and Crop Productivity: Mechanisms, Challenges, and Pathways to Climate-Resilient Agriculture. *Agronomy (Basel)*, *15*(11), 2669. Advance online publication. DOI: 10.3390/agronomy15112669

Khan, S. A., Sharma, G. K., Malla, F. A., Kumar, A., & Gupta, N. (2019). Microalgae based biofertilizers: A biorefinery approach to phycoremediate wastewater and harvest biodiesel and manure. *Journal of Cleaner Production*, *211*, 1412–1419. DOI: 10.1016/j.jclepro.2018.11.281

Kupriyanova, E. V., Pronina, N. A., & Los, D. A. J. P. (2023). Adapting from low to high: an update to CO2-concentrating mechanisms of cyanobacteria and microalgae. *12*(7), 1569.

Lizárraga, G. L. L. (2016). Considerations for photobioreactor design and operation for mass cultivation of microalgae. *Algae: Organisms for Imminent Biotechnology*, 55.

Malik, S., Shahid, A., Betenbaugh, M. J., Liu, C.-G., & Mehmood, M. A. (2022). A novel wastewater-derived cascading algal biorefinery route for complete valorization of the biomass to biodiesel and value-added bioproducts. *Energy Conversion and Management*, *256*, 115360. DOI: 10.1016/j.enconman.2022.115360

Novoveská, L., Nielsen, S. L., Eroldoğan, O. T., Haznedaroglu, B. Z., Rinkevich, B., Fazi, S., Robbens, J., Vasquez, M., & Einarsson, H. (2023). Overview and challenges of large-scale cultivation of photosynthetic microalgae and cyanobacteria. *Marine Drugs*, *21*(8), 445. DOI: 10.3390/md21080445 PMID: 37623726

Nugroho, Y. T., Hardianto, A., Syihab, A. B. M. I., Sidquni, S. A., Nurjannah, I., Adinisa, L., & Subroto, T. J. T. S. (2025). Carbon Capture at the Point Source: CO2 Gas Bubbling vs. *Bicarbonate Supplementation in Microalgal Cultivation.*, *22*(11), 10991–10991.

Nur, M. M. A. Mahreni, Murni, S. W., Setyoningrum, T. M., Hadi, F., Widayati, T. W., Jaya, D., Sulistyawati, R. R. E., Puspitaningrum, D. A., Dewi, R. N., Hadiyanto, & Hasanuzzaman, M. (2025). Innovative strategies for utilizing microalgae as dual-purpose biofertilizers and phycoremediators in agroecosystems. *Biotechnology Reports, 45*, e00870. 10.1016/j.btre.2024.e00870

Ördög, V., Stirk, W., Takács, G., Pőthe, P., Illés, A., Bojtor, C., Széles, A., Tóth, B., Van Staden, J., & Nagy, J. (2021). Plant biostimulating effects of the cyanobacterium Nostoc piscinale on maize (Zea mays L.) in field experiments. *South African Journal of Botany*, *140*, 153–160. DOI: 10.1016/j.sajb.2021.03.026

Panday, D., Bhusal, N., Das, S., & Ghalehgolabbehbahani, A. (2024). Rooted in nature: The rise, challenges, and potential of organic farming and fertilizers in agroecosystems. *Sustainability (Basel)*, *16*(4), 1530. DOI: 10.3390/su16041530

Parmar, P., Kumar, R., Neha, Y., & Srivatsan, V. (2023). Microalgae as next generation plant growth additives: Functions, applications, challenges and circular bioeconomy based solutions. *Frontiers in Plant Science*, *14*, 1073546. DOI: 10.3389/fpls.2023.1073546 PMID: 37063190

Penfold, C. M., Miyan, M., Reeves, T., & Grierson, I. T. (1995). Biological farming for sustainable agricultural production. *Australian Journal of Experimental Agriculture*, *35*(7), 849–856. DOI: 10.1071/EA9950849

Piyatilleke, S., Thevarajah, B., Nimarshana, P., & Ariyadasa, T. U. (2025). Microalgal biofuels: Challenges and prospective in the framework of circular bioeconomy. *Energy Nexus*, *17*, 100338. DOI: 10.1016/j.nexus.2024.100338

Popa, M. D., Simionov, I.-A., Petrea, S. M., Georgescu, P.-L., Ifrim, G. A., & Iticescu, C. (2025). Efficiency of microalgae employment in nutrient removal (nitrogen and phosphorous) from municipal wastewater. *Water (Basel)*, *17*(2), 260. DOI: 10.3390/w17020260

Renuka, N., Prasanna, R., Sood, A., Ahluwalia, A. S., Bansal, R., Babu, S., Singh, R., Shivay, Y. S., & Nain, L. (2016). Exploring the efficacy of wastewater-grown microalgal biomass as a biofertilizer for wheat. *Environmental Science and Pollution Research International*, *23*(7), 6608–6620. DOI: 10.1007/s11356-015-5884-6 PMID: 26638970

Ronga, D., Biazzi, E., Parati, K., Carminati, D., Carminati, E., & Tava, A. (2019). Microalgal Biostimulants and Biofertilisers in Crop Productions. *Agronomy (Basel)*, *9*(4), 192. https://www.mdpi.com/2073-4395/9/4/192. DOI: 10.3390/agronomy9040192

Sarma, U., Hoque, M. E., Thekkangil, A., Venkatarayappa, N., & Rajagopal, S. (2024). Microalgae in removing heavy metals from wastewater–An advanced green technology for urban wastewater treatment. *Journal of Hazardous Materials Advances*, *15*, 100444. DOI: 10.1016/j.hazadv.2024.100444

Schreiber, C., Schiedung, H., Harrison, L., Briese, C., Ackermann, B., Kant, J., Schrey, S. D., Hofmann, D., Singh, D., & Ebenhöh, O. J. J. o. a. p. (2018). Evaluating potential of green alga Chlorella vulgaris to accumulate phosphorus and to fertilize nutrient-poor soil substrates for crop plants. *30*(5), 2827-2836.

Shabbirahmed, A. M., Jacob, A., Dey, P., Somu, P., & Haldar, D. (2025). Biomass as eco-friendly adsorbents for the removal of emerging pollutants from wastewater: A review. *Discover Applied Sciences*, *7*(7), 771. DOI: 10.1007/s42452-025-07463-7

Singh Chauhan, D., Sahoo, L., & Mohanty, K. (2022). Maximize microalgal carbon dioxide utilization and lipid productivity by using toxic flue gas compounds as nutrient source. *Bioresource Technology*, *348*, 126784. DOI: 10.1016/j.biortech.2022.126784 PMID: 35104656

Solomon, W., Mutum, L., Janda, T., & Molnár, Z. (2023). Potential benefit of microalgae and their interaction with bacteria to sustainable crop production. *Plant Growth Regulation*, *101*(1), 53–65. DOI: 10.1007/s10725-023-01019-8

Stiles, W. A., Styles, D., Chapman, S. P., Esteves, S., Bywater, A., Melville, L., Silkina, A., Lupatsch, I., Grünewald, C. F., & Lovitt, R. (2018). Using microalgae in the circular economy to valorise anaerobic digestate: Challenges and opportunities. *Bioresource Technology*, *267*, 732–742. DOI: 10.1016/j.biortech.2018.07.100 PMID: 30076074

Sukačová, K., Vícha, D., & Dušek, J. (2020). Perspectives on microalgal biofilm systems with respect to integration into wastewater treatment technologies and phosphorus scarcity. *Water (Basel)*, *12*(8), 2245. DOI: 10.3390/w12082245

Sun, Z., Bo, C., Cao, S., & Sun, L. (2025). Enhancing CO(2) Fixation in Microalgal Systems: Mechanistic Insights and Bioreactor Strategies. *Marine Drugs*, *23*(3), 113. Advance online publication. DOI: 10.3390/md23030113 PMID: 40137299

Tredici, M. R., Rodolfi, L., Biondi, N., Bassi, N., & Sampietro, G. (2016). Techno-economic analysis of microalgal biomass production in a 1-ha Green Wall Panel (GWP®) plant. *Algal Research*, *19*, 253–263. DOI: 10.1016/j.algal.2016.09.005

Udaypal, G. R. K., Meharıya, S., & Verma, P. (2024). Advances in microalgae-based carbon sequestration: Current status and future perspectives. *Environmental Research, 249*, 118397. 10.1016/j.envres.2024.118397

Velásquez-Orta, S. B., Yáñez-Noguez, I., Ramírez, I. M., & Ledesma, M. T. O. (2024). Pilot-scale microalgae cultivation and wastewater treatment using high-rate ponds: A meta-analysis. *Environmental Science and Pollution Research International, 31*(34), 46994–47021. DOI: 10.1007/s11356-024-34000-7 PMID: 38985422

Wang, L., & Zhang, B. (2022). Cultivation of microalgae on agricultural wastewater for recycling energy, water, and fertilizer nutrients. In *Integrated Wastewater Management and Valorization Using Algal Cultures* (pp. 235–264). Elsevier. DOI: 10.1016/B978-0-323-85859-5.00006-3

Webster, L. J., Villa-Gomez, D., Brown, R., Clarke, W., & Schenk, P. M. (2024). A synthetic biology approach for the treatment of pollutants with microalgae. *Frontiers in Bioengineering and Biotechnology, 12*, 1379301. DOI: 10.3389/fbioe.2024.1379301 PMID: 38646010

Wu, G., Chong, J. W. R., Khoo, K. S., Tang, D. Y. Y., & Show, P. L. (2025). Upcycling food waste for microalgae cultivation toward lipid production in a closed-loop and system-integrated circular bioeconomy. *Biotechnology for Biofuels and Bioproducts, 18*(1), 74. DOI: 10.1186/s13068-025-02679-6 PMID: 40646635

Wu, G., Tham, P. E., Chew, K. W., Munawaroh, H. S. H., Tan, I. S., Wan, W. A. A. Q. I., Sriariyanun, M., & Show, P. L. (2023). Net zero emission in circular bioeconomy from microalgae biochar production: A renewed possibility. *Bioresource Technology, 388*, 129748. DOI: 10.1016/j.biortech.2023.129748 PMID: 37714493

Wu, G., Tham, P. E., Chew, K. W., Munawaroh, H. S. H., Tan, I. S., Wan-Mohtar, W. A. A. Q. I., Sriariyanun, M., & Show, P. L. (2023). Net zero emission in circular bioeconomy from microalgae biochar production: A renewed possibility. *Bioresource Technology, 388*, 129748. 10.1016/j.biortech.2023.129748. DOI: 10.1016/j.biortech.2023.129748 PMID: 37714493

Yang, Y., Tang, S., & Chen, J. P. (2024). Carbon capture and utilization by algae with high concentration CO(2) or bicarbonate as carbon source. *The Science of the Total Environment, 918*, 170325. DOI: 10.1016/j.scitotenv.2024.170325 PMID: 38278265

Yu, B. S., Pyo, S., Lee, J., & Han, K. (2024). Microalgae: A multifaceted catalyst for sustainable solutions in renewable energy, food security, and environmental management. *Microbial Cell Factories, 23*(1), 308. DOI: 10.1186/s12934-024-02588-7 PMID: 39543605

Zhang, X., An, L., Tian, J., Ji, B., Lu, J., & Liu, Y. J. B. T. (2023). Microalgal capture of carbon dioxide: A carbon sink or source?, *390*, 129824.

Zhou, W., Wang, J., Chen, P., Ji, C., Kang, Q., Lu, B., Li, K., Liu, J., & Ruan, R. (2017). Bio-mitigation of carbon dioxide using microalgal systems: Advances and perspectives. *Renewable & Sustainable Energy Reviews*, *76*, 1163–1175. 10.1016/j.rser.2017.03.065. DOI: 10.1016/j.rser.2017.03.065

Chapter 2

Biological CO_2 Sequestration in Microalgal Cultivation Systems:
Mechanisms and Process Dynamics

Brindha Devi Parthiban
http://orcid.org/0000-0003-2859-4659
Vels Institute of Science, Technology, and Advanced Studies, India

Yagna Priya
http://orcid.org/0009-0004-7133-104X
Vels Institute of Science, Technology, and Advanced Studies, India

Sangeetha Menon
http://orcid.org/0000-0001-6427-0200
Kristu Jayanti (Deemed to be) University, India

ABSTRACT

The study of Biological CO_2 Sequestration Mechanisms in Microalgal systems shows that the continuous rise of atmospheric carbon dioxide (CO_2) during the past several decades has become one of the most critical environmental problems which the world currently endeavors to solve. . The result has been rising global temperatures which now cause unpredictable weather patterns and ice cap melting and sea level rise and severe ecological damage. The existing carbon seizing and storage methods which include chemical absorption and geological storage and physical scrubbing

DOI: 10.4018/979-8-3373-7630-1.ch002

systems face multiple operational restrictions. Microalgae represent one of the top biological systems which scientist study for their potential to store carbon over extended periods. Microalgae represent tiny photosynthetic organisms which scientists discover in both marine and freshwater ecosystems, which they use because of their exceptional ability to absorb CO_2 and turn it into organic matter at a rapid pace.

1. INTRODUCTION

The worldwide presence of elevated atmospheric carbon dioxide (CO_2) levels has emerged as a significant problem because industrial activities have expanded and fossil fuel resources continue to be used. Carbon dioxide functions as one of the primary greenhouse gases which contribute to both global warming and climate change. The increasing levels of CO_2 have produced observable results which include rising global temperatures and altered weather patterns and glacier melting and ocean acidification. The global community has made it their top priority to find permanent solutions which will effectively decrease CO_2 levels in the atmosphere (Tang et al.,2020 & Wang et al 2022).

Current methods for controlling CO_2 emissions primarily aim to decrease fossil fuel usage while promoting the transition to renewable energy technologies. The existing methods serve as necessary solutions, but they lack the capacity to eliminate the total carbon pollution which has built up in the atmosphere since then. The development of carbon capture and storage (CCS) technologies functions as a solution to this problem. The methods capture CO_2 emissions from industrial facilities before they proceed to underground storage or alternative forms of containment(Singh & Gu,2010 & Wang et al 2022). The standard CCS methods face multiple challenges because they require high expenditure and consume large amounts of energy while creating doubts about their capacity to protect the environment over extended periods. The research community currently investigates sustainable solutions which provide more effective results than existing methods.

Carbon dioxide levels can be reduced through biological carbon sequestration which scientists view as a natural solution that protects the environment. Photosynthesis enables biological systems to transform carbon dioxide into organic matter. Microalgae has emerged as the most effective biological method for carbon dioxide sequestration among all the available biological methods. Microalgae are tiny aquatic microorganisms that uses sunlight for photosynthesis. They play a vital role in worldwide primary productivity while supporting the complete natural carbon cycle. Microalgae is known for its effective carbon-fixing organisms because they possess high photosynthetic efficiency. Microalgae show greater efficient conversion than terrestrial plants when converting sunlight into biomass because they can produce more sunlight-based energies. Microalgal species show rapid growth patterns as they can double their body masses within short time when environmental conditions allow. The plants achieve a higher rate of carbon dioxide absorption because they can grow at an exceptionally fast rate.

The cultivation of microalgae enables their growth outside areas designated for agricultural use. The microalgae grow in seawater and saline water and wastewater because they do not need fertile soil which traditional crops require to grow. The oil from microalgae serves as a feedstock for biodiesel production because it grows efficiently in ocean waters without using agricultural land or food resources. Marine microalgae show high economic value because they can grow successfully in various salt concentrations and different environmental conditions. Microalgae serve two main purposes since they capture carbon and they generate renewable energy. The biomass which these microalgae produce contains valuable components of lipids carbohydrates and proteins. Microalgal species accumulate lipids which serve us as biofuel production precursors while their carbohydrates transform into bioethanol and biogas through conversion processes. The dual function of microalgae for carbon capture and renewable energy production creates a promising solution to maintain sustainable environmental practices(Chisti, 2007 & Wijffels & Barbosa, 2010).

Microalgae show their capacity to absorb carbon dioxide from the atmosphere through their various biological processes and chemical transformations. The special adaptive traits which microalgae possess allow them to achieve their maximum capacity for carbon dioxide absorption. Microalgae main adaptation method involves carbon concentrating mechanisms which help them achieve higher internal carbon dioxide levels needed for their carbon fixation activities. The mechanism enables microalgae to sustain their photosynthetic activities under conditions when external CO_2 levels drop(Field et al 1998.,) Scientists study these mechanisms to create improved techniques that use microalgae for CO_2 emission capture.

Microalgae use environmental conditions to control their CO_2 fixation process which depends on light intensity and temperature and nutrient availability and pH and CO_2 supply. The cultivation system design directly impacts the total productivity of a given operation. Open pond systems and closed photobioreactors serve as the two primary methods for growing microalgae. Open pond systems provide easy operation at minimal expenses but these systems face higher risks of contamination while experiencing environmental changes. Closed photobioreactors provide enhanced control over growth parameters but they require higher expenses for building and servicing. The selection of a suitable cultivation system becomes critical for achieving successful CO_2 sequestration(Falkowski et al.,1998 & Beardall et al., 2009).

Recent advances in molecular biology and computational biology have established new scientific knowledge about microalgal carbon fixation. The combination of genome analysis and metabolic pathway modeling and systems biology tools enables researchers to find essential genes and pathways which control CO_2 fixation and biomass creation. The approaches generate valuable results which provide direction for experimental research and assist in achieving higher efficiency through deduction of unnecessary testing.

The potential of microalgae to sequester CO_2 requires additional solutions before scientists can implement their use at industrial scale. The main obstacles to progress remain because scientists need to resolve two essential issues which include the cost of cultivation and the efficiency of harvesting and the processes used after harvesting. The ongoing research and growth of technology in the field indicate that microalgae will become a key component of future methods to manage carbon emissions (Quinn & Davis, 2015; Slade & Bauen, 2013; Ruiz et al., 2016).

The aim of this paper is to discuss the biological mechanisms involved in CO_2 sequestration by microalgal systems. The study investigates photosynthesis and carbon concentrating mechanisms as essential physiological processes which affect their efficiency and their different cultivation methods and bioenergy production systems. Recent studies demonstrate that microalgae function as sustainable solutions for achieving atmospheric CO_2 reduction through environmentally friendly methods according to this research.

2. MICROALGAE IN GLOBAL CARBON CYCLING

Microalgae contribute to the global carbon cycle because of their small size which leads people to underestimate them. These organism serves as the main energy source for aquatic systems while its ability to transform the atmospheric carbon dioxide into biomass makes it vital for sustaining ecosystem process. The scientists estimates that microalgae together with phytoplankton constitute approximately 50 percent of the Earths total photosynthetic carbon uptake which makes them as vital as terrestrial vegetation for maintaining Earths carbon balance. Microalgae operate as extensive CO_2 biological sinks since they grow in all ocean and lake and aquatic environments. Microalgae use dissolved carbon dioxide from water sources during photosynthesis to create organic carbon in natural ecosystems(Falkowski et al 1998). The dissection process creates algal biomass which enables carbon movement through the food chain after algae become food for other organisms. Some of this carbon that researchers fixed will eventually descend to the deep sections of water bodies and sediments which will enable it to be kept there for extended durations. The biological carbon pump functions as the primary system which enables marine environments to store carbon permanently in their oceanic waters(Field et al.,1998).

Microalgae demonstrate higher carbon cycling efficiency compared to terrestrial plants. The organism demonstrates two distinct growth patterns which differentiate it from other species. Microalgae demonstrate rapid growth because their species can reach full development within hours while terrestrial plants need extended periods from weeks to months to achieve notable growth(Raven & Falkowski,1999 & Falkowski et al.,1998). Microalgae demonstrate rapid growth because they can adapt to environmental changes through carbon capture processes which increase during times of high CO_2 levels. Microalgae serve as efficient biological systems which achieve carbon fixation through their natural processes. The second essential component of their research involves studying photosynthetic efficiency. Microalgae demonstrate superior photosynthetic efficiency when compared to most terrestrial plants. Aquatic organisms can achieve better light distribution and nutrient uptake and gas exchange because they reside in underwater habitats. The microalgal species demonstrate higher efficiency for converting absorbed light energy into biomass chemical energy which they store. The global carbon cycle benefits from their activities because they occupy less space than forests and agricultural land (Beardall et al.,2009).

The marine microalgae species constitute the primary component of worldwide carbon cycling activities. The Earth has more than 70% of its surface covered by oceans which depend on marine microalgae as the fundamental component of their food webs(Raven & Falkowski,1999). The organisms continually take in carbon dioxide which exists in seawater that maintains a balance with atmospheric carbon dioxide levels. As atmospheric carbon dioxide levels rise, seawater absorbs more carbon dioxide which then becomes accessible to marine microalgae. The microalgae function as natural carbon buffers which help to stabilize atmospheric carbon dioxide concentrations(Falkowski et al.,1998).

The carbon handling methods of microalgae occur differently from terrestrial plants because of their reaction to different environmental conditions. Microalgal species possess the ability to alternate between dissolved carbon sources which include CO_2 and bicarbonate ions and other inorganic carbon sources. The organisms maintain their photosynthetic functions because of their capability to operate under different CO_2 concentration levels. Aquatic systems require this capacity because their carbon levels can be impacted by temperature changes and pH shifts and water movement patterns.

Researchers study microalgae in both natural ecosystems and controlled cultivation systems because microalgae play an essential role in carbon dioxide cycling. Microalgae organisms receive concentrated industrial CO_2 from exhaust gases when they grow in artificial ponds and photobioreactors. The method reduces CO_2 emissions while it increases algae growth and their ability to create biomass. The systems use algae to convert carbon into biomass which humans can collect for multiple applications such as bioenergy generation(Converti et al.,2009 & Griffiths & Harrison,2009).

Microalgae contribute to carbon cycling when their environmental nutrient levels permit them to grow. Algae need nitrogen and phosphorus and trace elements to grow and convert carbon dioxide into biomass. Microalgae exhibit rapid carbon uptake and biomass production when they grow in environments that provide them with abundant nutrients. Some species under nutrient-limited conditions will redirect their fixed carbon to create storage compounds which include lipids and carbohydrates. Microalgae demonstrate their significance in carbon cycling and energy storage through their ability to distribute carbon inside their cells.

Another important aspect of microalgal carbon cycling is their interaction with other microorganisms. Microalgae exist together with bacteria and other microbes in natural environments which affects the processes that transform carbon. Algae release organic carbon which bacteria can consume and bacteria provide nutrients that benefit algae. The interactions between these two groups of microbes create complex systems that control carbon movement through ecosystems. The design of artificial systems for microalgae-based CO_2 sequestration needs researchers to understand the interactions between different systems(Park et al.,2011 & Razzak et al.,2013).

Microalgae function as crucial components of the global carbon cycle because they demonstrate high productivity and rapid growth and inhabit all types of water bodies which exist on Earth. The combination of their effective carbon dioxide capture capabilities with their environmental resilience and their function as permanent carbon storage solutions makes them essential components that maintain atmospheric CO_2 balance. Scientists who study microalgae because of their role in global carbon cycling will discover their ability to help develop carbon sequestration technologies. This research establishes a framework for scientists to study how microalgae capture carbon dioxide through their unique biological processes.

3. BIOLOGICAL MECHANISMS OF CO_2 SEQUESTRATION IN MICROALGAE

3.1 Photosynthetic CO_2 obsession in Microalgae

Photosynthesis is the primary natural process through which microalgae prisoner carbon dioxide from their terrain and convert it into organic composites. analogous to advanced shops, microalgae use light energy to drive a series of biochemical responses that affect in the obsession of CO_2 into carbohydrates. still, microalgae are frequently more effective than terrestrial shops due to their simpler cellular structure and better access to dissolved inorganic carbon in submarine surroundings(Beardall et al.,2009).

In submarine systems, carbon dioxide exists substantially in dissolved form, along with bicarbonate and carbonate ions. Microalgae absorb dissolved CO_2 directly from the girding water during photosynthesis. Once inside the cell, CO_2 enters the photosynthetic pathway where it's converted into organic motes through a series of enzyme- driven responses. This process not only supports algal growth but also reduces the quantum of free CO_2 in the terrain(Chisti,2007).

The research team conducted studies on microalgae at their maximum capacity of 70mm throughout their experimental period. The technical structure of microalgae contains specialized organelles called chloroplasts which conduct photosynthesis through their chlorophyll pigments that absorb light energy. The light-dependent responses create energy-rich motes which include ATP and NADPH when light Energy undergoes absorption. The light-independent reactions utilize these molecules to convert carbon dioxide into organic compounds through their chemical processes. The fixed carbon is latterly employed for cell growth, conservation, and storehouse. Microalgal photosynthesis achieves its optimal efficiency through its main ability to function at maximum potential under first-rate environmental conditions(Beardall et al.,2009). Microalgal species demonstrate their ability to perform photosynthesis during both low-light conditions and changing environmental patterns. The organisms have developed this ability to sustain their carbon fixation processes in areas where terrestrial plants experience challenges. The microalgae exhibit a rapid ability to boost their photosynthetic performance when CO_2 levels rise which results in increased biomass production(Long et al.,2007).

Microalgae photosynthetic CO_2 fixation gets affected by environmental factors which include light intensity and temperature and pH levels and nutrient availability. Plants need sufficient light to perform photosynthesis, and they require specific temperature ranges to maintain their enzymatic activities. Cells require nitrogen and phosphorus nutrients to support their growth and metabolic functions. The equilibrium between these factors determines the efficiency of photosynthesis and the rate of carbon fixation.

Microalgae can gain another benefit because they can use various types of inorganic carbon sources. Many microalgal species have the ability to absorb dissolved CO_2 and bicarbonate ions, which they can use to produce CO_2 through their intracellular conversion process. The ability to use multiple carbon sources lets organisms increase their carbon absorption process because they can access free CO_2 at low concentrations in their aquatic habitats. Microalgae can maintain their ability to perform photosynthesis because they have various environmental conditions that support their growth(Singh & Gu,2010).

The process of photosynthetic CO_2 fixation establishes the basis for carbon sequestration in microalgal systems. Microalgae function as natural carbon sinks because they effectively transform both atmospheric and dissolved CO_2 into biomass. The basic photosynthetic process must be understood first before scientists can study advanced mechanisms which include carbon concentrating systems and specialized cellular structures that improve carbon fixation efficiency(Meyer et al.,2017).

3.2 Carbon Concentrating Medium (CCM)

Carbon Concentrating Medium (CCM) in Microalgae in submarine surroundings, the vacuity of free carbon dioxide is frequently limited due to slow gas prolixity in water. To overcome this limitation, numerous microalgae have developed a technical adaption known as the carbon concentrating medium (CCM). The main purpose of the CCM is to increase the attention of CO_2 around the carbon- fixing enzyme inside the cell, thereby perfecting photosynthetic effectiveness indeed when external CO_2 situations are low. The carbon concentrating medium workshop by laboriously transporting inorganic carbon into the algal cell. rather of counting only on dissolved CO_2, microalgae can take up bicarbonate ions (HCO_3^-) from the girding water using specific transport proteins located on the cell membrane and chloroplast membrane. Once inside the cell, bicarbonate is converted into CO_2 near the point of carbon obsession. This process ensures a nonstop and sufficient force of CO_2 for photosynthesis. One of the major advantages of the CCM is that it reduces the loss of carbon through photorespiration. In photosynthesis, the enzyme responsible for carbon obsession can occasionally reply with oxygen rather of CO_2, leading to energy loss. By adding the attention of CO_2 around this enzyme, the CCM minimizes oxygen hindrance and improves overall carbon obsession effectiveness. This makes microalgae largely effective indeed under conditions where CO_2 vacuity is limited. The exertion of the carbon concentrating medium is told by environmental conditions. Under low CO_2 conditions, the CCM becomes more active, with increased expression of carbon transporters and affiliated enzymes. When CO_2 is abundant, the CCM exertion may drop as unresistant prolixity becomes sufficient. This capability to regulate the CCM grounded on environmental conditions allows microalgae to acclimatize to shifting carbon vacuity in submarine ecosystems(Giordano et al.,2005 & Badger et al.,2006).

CCM functions as a crucial component which improves microalgae growth and productivity in artificial cultivation systems. The photobioreactor and open pond systems require effective carbon utilization methods which enable researchers to achieve maximum biomass output. The CCM maintains high internal CO_2 concentrations which enable plants to continue photosynthesis while experiencing increased growth speed. This feature becomes essential when microalgae are applied to carbon sequestration and bioenergy production purposes.

The CCM research field establishes knowledge which scientists use to enhance microalgal productivity. Researchers study carbon transport and conversion processes through gene and protein analysis to discover optimal methods for improving carbon fixation through different cultivation techniques and genetic methods. Scientists increasingly employ in-silico studies and metabolic modelling to study CCM pathways and forecast microalgae behavior under varying CO_2 levels(Raven et al.,2012).

Microalgae survive in carbon-limited environments because their primary physiological adaptation is the carbon concentrating mechanism. The photosynthetic efficiency of microalgae increases because they can capture and concentrate inorganic carbon which leads to crucial biological carbon dioxide sequestration. The mechanism establishes an essential connection between environmental carbon availability and the efficiency of cellular carbon fixation(Ho et al.,2011).

3.3 Role of Rubisco in Carbon Fixation

Rubisco functions as the main enzyme through which microalgae and higher plants accomplish their carbon fixation activities. The enzyme starts the Calvin cycle which transforms carbon dioxide into an organic compound. Scientists consider Rubisco control methods essential for effective carbon dioxide fixation because the enzyme exists in Earth's ecosystems at high concentrations but operates at reduced speeds with minimal output. Rubisco exists as a dual-functioning enzyme because it can establish connections with both carbon dioxide and oxygen molecules. The relationship between Rubisco and CO_2 enables effective carbon capture which results in plant growth through increased biomass production. The process of photorespiration in plants occurs when Rubisco interacts with oxygen which results in energy loss and a reduction of carbon fixation capacity. Aquatic environments become particularly affected by this restriction because water-dwelling organisms experience fluctuations in CO_2 concentrations. Microalgae have developed methods to enhance Rubisco activity through the creation of higher CO_2 environments that surround the enzyme(Long et al.,2007).

Rubisco exists in microalgae chloroplasts because they contain this enzyme, which occurs at distinct cellular locations. The enzyme operates through its connection to the carbon concentrating mechanism (CCM) system, which delivers high CO_2 concentration to the enzyme. Microalgae maintain high CO_2 levels in their environment because this process decreases the possibility of Rubisco binding to oxygen, which results in photorespiration, thus enhancing photosynthetic efficiency (Converti et al.,2009).

Rubisco performance depends on environmental conditions which include light levels and temperature and nutrient supply. Rubisco operates at its highest efficiency under ideal environmental conditions which enable fast carbon capture and plant development. The presence of low nutrients or excessive light can impact Rubisco performance during stressful situations. Microalgae use metabolic adjustments to maintain energy balance while fixing carbon which enables them to endure harsh environmental conditions.

Researchers examine the microalgal Rubisco structure and its regulatory mechanisms to improve carbon sequestration techniques. The differences in Rubisco composition between microalgae and higher plants indicate that algal Rubisco possesses specific advantages because it can better adapt to different CO_2 concentrations. Microalgae have become a crucial model organism for researchers who want to investigate molecular mechanisms of carbon fixation(Khan et al 2018).

The researchers perform in-silico analysis to investigate Rubisco-related biological pathways. The computational methods enable researchers to identify regulatory genes and enzyme interactions and the metabolic bottlenecks that occur in carbon fixation. The studies show that researchers can enhance Rubisco activity through better carbon supply and cellular control mechanisms instead of direct enzyme alterations.

Rubisco functions as the main mechanism which microalgae use to absorb CO_2 through its function as the primary carbon entry point for their cellular metabolism. Microalgae have developed systems which help them to utilize microalgal enzymes even though these enzymes have built-in performance constraints. The study of Rubisco functions serves two main purposes: it helps advance carbon dioxide removal methods and it enables researchers to create sustainable microalgal systems.

3.4 Pyrenoid Structure and Function in Microalgae

The pyrenoid functions as a specialized subcellular component which exists within the chloroplasts of various microalgal species and controls carbon dioxide fixation. The carbon concentrating mechanism (CCM) system depends on this component which creates a weighty impact during Rubisco enzyme activities. The establishment of pyrenoids in microalgae establishes a competitive edge which enables these organisms to absorb carbon dioxide under conditions of low CO_2 availability. The pyrenoid exists as an extensive chloroplast area which holds a dense concentration of Rubisco protein. Microalgae use this site to achieve efficient carbon fixation because they can store their enzymes in compact areas. The pyrenoid area receives inorganic carbon through thylakoid membranes and tubules which serve as transport channels. The pyrenoid and thylakoids maintain a close relationship which enables continuous CO_2 supply to Rubisco(Engel et al.,2015).

The pyrenoid functions as a mechanism which transforms bicarbonate ions that enter chloroplasts into carbon dioxide. The conversion process depends on enzymes which function similarly to carbonic anhydrase. The pyrenoid delivers carbon dioxide to Rubisco which enables the enzyme to interact with carbon dioxide instead of oxygen. This process decreases photorespiration while it boosts the complete efficiency of photosynthesis(Meyer et al.,2017).

Environmental factors determine both the shape and operation of the pyrenoid system. The activated CCM system shows increased pyrenoid shape when CO_2 levels decline. When CO_2 levels are high blood CCO^2 levels show decreased need for the pyrenoid system. Microalgae use this adaptive control system to quickly adapt their carbon intake process to different oceanic surface conditions.

Recent exploration has shown that the pyrenoid is n't a stationary structure but a dynamic bone. Its size, shape, and association can change depending on growth conditions, light intensity, and nutrient vacuity. This rigidity further highlights its significance in maintaining effective carbon obsession under varying environmental stresses(Giordano et al.,2005 & Badger et al.,2006).

From an applied perspective, understanding pyrenoid structure and function is important for perfecting microalgal CO_2 insulation systems. perceptivity into pyrenoid- associated proteins and nonsupervisory mechanisms can help identify strategies to enhance carbon obsession effectiveness. In- silico studies and molecular analysis have decreasingly concentrated on pyrenoid-affiliated pathways to prognosticate how changes at the cellular position may impact overall productivity.

In summary, the pyrenoid plays a central part in microalgal carbon insulation by concentrating Rubisco and easing effective CO_2 delivery. Together with the carbon concentrating medium and photosynthetic ministry, the pyrenoid enables microalgae to fix carbon effectively indeed under limiting conditions. This makes microalgae largely suitable organisms for natural CO_2 insulation and related sustainable operations (Wang et al., 2022).

4. FACTORS INFLUENCING CO_2 SEQUESTRATION IN MICROALGAL SYSTEMS

4.1 Light Intensity and Photoperiod

Light acts as one of the main elements which determines carbon insulation capacity in microalgal systems because it serves as the primary energy source for photosynthetic processes. Microalgae need light energy to transform carbon dioxide into organic materials, but any alteration in light conditions will affect their ability to grow and their capacity to capture carbon dioxide. Algal productivity depends on both light intensity and the duration of light exposure which scientists call photoperiod. Microalgae reach their highest photosynthetic output at optimal light intensity while their CO_2 absorption capacity operates at peak efficiency. The presence of low light conditions prevents photosynthesis from occurring because plants lack sufficient energy resources which results in decreased biomass production. Extremely high light conditions result in photoinhibition, which occurs when excess light damages the photosynthetic equipment and reduces the system's ability to absorb carbon. Proper light intensity maintenance functions as a fundamental requirement which supports continuous microalgal development(Beardall et al.,2009 & Tang et al.,2020).

Photoperiod, which refers to the cycle of light and dark exposure, also influences algal metabolism. numerous microalgal species perform better under controlled light dark cycles rather than nonstop illumination. During the light phase, photosynthesis laboriously fixes CO_2 and produces energy-rich composites. The dark phase allows cells to use stored energy for growth, form, and metabolic balance. A balanced photoperiod helps maintain steady growth and effective carbon application. In laboratory and civilization setups, artificial lighting systems are frequently used to give controlled illumination. Cool white fluorescent or LED lights are generally preferred due to their stable intensity and lower heat generation. Proper positioning of the light source ensures invariant exposure and prevents shadowing goods, which can else limit photosynthesis in thick societies (Converti et al.,2009).

Light vacuity also affects the distribution of carbon within microalgal cells. Under sufficient light, further carbon is directed toward biomass product. Under limited or shifting light conditions, microalgae may shift their metabolism to store carbon in the form of lipids or carbohydrates. This response is particularly important when microalgae are studied for both CO_2 insulation and bioenergy operations. Overall, light intensity and photoperiod explosively impact microalgal photosynthesis and carbon insulation. Optimizing these parameters is pivotal for perfecting CO_2 obsession effectiveness and icing stable growth in both experimental and large- scale civilization systems (Ugwu et al.,2008 & Posten, 2009).

4.2 CO_2 attention and Carbon Vacuity

Carbon dioxide attention is a major factor impacting the growth of microalgae and their capability to fix carbon efficiently. Since CO_2 serves as the primary carbon source for photosynthesis, its vacuity directly affects biomass product and carbon insulation rates. In submarine systems, CO_2 is present in dissolved form along with bicarbonate and carbonate ions, and microalgae use these inorganic carbon forms depending on environmental conditions. Under low CO_2 conditions, photosynthetic exertion can come limited, performing in slower growth and reduced carbon obsession. To overcome this, numerous microalgal species calculate on carbon concentrating mechanisms that help them use bicarbonate efficiently. still, indeed with these mechanisms, adding the external CO_2 attention generally leads to enhanced photo-synthesis and biomass accumulation, especially in controlled civilization systems.

In artificial civilization setups, CO_2 is frequently supplied externally to ameliorate algal productivity. This can be achieved by washing CO_2 fortified air into the culture medium. Controlled CO_2 delivery not only provides sufficient carbon for photosynthesis but also helps maintain pH stability in the medium. still, inordinate CO_2 force can lower the pH too important and negatively affect algal growth. thus, maintaining an optimal CO_2 attention is essential. Carbon vacuity also influences the metabolic pathways of microalgae. When sufficient CO_2 is available, microalgae primarily direct carbon toward growth and biomass conformation. Under carbon-limited conditions, growth slows down and cellular metabolism may shift toward survival strategies. This inflexibility allows microalgae to acclimatize to shifting carbon situations in natural and artificial surroundings(Ho et al.,2011 & Razzak et al.,2013).

In CO_2 insulation studies, measuring carbon vacuity and uptake is important for assessing system effectiveness. circular pointers similar as pH change, bubble rate, and biomass accumulation are generally used to assess CO_2 transfer and application These measures help determine how effectively carbon is being supplied and fixed by microalgal societies. Overall, CO_2 attention and carbon vacuity play a critical part in determining the success of microalgal carbon insulation systems. Proper regulation of CO_2 input ensures effective photosynthesis, stable growth, and bettered carbon obsession performance (Tang et al.,2020 & Ho et al.,2011).

4.3 Nutrient Availability and Limitation

The nutrient supply in the environment controls the capacity of microalgae to grow and their potential for carbon capture. Microalgae require both nitrogen and phosphorus together with trace nutrients to grow and complete their photosynthesis and essential cellular operations. The highest growth rates and peak carbon capture of microalgae occur when they receive the necessary nutrients in exact balanced proportions. Microalgae require nitrogen as an essential nutrient which enables their growth through protein and enzyme production and chlorophyll synthesis. Microalgae use nitrogen to support their growth process when it exists in sufficient quantities because they need it for cell division and biomass production. Organisms display reduced growth rates when their nitrogen levels drop below normal limits. Microalgal species develop the ability to accumulate storage materials through carbon redirection when they encounter nitrogen stress. Researchers who study biofuel production should pay special attention to this particular shift (Chisti,2007).

Phosphorus serves as an essential nutrient which enables energy transfer processes and nucleic acid production and membrane development. Phosphorus deficiency in the soil causes decreased plant growth and lower photosynthetic performance. Phosphorus deficiency leads to changes in carbon distribution inside the cell which resemble the effects of nitrogen stress. The metabolic changes that occur due to nutrient deficiencies demonstrate that organizations need to manage their nutrient intake to achieve specific goals which include maximum biomass production and better storage compound development. Trace elements such as iron, magnesium, and zinc are required in smaller amounts but play critical roles in enzymatic reactions and photosynthetic processes. The absence of these essential micronutrients results in adverse effects on the ability to perform photosynthesis and carbon fixation. The presence of these elements, which need to be taken in small quantities, plays a vital role in sustaining the health of microalgal cultures(Griffiths & Harrison,2009).

Natural environments experience changes in nutrient availability because water mixing and pollution and seasonal variations alter the environment. Microalgae developed adaptive mechanisms which enable them to survive in nutrient-limited environments by changing their metabolic processes and resource management. Their ability to adapt allows them to grow in different environments which include wastewater treatment facilities. Controlled cultivation systems allow researchers to modify nutrient levels which leads to better microalgal growth results. The initial growth stage requires adequate nutrient supplies which help to increase biomass development. The later phases of the process require researchers to implement nutrient restrictions which will boost production of lipids and carbohydrates. The two-stage cultivation method serves as a main research focus in studies about carbon sequestration and bioenergy research through microalgae. Nutrient availability determines how microalgae grow their carbon fixation capacity and their metabolic pathways. Researchers need to learn about nutrient conditions because it helps them create systems which use microalgae for effective CO_2 sequestration and their associated functions(Park et al,2011).

4.4 Temperature and pH

Temperature and pH are important environmental factors that directly affect the growth, metabolism, and carbon sequestration ability of microalgae. Since microalgae are sensitive to changes in their surrounding conditions, maintaining suitable temperature and pH ranges is necessary for stable growth and efficient CO_2 fixation. Temperature influences enzyme activity, photosynthesis, and overall metabolic processes in microalgal cells. Each microalgal species has an optimal temperature range where growth and carbon fixation are highest. When the temperature falls below this range, metabolic reactions slow down, leading to reduced photosynthesis and biomass formation. On the other hand, temperatures above the optimal range can cause stress, damage cellular components, and reduce photosynthetic efficiency. Therefore, maintaining moderate and stable temperatures is important, especially in controlled cultivation systems(Converti et al.,2009)

The pH level controls carbon distribution in aquatic ecosystems through its effect on carbon dissolved in water. The culture medium's pH level determines the equilibrium among dissolved CO_2, bicarbonate, and carbonate ions. Microalgae can use dissolved CO_2 as their primary carbon source when pH values reach their lowest levels. The rising pH levels cause carbon content to transform into bicarbonate and carbonate compounds which create different availability patterns for various organisms. Sudden pH shifts together with strong pH alterations make it difficult for algae to grow and for them to carry out their photosynthesis process. Microalgae use CO_2 for photosynthesis which results in pH levels rising gradually over time. The addition of external CO_2 to the medium causes its pH level to decrease. The continuous carbon supply requires stable pH levels because pH fluctuations create conditions that result in stress for organisms. Laboratories commonly use pH monitoring which uses indirect methods through CO_2 control to adjust pH instead of chemical solutions(Carvalho et al.,2006 & Ugwu et al.,2008).

The interaction between temperature and pH creates an impact on both nutrient uptake and enzyme activity. The presence of suboptimal conditions leads to decreased nutrient availability together with a reduction in the speed of cellular processes responsible for carbon fixation. Microalgae that can tolerate a wide range of temperature and pH conditions are particularly valuable for carbon sequestration studies because they can maintain performance under varying environmental conditions. Microalgal physiology together with carbon sequestration efficiency experience their strongest effects from temperature and pH changes. The system requires proper parameter control and monitoring to maintain healthy cultures while achieving effective CO_2 fixation in both experimental and applied systems.

4.5 Cultivation Conditions and System Design

The efficiency of CO_2 sequestration in microalgal systems depends on both the design of their systems and the conditions used for their cultivation. The presence of suitable biological and environmental conditions will not lead to growth and carbon fixation when system design defects exist. The proper cultivation setup selection process stands as the key requirement for achieving successful microalgal cultivation results. Open systems and closed systems together represent the two primary system types used for growing microalgae. Raceway ponds operate as open systems which provide a straightforward and economical solution for algae cultivation. The system enables large-scale algae production through the utilization of natural sunlight and ambient environmental conditions. The open systems suffer from high vulnerability because environmental conditions and contaminants can enter the system while water through evaporation gets lost. The combination of these elements leads to decreased system performance which results in unstable cultures and lower carbon capture abilities during extended periods.

Photobioreactors act as closed systems which enable researchers to regulate all aspects of their experiments through complete control over light intensity and temperature and pH levels and CO_2 supply. The systems enable better carbon dioxide utilization because they prevent contamination risks which help maintain operational efficiency. Research and controlled studies which require consistent and accurate results prefer photobioreactors despite their higher initial and operational costs. The design of cultivation systems requires both aeration and mixing to function effectively. The process of proper mixing guarantees that nutrients and light and CO_2 resources will spread evenly throughout the entire culture. The system prevents cell settling while it improves gas exchange. Laboratory-scale setups use air pumps and bubbling systems to provide aeration which also delivers carbon dioxide to the culture medium.

The system requires proper light distribution because it serves as a fundamental element of its operation. Plants in dense cultures experience reduced photosynthesis because of their inability to receive adequate light. The combination of transparent containers and proper light placement and controlled light cycles enables plants to receive constant light, which leads to improved their photosynthesis rates. Natural lighting conditions can be recreated in small-scale experiments through the use of artificial lights that operate with light and dark cycles. Microalgal cultivation system design must consider scalability as an essential aspect for effective CO_2 sequestration. Laboratory systems provide valuable insights into how organisms grow and how they convert carbon into their biological systems, but actual production needs systems that operate at affordable costs while protecting the environment. The research work should concentrate on developing affordable basic systems that can advance into bigger operating systems. Microalgal performance depends on two main factors which are the cultivation conditions and system design. The systems which have been designed well achieve three main objectives by maximizing CO_2 usage and establishing constant growth conditions and enabling effective carbon capture. The laboratory results need these aspects to be optimized because they are required for developing practical solutions which can be implemented at scale (Beardall et al., 2009 & Wang et al., 2022).

5. CULTIVATION SYSTEMS FOR MICROALGAL CO_2 SEQUESTRATION

The success of microalgal CO_2 insulation largely depends on the civilization system in which the algae are grown. While microalgae naturally retain high photosynthetic effectiveness and carbon obsession capability, these advantages can only be completely employed when suitable growth conditions are handed. civilization systems control crucial factors similar as light exposure, carbon dioxide force, nutrient distribution, temperature, and mixing, all of which influence algal growth and carbon uptake. Different civilization systems have been developed to study and use microalgae for CO_2 insulation under laboratory and field conditions(Posten,2009). These systems range from simple open setups to largely controlled unrestricted reactors. Each system offers specific benefits and challenges in terms of productivity, cost, impurity threat, and ease of operation. thus, opting the applicable civilization system is an important step in microalgal exploration and operation(Ugwu et al.,2008).

In experimental studies, civilization systems are designed to allow precise control of growth parameters so that the goods of individual factors on carbon obsession can be easily understood. In applied or large-scale settings, the focus shifts toward scalability, profitable feasibility, and long- term stability. A balance between control and cost is thus needed when choosing a suitable system. This section provides an overview of the different civilization systems used for microalgal CO_2 insulation. The design, functional features, and limitations of open and unrestricted civilization systems are bandied to punctuate their applicability in both exploration and practical operations(Borowitzka,1999 & Park et al.,2011).

5.1 Open Cultivation Systems

Open civilization systems are one of the most generally used styles for growing microalgae, especially at large scale. These systems involve growing algae in open surroundings where societies are exposed directly to natural conditions similar as sun, temperature variations, and atmospheric feasts. Common exemplifications of open civilization systems include open ponds and waterway ponds. One of the main advantages of open systems is their low cost. They're easy to construct, bear minimum structure, and can be operated using natural sun, which reduces energy input. Because of these factors, open systems are frequently considered suitable for large- scale operations similar as wastewater treatment and bulk biomass product. The simplicity of operation makes them seductive, especially in regions with high sun vacuity.

In open systems, mixing is generally achieved using paddle bus or simple aeration. Mixing helps distribute nutrients unevenly, prevents algal cells from settling, and improves exposure to light. Carbon dioxide is supplied either from the atmosphere or through external bubbling, although precise control of CO_2 attention is delicate in these systems. Despite their advantages, open civilization systems have several limitations. One major issue is impurity by unwanted microorganisms similar as bacteria, fungi, or other algal species. Since the system is open to the terrain, maintaining a pure culture becomes gruelling. Environmental oscillations similar as temperature changes, downfall, evaporation, and dust can also affect algal growth and stability(Carvalho et al.,2006).

Another limitation is lower productivity compared to unrestricted systems. Light penetration is frequently uneven, especially in deeper ponds, and CO_2 application effectiveness is fairly low. As a result, carbon obsession rates and biomass yields may vary over time. Overall, open civilization systems are suitable for low- cost, large- scale microalgal civilization where strict control is n't needed. While they may not offer the loftiest effectiveness, they give a practical option for studying microalgal growth and carbon insulation under near-natural conditions.

5.2 Closed Cultivation Systems (Photobioreactors)

Photobioreactors which function as closed cultivation systems enable scientists to regulate microalgae growth conditions through their enclosed design. Photobioreactors operate as closed systems which protect algae cultures from both external contamination and changes in environmental conditions. The systems find extensive application in research laboratories and controlled settings which study CO_2 capture processes. Photobioreactors enable facilities to control multiple systems which include light intensity and temperature and pH and nutrient levels and CO_2 delivery systems. The controlled environment enables microalgae to achieve their highest growth performance which results in increased biomass production and better carbon fixation efficiency. The system delivers CO_2 through controlled amounts which improves carbon usage efficiency while decreasing carbon emissions(Carvalho et al.,2006).

Photobioreactors use transparent materials to create systems which allow light to reach their biological cultures. Researchers developed three different photobioreactor designs which include tubular and flat panel and column systems to achieve better light distribution and gas exchange in their systems. The process of mixing with either aeration or mechanical stirring creates conditions which allow every part of the system to receive equal light and nutrient distribution while stopping cells from settling down. Closed systems provide another major benefit because they limit the risk of contamination. The enclosed culture system prevents unwanted microorganisms from entering which helps researchers maintain pure algal cultures. The research needs this element because it focuses on particular species and requires extended testing(Posten,2009).

Photobioreactors face specific limitations. The system demands greater financial investment and operational expenses than open systems. The total operational costs raise expenses because of energy needs for lighting and temperature management and aeration systems. The enclosed design makes maintenance and cleaning tasks more challenging because of its restricted access points. Closed cultivation systems present research and pilot-scale studies with a solution that maintains essential accuracy while providing reproducible results. The system enables researchers to study microalgal carbon sequestration mechanisms through its ability to maintain controlled CO_2 delivery and stable growth conditions(Ho et al.,2011 & Wang et al.,2022).

5.3 Laboratory Scale Cultivation Systems

The research studies use laboratory scale cultivation systems to study how microalgae grow and fix carbon and show their physiological characteristics under controlled environmental conditions. The experimental systems operate at smaller scale than their original design but they produce important research results which scientists can repeatedly test. Microalgae cultivation in laboratory environments uses conical flasks and culture bottles and small transparent containers as their primary growing equipment. The systems enable researchers to precisely manage all operational aspects including light dark cycle patterns and air circulation and nutrient levels and carbon dioxide (CO_2) exposure. Researchers use artificial light sources to deliver steady light output while air or CO_2 bubbling keeps the system mixed and enables gas transfer(Posten,2009 & Ho et al.,2011).

Laboratory scale systems offer their primary benefit because they provide researchers with operational flexibility to test different experimental conditions. Researchers can conduct experiments to assess different growth conditions by changing one specific parameter at a time which includes light intensity and nutrient levels and CO_2 supply. The laboratory systems provide researchers with a solution that enables them to optimize their work while they conduct research to compare different algal strains. The researchers can use the systems to monitor the initial growth period and the organisms' process of adapting to new environments and their reaction to stress situations. Scientists can achieve monitoring purposes with greater convenience. The system allows for continuous measurement of optical density and pH and temperature and biomass without interrupting the growth process. The use of small culture volumes enables scientists to conduct multiple tests while using less materials which helps them obtain precise results in their research(Griffiths & Harrison,2009).

The laboratory-scale systems which exist today lack the ability to operate at larger production capacities. The small flask experimental conditions do not accurately simulate the actual conditions which occur during large-scale cultivation processes. The three elements which include light penetration and mixing efficiency and gas transfer exhibit different behaviour for systems which operate at larger volumes. The laboratory results need to undergo detailed analysis before researchers can use them for their study of large-scale systems. Microalgal CO_2 sequestration research begins at laboratory-scale cultivation systems which provide essential foundational research. The systems enable researchers to determine growth conditions which serve as basic standards while they assess system performance and design larger systems which are more effective for actual operations(Tang et al.,2020).

5.4 Significance of Cultivation System Selection

The selection of the appropriate cultivation system represents a crucial milestone for research concerning microalgae-based CO_2 sequestration. The experimental outcomes depend on two factors which include the chosen system and its impact on algal development and carbon absorption efficiency. The growth conditions established through the cultivation system determine whether a highly effective algal species will succeed or fail. Each operational system provides organizations with unique benefits because of its different control capabilities and expenses and expansion potential. Organizations can conduct operations at low expenses through open systems which enable them to achieve extensive production but provide minimal control over their operational environment. Photobioreactors function as closed systems which enable operators to manage their cultivation conditions more effectively but require higher operational expenses. Researchers can study systems at laboratory scale to conduct experiments but these systems do not accurately reflect actual environmental conditions. Researchers should select their systems based on their research objectives(Borowitzka,1999).

For exploration purposes, controlled systems are frequently preferred to understand physiological and biochemical mechanisms easily. For applied studies and unborn scale- up, factors similar as profitable feasibility, energy input, and long- term stability come inversely important. Balancing these aspects is necessary to design effective microalgal CO_2 insulation strategies. In conclusion, civilization system selection plays a major part in determining the success of microalgal carbon insulation. A well- chosen system ensures optimal growth conditions, effective carbon application, and meaningful experimental issues, forming the base for unborn development and operation(Stephens et al.,2010).

6. CHALLENGES AND LIMITATIONS OF MICROALGAL CO_2 SEQUESTRATION

Indeed, though microalgae show strong eventuality for carbon dioxide insulation, there are several challenges and limitations that need to be addressed before these systems can be extensively applied. Understanding these limitations is important to really estimate the feasibility of microalgal CO_2 insulation and to design better strategies for future operations. One of the major challenges is the cost of civilization. Growing microalgae requires inputs similar as nutrients, light, aeration, and CO_2 force. While laboratory scale civilization is manageable, spanning up these systems significantly increases functional costs. Media factors, energy for lighting and mixing, and conservation of civilization systems contribute to the overall expenditure, which can limit large- scale relinquishment. Another important limitation is harvesting and downstream processing. Microalgal cells are veritably small and generally present in dilute attention in culture media. Energy requires forceful methods like centrifugation and filtration and flocculation to separate biomass from large amounts of water. The process needs proper optimization because it results in higher energy usage which makes the process less sustainable. The process needs to develop better CO_2 transfer systems because current methods still face difficulties. Microalgae use CO_2 efficiently, but their culture only absorbs part of the CO_2 supply. During aeration and washing processes, a major part of CO_2 emissions escapes into the environment, which decreases the total insulation capacity. The process needs specialized methods to maintain proper CO_2 balance between force and uptake because open systems and semi-controlled systems create operational difficulties (Tang et al.,2020).

Environmental perceptivity represents an additional restriction. The microalgal growth process depends on environmental variables which include temperature and light and pH and nutrient vacuity. Changes to these parameters which occur without warning will create harmful effects on both growth and carbon fixation. The open systems face difficulty in controlling their environmental changes because of their high sensitivity which results in unstable operational results (Khan et al.,2018). Open civilization systems create a widespread problem because they face challenges with contamination. The presence of unwanted microorganisms can compete with algal growth while obstructing their development which leads to lower productivity. Researchers encounter difficulties when they try to keep pure cultures for extended periods because this impacts their ability to conduct experiments and execute their large- scale work. The process of scaling presents another obstacle. The results from laboratory studies do not always transfer to results found in large industrial operations(Molina Grima et al.,2003). The three factors which need assessment include light penetration and mixing effectiveness and gas transfer efficiency at advanced volumes(Slade & Bauen,2013). The process requires both optimization and implementation through pilot-scale studies before actual execution. The current public awareness combined with absent standardization for natural carbon insulation programs hinders the progress of microalgal technology development and commercialization. The implementation of large- scale projects needs additional research together with experimental support and regulatory testing(Quinn & Davis,2015).

In summary, while microalgae offer a sustainable and biologically effective approach for CO_2 insulation, several specialized, profitable, and functional challenges must be addressed. Feting these limitations helps in relating exploration gaps and developing bettered strategies for future operations(Uduman et al.,2010).

7. FUTURE PROSPECTS AND APPLICATIONS OF MICROALGAL CO_2 SEQUESTRATION

Microalgal CO_2 sequestration has strong potential to become an important component of future climate mitigation strategies. The world currently needs environmentally friendly solutions because global warming and carbon emissions have developed into major environmental problems. Microalgae provide unique benefits through their ability to capture carbon while producing biomass that supports various uses. Industrial CO_2 sources will become the main application for microalgal systems according to their future developments. Microalgal cultivation uses flue gases from power plants and cement industries and other carbon-intensive sectors as carbon sources. The method reduces CO_2 emissions at their source while supplying additional carbon for algal growth which leads to improved growth rates. The integrated systems lead to improved efficiency and decreased expenses for CO_2 delivery.

Microalgae demonstrate strong potential to produce renewable energy sources. The biomass generated from CO_2 sequestration can be transformed into biofuels and biogas and other valuable products. The system functions to capture carbon while it generates renewable energy resources for use. Future biomass yield and energy recovery improvements will result from better cultivation methods and new strain selection techniques. The wastewater treatment process serves as an essential function. Microalgae use wastewater nutrients to grow while they capture CO_2 thus solving two environmental issues. The integrated method decreases nutrient pollution and reduces treatment expenses while generating valuable biomass. The systems have special importance for developing nations which require affordable and environmentally friendly solutions. Scientific research into genetics combined with computational modeling will help scientists advance microalgal research about its performance. Scientists can use their understanding of metabolic pathways that drive carbon fixation and biomass growth to identify methods for enhancing productivity without conducting extensive experimental work. The present approaches will direct research toward developing systems that optimize efficiency through targeted system improvements.

Future research needs to develop better system designs which will decrease cultivation expenses and boost CO_2 transfer efficiency. The development of low-cost cultivation media together with easy-to-use aeration systems and expandable microalgal systems will make these technologies accessible to more people (Cuellar-Bermudez et al.,2015). The transition from laboratory research to practical application requires pilot-scale studies and field studies which serve as vital research. Microalgae serve as reliable biological systems which can effectively capture CO_2 emissions and support sustainable environmental management according to research findings. Social scientists are working to establish proof that microalgae systems can deliver measurable carbon reduction results which will support the shift to an environmentally sustainable future(Owende,2010 & Wijffels et al.,2013).

8. CONCLUSION

The research shows that microalgae serve as an eco-friendly method to capture carbon dioxide because of their natural biological processes. Microalgae use their photosynthetic capabilities to absorb carbon dioxide from both the atmosphere and water and transform it into biomass. The environmental impact of microalgal systems enables them to provide a carbon capture method that better supports sustainable development goals when compared to traditional carbon capture methods. The review shows how microalgae contribute to global carbon cycling while explaining the biological pathways which enable them to fix CO_2 through their photosynthetic processes and carbon concentrating mechanisms and Rubisco enzymatic activity and pyrenoid structure. Microalgae use these mechanisms to achieve effective carbon fixation throughout different environmental conditions.

The researchers discovered that five environmental elements including light and carbon and nutrient and temperature and pH levels together impacted microalgal growth and their ability to sequester carbon dioxide. The researchers found that selecting a specific cultivation system determined the total operational success of their process. The advantages and disadvantages of open systems and closed photobioreactors and laboratory-scale setups become evident when their specific application needs are evaluated. The process of microalgae CO_2 sequestration shows great potential yet it still experiences existing obstacles which impact its cost and ability to scale and collect materials and operate efficiently. The future development process needs to solve these limitations through better system design together with optimization methods and combined system solutions.

Microalgae provide an effective biological method to decrease atmospheric CO_2 emissions while enabling sustainable processes for bioenergy generation and wastewater management. The continuing research efforts combined with technological progress will enable microalgal systems to deliver substantial assistance in fighting climate change and achieving environmental sustainability goals.

REFERENCES

Badger, M. R., Price, G. D., Long, B. M., & Woodger, F. J. (2006). The environmental plasticity and ecolcgical genomics of the cyanobacterial CO_2 concentrating mechanism. *Journal of Experimental Botany*, *57*(2), 249–265. DOI: 10.1093/jxb/eri286 PMID: 16216846

Beardall, J., Raven, J. A., & Giordano, M. (2009). Environmental regulation of photosynthesis in microalgae. *Journal of Experimental Botany*, *60*(10), 2491–2507.

Borowitzka, M. A. (1999). Commercial production of microalgae: Ponds, tanks, tubes and fermenters. *Journal of Biotechnology*, *70*(1–3), 313–321. DOI: 10.1016/S0168-1656(99)00083-8

Carvalho, A. P., Meireles, L. A., & Malcata, F. X. (2006). Microalgal reactors: A review of enclosed system designs and performances. *Biotechnology Progress*, *22*(6), 1490–1506. DOI: 10.1002/bp060065r PMID: 17137294

Carvalho, A. P., Meireles, L. A., & Malcata, F. X. (2006). Microalgal reactors: A review of enclosed system designs and performances. *Biotechnology Progress*, *22*(6), 1490–1506. DOI: 10.1002/bp060065r PMID: 17137294

Chisti, Y. (2007). Biodiesel from microalgae. *Biotechnology Advances*, *25*(3), 294–306. DOI: 10.1016/j.biotechadv.2007.02.001 PMID: 17350212

Converti, A., Casazza, A. A., Ortiz, E. Y., Perego, P., & Del Borghi, M. (2009). Effect of temperature and nitrogen concentration on the growth and lipid content of Nannochloropsis oculata and Chlorella vulgaris. *Chemical Engineering and Processing*, *48*(6), 1146–1151. DOI: 10.1016/j.cep.2009.03.006

Cuellar-Bermudez, S. P., Garcia-Perez, J. S., Rittmann, B. E., & Parra-Saldivar, R. (2015). Photosynthetic bioenergy utilizing CO_2: An approach on flue gases utilization for microalgae cultivation. *Journal of Cleaner Production*, *98*, 53–65. DOI: 10.1016/j.jclepro.2014.03.034

Engel, B. D., Schaffer, M., & Cuellar, L. K.. (2015). Native architecture of the Chlamydomonas chloroplast revealed by in situ cryo-electron tomography. *eLife*, *4*, e04889. DOI: 10.7554/eLife.04889 PMID: 25584625

Falkowski, P. G., Barber, R. T., & Smetacek, V. (1998). Biogeochemical controls and feedbacks on ocean primary production. *Science*, *281*(5374), 200–206. DOI: 10.1126/science.281.5374.200 PMID: 9660741

Field, C. B., Behrenfeld, M. J., Randerson, J. T., & Falkowski, P. (1998). Primary production of the biosphere: Integrating terrestrial and oceanic components. *Science*, *281*(5374), 237–240. DOI: 10.1126/science.281.5374.237 PMID: 9657713

Giordano, M., Beardall, J., & Raven, J. A. (2005). CO_2 concentrating mechanisms in algae: Mechanisms, environmental modulation, and evolution. *Annual Review of Plant Biology*, *56*(1), 99–131. DOI: 10.1146/annurev.arplant.56.032604.144052 PMID: 15862091

Griffiths, M. J., & Harrison, S. T. L. (2009). Lipid productivity as a key characteristic for choosing algal species for biodiesel production. *Journal of Applied Phycology*, *21*(5), 493–507. DOI: 10.1007/s10811-008-9392-7

Ho, S. H., Chen, C. Y., Lee, D. J., & Chang, J. S. (2011). Perspectives on microalgal CO_2-emission mitigation systems. *Biotechnology Advances*, *29*(2), 189–198. DOI: 10.1016/j.biotechadv.2010.11.001 PMID: 21094248

Khan, M. I., Shin, J. H., & Kim, J. D. (2018). The promising future of microalgae: Current status, challenges, and optimization of a sustainable and renewable industry for biofuels, feed, and other products. *Microbial Cell Factories*, *17*(1), 36. DOI: 10.1186/s12934-018-0879-x PMID: 29506528

Long, B. M., Badger, M. R., Whitney, S. M., & Price, G. D. (2007). Analysis of Rubisco biogenesis in the chloroplast reveals alternative pathways for protein targeting. *Proceedings of the National Academy of Sciences of the United States of America*, *104*(6), 2178–2183.

Meyer, M. T., Griffiths, H., & Olofsson, L. (2017). The pyrenoid: A microcompartment for carbon-concentrating mechanisms in algae. *Plant Physiology*, *173*(1), 1–10. PMID: 28049854

Molina Grima, E., Belarbi, E. H., Acién Fernández, F. G., Robles Medina, A., & Chisti, Y. (2003). Recovery of microalgal biomass and metabolites: Process options and economics. *Biotechnology Advances*, *20*(7–8), 491–515. DOI: 10.1016/S0734-9750(02)00050-2 PMID: 14550018

Park, J. B. K., Craggs, R. J., & Shilton, A. N. (2011). Wastewater treatment high rate algal ponds for biofuel production. *Bioresource Technology*, *102*(1), 35–42. DOI: 10.1016/j.biortech.2010.06.158 PMID: 20674341

Posten, C. (2009). Design principles of photo-bioreactors for cultivation of microalgae. *Engineering in Life Sciences*, *9*(3), 165–177. DOI: 10.1002/elsc.200900003

Quinn, J. C., & Davis, R. (2015). The potentials and challenges of algae based biofuels: A review of the techno-economic, life cycle, and resource assessment modeling. *Bioresource Technology, 184*, 444–452. DOI: 10.1016/j.biortech.2014.10.075 PMID: 25453439

Raven, J. A., & Falkowski, P. G. (1999). Oceanic sinks for atmospheric CO_2. *Plant, Cell & Environment, 22*(6), 741–755. DOI: 10.1046/j.1365-3040.1999.00419.x

Raven, J. A., Giordano, M., Beardall, J., & Maberly, S. C. (2012). Algal and aquatic plant carbon concentrating mechanisms in relation to environmental change. *Photosynthesis Research, 109*(1–3), 281–296. PMID: 21327536

Razzak, S. A., Hossain, M. M., Lucky, R. A., Bassi, A. S., & de Lasa, H. (2013). Integrated CO_2 capture, wastewater treatment and biofuel production by microalgae culturing. *Renewable & Sustainable Energy Reviews, 27*, 622–653. DOI: 10.1016/j.rser.2013.05.063

Ruiz, J., Olivieri, G., de Vree, J., Bosma, R., Willems, P., Reith, J. H., Eppink, M. H. M., Kleinegris, D. M. M., Wijffels, R. H., & Barbosa, M. J. (2016). Towards industrial products from microalgae. *Energy & Environmental Science, 9*(10), 3036–3043. DOI: 10.1039/C6EE01493C

Singh, J., & Gu, S. (2010). Commercialization potential of microalgae for biofuels production and CO_2 mitigation. *Renewable & Sustainable Energy Reviews, 14*(9), 2596–2610. DOI: 10.1016/j.rser.2010.06.014

Slade, R., & Bauen, A. (2013). Micro-algae cultivation for biofuels: Cost, energy balance, environmental impacts and future prospects. *Biomass and Bioenergy, 53*, 29–38. DOI: 10.1016/j.biombioe.2012.12.019

Stephens, E., Ross, I. L., King, Z., Mussgnug, J. H., Kruse, O., Posten, C., Borowitzka, M. A., & Hankamer, B. (2010). An economic and technical evaluation of microalgal biofuels. *Nature Biotechnology, 28*(2), 126–128. DOI: 10.1038/nbt0210-126 PMID: 20139944

Tang, D. Y. Y., Khoo, K. S., & Chew, K. W.. (2020). CO_2 biofixation by microalgae and its potential contribution to climate change mitigation. *Bioresource Technology, 299*, 122635.

Uduman, N., Qi, Y., Danquah, M. K., Forde, G. M., & Hoadley, A. (2010). Dewatering of microalgal cultures: A major bottleneck to algae-based fuels. *Journal of Renewable and Sustainable Energy, 2*(1), 012701. DOI: 10.1063/1.3294480

Ugwu, C. U., Aoyagi, H., & Uchiyama, H. (2008). Photobioreactors for mass cultivation of algae. *Bioresource Technology*, *99*(10), 4021–4028. DOI: 10.1016/j.biortech.2007.01.046 PMID: 17379512

Wang, J., Li, J., & Wang, X.. (2022). Microalgae-based carbon capture, utilization and storage (CCUS): A critical review. *Journal of Cleaner Production*, *344*, 131068.

Wijffels, R. H., & Barbosa, M. J. (2010). An outlook on microalgal biofuels. *Science*, *329*(5993), 796–799. DOI: 10.1126/science.1189003 PMID: 20705853

Wijffels, R. H., Kruse, O., & Hellingwerf, K. J. (2013). Potential of industrial biotechnology with cyanobacteria and eukaryotic microalgae. *Current Opinion in Biotechnology*, *24*(3), 405–413. DOI: 10.1016/j.copbio.2013.04.004 PMID: 23647970

Zhu, L. (2015). Biorefinery as a promising approach to promote microalgae industry: An innovative framework. *Renewable & Sustainable Energy Reviews*, *41*, 1376–1384. DOI: 10.1016/j.rser.2014.09.040

Chapter 3
Selection and Evaluation of Microalgal Strains for Biological CO_2 Sequestration

Salma Hameed
Independent Researcher, Pakistan

Muhammad Saqlain
http://orcid.org/0009-0008-1822-3970
University of Jhang, Pakistan

Nadia Tehseen
Independent Researcher, Pakistan

Qudrat Ullah
http://orcid.org/0000-0003-3881-0917
Independent Researcher, Thailand

Tauqeer Ahmed Qadri
http://orcid.org/0000-0003-2128-1633
Independent Researcher, Pakistan

Rashid Abbas Khan
http://orcid.org/0000-0002-5438-6274
University of South Africa, South Africa

ABSTRACT

Microalgae is a disruptive technology for Carbon Capture and Utilization in a circular bioeconomy, showing volumetric productivity and CO_2 fixation rates much higher than terrestrial plants. These organisms act as efficient biological carbon sinks due to specialized Carbon Concentrating Mechanisms that enhance inorganic carbon concentration at RuBisCO to maximize the Calvin–Benson cycle. Reported sequestration rates among microalgal strains range from 0.06 to 2.57 g L^{-1} d^{-1}, with Haematococcus sp. at the upper end. Genetic and metabolic engineering improved performance, Chlamydomonas reinhardtii manipulation increased carbon fixation by 35%, while transgenic Chlorella sorokiniana achieved 88.5% carbon capture. Coupling microalgal systems with wastewater treatment enables nutrient recovery.

DOI: 10.4018/979-8-3373-7630-1.ch003

removing up to 93% nitrates and 83% phosphates, yielding 3.21 g L^{-1} biomass. Optimal growth occurs at ~2000 lux and up to 35 °C. Although 5 vol% CO_2 enhances biomass, 9 vol% can cause culture collapse. High PBR capital costs and energy-intensive harvesting remain major commercialization barriers.

1. INTRODUCTION

One of the most urgent environmental issues of the twenty first century is global climate change that is caused by anthropogenic emission of greenhouse gases. The successive elevation of carbon dioxide (CO_2) concentration in the atmosphere has aggravated the quest to develop scalable, sustainable, and carbon capture and utilization (CCU) technologies that can help mitigate the effects of climate on the planet and integrate into circular bioeconomy systems (Almomani et al., 2023). Microalgae have become one of the most recent solutions to natural problems as biological carbon sinks effectively able to fix CO_2 through photosynthesis and at the same time, solve other related environmental issues such as wastewater treatment and nutrient recycling (Kirbayeva et al., 2025). The integration of the carbon sequestration and resource valorization makes the microalgal systems a multitalent platform, which facilitates the congruency between climate mitigation and sustainable approaches to industrialization. Unicellular photosynthetic organisms (microalgae) are of interest as biological carbon capture because they have significantly higher volumetric productivity and CO_2 fixation rates than terrestrial plants (Sarmah et al., 2026). Systematic reviews in the recent past have reported 0.06 to 2.57 g L^{-1} d^{-1} as carbon sequestration rates in the various microalgal strains, with *Haematococcus sp.* having fixed rates at the higher end of the scale (Sarmah et al., 2026). They have specialized carbon concentrating mechanisms (CCMs) to increase inorganic carbon supply at the location of ribulose-1,5-bisphosphate carboxylase/oxygenase (RuBisCO), which increases the efficiency of photosynthesis and minimizes photorespiratory losses (Priyadharsini et al., 2022). Moreover, micro algae are also able to endure low CO_2 levels, and adulterants that are found in industrial flue gases which is why they can be applied to coal-fired power plants, cement manufacturing plants, and other point source emitters (Almomani et al., 2023). Photosynthetic carbon fixation via the Calvin-Benson cycle depends upon the biological processes of CO_2 reduction into organic carbon and then into biomolecular pathways of lipid, carbohydrate and protein synthesis to form biomass is the biological basis of microalgal CO_2 sequestration. Such a process is complex in its regulation, with an interaction between environmental and physiological factors: the intensity of light, temperature, pH, nutrient availability, and CO_2 concentration. Genetic improvement methods such as mutagenesis, adaptive laboratory evolution, and directed genetic engineering have shown potentials of great importance to improve carbon fixation activity. An example is genetic manipulations in *Chlamydomonas reinhardtii* which were reported to produce about 35% and 25% carbon fixation capacity and biomass productivity enhancement, respectively (Sarmah et al., 2026). These developments highlight the effectiveness of metabolic engineering to streamline microalgal strains to use as an industrial scale source of carbon capture. A unique opportunity of microalgal CO_2 sequestration is the opportunity to synergistically integrate with wastewater treatment and valorization systems. Microalgae could absorb nitrogen, phosphorus, and other nutrients in municipal, agricultural and industrial effluents, and at the same time eliminate nutrients and produce biomass (Li,S et al., 2023). The same dual functionality will not only minimize the use of artificial fertilizers, but will also convert the waste streams into biomass feed stock to biofuels, animal feed, bioplastics and other bioproducts (Kirbayeva et al., 2025). The circular bioeconomy approach puts the microalgal systems in the context of closed-loop systems, in which CO_2 capture is linked to nutrient recovery and biomass valorization thus enhancing the process economics and environmental sustainability (Frongia et al., 2023). Pilot scale experiments have shown that integrated microalgae-wastewater systems are feasible, strain screening experiments have shown microalgae with strong capacity of carbon-fixing in intricate wastewater matrices (Li,S et al., 2023). Although microalgal CO_2 sequestration has a good potential, they are still faced with a number of technical and economic challenges. The existing photosynthetic efficiency in the industrial condition is not always optimal, and the costs of both capitals and processes involve in growing the large-scale cultivation and harvesting of biomass are a big obstacle to commercialization (Li et al., 2023). Moreover, the carbon equilibrium of microalgal systems in the long term is also determined by the downstream biomass utilization mechanisms and the degree to which carbon is immobilized in long-term products or storage in soil matrix. It has been reported in recent studies on edaphic (soil-applied) microalgae that soil organic carbon increased by 15-30% and crop yields increased by up to 20%, demonstrating the potential of application in agronomic studies as yet another carbon store (Swami et al., 2025). This chapter will give a detailed analysis of the biological processes, environmental regulations, and technological measures that support the process of CO_2 sequestration in microalgae. It examines carbon capture in combination with wastewater valorization as part of the circle bioeconomy, the evaluation of cultivation system and process optimization strategies, and the evaluation of new advances in genetic engineering and hybrid biological systems. The chapter will inform the future of microalgal technology in terms of the scaling to help play a significant role in global climate mitigation and sustainable management of the resource by synthesizing existing knowledge, and establishing gaps in research (Krishnamoorthy et al., 2024).

2. MICROALGAE AS EFFICIENT BIOLOGICAL CARBON SINKS

2.1 Photosynthetic Efficiency Compared with Terrestrial Plants

Microalgae can be described as one of the most effective photosynthetic organisms in nature with a much higher carbon fixation rate than the terrestrial plants when they are grown in ideal conditions (Nayak et al., 2025),(Prasad et al., 2021). Whereas land-based vegetation needs a large land mass and seasonal growth, microalgae have greater areal productivity and greater biomass turnover especially in regulated, high-CO_2 settings (Krishnamoorthy et al., 2024). This superiority in photosynthesis has been demonstrated in recent empirical studies. As an example, laboratory cultures using *Microchloropsis gaditana* were able to show that the increase in CO_2 levels in ambient conditions (around 300 ppm) to enriched conditions (30,000 ppm, 3% v/v) led to a 2.0-fold increase in the biomass productivity and the concomitant diversion of fixed carbon into long-chain polyunsaturated fatty acids (PUFAs) (Kareya et al., 2020). This great CO_2 sensitivity highlights the practical utility of microalgae in capturing carbon. The improved photosynthetic performance of microalgae is as a result of a number of natural benefits. Their unicellular or colonial form means that practically all the cells have direct access to light, CO_2 and nutrients and not the energy is expended in supporting tissues not involved in photosynthesis that are present in terrestrial plants (Nayak et al., 2025),(Wang et al., 2025). Moreover, microalgae have short doubling times of only a few hours as opposed to days or even weeks and seasonal production of biomass by continuously growing microalgae can be achieved under controlled environmental conditions (Prasad et al., 2021),(Krishnamoorthy et al., 2024). All these attributes boil down to an increased rate of photosynthetic carbon uptake per cultivated hectare as compared to traditional land crops, which makes microalgae of interest in any intensive carbon sequestration plan as part of climate mitigation approach (Nayak et al., 2025),(Vishwakarma et al., 2025).

2.2 Major Microalgal Groups Involved in CO_2 Sequestration

The process of selecting the right species of microalgae to be used as a CO_2 sequestration agent is based on various factors such as the rate of growth, environmental factors, the capacity to sequester CO_2 and the possibility to valorise biomass (Prasad et al., 2021),(Krishnamoorthy et al., 2024). A number of genera have become candidates to be used in industrial-scale carbon capture operations:

a) The *Chlorella* species has been widely studied and used in the CO_2 sequestration process based on their high growth rates, robustness in the fresh water system and demonstrated capability in the demonstration of the biomass production process (Prasad et al., 2021),(Krishnamoorthy et al., 2024). They can be used in research as well as in commercial activities as they are fully adaptable to different cultivation systems.

b) The presence of high protein and exceptional alkalinity tolerance, making it possible to implement in high-density cultivation systems where other species may not, has made spirulina (*Arthrospira*) especially useful (Prasad et al., 2021),(Wang et al., 2025). Several genera have emerged as prime candidates for industrial-scale carbon capture applications:

c) *Dunaliella* species have certain benefits on growth in saline and hypersaline waters, enabling the capture of CO_2 activities on non-arable lands using seawater or brackish water sources (Prasad et al., 2021),(Wang et al., 2025). Their light tolerance also increases the productivity of the outdoor cultivation systems.

d) The *Nannochloropsis* species and closely related *Microchloropsis* species have become popular as marine oleaginous species as a source of biofuel feedstock, due to their ability to store large lipid reserves (Krishnamoorthy et al., 2024),(Kareya et al., 2020). The genus *Microchloropsis gaditana* in particular has proven to be an outstanding CO_2 responsive organism with metabolic flexibility to divert captured carbon to high value products (Kareya et al., 2020).

e) The commonly used cyanobacteria (such as filamentous ones) are often chosen due to its fast growth rate, ease of genetic manipulation, and successful application in CO2 sequestration and biofuel studies (Krishnamoorthy et al., 2024),(Wang et al., 2025). They are also prokaryotes, which makes it easy to use genetic engineering strategies to improve on the carbon fixation pathways.

2.3 Carbon Concentrating Mechanisms (CCMs)

The use of advanced carbon concentrating mechanisms (CCMs) is the mechanism that distinguishes microalgae among the majority of terrestrial vegetation, as it increases the efficiency of photosynthesis (Kareya et al., 2020),(Vishwakarma et al., 2025),(Leech et al., 2025). They mitigate low CO_2 affinity of Rubisco and inhibit photorespiration by accumulating CO_2 and bicarbonate (HCO3-) actively around Rubisco up to 1000-fold (Vishwakarma et al., 2025),(Leech et al., 2025). This guarantees a high-level of Rubisco saturation, which enhances the efficacy of carboxylation (Nayak et al., 2025),(Kareya et al., 2020). Biophysical CCMs are based on bicarbonate transporters and carbonic anhydrases, and Rubisco is either localized in pyrenoids in green algae or in carboxysomes in cyanobacteria (Kareya et al., 2020),(Vishwakarma et al., 2025),(Leech et al., 2025). Biochemical CCMs are similar to C4 photosynthesis, but use metabolic shuttles to concentrate carbon but are less abundant (Vishwakarma et al., 2025),(Leech et al., 2025). Localized Rubisco, membrane-spanning transporters, and carbonic anhydrases are some of the main CCM elements (Kareya et al., 2020),(Vishwakarma et al., 2025). CCMs evolutionary acquisition enhances the CO_2 constraints of Rubisco which allows microalgae to sustain a high carbon fixation rate in changing ambient conditions (Vishwakarma et al., 2025).

2.4 Advantages within Climate Mitigation Frameworks

Microalgae can have a significant climate mitigation potential based on combined strategies. They make use of industrial flue gas CO_2 directly converting it to biomass and decreasing greenhouse gases (Nayak et al., 2025),(Krishnamoorthy et al., 2024),(Sridevi et al., 2026). Coupling with wastewater treatment allows the sequestration of CO 2, removal of nutrients, and biofuel and bioproduct generation (Nayak et al., 2025),(Krishnamoorthy et al., 2024),(Sridevi et al., 2026). Scalability is system-design based, closed photobioreactors, and open raceway ponds, but there are light penetration and mixing issues, cost of harvesting, and seasonality (Nayak et al., 2025),(Leech et al., 2025). The biomass is used to support the biomass valorization to fuels, feeds, nutraceuticals, pharmaceuticals, and bioplastics, thereby enhancing the economic viability (Krishnamoorthy et al., 2024),(Wang et al., 2025),(Sridevi et al., 2026),(Leech et al., 2025). The ongoing studies are concerned with the optimization of CCMs, strain engineering, and process integration (Nayak et al., 2025),(Prasad et al., 2021),(Krishnamoorthy et al., 2024),(Leech et al., 2025).

3. BIOLOGICAL MECHANISMS OF CO_2 SEQUESTRATION

3.1 Photosynthetic CO_2 Fixation Pathways: The Calvin-Benson Cycle

The Calvin-Benson (CB) cycle is the biochemical route of converting atmospheric CO_2 to organic carbon compounds by the photosynthetic organisms. This metabolic pathway exists in three phases carboxylation, reduction, and regeneration that are cyclic in nature (Raines et al., 2022). In carboxylation, the CO_2 acceptor molecule is ribulose-1,5-bisphosphate (RuBP) and the product of the reaction is 3-phosphoglycerate (3-PGA) which is catalyzed by ribulose-1,5-bisphosphate carboxylase/oxygenase (RuBisCO). Reduction step then transforms 3-PGA into glyceraldehyde-3-phosphate (G3P) which utilizes the ATP and NADPH produced by the light-dependent photosynthetic reactions. Lastly, the regeneration stage makes use of a complicated chain of enzymatic reactions to replenish RuBP, thus being able to restart the continuous process of CO_2 assimilation (Raines et al., 2022). Recent developments have determined that there are several enzymatic bottlenecks in the CB cycle that could limit the efficiency of overall carbon fixation. In addition to RuBisCO, rate limiting enzymes have been identified during the regeneration phase as sedoheptulose-1,7-bisphosphatase (SBPase) and fructose-1,6-bisphosphate aldolase (FBA) (Raines et al., 2022). The current engineering approaches focus on pathway-based interventions, such as expression of these restraining enzymes and incorporation of CO_2 concentrating mechanisms (CCMs) based on microalgae and cyanobacteria (Yamaoka et al., 2025),(Prasad et al., 2021). Microalgal system research has shown that organisms with native CCMs have far greater rates of carbon fixation, which can be used as useful models of engineering more efficient photosynthetic processes in plants on earth (Prasad et al., 2021),(Bongirwar et al., 2023).

3.2 Role of RuBisCO and Enzymatic Regulation

RuBisCO is at the center of the world carbon fixation process and has been identified as the largest enzyme in the planet (Nayak et al., 2025). Although it is common, RuBisCO has fundamental catalytic defects, such as a low turnover rate (k_cat ~3-10 s^{-1}) and ineffective selection between CO_2 and O_2 as substrates (Nayak et al., 2025). This oxygenase action triggers the process of photorespiration, which is an inefficient process that lowers the net carbon gain by about 25-30% in normal atmospheric conditions (Raines et al., 2022),(Zhu et al., 2025). In a majority of plants and green algae, the enzyme exists as an eight large (RbcL) and eight small (RbcS) multimeric holoenzymes. Its assembly involves complex chaperone systems and the knowledge of these biosynthetic regulatory processes in microalgae has informed the strategy to be employed in engineering in both plant and algal systems (Zhu et al., 2025),(Ganesan et al., 2022). The regulatory processes of the activity of RuBisCO involve activation by RuBisCO activase, which removes the inhibitor sugar phosphates in the active site, and the activation or suppression by carbamylation of a highly conserved lysine residue (Nayak et al., 2025). The recent engineering strategies have concentrated on three major strategies of improving performances of RuBisCO. First, experiments that involve variant transfer are carried out by substituting native plant RuBisCO with more efficient microalgae or cyanobacteria isoforms with a higher catalytic specificity or catalytic turnover rate (Zhu et al., 2025),(Qin et al., 2025). Second, condensation and compartmentalization approaches are supposed to convert RuBisCO to liquid-liquid phase-separated microcompartments in chloroplasts, akin to the natural pyrenoid organelles in the algae or bacterial carboxysomes (Nayak et al., 2025),(Chen et al., 2025). These condensates produce local CO_2 enrichments surrounding RuBisCO, which essentially enhances substrate accessibility and reduces oxygenase activity. Third, metabolic tuning strategies maximize the expression of the RuBisCO and co-factors, in order to coordinate the overall flux improvement with other enzymes of the CB cycle (Yamaoka et al., 2025),(Ganesan et al., 2022)-).

3.3 Intracellular Carbon Transport and Storage Pathways

The triose phosphate intermediates formed after fixation at the CB cycle take over several compartments and metabolic routes within the cell. A part of fixed carbon is instantaneously stored as starch in the chloroplasts as a temporary store of reserve (Prasad et al., 2021). The other triose phosphates are also exported to the cytosol through triose phosphate/phosphate translocator that, in turn, supplies the triose phosphates into the biosynthesis of sucrose and other metabolic pathways (Raines et al., 2022),(Prasad et al., 2021). The distribution of fixed carbon between the different fates depends on the availability of light, supply of ATP/NADPH, and Metabolic sink demand that is highly regulated (Prasad et al., 2021). CCM functioning creates extra ATP expenses in microalgae and cyanobacteria, but the CO_2 concentration rise nearby RuBisCO is significantly better (Prasad et al., 2021),(Bongirwar et al., 2023). Research has stressed that high throughput of the CB cycle and efficient transport of carbon to storage compartments require optimum regimes of irradiance (Prasad et al., 2021). It has been shown through engineering that developing or improving metabolic sinks decreases the inhibition of feedback on the CB cycle, resulting in more net carbon assimilation (Bongirwar et al., 2023),(Liu, Y et al., 2021). Engineered cyanobacterial systems with enhanced product pathways have a higher rate of CO_2 fixation since there is a reduction in bottlenecks in carbon partitioning (Bongirwar et al., 2023). The results of these findings underscore the role of equal flow of carbon through fixation and storage as a measure of the total sequestration capacity.

Figure 1. Enhanced biological co₂ sequestration: from fixation to stable sinks

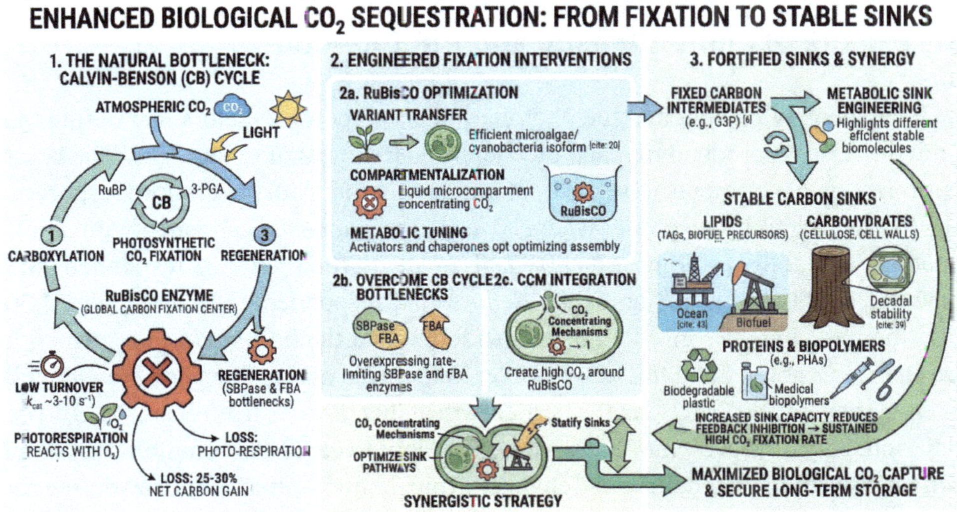

3.4 Biomolecule Synthesis as Carbon Sinks

The biological sequestration of carbon is based on the fixation of a fixed carbon content in stable biomolecules, primarily carbohydrates, lipids, and proteins (Prasad et al., 2021),(Bongirwar et al., 2023). Primary storage products are carbohydrates, including starch as short- to medium-term storage and structural polysaccharides including cellulose and hemicellulose as having longer residence periods of carbon (Prasad et al., 2021). Triacylglycerides (TAGs) are lipids that are extremely stable and energetically dense carbon stores (Bongirwar et al., 2023),(Ganesan et al., 2022). Microalgae are able to store lipids (up to 20-50% of dry weight) under stress, making it an effective platform to engineered sequestration (Prasad et al., 2021),(Bongirwar et al., 2023). Lipid biosynthesis is promoted using strategies like overexpressing the Kennedy pathway enzymes, inhibiting other pathways and optimization of growth conditions (Ganesan et al., 2022). Growing the carbon diversion to lipids enhances sequestration as it produces biofuel precursors (Ganesan et al., 2022),(Liu, Y et al., 2021). Proteins (including polyhydroxyalkanoates (PHAs)) and biopolymers could also be used as carbon sinks and value-added products (Liu, Y et al., 2021). The CO_2 fixation is stimulated through augmenting metabolic sinks by lessening the Calvin-Benson cycle feedback inhibition (Yamaoka et al., 2025),(Bongirwar et al., 2023). Combining sink pathway engineering with carbon concentrating mechanisms can be used to achieve synergies in sequestration capacity (Yamaoka et al., 2025),(Nayak et al., 2025),(Ganesan et al., 2022).

4. ENVIRONMENTAL AND PHYSIOLOGICAL CONTROLS

4.1 Effects of Light Intensity and Photoperiod

The intensity of light and the photoperiod are important factors that define the growth, biomass productivity, and biochemical structures of microalgae. The latest optimization experiments in vertical tubular photobioreactors also showed species-specific productivity response, where *Arthrospira platensis* responded with a productivity of 4.24 gL^{-1} biomass and 0.72 $gL^{-1}d^{-1}$, *Chlorella vulgaris* responded with 2.37 gL^{-1} and 0.36 $gL^{-1}d^{-1}$, and *Dunaliella salina* responded with the highest 4.50 gL^{-1} biomass and 0.77 $gL^{-1}d^{-1}$. It was also found that the best strain to screen with was native as about 2000 lux was considered to be the optimal level of growth with *Scenedesmus* AIN01 showing the best growth at this level (Najeeb et al., 2023). The light intensity, however, may cause excessive light to cause photoinhibition, which activates the lipid remodelling and changes in biochemical profiles and lowers the net growth rates (Matinha-Cardoso et al., 2022). A tradeoff between areal productivity and desired biochemical composition necessitates a close balance between species selection in relation to illumination regimes, as evidenced by the large productivity variation among taxa under the same regimes of light (Hawrot-Paw et al., 2023).

4.2 Temperature, pH, and Nutrient Availability

Temperature is a species-specific control of enzymatic rates, cell division kinetics and cell responses to stress. It was demonstrated through a thorough 10 strain optimization study that 35°C was the optimum growth temperature of 50% of the isolates studied and Scenedesmus AIN01 was much better able to grow at the high temperature of 35°C than other strains (Najeeb et al., 2023). On the other hand, the moderate temperature (22°C) and nitrate supplementation (160 mgL^{-1}) yielded a 1.3-fold increase in biomass at pH 8, which demonstrates complicated relations between temperature, nutrients, and pH (Polat et al., 2023). Thermal stress TSS causes a reduction in photosynthetic machinery, which selectively favors carbon compounds associated with stress over growth, and also in species-specific optima results in an increase in carbon partitioning (Matinha-Cardoso et al., 2022),(Polat et al., 2023). The pH has a critical impact on the inorganic carbon speciation, enzyme activity, and bioavailability of nutrients. In recent screening studies, most freshwater isolates (8 of 10 strains) grew best at pH 7 (Najeeb et al., 2023), with exceptions of alkaliphilic species like *Arthrospira* which naturally grow best in higher pH. The CO_2 relative to photosynthetic uptake determined dynamic pH responses in the course of cultivation; culture of *Arthrospira platensis* with 1 and 5 vol% CO_2 exhibited increasing pH, and 9 vol% CO_2 led to a decrease in pH between pH 9.2

and 8.8 during the experimental time (Chunzhuk et al., 2023). These pH changes cause a direct change in the ratio of CO_2, HCO_3^-, and CO_3^{2-}, which changes the supply of inorganic carbon to photosynthesis (Doppler et al., 2021). Growth rates and biomass yields are in fact influenced by nitrogen and phosphorus availability which is a macronutrient. The best nitrogen level was found to be 40 mM $NaNO_3$ in 70% of the examined autotrophic isolates, and phosphorus optima were 1.5 2 mM phosphate in 50% of the strains (Najeeb et al., 2023). The change in biomass in response to the addition of nitrogen is large; a controlled experiment found that the addition of 160 mgL^{-1} nitrate resulted in a 1.3-fold increase in biomass under controlled conditions of temperature and pH (Polat et al., 2023). It is important to note that nutrient deficiency, particularly nitrogen, activates a metabolic conversion of growth to neutral lipid storage, which is a popular biofuel-based farming practice that requires two-step production regimens to trade-off biomass productivity against lipid production (Matinha-Cardoso et al., 2022),(Polat et al., 2023).

4.3 CO_2 Concentration and Gas Transfer Dynamics.

The CO_2 level has a severe effect on the rate of growth, the culture pH, and biochemical distribution. Experiments of controlled photobioreactor with *Arthrospira platensis* at an initial level of CO_2 contents of 1, 5 and 9 vol% results in a growth rate of the biomass density of 79.4, 76.3 and 48.4 $mgL^{-1}day^{-1}$. The maximal recorded rate of CO_2 uptake was about 0.10 vol percent CO_2/day in the cultures that were inoculated with 5 vol% CO_2. Most importantly, although cultures were viable and yielded high-quality biomass at 1 and 5 vol% CO_2, the face of 9 vol% CO_2 led to the death of the cells in large numbers and acidified the medium in the middle, proving the presence of species-specific CO_2 tolerance limits (Chunzhuk et al., 2023). The elevated CO_2 also changed biochemical composition with a resultant reduction in lipid and protein content with an increase of CO_2 concentration--the opposite effect of lipid accumulation normally achieved by nutrient stress (Chunzhuk et al., 2023). The dynamics of gas transfer and the local CO_2 concentration at cells is closely related to the mass-transfer peculiarities of the reactor, which highlights the significance of the engineering parameters in the realization of effective autotrophic growth (Doppler et al., 2021).

4.4 Stress Physiology and Adaptive Responses.

Microalgae have adaptive responses to environmental stressors which are preserved, such as reduced growth rates, membrane reorganization, pigment changes and changes in carbon storage. The deficiency of nutrients, extreme irradiance, and temperature changes usually provoke the neutral lipid deposition and alter the fatty acids composition, however, usually at the cost of slower growth rates (Matinha-Cardoso et al., 2022),(Polat et al., 2023). But stress responses do not follow a consistent pattern, lipid and protein status reduction during CO_2 elevation found in *Arthrospira platensis* (Chunzhuk et al., 2023) is in contrast to lipid-accumulation response to nutrient limitation, which supports species-specific and stressor-specific responses. When subjected to environmental stress, microalgae shift carbon and energy of growth to storage compounds, stress-protective metabolites, and pigment synthesis (Matinha-Cardoso et al., 2022),(Polat et al., 2023). The adaptive management approaches are becoming more and more utilitarian of two-stage cultivation protocols; an initial nutrient-saturated growth followed by a controlled stress to achieve the required biochemical accretion to balance the biomass productivity and product quality goals (HD et al., 2025). Early warning of stress onset under the form of monitoring pH shifts, optical density changes and biomass composition offers early warning of the occurrence of a culture collapse; as with *Arthrospira* (Chunzhuk et al., 2023), the onset of culture collapse at 9 vol% CO_2 was predicted by a rapid decline in pH and loss of productivity (Doppler et al., 2021).

5. SPECIES SELECTION AND STRAIN IMPROVEMENT

When choosing microalgal species to be very effective in CO_2 fixation and biomass generation, it is necessary to perform a systematic screening of various physiological, biochemical, and genetic parameters. Species selection criteria should be able to strike a balance between laboratory performance metrics and industrial scalability such that high-performing strains can be able to endure the real-world cultivation conditions such as exposure to flue gas contaminations (Liang et al., 2025). The main selection criteria are as follows: CO_2 fixation capacity in the form of instantaneous or integrated carbon uptake rates (mg CO_2 L^{-1} d^{-1}), biomass productivity in the form of g L^{-1} d^{-1}, photosynthetic efficiency in terms of PSII maximum quantum yield ratio (Fv/Fm) and CO 2 tolerance between 13 and 20% v/v under potential realistic flue-gas conditions containing NO_x and SO_x (Ma, Z et al., 2022). Also, carbon partitioning profiles which measure the proportion of fixed carbon in carbohydrates, lipids, and proteins are important to understand the potential of metabolic capacity and optimization of products in the future (Shang et al., 2022). The enzymatic parameters of performance, specifically the activity of RuBisCO, the activity of photosynthetic antenna complex, etc., are the direct predictors of CO_2 assimilation capacity and the main focus of genetic improvement (Ganesan et al., 2022). The genetic tractability such as cell wall properties that can be transformed with modern tools and the availability of molecular tools have become critical requirements to modern strain development programs (Li, X et al., 2021).

Comparative studies and screening programs have identified the high-performance microalgal species with the best CO_2 fixation capability. The photoautotrophic production in high CO_2 condition increased threefold in chlamydomonas reinhardtii engineered with a universal plasma membrane H+-ATPase which was able to grow under simulated conditions of flue gases of 13 vol% CO_2, 20 ppm NO_x and 32 ppm SO_x resulting in twofold more CO_2-based bioproducts compared to the wild type (Choi et al., 2021). Transgenic Chlorella sorokiniana strains overexpressing pyridoxal kinase exhibited elevated carbon capturing efficacy attaining 88.5 percent with 34 percent growth speed increase in dynamic CO_2 concentrations indicating the promise of vitamin B6 metabolism regulation in enhancing CO_2 assimilation (Liang et al., 2025). The nuclear mutagenesis and polyethylene glycol (PEG) adaptive screening mutant M800 had a maximum CO_2 fixation rate that was 1.47 times greater than the wild type strain and reached a biomass dry weight of 2.31 g L^{-1} in the presence of PEG (1 mM) supplementation (Xin et al., 2024). A mutant of Selenastrum capricornutum selected with UV mutagenesis and 96-well microplate screening SDEC-2M exhibited superior tolerance of 15% CO_2 with a photosynthetic efficiency of 14.81%, carbohydrate productivity of 11.22 mg L^{-1} d^{-1}, and PSII maximal quantum yield of 70.14% (Shang et al., 2022). These show the variety of high-performance species that are available to be used in CO_2 capture applications and the considerable increase that can be made by developing strains with specific characteristics.

Genetic diversity screening and strain selection systems entail a combination of traditional mutagenesis and cutting-edge high-tech phenotyping systems to determine better variants. This can be done using UV mutagenesis combined with 96-well microplate screening to screen hundreds of mutants quickly as it was in the isolation of Selenastrum capricornutum SDEC-2M that grew and was more productive in high CO_2 environments (Shang et al., 2022). Adaptive evolution coupled with nuclear mutagenesis through ^{60}Co y-ray irradiation, followed by specific stressor, polyethylene glycol, has turned out to be useful in deriving mutants that had a much better CO_2 fixation capacity and biomass accumulation (Xin et al., 2024). Phenotypic screening measures usually consist of measures of biomass dry weight, target compound specific productivities (carbohydrates, lipids, proteins), photosynthetic efficiency percentages, enzymatic activities (RuBisCO assays) as quantitative endpoints (Shang et al., 2022)(Xin et al., 2024). The methodological models are becoming more focused on complementing high-throughput physiological assays with downstream genetic characterization and transformation-assays allowing priorities to be given to the strains that display both high performance and genetically amenable to further engineering (Ma, Z et al., 2022)(Li, X et al., 2021).

Genetic engineering and metabolic enhancement strategies have come a long way with various pathways showing a significant enhancement on CO_2 fixation and biomass production. The growth under acidic and high-CO_2 conditions, which enhances pH homeostasis and CO_2 tolerance, is achieved by expression of membrane transport proteins, especially the plasma membrane H+-ATPase, and the single-gene expression of which yields 3.2-fold increases in photoautotrophic production and doubles the quantity of products under flue gas cultivation (Choi et al., 2021). Regulation of cofactor biosynthesis pathways such as overexpression of pyridoxal kinase in Chlorella sorokiniana promotes effective overall cellular metabolism and carbon capture efficiency of 88.5% and 34% growth increase respectively (Liang et al., 2025). Optimization of the photosynthetic machinery by maximizing the use of RuBisCO, size reduction of light-harvesting antenna to minimize photoinhibition and lengthening of the utilization of photosynthetically active radiation is a holistic approach to enhancing net carbon fixation (Ganesan et al., 2022)(Botelho et al., 2021). Gene editing with CRISPR-Cas9, specifically with the use of Cas9 protein-gRNA ribonucleoproteins (RNPs) which do not require any vector constructs, provides high accuracy in genome editing, although cell wall entry by enzymatic digestion, or selection of inherently transformable species, is essential to achieve successful transformation (Li, X et al., 2021). The engineering of metabolic sinks that divert the fixed carbon into stable storage compounds or high-value products is the most effective way to maximize the carbon capture potential, and fast-growing cyanobacterial species are promising chassis of flux-balance modeling coupled synthetic biology (Kim et al., 2024). Such genetic engineering approaches coupled with conventional mutagenesis and screening technologies enable a powerful suite of tools in the creation of next-generation microalgal strains that have been optimized to serve industrial CO_2 capture and utilization purposes (Titirici et al., 2013).

Table 1. Species Selection and Strain Improvement for Enhanced CO_2 Fixation and Biomass Generation.

Microalgal Species / Strain	*Improvement Strategy*	*Key Performance Results & Metrics*
Chlamydomonas reinhardtii	Genetic Engineering: Expression of universal plasma membrane H^+-ATPase.	3-fold increase in photoautotrophic production; 2-fold increase in CO_2-based bioproducts; tolerance to 13 vol% CO_2, 20 ppm NO_x, and 32 ppm SO_x.
Chlorella sorokiniana	Genetic Engineering: Overexpression of pyridoxal kinase (Vitamin B6 metabolism).	88.5% carbon capture efficacy; 34% increase in growth speed under dynamic CO_2 concentrations.
Euglena gracilis (Mutant M800)	Nuclear mutagenesis and PEG (polyethylene glycol) adaptive screening.	CO_2 fixation rate 1.47 times greater than wild type; biomass dry weight reached 2.31 g L^{-1}.
Selenastrum capricornutum (SDEC-2M)	UV mutagenesis combined with 96-well microplate screening.	15% CO_2 tolerance; 14.81% photosynthetic efficiency; 11.22 mg L^{-1} d^{-1} carbohydrate productivity; 70.14% Fv/Fm.
Cyanobacterial species	Flux-balance modeling and synthetic biology to engineer metabolic sinks.	Effective diversion of fixed carbon into stable storage compounds or high-value products.
General Strains (Photosynthetic focus)	Optimization of RuBisCO activity and reduction of light-harvesting antenna size.	Minimized photoinhibition; increased net carbon fixation; lengthened utilization of photosynthetically active radiation.

6. INTEGRATION WITH WASTEWATER VALORIZATION SYSTEMS

Combining microalgae-based systems with wastewater treatment is a paradigm shift in ensuring sustainable carbon capture and at the same time tackle the issue of nutrient pollution. This synergetic strategy not only turns wastewater into a disposal problem but also makes it a valuable resource, which is the principle of a circular bioeconomy, with the simultaneous elimination of nitrogen and phosphorus, fixation of CO_2, and biomass that can be further used to get another value, which can be used in the valorization of wastewater.

6.1 Wastewater as Nutrient Source for Carbon Capture.

Municipal and industrial effluents have high levels of nitrogen, phosphorus and organic carbon which can promote photoautotrophic growth of microalgae and facilitate CO_2 capture. Recent laboratory studies have shown that the technical viability of this dual-purpose approach has quantifiable performance measures.

Growth of *Chlorella sorokiniana* pa.91 in a flat-plate photobioreactor with real settled municipal wastewater (COD = 211 mg L^{-1}) at 200 mL prior to eight days in this system, with a peak biomass of 3.21 g L^{-1} and a CO_2 bio-fixation rate of 0.59 mg L^{-1} d^{-1} and removal of 35.6% (Yaqoubnejad et al., 2021). At the same time, the system had a high level of nutrient removals NH_4^+ (74%), NO_3^- (93%), PO_4^{3-} (83%), and COD (76%) (Yaqoubnejad et al., 2021). Supplemental CO_2 is also provided, which only increases the nutrient removal and biomass productivity. The *Scenedesmus dimorphus* that grew in tertiary municipal effluent where CO_2 was added to the medium 2% had its biomass increase of 567 mg L^{-1} (air only) per 733 mg L^{-1}, whereas total nitrogen removal increased 28-100% without CO_2 addition to the medium and 60-100% when CO_2 was added (Faruque et al., 2023). Complete removal of phosphorus in an original concentration of 13mg/L ranged between 37-57%, and lipid content was 29-percent of dry weight with CO_2 enriched conditions (Faruque et al., 2023). These results highlight the importance of carbon supply to the nutrient uptake and bioproducts growth. Wastewater of industries can also be used as good sources of nutrients. The effectively grown *Scenedesmus sp.* directly in the industrial effluents with simultaneous CO_2 biocapture, which evidences that high-strength effluents could be used to facilitate high levels of microalgal growth under appropriate conditions (López-Pacheco et al., 2021). To high-strength slurries like biogas digestate, the microalgae-based purification systems reviewed and reported economic values of pollutant removal and CO_2 removal 14.2-39.0% USD^{-1} and 19.9-23.3% USD^{-1}, respectively (Zhang et al., 2020).

6.2 Coupling CO_2 Sequestration with Nutrient Removal.

Photosynthetic assimilation is the mechanistic relationship that facilitates the sequestration of carbon and removal of nutrients where micro algae are able to fix inorganic carbon (CO_2) and at the same time the cellular biomass accumulates dissolved nitrogen and phosphorus. Such a combined metabolic mechanism generates operation synergies that improve the overall system performance in contrast to single purpose technologies.

Experimental results prove that CO_2 enrichment does not only promote the decrease in the rate of carbon fixation, but it also promotes the uptake of nutrients due to the increase in photosynthetic activity and biomass formation (Faruque et al., 2023). This synergy is especially strong in systems that are limited in nitrogen, in which high carbon abiotic conditions cause metabolic flux to switch towards lipid biosynthesis, without impairing high nutrient polishing abilities (Faruque et al., 2023). The light intensity, temperature, N:P ratio, and CO_2 concentration are the key parameters important to the operational conditions that determine the balance between the efficiency of nutrient removal, biomass yield, and the formation of bioproducts (Yaqoubnejad et al., 2021)(Faruque et al., 2023)(Zhang et al., 2020). The consortia-based methods also maximize efficiency in coupling. Symbioses between microalgae and bacteria or microalgae and fungi take advantage of differing metabolic activities, the bacteria mineralizing organic material and the supply of CO_2 to the algae photosynthetic process, and the algae respirative metabolism with oxygen. It has been reported that the symbiosis of microalgae and fungi is more energy-efficient compared to monocultures when applied in slurry purification and biogas upgrading systems (Zhang et al., 2020). These biological synergies have the effect of decreasing aeration and increasing process stability especially in high strength or variable wastes.

6.4 Case Studies and Pilot-Scale Implementations.

A number of laboratory and pilot tests confirm the integrated system of wastewater-carbon capture at various levels. The above mentioned flat-plate photobioreactor experiment is a good controlled laboratory scale demonstration that has a detailed performance characterization (Yaqoubnejad et al., 2021). Likewise, the tertiary effluent trials which have been performed and which methodically assessed the influence of CO_2 supplementation on nutrient removal and lipid production (Faruque et al., 2023). The German ATS system studied at the pilot scale that seasonally operated at 50°54′19″N, 6°24′55″E and generated algal-bacterial biofilm that was useful in anaerobic digestion and recovery of energy (Cheenakula et al., 2023) This experimental pilot showed that it was feasible throughout the year in temperate climates and scaled what could be made in a full size design. Piloting of industrial wastewater applications has been done also. The grown microalgae on actual industrial effluent and testified strength in the fluctuation of loading conditions and CO_2 biocapture capability in non-municipal systems (López-Pacheco et al., 2021) Systematic reviews are used in synthesising several pilot investigations on biogas slurry treatment with biogas upgrading in quantifying ranges of energy and economic performance through various implementations (Zhang et al., 2020)(Pant et al., 2025). All of these case studies exhibit reproducible nutrient removal (usually >80-90% in the case of nitrogen and phosphorus), quantifiable CO_2 fixation (tens of percent removal efficiency), and plausible energy recovery pathways. Nevertheless, they also emphasize the importance of site-specific optimization of reactor design, hydraulic residence time, light delivery, and downstream processing in the context of large-scale delivery of the outcomes of circular bioeconomy.

7. CULTIVATION SYSTEMS AND PROCESS OPTIMIZATION

The biomass production of microalgae depends on the choice, as well as the optimization of cultivation systems in terms of their productivity and costs, as well as operational simplicity. Determinism of CO_2 delivery strategies, the final decision between open ponds and closed photobioreactors (PBRs), selection of a metabolic mode and scale-up engineering critically affect technical and economic viability of industrial microalgae bioprocesses.

7.1 Open Ponds versus Photobioreactors.

Two radically different methods of microalgal growth culturing include open pond systems and photobioreactors, which have their own benefits and shortcomings. Open ponds commonly in the form of raceways have the main benefit of reduced capital and operation costs as they are less complicated to build and do not need to have active environmental control mechanisms (Yang et al., 2023). Nevertheless, such systems have a series of shortcomings such as low light distribution, low volumetric productivity, low mass transfer efficiency of gases-liquid, and high susceptibility to contamination by colonizing microorganisms and predators (Yahaya et al., 2025). Photobioreactors on the contrary allow better control over such important cultivation parameters as light intensity, temperature, pH and dissolved gas concentrations. Closed PBR designs Tubular, flat-plane and bubble-column designs permit a higher cell density and uniform quality products. As an example, bubble-column photobioreactors cultivating *Chlorella sorokiniana* have proven to have a rise in biomass concentration of 1.05 g L^{-1} to 1.85 g L^{-1} and volumetric productivity of 0.33 g L^{-1} d -1 to 0.48 g L^{-1} d -1 with a rise in CO_2 concentration to 3%. (Nair et al., 2020). In spite of these productivity benefits, photobioreactors have a much higher capital cost and operational expenses in terms of materials and pumping, temperature control, as well as advanced gas handling systems (Fernández et al., 2024). The recent comparative analyses of S-shaped photobioreactors have reported volumetric productivities of 0.23 g L^{-1} d -1 at mixotrophic conditions and 0.43 g L^{-1} d -1 with optimized culture modes, demonstrating the significant role of the reactor design and the operational mode on the system performance (Yang et al., 2023). The choice made between open and closed systems will finally be WT determination of the target product value, purity required, and the scale of production with high-value metabolites usually warranting the extra investment in closed photobioreactor technology.

7.2. CO_2 Delivery Strategies and Mass Transfer Optimization.

Mass transfer and carbon dioxide delivery processes still represent the major bottlenecks in the growth of photoautotrophic and mixotrophic microalgae. Effective supply of CO_2 should be able to balance carbon supply and constraints like low solubility of gas, pH change and oxygen build-up. Direct gas sparging that uses diffusers, bubble column aeration, and membrane-based contactors, are predominantly applied in industrial systems (Sun et al., 2025). The fine bubble diffusion uses a lot of bubbles to increase the interfacial area and contact time at the risk of rising shear stress, and cell damage. It is demonstrated by the bubble column photobioreactors that a high level of feed CO_2 concentration to 3-10 percent of the air volume helps to enhance biomass and lipid accumulation dramatically (Nair et al., 2020). Nevertheless, dissolved CO_2 in high amounts may suppress metabolism and can lead to drastic changes in pH and should be well controlled. With the gas-permeable membranes, membrane contactors can offer enhanced mass transfer with a minimal shearing force and reduction of CO_2 caused by shearing stress, allowing the precise control of dissolved CO_2 (Zhang, B et al., 2025). Reactor hydrodynamics, mixing, and residence time distribution are also required to optimize mass transfer (Sun et al., 2025). The elimination of oxygen is also important, since the cultures may contain 200 percent or more of oxygen, which results in photoinhibition. It is feasible to combine CO2 injection with oxygen removal so that the systems will be stable and highly productive (Yang et al., 2025).

7.3 Mixotrophic and Heterotrophic Cultivation.

Metabolic flexibility is a strong tool of productivity increase of microalgal systems. Hybrid photoautotrophic systems (that are mixotrophic, that is, capable of CO_2 fixation alongside heterotrophic uptake of organic carbon substrates) are capable of yielding much higher biomass and lipid concentrations than autotrophic systems alone (Nair et al., 2020). Organic substrates, which are usually acetate, glucose, or glycerol, are determined by strain-specific metabolic abilities and price factors of organic substrates. Combination of acetic acid with *Chlorella sorokiniana* a high CO_2 concentration (3%) and an optimized light conditions resulted in high productivity gains: four-fold increase in biomass yield (0.7 to 2.79 g L^{-1}), five-fold increase in productivity (0.124 to 0.593 g L^{-1} d^{-1}), and six-fold increase in total lipid yield (0.12 to 0.75 g L^{-1}) over control photoautotrophic conditions (Nair et al., 2020). Complementary working of the photosynthetic and respiratory pathways give rise to these synergistic effects which minimize biomass losses to the night, and enhance the total efficiency of carbon conversion. Thermoacidophilic *Galdieria sulphuraria* glucose-based mixotrophy has been successfully demonstrated in tubular photobioreactors systems at pilot scale through the use of model-driven control mechanisms to achieve oxygen homeostasis. Researchers had reached a stable high-density mixotrophic growth by dynamically controlling the glucose supply using real-time measurement of dissolved oxygen and prevented the accumulation of oxygen leading to inhibition of photosynthetic activity (Fernández et al., 2024)(Moñino Fernández et al., 2025). This O_2-based mixotrophy scheme is a feasible industrial implementation strategy. Purely heterotrophic culture removes the need of light, and may obtain very high cell densities, but requires sterile culture, continuous aeration, and inexpensive organic substrates. The decision between photoautotrophic, mixotrophic or heterotrophic mode is eventually a trade-off between productivity and substrate costs and the complexity of the processes (Yahaya et al., 2025).

8. BIOMASS UTILIZATION AND LONG-TERM CARBON STORAGE

8.1 Biofuels, Bioproducts, and Biorefineries

Biomass conversion to biofuels and bioproducts via integrated biorefinery systems has shown to have a high potential of curbing greenhouse gas emissions as opposed to fossil fuel counterparts. In recent life cycle assessment of new thermochemical-biochemical biomass-to-liquid routes, the GHG reductions of 60-86% relative to fossil aviation and marine fuels, and non-renewable energy consumption reductions of 56-83% were reported under different conditions (Kourkoumpas et al., 2024). These scale benefits are extremely dependent on the choice of feedstock, logistics, and the renewable electricity mix used in the change method and the anticipated decarbonization of the grid could reduce pathway GHG emissions by an extra 26-68% (Kourkoumpas et al., 2024).

Life cycle analysis of the various bio-jet fuel production pathways indicates some different intensities of carbon: the aqueous phase reforming (APR) has shown the lowest global warming potential of 29.2 gCO_2e/MJ, then hydro processed esters and fatty acids (HEFA) with 43.6 gCO_2e/MJ, and hydrothermal liquefaction (HTL) with 51.2 gCO_2e/MJ (Zhu, X et al., 2022). These values are great additions to the traditional jet fuel when considered in a cradle-to-grave analysis. This is shifting to modern biorefineries, where albeit in a less advanced form than petroleum refineries, they are diversifying their product range to include a share on fuels of 85-90% and chemicals of 10-15% (Asghar et al., 2022). This is a product-mix optimization strategy that is necessary to enhance economic feasibility and competitiveness in the market. Valorization of co-products such as biochar, biochemicals and heat have long been found to be a key towards increasing the profit margins as well as the overall life cycle environmental performance at large (Asghar et al., 2022).

8.2 Carbon Stabilization Pathways.

Biomass stabilization of carbon entails various processes where carbon is transferred to long term storage pools in the atmosphere. Application of biochar is one of the most measurable stabilization processes. The carbon intensity of the resulting biofuel can also be reduce approximately 40.8–41.2 gCO_2e/MJ to 19.0–19.7 gCO_2e/MJ (Lan et al., 2021).. This is the dramatic decrease of the carbon sequestration benefit of biochar retention in agricultural soils. Long-term stabilization of carbon and also the replacement of fossil fuel use has been shown as environmental credits that may be derived by the thermochemical conversion processes especially torrefaction and pyrolysis (Chen et al., 2025). The size of these credits is very much reliant on the limits of systems, ultimate applications and the methodological decisions in life cycle accounting. Specific cellulosic energy crops and afforestation plans have the benefit of providing carbon additionality as they amplify quantifiable carbon fluxes to the atmosphere-ecosystem, without competing with food production systems (Field et al., 2021). Extensive deliberations on the biomass residue-to-biochar systems report tremendous environmental and energy advantages, and application-specific analysis is advised to maximize the stabilization potential and net carbon credits (Lee et al., 2020). It should be noted, however, that although near-term carbon benefits have been widely reported, there are still strong quantitative estimates of century-scale carbon permanence across a variety of systems that have not been well reported in the literature and, as such, long-term monitoring studies are needed.

8.3 Life Cycle Assessment of Carbon Balance.

Biomass carbon balance shows significant nuances in the framework of dynamic life cycle assessment methodologies that take into account the temporal accounting of carbon. Methods that use time discounting LCA may be expected to inflate estimated carbon intensities relative to non-time-discounted analyses- such as, when the timing of forest regeneration and the payback of carbon gains are explicitly accounted using time-discounted LCA, the carbon intensity on fast pyrolysis biofuel would increase, since the intensity is estimated at 41 gCO_2e/MJ (static) rather than 51-65 gCO_2e/MJ (time-discounted). On the other hand, biochar soil addition significantly lowers time-discounted carbon intensities which can be seen to highlight the paramount essence of selectivity in end-use pathway. Extensive integrated tests involving techno-environmental studies along with carbon capture and storage (CCS) methods have a prospect of producing almost no or negative carbon emission by biomass gasification systems involving woody biomass (Antonini et al., 2021). These studies measure trade-offs in production-use and to define the use of CO_2 capture of syngas as a possible route to low-carbon hydrogen and synthetic fuel production. The integrated LCA modeling using agents indicates that bio-energy adoption may have counterintuitive and even complex impacts when policy incentives are used. The adoption of perennial energy grasses on 25 percent of high yielding agricultural areas on a simulated basis showed that the effect of crop displacement and structural land use can significantly affect net performance of GHG life cycles, and that the application of spatially explicit policies should be evaluated as an important issue (i Losada et al., 2024).

8.4 Role in Sustainable Bioeconomy

The use of biomass is a key element in the new sustainable bioeconomy in various ways. Circular economy ideals, such as giving meaning to waste streams, co-production of chemicals and materials and cascading use of biomass should be integrated to meet both the economic and environmental sustainability goals (Asghar et al., 2022). There should be policy frameworks, which consider regional land-use relationship, the impact of soil organic carbon and those of indirect market effects to prevent unintended consequences like displacement of crops or destruction of ecosystems (i Losada et al., 2024). A shift to a renewable electricity grid is a significant economic and environmental lever, and has the potential to radically boost the carbon footprint and cost-competitiveness of biorefinery processes (Kourkoumpas et al., 2024). The concept of integrated life cycle assessment and techno-economic analysis systems are becoming recognized as essential instruments that need to be used when developing technologies, designing policies, and making investment choices in the bioeconomy sector (Asghar et al., 2022)(Chen et al., 2025)(Lee et al., 2020)

9. EMERGING TECHNOLOGIES AND INNOVATIONS

Synthetic biology, computational intelligence and systems integration are changing the nature of algal biotechnology to precision-based industrial platform. Specific genetic engineering, machine learning-based growing, designed microbial consortia, and industrial carbon capture have become a reality, responding to productivity and economic limitations. Synthetic biology allows specific editing of the genome and control of metabolic pathways through the CRISPR platform and completely modular cloning systems (Ng et al., 2020). This possible factor was demonstrated by strategic enzyme sourcing whereby the red-alga enzyme of CO_2/O_2 with red alga, Griffithsia monilis, has a factor of 167, much greater than that of normal green C3 leaf enzymes, which enhances photosynthetic efficiency (Naduthodi et al., 2021). In addition to the single-enzyme optimization, further combined metabolic engineering and machine learning realized the biomass productivity of 0.1 g L^{-1} h^{-1} and the productivity of limonene of 0.2 mg L^{-1} h^{-1} in photobioreactors (Long et al., 2022). Given the scaling to the outdoor raceway ponds, the biomass productivity of 43.3 g m^{-2} day^{-1} and the minimum selling price of an estimated $281 per tonne were observed, which indicated the economic scalability (Long et al., 2022). The IoT technologies and artificial intelligence allow predicting the amount of light exposure, nutrient delivery, and harvesting (Long et al., 2022). The behavior of hybrid AI and fuzzy logic models optimized through genetic algorithmic methods forecast the CO_2 fixation in different temperature, pH, and nutrient conditions (Kushwaha et al., 2023). High-throughput phenotyping and sensor networks develop closed-loop systems, which make yield more stable and generate strain development based on data (Fabris et al., 2020). Specific algae-bacteria consortia are also better designed to enhance nutrient recycling, contamination resistance and stability of processes by means of metabolic complementarity (Nneoma et al., 2025)(Fabris et al., 2020). Even though encouraging, standardized industrial scale validation is scarce (Nneoma et al., 2025). It has been established that the integration of carbon capture in industries has a productivity of outdoor biomass of 43.3 g m^{-2} day^{-1}, and the AI-based models indicate site-specific optimization and techno-economic analysis (Long et al., 2022) (Kushwaha et al., 2023). Nevertheless, unified use of CO_2 data and costing are still research areas of commercial implementation.

Table 2. Key Emerging Technologies in Algal Biomanufacturing

Technology	Key Innovation	Primary Impact
Synthetic Biology	CRISPR-based editing and modular cloning.	Precision control of carbon fixation and lipid pathways.
Enzyme Engineering	Sourcing high-efficiency RuBisCO (e.g., from Griffithsia monilis).	Potential for 2–4x higher CO2/O2 specificity than plants.
AI & Machine Learning	Hybrid AI-fuzzy logic and predictive growth modeling.	Real-time optimization of light, nutrients, and CO2 fixation.
IoT & Automation	Sensor networks and closed-loop control systems.	Enhanced yield consistency and data-driven strain screening.
Microbial Consortia	Designed algae-bacteria "synthetic microbiomes".	Improved nutrient recycling and resistance to contamination.
Industrial Integration	Direct capture from flue gas (power/cement plants).	Economically viable biomass production (43.3 g m^{-2} d^{-1}).

10. CHALLENGES, LIMITATIONS, AND POLICY CONTEXT

Although the use of microalgae as a sustainable biofuel feedstock holds a lot of potential, the research-to-commercial-scale transition remains limited by a host of technical, economic, environmental, and regulatory challenges. These issues are solved only with the coordinated innovation of cultivation systems, harvesting technologies, integration of the process, and the supportive policy frameworks.

10.1 Technical Barriers

The technical barrier to development of microalgae biofuel still takes a pre-eminent role. Poor lipid productivity in scaleable production cultures is an ongoing issue, with most promising strains that were optimized in laboratory cultures not continuing to be highly productive in lipids when scaled into outdoor, fluctuating production cultures (Tang et al., 2020)(Babu et al., 2022). Bacterial, fungal, grazers, competing microalgae species contamination presents intense threats to culture stability, which regularly causes disastrous culture crashes to affect yields and require expensive remedial action (Tang et al., 2020)(Kashyap et al., 2023). This weakness is especially severe in open-pond systems, in which it is not cost-effective to maintain axenic or near-axenic cultures. The problem of strain optimization is also made more difficult by genetic instability, regulatory restrictions on genetically modified organisms, and inevitability of trade-offs between high lipid accumulation and rapid growth rates (Babu et al., 2022)(Geng et al., 2025). Most recent developments of genetic and metabolic engineering, adaptive laboratory evolution, and hybrid cultivation solutions, using closed photobioreactors to produce inoculums with open ponds to scale up, provide encouraging directions in reducing the risk of contamination and the cost control (Babu et al., 2022)(Geng et al., 2025). Early culture perturbation monitoring through AI and IoT-based systems has become a new area of operation increasing operational resilience (Geng et al., 2025).

10.2 Economic Barrier and Cost Competitiveness.

Economic viability of the microalgae biofuels is hampered by the high capital and operating costs that are at the present exceeding the competitiveness to first-generation biofuels and fossil fuels in the market. Downstream processing (mainly harvesting, dewatering and drying) which consumes a lot of energy is the main contributor to the high production costs and may constitute a major part of the overall operation costs (Subhash et al., 2022)(Kumar et al., 2024). Major investments in capital infrastructure such as photobioreactor, mixing, and downstream processing facilities, restrict the investment of the private sector without policy support or diversification of income streams (Geng et al., 2025)(Bello et al., 2024). Inadequate standardization of techno-economic assessment (TEA) protocols and life cycle assessment (LCA) techniques add to the complexity of making investment decisions and comparing systems and studies without transparency (Geng et al., 2025). To address this disparity in the economy, the new reviews have recommended combined biorefinery models that co-produces high-value products such as pigments, proteins, nutraceuticals and biopolymers along with the fuels, in order to enhance the total revenue streams and spread the cost over a variety of product lines (Geng et al., 2025)(Kumar et al., 2024). Such policy tools as specific subsidies, algal biofuels feed-in tariffs, carbon credits and certification schemes of sustainable feedstocks are suggested as indispensable tools to de-risk investment and spur commercialization (Bello et al., 2024)(Zhang et al., 2020).

10.3 Harvesting and Scalability Issues

The small cell sizes of most productive species and the dilute characteristics of the microalgae cultures (usually 0.5-2 g/L biomass concentration) continue to be bottlenecks in harvesting and dewatering the microalgae cultures (Babu et al., 2022)(Subhash et al., 2022). Traditional mechanical techniques like centrifugation, albeit good, are energy consuming and would be costly in large-scale commercial applications. Chemical flocculation decreases the energy consumption but creates issues about the chemical residues and inconsistent specific species and water chemistry performance (Babu et al., 2022)(Subhash et al., 2022). New low-energy technologies such as bioflocculation, electroflocculation and integrated harvesting chains of gentle aggregation with passive separation by gravity, have potential in minimizing energy penalties (Geng et al., 2025)(Rajpoot et al., 2025). Most of these technologies are however in their infancy stages and there is no certainty of the scale up and consistency of the technology with fluctuating field conditions (Rajpoot et al., 2025). The transition of a laboratory to a commercial level creates more complexity. Mass transfer and light penetration problems in high-density cultures impair photosynthetic efficiency and it becomes increasingly challenging and costly to maintain even mixing and temperature as mass scale increases (Babu et al., 2022)(Geng et al., 2025). These limitations in engineering combined with decreased areal productivity and higher variability of operation weaken economic estimates on the basis of lab-scale operation (Geng et al., 2025).

10.4 Environmental Variability and Site-Specific Constraints.

Cultivation systems used outdoors are automatically vulnerable to the changes in the environment that influence productivity and consistency. The influence of temperature on metabolic rates and the potential to alter the cellular structure to disfavor lipid accumulation, which is dependent on the geographic position and time of the year (Babu et al., 2022)(Kashyap et al., 2023)(Geng et al., 2025). Light variability during the day and season affects the photosynthetic efficiency and requires properly decided design considerations on depth of culture, mixing regimes, and harvesting times as a way of maximizing light use (Geng et al., 2025). The availability of nutrients, be it through synthetic inputs or wastewater integration, needs a careful management to prevent the loss of productivity or subsequent pollution (Bello et al., 2024). Feasibility, however, depends on site selection, based on local climatic conditions and availability of resources and is thus constrained by site-specific environmental factors, making it difficult to scale-up microalgae biofuel production worldwide (Bello et al., 2024).

11. FUTURE PERSPECTIVES AND RESEARCH DIRECTIONS

To scale-up the implementation of microalgae innovations in the laboratory to an industrial scale, microalgae translations would require the coordination of climate policies, alignment of infrastructures, targeted research, and scale-up policies. Microalgae can be used to produce biofuels and value-added products, which help in the sequestration of CO_2, and aid carbon capture and conversion to further consumption in climate mitigation (Miranda et al., 2022)(Nagi et al., 2021). They have the potential to greatly decrease the carbon footprint of the lifecycle, especially when combined with point-source CO_2 and downstream biorefineries, especially when technological and technoeconomic enhancements become possible (Sarker et al., 2024)(Maliha et al., 2023). Biofuel and bioproducts generation coupled with CO_2 capture maximize mitigation by substituting fossil feedstocks and diversifying revenue sources (Nagi et al., 2021)(Maliha et al., 2023). Strong lifecycle analysis and pilot tests are required to confirm greenhouse gas emissions and lead policy frameworks (Miranda et al., 2022)(Sarker et al., 2024). Wastewater integration has good economic and circularity advantages. Municipal and industrial effluents are a source of nutrients, they decrease the freshwater usage and allow simultaneous bioremediation (Ummalyma et al., 2023)(Zielinski et al., 2023). Treatment facilities co-location reduces the costs of cultivation, and helps to comply with regulations, but there are several challenges such as the variability of the wastewater composition, competition among microbes, costs of harvesting, and regulatory limitations (Ummalyma et al., 2023)(Zielinski et al., 2023). The barriers to commercialization consist of high cultivation cost, downstream cost, strain inefficient harvesting and scarce pilot scale validation (Nagi et al., 2021)(Ummalyma et al., 2023)(Olabi et al., 2022)(Veeramuthu et al., 2020). It prioritizes genetic and process engineering, low-energy collection, low-cost dewatering, integrated conversion, and larger pilot demonstrations with industrial CO_2 and wastewater feeds (Nagi et al., 2021)(Olabi et al., 2022)(Ummalyma et al., 2023)(Veeramuthu et al., 2020)(Sarker et al., 2024)(Zielinski et al., 2023).

CONCLUSION

Microalgae is an efficient and high productivity intervention on carbon capture and utilization (CCU) in a closed bioeconomy. Their high efficiency in photosynthesis, which is made possible through carbon concentrating mechanisms (CCMs) and the optimization of enzymatic processes, enables much higher CO 2 fixation rates than are possible in terrestrial plants. These biological sinks have since been improved by innovation in genetic engineering and optimization of metabolic sinks with fixation efficiencies reaching 88.5%. Microalgal cultivation applied together with wastewater treatment forms a synergistic process that reclaims important nutrients at the expense of reducing industrial emissions. Although these improvements have been made, technical harvesting and scaling barriers must be overcome in order to make the process commercially viable. To achieve future success, policy support should be coordinated, the processes should be streamlined with AI, and high-value co-products need to be created to de-risk investments in mitigating global climate.

REFERENCES

Almomani, F., Abdelbar, A., & Ghanimeh, S. (2023). A review of the recent advancement of bioconversion of carbon dioxide to added value products: A state of the art. *Sustainability (Basel), 15*(13), 10438. DOI: 10.3390/su151310438

Antonini, C., Treyer, K., Moioli, E., Bauer, C., Schildhauer, T. J., & Mazzotti, M. (2021). Hydrogen from wood gasification with CCS–a techno-environmental analysis of production and use as transport fuel. *Sustainable Energy & Fuels, 5*(10), 2602–2621. DOI: 10.1039/D0SE01637C

Asghar, A., Sairash, S., Hussain, N., Baqar, Z., Sumrin, A., & Bilal, M. (2022). Current challenges of biomass refinery and prospects of emerging technologies for sustainable bioproducts and bioeconomy. *Biofuels, Bioproducts & Biorefining : Biofpr, 16*(6), 1478–1494. DOI: 10.1002/bbb.2403

Babu, S. S., Gondi, R., Vincent, G. S., JohnSamuel, G. C., & Jeyakumar, R. B. (2022). Microalgae biomass and lipids as feedstock for biofuels: Sustainable biotechnology strategies. *Sustainability (Basel), 14*(22), 15070. DOI: 10.3390/su142215070

Bello, M., Modu, A. K., Boryo, D. E. A., & Ranganathan, P. (2024). Sustainable algal biorefinery: A review on current perspective on technical maturity, supply infrastructure, business and industrial opportunities. *Journal of Environmental Management, 370*, 122208. DOI: 10.1016/j.jenvman.2024.122208 PMID: 39243640

Bongirwar, R., & Shukla, P. (2023). Metabolic sink engineering in cyanobacteria: Perspectives and applications. *Bioresource Technology, 379*, 128974. DOI: 10.1016/j.biortech.2023.128974 PMID: 36990331

Botelho, D. F., Dias, B. H., de Oliveira, L. W., Soares, T. A., Rezende, I., & Sousa, T. (2021). Innovative business models as drivers for prosumers integration-Enablers and barriers. *Renewable & Sustainable Energy Reviews, 144*, 111057. DOI: 10.1016/j.rser.2021.111057

Cheenakula, D., Hoffstadt, K., Krafft, S., Reinecke, D., Klose, H., Kuperjans, I., & Grömping, M. (2023). Anaerobic digestion of algal–bacterial biomass of an Algal Turf Scrubber system. *Biomass Conversion and Biorefinery, 13*(15), 13605–13619. DOI: 10.1007/s13399-022-03236-z

Chen, T., Hojka, M., Davey, P., Sun, Y., Zhou, F., Lawson, T., Nixon, P. J., Lin, Y., & Liu, L. N. (2025). Engineering Rubisco condensation in chloroplasts to manipulate plant photosynthesis. *Plant Biotechnology Journal, 23*(6), 2140–2149. DOI: 10.1111/pbi.70047 PMID: 40087764

Chen, W. H., Biswas, P. P., Lee, K. T., Chang, J. S., Lin, K. Y. A., Petrissans, M., & Petrissans, A. (2025). Integrating comparative life cycle assessment and applications of torrefaction and pyrolysis: A comprehensive review of net carbon and environmental credits. *Chemical Engineering Journal*, •••, 172106.

Choi, H. I., Hwang, S. W., Kim, J., Park, B., Jin, E., Choi, I. G., & Sim, S. J. (2021). Augmented CO2 tolerance by expressing a single H+-pump enables microalgal valorization of industrial flue gas. *Nature Communications*, *12*(1), 6049. DOI: 10.1038/s41467-021-26325-5 PMID: 34663809

Chunzhuk, E. A., Grigorenko, A. V., Chernova, N. I., Kiseleva, S. V., Ryndin, K. G., Popel, O. S., Malaniy, S. Y., Slavkina, O. V., de Farias Neves, F., Leng, L., Kumar, V., & Vlaskin, M. S. (2023). Direct study of CO2 capture efficiency during microalgae Arthrospira platensis cultivation at high CO2 concentrations. *Energies*, *16*(2), 822. DOI: 10.3390/en16020822

Doppler, P., & Spadiut, O. (2021). Introduction to autotrophic cultivation of microalgae in photobioreactors. In Kourist, R., & Schmidt, S. (Eds.), *The Autotrophic Biorefinery* (pp. 113–130). DOI: 10.1515/9783110550603-005

Fabris, M., Abbriano, R. M., Pernice, M., Sutherland, D. L., Commault, A. S., Hall, C. C., Labeeuw, L., McCauley, J. I., Kuzhiuparambil, U., Ray, P., Kahlke, T., & Ralph, P. J. (2020). Emerging technologies in algal biotechnology: Toward the establishment of a sustainable, algae-based bioeconomy. *Frontiers in Plant Science*, *11*, 279. DOI: 10.3389/fpls.2020.00279 PMID: 32256509

Faruque, M. O., Hossain, M. M., Farooq, W., & Razzak, S. A. (2023). Phototrophic bioremediation of municipal tertiary wastewater coupling with lipid biosynthesis using Scenedesmus dimorphus: Effect of nitrogen to phosphorous ratio with/without CO2 supplementation. *Sustainability (Basel)*, *15*(2), 1409. DOI: 10.3390/su15021409

Fernández, P. M., Buscaglia, L., Jordaan, D., Barbosa, M., & Janssen, M. (2024). Model-driven design and validation of glucose supply control in a tubular photobioreactor operated under oxygen-balanced mixotrophy. *Chemical Engineering Journal*, *499*, 155718. DOI: 10.1016/j.cej.2024.155718

Field, J. L. (2021). Revisiting "Additional Carbon": Tracking atmosphere–ecosystem carbon exchange to establish mitigation and negative emissions from bio-based systems. *Frontiers in Climate*, *3*, 603239. DOI: 10.3389/fclim.2021.603239

Frongia, F., Cramarossa, M., Arru, L., & Forti, L. (2023). Integrated wastewater treatment and CO2 capture by microalgae-based system. *Journal of Biotech Research*, *14*, 227–233.

Ganesan, A., Nawkarkar, P., Ali, S., Kajlaand, S., & Kumar, S. (2022). Genetic engineering of microalgae for enhanced photosynthetic efficiency, CO2 fixation, and fuel-based products.

Geng, Y., Shaukat, A., Azhar, W., Raza, Q. U. A., Tahir, A., Abideen, M. Z. U., Zia, M. A. B., Bashir, M. A., & Rehim, A. (2025). Microalgal biorefineries: A systematic review of technological trade-offs and innovation pathways. *Biotechnology for Biofuels and Bioproducts*, *18*(1), 93. DOI: 10.1186/s13068-025-02694-7 PMID: 40817213

Hawrot-Paw, M., & Sąsiadek, M. (2023). Optimization of microalgal biomass production in vertical tubular photobioreactors. *Energies*, *16*(5), 2429. DOI: 10.3390/en16052429

HD. A., & Vijayan P. U, H. (2025). Bio-green Evolution: The Journey of Microalgae Products from Lab to Market—Considerations and Recommendations—Part II. In *Microalgae Horizons: Fundamentals, Innovations, and Industrial Applications* (pp. 277-299). Cham: Springer Nature Switzerland.

Kareya, M. S., Mariam, I., Shaikh, K. M., Nesamma, A. A., & Jutur, P. P. (2020). Photosynthetic carbon partitioning and metabolic regulation in response to very-low and high CO2 in Microchloropsis gaditana NIES 2587. *Frontiers in Plant Science*, *11*, 981. DOI: 10.3389/fpls.2020.00981 PMID: 32719702

Kashyap, M., Chakraborty, S., Kumari, A., Rai, A., Varjani, S., & Vinayak, V. (2023). Strategies and challenges to enhance commercial viability of algal biorefineries for biofuel production. *Bioresource Technology*, *387*, 129551. DOI: 10.1016/j.biortech.2023.129551 PMID: 37506948

Kim, D. S., Moreno-Cabezuelo, J. Á., Schulz, E. N., Lea-Smith, D. J., & Sagaram, U. S. (2024). Recent advances in engineering fast-growing cyanobacterial species for enhanced CO2 fixation. *Frontiers in Climate*, *6*, 1412232. DOI: 10.3389/fclim.2024.1412232

Kirbayeva, D. K., Balouch, H., Xu, T., Bauenova, M. O., Zaletova, D. E., & Kossalbayev, B. D. (2025). The role of microalgae in circular carbon capture systems. *Functional Plant Biology : FPB*, *52*(9), FP25195. DOI: 10.1071/FP25195 PMID: 40977150

Kourkoumpas, D. S., Bon, A., Sagani, A., Atsonios, K., Grammelis, P., Karellas, S., & Kakaras, E. (2024). Life cycle assessment of novel thermochemical–biochemical biomass-to-liquid pathways for sustainable aviation and maritime fuel production. *Bioresource Technology*, *393*, 130115. DOI: 10.1016/j.biortech.2023.130115 PMID: 38013031

Krishnamoorthy, S., Kuppam, C., & Mamilla, R. C. R. (2024). Evaluation of carbon capture methodologies, mechanisms, and improvements for sustainable carbon dioxide mitigation using microalgae. *Industrial Biotechnology (New Rochelle, N.Y.)*, *20*(5), 186–203. DOI: 10.1089/ind.2024.0015

Kumar, L., & Bharadvaja, N. (2024). A review on microalgae biofuel and biorefinery: Challenges and way forward. *Energy Sources. Part A, Recovery, Utilization, and Environmental Effects*, *46*(1), 16599–16622. DOI: 10.1080/15567036.2020.1836084

Kushwaha, O. S., Uthayakumar, H., & Kumaresan, K. (2023). Modeling of carbon dioxide fixation by microalgae using hybrid artificial intelligence (AI) and fuzzy logic (FL) methods and optimization by genetic algorithm (GA). *Environmental Science and Pollution Research International*, *30*(10), 24927–24948. DOI: 10.1007/s11356-022-19683-0 PMID: 35349067

Lan, K., Ou, L., Park, S., Kelley, S. S., Nepal, P., Kwon, H., Cai, H., & Yao, Y. (2021). Dynamic life-cycle carbon analysis for fast pyrolysis biofuel produced from pine residues: Implications of carbon temporal effects. *Biotechnology for Biofuels*, *14*(1), 191. DOI: 10.1186/s13068-021-02027-4 PMID: 34587989

Lee, M., Lin, Y. L., Chiueh, P. T., & Den, W. (2020). Environmental and energy assessment of biomass residues to biochar as fuel: A brief review with recommendations for future bioenergy systems. *Journal of Cleaner Production*, *251*, 119714. DOI: 10.1016/j.jclepro.2019.119714

Leech, D., Previtera, D. A., Zhang, Y., Botella, J. R., & Crisp, P. A. (2025). Precision plant epigenome editing: What, how, and why. *Trends in Plant Science*. PMID: 40915965

Li, G., Xiao, W., Yang, T., & Lyu, T. (2023). Optimization and process effect for microalgae carbon dioxide fixation technology applications based on carbon capture: A comprehensive review. *C*, *9*(1), 35.

Li, S., & Zhu, L. (2023). Copper regulates degradation of typical antibiotics by microalgal-fungal consortium in simulated swine wastewater: Insights into metabolic routes and dissolved organic matters. *Water Research*, *245*, 120654. DOI: 10.1016/j.watres.2023.120654 PMID: 37778083

Li, X., Sun, X., Gao, L., Xu, J., & Gao, G. (2021). Effects of periodical dehydration on biomass yield and biochemical composition of the edible red alga Pyropia yezoensis grown at different salinities. *Algal Research*, *56*, 102315. DOI: 10.1016/j.algal.2021.102315

Liang, R. X., Hung, J. C., Diankristanti, P. A., Chen, Y. T., Chung, C. W., & Ng, I. S. (2025). Enhanced carbon capture and utilization in transgenic Chlorella sorokiniana harboring pyridoxal kinase under dynamic carbon dioxide levels. *Bioresource Technology*, *425*, 132315. DOI: 10.1016/j.biortech.2025.132315 PMID: 40023335

Liu, Y., Cruz-Morales, P., Zargar, A., Belcher, M. S., Pang, B., Englund, E., Dan, Q., Yin, K., & Keasling, J. D. (2021). Biofuels for a sustainable future. *Cell*, *184*(6), 1636–1647. DOI: 10.1016/j.cell.2021.01.052 PMID: 33639085

Long, B., Fischer, B., Zeng, Y., Amerigian, Z., Li, Q., Bryant, H., Li, M., Dai, S. Y., & Yuan, J. S. (2022). Machine learning-informed and synthetic biology-enabled semi-continuous algal cultivation to unleash renewable fuel productivity. *Nature Communications*, *13*(1), 541. DOI: 10.1038/s41467-021-27665-y PMID: 35087023

López i Losada, R., Rosenbaum, R. K., Brady, M. V., Wilhelmsson, F., & Hedlund, K.i Losada. (2024). Agent-Based Life Cycle Assessment enables joint economic-environmental analysis of policy to support agricultural biomass for biofuels. *The Science of the Total Environment*, *916*, 170264. DOI: 10.1016/j.scitotenv.2024.170264 PMID: 38253104

López-Pacheco, I. Y., Castillo-Vacas, E. I., Castañeda-Hernández, L., Gradiz-Menjivar, A., Rodas-Zuluaga, L. I., Castillo-Zacarías, C., Sosa-Hernández, J. E., Barceló, D., Iqbal, H. M. N., & Parra-Saldívar, R. (2021). CO2 biocapture by Scenedesmus sp. grown in industrial wastewater. *The Science of the Total Environment*, *790*, 148222. DOI: 10.1016/j.scitotenv.2021.148222 PMID: 34380253

Ma, Z., Cheah, W. Y., Ng, I. S., Chang, J. S., Zhao, M., & Show, P. L. (2022). Microalgae-based biotechnological sequestration of carbon dioxide for net zero emissions. *Trends in Biotechnology*, *40*(12), 1439–1453. DOI: 10.1016/j.tibtech.2022.09.002 PMID: 36216714

Maliha, A., & Abu-Hijleh, B. (2023). A review on the current status and post-pandemic prospects of third-generation biofuels. *Energy systems*, *14*(4), 1185-1216.

Matinha-Cardoso, J., Santos, T., Pereira, H., Varela, J., Tamagnini, P., & Mota, R. (2023). Pilot scale production of Crocosphaera chwakensis CCY0110 and evaluation of its biomass nutritional potential. *Algal Research*, *69*, 102939. DOI: 10.1016/j.algal.2022.102939

Miranda, A. M., Hernandez-Tenorio, F., Ocampo, D., Vargas, G. J., & Sáez, A. A. (2022). Trends on CO2 capture with microalgae: A bibliometric analysis. *Molecules (Basel, Switzerland)*, *27*(15), 4669. DOI: 10.3390/molecules27154669 PMID: 35897845

Moñino Fernández, P., López Morales, M., de Winter, A., van den End, F., Janssen, M., & Barbosa, M. (2025). Pilot-scale oxygen-balanced mixotrophic cultivation of Galdieria sulphuraria. *ACS Sustainable Chemistry & Engineering*, *13*(5), 2132–2140. DOI: 10.1021/acssuschemeng.4c09186 PMID: 39950107

Naduthodi, M. I. S., Claassens, N. J., D'Adamo, S., van der Oost, J., & Barbosa, M. J. (2021). Synthetic biology approaches to enhance microalgal productivity. *Trends in Biotechnology*, *39*(10), 1019–1036. DOI: 10.1016/j.tibtech.2020.12.010 PMID: 33541719

Nagi, G. K., Minhas, A. K., Gaur, S., Jain, P., & Mandal, S. (2021). Integration of algal biofuels with bioremediation coupled industrial commodities towards cost-effectiveness. *Frontiers in Energy Research*, *9*, 735141. DOI: 10.3389/fenrg.2021.735141

Nair, A., & Chakraborty, S. (2020). Synergistic effects between autotrophy and heterotrophy in optimization of mixotrophic cultivation of Chlorella sorokiniana in bubble-column photobioreactors. *Algal Research*, *46*, 101799. DOI: 10.1016/j.algal.2020.101799

Najeeb, M. I., Anjum, A. A., Ahmad, M. U. D., Ali, T., Maqbool, A., & Manzoor, R. (2023). Optimization of physicochemical factors for microalgal biomass using indigenously isolated microalgae. *JAPS: Journal of Animal & Plant Sciences*, *33*(6)

Nayak, S. K., Bhattacharyya, P., Padhy, S. R., Das, A., Parida, S. P., Rath, M., & Nayak, A. (2025). Harnessing the potential of microalgae for carbon sequestration to achieve net-zero emissions. *Mitigation and Adaptation Strategies for Global Change*, *30*(7), 58. DOI: 10.1007/s11027-025-10249-2

Ng, I. S., Keskin, B. B., & Tan, S. I. (2020). A critical review of genome editing and synthetic biology applications in metabolic engineering of microalgae and cyanobacteria. *Biotechnology Journal*, *15*(8), 1900228. DOI: 10.1002/biot.201900228 PMID: 32080963

Nneoma, U. C., Chukwudi, O. F., Nnenna, U. J., & Paul-Chima, U. O. (2025). Hybrid biofactories: Integrating microalgae and engineered microbiomes for enhanced biofuel production in circular carbon systems. *Frontiers in Energy Research*, *13*, 1654079. DOI: 10.3389/fenrg.2025.1654079

Olabi, A. G., Alami, A. H., Alasad, S., Aljaghoub, H., Sayed, E. T., Shehata, N., Rezk, H., & Abdelkareem, M. A. (2022). Emerging technologies for enhancing microalgae biofuel production: Recent progress, barriers, and limitations. *Fermentation (Basel, Switzerland)*, *8*(11), 649. DOI: 10.3390/fermentation8110649

Pant, S., Goel, M., Sahoo, N. K., Yuan, Q., & Rout, P. R. (2025). Microalgae-based treatment for wastewater management and valorisation. *Current Opinion in Environmental Science & Health*, *48*, 100662. DOI: 10.1016/j.coesh.2025.100662

Polat, E., & Altınbaş, M. (2023). A review on microalgal growth stress parameters for sustainable biofuel production. *Clean Technologies and Environmental Policy*, *25*(8), 2469–2487. DOI: 10.1007/s10098-023-02535-x

Prasad, R., Gupta, S. K., Shabnam, N., Oliveira, C. Y. B., Nema, A. K., Ansari, F. A., & Bux, F. (2021). Role of microalgae in global CO2 sequestration: Physiological mechanism, recent development, challenges, and future prospective. *Sustainability (Basel)*, *13*(23), 13061. DOI: 10.3390/su132313061

Priyadharsini, P., Nirmala, N., Dawn, S. S., Baskaran, A., SundarRajan, P., Gopinath, K. P., & Arun, J. (2022). Genetic improvement of microalgae for enhanced carbon dioxide sequestration and enriched biomass productivity: Review on CO2 bio-fixation pathways modifications. *Algal Research*, *66*, 102810. DOI: 10.1016/j.algal.2022.102810

Qin, K., Ye, X., Luo, S., Fernie, A. R., & Zhang, Y. (2025). Engineering carbon assimilation in plants. *Journal of Integrative Plant Biology*, *67*(4), 926–948. DOI: 10.1111/jipb.13825 PMID: 39783795

Raines, C. A., Cavanagh, A. P., & Simkin, A. J. (2022). Improving carbon fixation. In *Photosynthesis in action* (pp. 175–192). Academic Press. DOI: 10.1016/B978-0-12-823781-6.00009-5

Rajpoot, A. S., Shende, V., Chelladurai, H. M., Dwivedi, G., Verma, T. N., & Choudhary, T. (2025). Exploring the potential and progress of microalgae biomass production, harvesting, and pre-treatment for sustainable biofuel production: A comprehensive review. *Environment, Development and Sustainability*, •••, 1–51. DOI: 10.1007/s10668-025-05984-0

Rossi, S. (2025). Bioeconomy analysis of microalgal products and process: the role of microalgae technology in the blue economy. In *Microalgae as Promising Source of Commercial Bioproducts: Advances and Emerging Technologies in Bioactive Compounds, Biofuels, Food and Biorefineries* (pp. 343–352). Springer Nature Switzerland. DOI: 10.1007/978-3-031-86433-9_23

Sarker, N. K., & Kaparaju, P. (2024). Microalgal bioeconomy: A green economy approach towards achieving sustainable development goals. *Sustainability (Basel)*, *16*(24), 11218. DOI: 10.3390/su162411218

Sarmah, M., & Panda, M. R. (2026). Systematic review on proliferation of microalgae and carbon sequestration potential for sustainability. *World Journal of Microbiology & Biotechnology*, *42*(3), 100. DOI: 10.1007/s11274-026-04849-6 PMID: 41731188

Shang, Z., Qi, F., Mu, R., Ren, Y., Ma, G., & Yin, M. (2022). Enhanced potential of a Selenastrum capricornutum mutant for CO2 fixation and biofuels production. *Environmental Progress & Sustainable Energy*, *41*(5), e13838. DOI: 10.1002/ep.13838

Song, Y., Hu, Z., Yang, X., An, Y., & Lu, Y. (2025). Synergistic microalgae-duckweed systems for enhanced aquaculture wastewater treatment, biomass recovery, and CO2 sequestration: A novel approach for sustainable resource recovery. *Environmental Research*, *274*, 121271. DOI: 10.1016/j.envres.2025.121271 PMID: 40043930

Sridevi, V., Hamzah, H. T., Terapalli, A., Varma, J. V., Rao, C. S., & Ramesh, F. (2026). Net-Zero Emission and Microalgae Biotechnology Carbon Dioxide Sequestration. *Microalgae Biotechnology and Sustainable Development*, 131-155.

Subhash, G. V., Rajvanshi, M., Kumar, G. R. K., Sagaram, U. S., Prasad, V., Govindachary, S., & Dasgupta, S. (2022). Challenges in microalgal biofuel production: A perspective on techno economic feasibility under biorefinery stratagem. *Bioresource Technology*, *343*, 126155. DOI: 10.1016/j.biortech.2021.126155 PMID: 34673195

Sun, Z., Bo, C., Cao, S., & Sun, L. (2025). Enhancing CO2 fixation in microalgal systems: Mechanistic insights and bioreactor strategies. *Marine Drugs*, *23*(3), 113. DOI: 10.3390/md23030113 PMID: 40137299

Swami, S., Lallawmkimi, M. C., Singh, V., Yadav, K. K., Alagendran, S., Koushal, S., & Lairenjam, G. (2025). Harnessing edaphic microalgae for soil carbon sequestration and climate-smart agriculture. *Archives of Current Research International*, *25*(3), 155–171. DOI: 10.9734/acri/2025/v25i31105

Tang, D. Y. Y., Khoo, K. S., Chew, K. W., Tao, Y., Ho, S. H., & Show, P. L. (2020). Potential utilization of bioproducts from microalgae for the quality enhancement of natural products. *Bioresource Technology*, *304*, 122997. DOI: 10.1016/j.biortech.2020.122997 PMID: 32094007

Titirici, M. M. (2013). Green carbon. *Sustainable Carbon Materials from Hydrothermal Processes*, 1-36.

Ummalyma, S. B., Sirohi, R., Udayan, A., Yadav, P., Raj, A., Sim, S. J., & Pandey, A. (2023). Sustainable microalgal biomass production in food industry wastewater for low-cost biorefinery products: A review. *Phytochemistry Reviews*, *22*(4), 969–991. DOI: 10.1007/s11101-022-09814-3 PMID: 35431709

Veeramuthu, A., & Ngamcharussrivichai, C. (2020). Potential of microalgal biodiesel: challenges and applications. *Renewable Energy-Technologies and Applications*, 167-180.

Vishwakarma, R., Selvaraju, K., & Malik, A. (2025). Critical insights into the dynamics of carbon capture for enhancement of photosynthetic efficiency in microalgae. *Environmental Science and Pollution Research International, 32*(34), 20787–20797. DOI: 10.1007/s11356-025-36837-y PMID: 40813513

Wang, R., Wang, X., & Zhu, T. (2025). Research progress and application of carbon sequestration in industrial flue gas by microalgae: A review. *Journal of Environmental Sciences (China), 152*, 14–28. DOI: 10.1016/j.jes.2024.04.018 PMID: 39617540

Wicker, R. J. (2023). Bioprospecting for nordic photosynthetic consortia for wastewater treatment, carbon capture, and value creation.

Xin, K., Cheng, J., Guo, R., Qian, L., Wu, Y., & Yang, W. (2024). Nuclear mutagenesis and adaptive evolution improved photoautotrophic growth of Euglena gracilis with flue-gas CO2 fixation. *Bioresource Technology, 397*, 130497. DOI: 10.1016/j.biortech.2024.130497 PMID: 38408501

Yahaya, E., Yeo, W. S., Nandong, J., & Ngu, J. C. Y. (2025). CO2 fed microalgae cultivation in photobioreactor: Review on challenges and possible solutions. *Environmental Technology Reviews, 14*(1), 540–564. DOI: 10.1080/21622515.2025.2508945

Yamaoka, Y., Petroutsos, D., Je, S., Yamano, T., & Li-Beisson, Y. (2025). Light, CO2, and carbon storage in microalgae. *Current Opinion in Plant Biology, 84*, 102696. DOI: 10.1016/j.pbi.2025.102696 PMID: 39983365

Yang, H., & Xin, X. (2023). CO2 capture and lipid production performance of microalgae in the S-shaped photobioreactor under different culture modes. *Enzyme and Microbial Technology, 165*, 110194. DOI: 10.1016/j.enzmictec.2023.110194 PMID: 36682097

Yang, Q., Wu, J., & Li, G. (2025). Control of cell–cell communication and deciding when to grow. *Trends in Plant Science, 30*(6), 585–587. DOI: 10.1016/j.tplants.2025.01.003 PMID: 39893119

Yaqoubnejad, P., Aminirad, H., & Taghavijeloudar, M. (2021). Optimization of Cultivation Condition of Newly Isolated Strain Chlorella Sorokiniana pa. 91 for CO2 Bio-Fixation and Nutrients Removal From Real Municipal Wastewater: Impact of Temperature and Light Intensity.

Zhang, B. (2025). *Development and evaluation of a methanotroph-microalgae mixed-culture platform for integrated biogas utilization and valorisation* (Doctoral dissertation, Nanyang Technological University).

Zhang, M., Wang, Z., Zhang, X., Qian, X., & Shen, G. (2020). Biogas and quality fertilizer production from dry anaerobic digestion of rice straw with nitrogen addition. *Bioresource Technology Reports*, *11*, 100509. DOI: 10.1016/j.biteb.2020.100509

Zhang, W., Zhao, C., Cao, W., Sun, S., Hu, C., Liu, J., & Zhao, Y. (2020). Removal of pollutants from biogas slurry and CO2 capture in biogas by microalgae-based technology: A systematic review. *Environmental Science and Pollution Research International*, *27*(23), 28749–28767. DOI: 10.1007/s11356-020-09282-2 PMID: 32468373

Zhu, T., Ning, P., Liu, Y., Liu, M., Yang, J., Wang, Z., & Li, M. (2025). Knowledge of microalgal Rubiscos helps to improve photosynthetic efficiency of crops. *Planta*, *261*(4), 78. DOI: 10.1007/s00425-025-04645-w PMID: 40042639

Zhu, X., Xiao, J., Wang, C., Zhu, L., & Wang, S. (2022). Global warming potential analysis of bio-jet fuel based on life cycle assessment. *Carbon Neutrality*, *1*(1), 25. DOI: 10.1007/s43979-022-00026-4

Zieliński, M., Dębowski, M., Kazimierowicz, J., & Świca, I. (2023). Microalgal carbon dioxide (CO2) capture and utilization from the European Union perspective. *Energies*, *16*(3), 1446. DOI: 10.3390/en16031446

Chapter 4
Genetic Engineering and Synthetic Biology for Enhanced Carbon and Nutrient Capture

Rahul Kumar
http://orcid.org/0009-0000-1323-2220
Department of Microbiology, Graphic Era (Deemed to be) University, Dehradun, India

ABSTRACT

Microalgae are promising biotechnological tools for carbon sequestration and wastewater valorization, with the dual benefit of greenhouse gas emission mitigation and nutrient recovery from waste streams. However, the natural metabolic efficiency of microalgae for carbon fixation and nutrient uptake is suboptimal for large-scale applications. Recent technological breakthroughs in genetic engineering and synthetic biology have enabled the selective manipulation of photosynthetic, metabolic, and regulatory networks to improve the efficiency of carbon and nutrient capture. The present chapter discusses the molecular and synthetic biology tools used for optimizing critical pathways, such as the Calvin Benson cycle, carbon concentrating mechanisms, and nutrient assimilation systems for nitrogen, phosphorus, and sulfur. Advanced strategies, such as CRISPR/Cas-mediated genome editing, pathway rewiring, synthetic promoter design, and genome scale metabolic modeling, are report for creating superior algal strains.

DOI: 10.4018/979-8-3373-7630-1.ch004

INTRODUCTION

The advancements in the field of genetic engineering and synthetic biology have greatly altered the possibilities of biological systems to address some of the most pressing concerns of the twenty first century related to the environment (Goold et al., 2018). Notably, the increasing levels of global carbon emissions and nutrient imbalances found in the environment, leading to soil degradation, eutrophication, and the use of synthetic fertilizers, have resulted in an urgent demand from the scientific community for novel solutions that can be easily combined with natural systems. Taking advantage of newer technologies, such as CRISPR genome editing, metabolic engineering, and the construction of synthetic carbon fixation pathways, there is an opportunity to reengineer plants, microbes, and algae to convert carbon and nutrient inputs to novel endpoints with unprecedented levels of biological efficiency and effectiveness. Climate resilient, low input agro-systems comprising engineered crops with superior photosynthetic pathways, deep roots, and synthetic symbioses with nitrogen fixing microbes offer an unprecedented opportunity to counter the problem of climate change (Ullah et al., 2025). The synthetic microbes and cyanobacteria, which act as biological factories that convert CO_2 to fuel, bioplastics, and long lived soil carbon, offer an immediate opportunity to fulfill the national CO_2 sequestration targets that most nations face today because of the Kyoto Treaty to mitigate climate change impacts. The cutting edge of this technology, artificial metabolic cycles and the production of semi-synthetic organisms extend the boundaries of biological capabilities, overcoming the limits of natural evolution to achieve CO_2 fixation and specific biochemical conversions that exceed the levels found in natural systems (Singla, 2021). In this respect, rapid progress in this technology offers an unprecedented opportunity to change the future of biological sciences from a system that preserves nature to one that redesigns nature with clever technology to make the environment and the world safe, healthier, and even cleaner. Nonetheless, there is an immediate demand from the scientific and global communities to redefine the biosafety, use, and restraints of this modern scientific technology to mitigate the risks related to the biosphere, environment, and health, apart from the intellectual property concerns and equity that this technology offers to the world's developing nations (Forero-Pineda, 2006; Grama et al., 2022).

Role of Microalgae in Carbon Sequestration and Wastewater Valorisation

The major application areas for microalgae revolve around carbon capture and wastewater treatment, owing to their unique metabolic adaptability, fast growth rates, and tolerance to nutrient rich wastewater. The advanced photosynthetic organisms, microalgae possess a tremendous capacity to capture carbon dioxide from the atmosphere, far surpassing the capacity of plants to transform it into biomass by utilizing the Calvin cycle and specialized carbon fixation mechanisms (Prasad et al., 2021). Thus, they act as potent biological carbon capture technologies that can potentially offset carbon emissions from industrial flue gases within closed photo bioreactors and open pond reactors. Microalgae demonstrate unique properties in wastewater treatment because of their natural capacity to readily take up nitrogen, phosphorus, and organic compounds, thereby lowering biochemical oxygen demand and inhibiting the development of eutrophication in water bodies after wastewater discharge. The wastewater exists as an inexpensive and readily available nutrient supply for algal culture development, presenting a symbiotic relationship for both carbon capture and wastewater treatment technologies (Viswanaathan et al., 2022). The algal biomass obtained after carbon capture and wastewater treatment can then be converted into valuable products, such as biofuels, bioplastics, animal feed, biofertilizers, pigments, and medications, thereby opening new horizons for a biologically based recycling economy. The latest biotechnological breakthroughs encompassing CRISPR-assisted strain modifications for improving microalgal carbon fixation rates and protein/lipid synthesis could potentially increase the biotechnological utilization potential of microalgae based biorefineries for multiple applications (Ummalyma et al., 2025; Cao et al., 2023). In addition, the integration process using anaerobic digestion, carbon rich flue gas capture, and hydrothermal liquefaction technologies could potentially improve the sustainability and economy associated with carbon capture using microalgae, while promoting biorefineries for wastewater treatment and carbon capture applications. Therefore, the application potential for carbon capture using microalgae could act as a sustainable multi-faceted technological system addressing climate change issues, resource exploitation, and pollution management simultaneously using biotechnologies (Ravichandran et al., 2024).

Limitations of Natural Metabolic Efficiency in Microalgae

Microalgae have a remarkable ability to photosynthesize and biochemically transform substances; however, their natural metabolic efficiency is limited by various biological and environmental constraints, rendering them unsuitable for carbon sequestration and biological production on a larger scale. The major bottleneck in their metabolic efficiency is the slow rate of Rubisco, the major carbon fixing enzyme, which turns over relatively slowly and often involves oxygenation, contributing to the wastefulness of photorespiration (Busch, 2020). The photosynthetic apparatus of microalgae reaches saturation at lower light intensities, whereas the production of photo inhibition, along with the production of reactive oxygen species, causes cellular damage to their machinery, thus reducing photosynthetic efficiency. Nutrient limitation further poses metabolic constraints, as the natural photosynthetic metabolism of microalgae depends greatly on the right amounts of nitrogen, phosphorus, trace elements, and vitamins in their medium, making their growth conditions, especially those in wastewater or outdoors, highly susceptible to variations in nutrient concentrations. In addition, the use of carbon dioxide in photosynthetic activity may be hampered by the loss of expression of carbon-fixation mechanisms in particular environments, thereby reducing the efficiency of carbon use in photosynthetic assays (Ray et al., 2022). In the natural strains, their metabolic pathways may be optimized toward survival rather than toward high biomass or metabolic production, leading to a metabolic imbalance in the partitioning of their photosynthetic products, including lipids, carbohydrates, or proteins of high economic value to their biomass. Metabolite production, particularly lipids, through stress responses often requires inhibition of biomass production, thus rendering them production in a tradeoff between the two in continuous cultures. In addition, various biological constraints, including daily variations, high levels of temperature, contamination, and competition, pose diverse additional metabolic inefficacies within the photosynthetic reactions of the microorganisms, particularly in the case of photosynthetic reactions in the open (Pan et al., 2021).

The Need for Genetic Engineering and Synthetic Biology to Overcome Biological Constraints

The natural biological limitations of living organisms, such as the inevitable suboptimization of photosynthetic efficiency, carbon fixation rates, nutrient utilization, and metabolic tradeoffs, highlight the increasing necessity of genetic engineering and synthetic biology to optimize the potential of living organisms and systems. Although natural selection has optimized plants, microalgae, and microbes for survival in certain natural environments, these systems are by no means optimized for exploitation and use in controlled environments (Amin et al., 2025). Rather, the most crucial enzymes involved in these systems, including Rubisco, work well below their theoretical efficiencies, while metabolic networks direct carbon resources to maintain the organism and utilize them to synthesize valuable compounds. Furthermore, carbon utilization systems and nutrient-uptake mechanisms are frequently under optimized and insufficient to support the rapid growth and utilization of these microorganisms. Genetic engineering addresses these limitations by providing the capability to make precise changes to microbial genomes and metabolic networks (Montaño López et al., 2022). Synthetic biology, however, takes a much more holistic and comprehensive approach that enables the design and construction of novel biological systems and pathways that are unachievable in natural biological systems. Synthetic biology will enable the design and implementation of novel metabolic and carbon fixation pathways, advanced light harvesting mechanisms, stress tolerant microorganisms, and microorganisms that produce commodity chemicals, such as fuels and soil-stabilizing polymers, on a scale that will meet global demands. Synthetic biology also provides the design and construction capabilities of novel chassis microorganisms programmed and optimized for a wide variety of functions, including the removal of nutrients from wastewater and the use of microbes to produce carbon negative goods (Liu et al., 2020). Moreover, the need to combine the best attributes of genetic engineering and synthetic biology underlines the optimization and design of these technologies as a prerequisite for optimizing and meeting the growth and utilization demands of these microbes on a scale that is commercially viable and will fulfill the demands and deficiencies of the world's energy and carbon utilization (Figure 1).

Figure 1. sources and synthetic biology development strategies

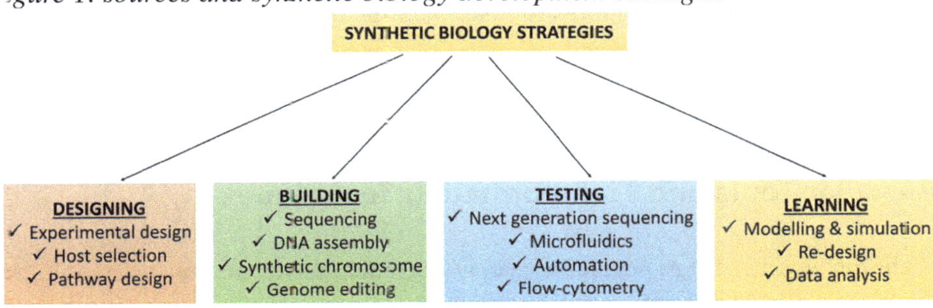

FUNDAMENTALS OF GENETIC ENGINEERING AND SYNTHETIC BIOLOGY IN MICROALGAE

Concepts and Distinctions Between Genetic Engineering and Synthetic Biology

Although genetic engineering and synthetic biology stand out as interrelated but conceptually distinct approaches, both provide specific modes of modifying biological systems. Genetic engineering broadly refers to the purposeful manipulation of genes within a host to achieve a definite functional objective through gene insertion, deletion, silencing, and modification (Lanigan et al., 2020). It acts within the natural biological context of an organism by modifying its individual components to enhance certain traits, such as metabolite production, stress tolerance, and growth efficiency. The synthetic biology has gone beyond manipulation of just a few genes to the systematic design and construction of new biological parts, pathways, and whole organisms. It incorporates engineering principles, standardization, modularity, and abstraction to produce programmable biological systems that may not exist in nature (Kim et al., 2025). Whereas genetic engineering is often done in increments and has been organism specific, synthetic biology has emphasized the building of synthetic gene circuits, pathways, and chassis organisms with defined biological modules like promoters, ribosome binding sites, and regulatory switches. Another point of divergence pertains to the scale and level of complexity; while traditional genetic engineering adjusts one or a few genes, synthetic biology performs large scale re-wiring that enables the construction of artificial carbon fixation cycles, bio-computational networks, or engineered microbial consortia. Synthetic biology uses computational modelling, high throughput design build test cycles, and automation to allow predictability and rational design (Matzko et al., 2024). Genetic engineering provides a fundamental set of tools for manipulating the genome, whereas synthetic biology provides a conceptual and technical framework for building novel biological functions. The convergence of these two creates transformational microalgae into efficient platforms for carbon capture, nutrient recycling, and bio manufacturing.

Molecular Tools and Transformation Techniques

The development of molecular biotechnology and transformation methodologies is crucial for the genetic engineering of microalgae to improve carbon sequestration, nutrient utilization, and bioproduct formation. The major molecular biotechnology tools used for this purpose are CRISPR-Cas systems, zinc-finger nucleases (ZFNs), and transcription activator like effector nucleases (TALENs), which enable precise editing of microalgae genomes by generating site specific mutations, insertions, or deletions (Zhang et al., 2019). Another technique, RNA interference (RNAi), and antisense RNA enable gene silencing at the post-transcriptional level, offering additional regulatory options for carbon and nutrient fluxes. Additional molecular biotechnology resources for genetic engineering of microalgae include reporter genes and regulatory elements and inducible promoter regulatory components for selecting and directing the expression of genes of interest (Ahmad Kamal et al., 2024). On the other hand, the methodology for cell transformation of microalgae has varied based on the species of interest and the target genomes, including the nucleus, chloroplast, and mitochondria. Cell transformation techniques for this purpose include electroporation and biolistic gel-bombardment, mechanical cell agitation using glass beads and silicon carbide fibers, and PEG-mediated transfection and liposome mediated vectors to improve cell uptake of DNA. Agrobacterium mediated transformation of microalgae has been successfully developed based on natural horizontal gene transfer

(Sharma et al., 2021). Another major advantage of chloroplast transformation in microalgae is the improved protein expression and minimization of gene silencing and transgene escape due to sexual reproduction. Furthermore, the development of synthetic biotechnology for this purpose has been crucial for designing optimal carbon and nutrient fluxes for photosynthesis, improving bioproduct synthesis, and developing artificial carbon-concentrating mechanisms (Table 1).

Table 1. synthetic biology frameworks for algal systems, including modular design, chassis optimization, and gene circuits

Agenda	Approaches	Applications in Algae	Advantages	Challenges
Modular Design	Standardized biological parts (promoters, ribosome binding sites, terminators) Modular metabolic pathways interchangeable regulatory elements	Assembly of synthetic carbon fixation pathways, Lipid or biofuel production, Stress responsive modules	Predictable and repeatable designs, Easy combination of modules, Rapid prototyping	Context dependent interactions may affect performance, Limited availability of well characterized algal parts
Chassis Optimization	Selection of robust algal species (*Chlamydomonas reinhardtii*, Nannochloropsis), Genome streamlining, Codon optimization, Organelle targeting (chloroplast, mitochondria)	Enhanced growth and biomass accumulation, Improved carbon capture and nutrient uptake, High level recombinant protein expression	Increased metabolic efficiency, Reduced metabolic burden, Stable transgene expression	Species specific limitations, Balancing growth vs metabolite production, Risk of ecological escape in open systems
Synthetic Gene Circuits	Inducible promoters and sensors, Toggle switches and logic gates, Feedback and feedforward loops, CRISPR based regulatory circuits	Controlled expression of carbon fixation genes, Stress or nutrient responsive metabolite production, Dynamic regulation of lipid or carbohydrate synthesis	Precise temporal and spatial control, Responsive to environmental cues, Integration with metabolic engineering strategies	Circuit complexity may reduce reliability, Potential off target effects, Burden on host metabolism
Integrated Synthetic Platforms	Combination of modular pathways, optimized chassis, and gene circuits, High throughput design build test learn cycles, Computational modeling and simulation	Industrial scale carbon sequestration, Wastewater remediation with nutrient recycling, Bio product synthesis (biofuels, bioplastics, pigments)	System level optimization, Predictable and scalable performance, Multifunctional algal systems	High engineering complexity, Need for advanced computational and experimental infrastructure, Regulatory and biosafety considerations

ENGINEERING CARBON FIXATION PATHWAYS

The potential application of engineered carbon fixation cycles in microalgae and other phototrophic model organisms has emerged as a promising direction for mitigating the inherent limitations and challenges of natural photosynthesis in carbon sequestration. The main enzyme involved in the assimilation of CO_2, RuBisCO, has limited turnover efficiency and sometimes performs oxygenation reactions that promote wasteful photorespiration cycles (Busch, 2020). To tackle these challenges, various synthetic biology tools have been developed, including the optimization of Rubisco kinetic properties, designer Rubisco variants isolated from cyanobacteria and microalgae, and the design of engineered chaperones that improve enzyme stability and assembly properties. Furthermore, the integration of carbon fixation mechanisms such as carboxysomes and bicarbonate transporters has improved the concentration of CO_2 near the RuBisCO enzyme, thereby increasing the overall carbon sequestration efficiency and preventing wasteful photorespiration (Jin et al., 2023). In addition to enhancing the efficiency of the Calvin-Benson-Bassham cycles, researchers have developed new fully synthetic routes such as the CETCH pathway, reversed TCA cycles, and hybrid metabolic pathways that could be far more optimum in efficiency and metabolic energy utilization (Rubin-Blum et al., 2019). Other metabolic engineering techniques allow the conversion of sequestered carbon into various valuable commodities, such as lipids, carbohydrates, or polymers, which can serve as valuable by products in linking biological carbon sequestration with biomanufactures. Advanced techniques involve combined computational models, genome-scale metabolic reconstructions, and CRISPR regulation mechanisms that help predict the efficiency of these metabolic pathways on larger scales by ensuring predictable control of these pathways (Cardiff et al., 2024). In conclusion, metabolic engineering of pathways involved in carbon sequestration in microalgae has vast potential in reducing CO_2 emissions and enhancing overall biomass productivity by establishing sustainable protocols that link carbon sequestration with bioprocess manufacturers that convert these manufacturers into biofuels, bioplastics, and nutrient recovery by filling the existing gaps between basic biological mechanisms of photosynthesis and its biotechnological applications.

Algal Photosynthetic Carbon Metabolism

A key component of algal photosynthetic carbon metabolism is the ability to reduce atmospheric CO_2 concentrations and covalently bond them into compounds that are essential for growth, biomass, and bioproduct formation. The underlying carbon fixation process is catalyzed by the Calvin-Benson-Bassham (CBB) cycle, which catalyzes the binding of CO_2 to 3-phosphoglycerate through the enzyme ribulose-1,5-bisphosphate carboxylase/oxygenase (Rubisco), which is subsequently converted through a series of reactions that regenerate ribulose-1,5-bisphosphate and yield triose phosphates (Meloni et al., 2023). RuBisCO is a widespread enzyme in algal photosynthesis; however, it is known for its low activity rate, tendency for oxidation, and involvement in photorespiration, which leads to energy losses in algal cells. To counter these shortcomings, most algal cells have implemented strategies designed as carbon-concentrating mechanisms (CCMs). These are biological methods that actively increase the concentration of CO_2 inside algal cells near Rubisco enzymes, which have limited activity at low CO_2 concentrations in the atmosphere under sunlight exposure (Zhang and Liu, 2021). These strategies include biologically active methods such as the bicarbonate transporter system, as well as the use of specialized compartments such as carboxysomes and pyrenoids, which are designed for Rubisco entrapment, ensuring higher concentrations of CO_2 in algal cells. These are in activating biologically active carbon fixation pathways in sunlight for enhanced algal growth, biomass, and bioactive compound yields in biotechnological applications such as bioproducts, bioplastics, bioliquids, biogases, and biocolloids in bioremediation applications in biotechnology (Akash et al., 2023). Moreover, these biologically active carbon fixation pathways in algal cells under sunlight illumination offer substantial platforms for understanding how algal cells utilize environmental signals such as sunlight, CO_2, nutrients, and temperature, hence controlling the biosynthesis pathways of lipids, carbohydrates, proteins, lipoproteins, and biocolloids through biologically active algal carbon metabolic pathways in critical biotechnological applications in bioremediation for enhancing bioproducts and bioplastics.

Strategies for Enhancing Carbon Fixation

Methods for improving carbon fixation in microalgae and other photoorganisms involve the use of metabolic, genetic, and synthetic biology tools to mitigate the innate drawbacks of natural photosynthetic efficiency. A principal approach is aimed at optimizing RuBisCO, the principal enzyme of carbon fixation, by introducing variants in cyanobacteria/algal RuBisCO enzymes, including chaperone enhancement of enzyme aggregation and stability (Chen et al., 2023). In a complementary effort, the maximization or addition of carbon concentrating mechanisms (CCMs), such as pyrenoids, carboxysomes, or transporters of bi-carbonates, increases the localized concentration of Rubisco in the synthesized photosynthetic pigments, thus opposing photorespiration and optimizing the efficiency of carbon fixation. In metabolic engineering, the focus is on the redirection of carbon metabolism to target biomass products, including lipids, starch, or valuable primary/secondary compounds, while ensuring minimal metabolic loss of carbon-rich products (Mipeshwaree Devi et al.,

2023). Synthetic biologists have recently been successful in adding the CETCH cycle or the rTCA cycle to the photosynthetic machinery of algae or other target organisms, potentially outperforming the efficiency of the Calvin-Benson-Bassham cycle under specific conditions (Feng et al., 2023). Additional approaches to boost photosynthetic efficiency include the optimization of the light absorption properties of photosystem antenna rings, stress tolerance, or modification of regulatory elements to maintain high rates of carbon fixation performance even under dynamic light response conditions. Aspects of photosynthetic engineering have been merged with or supplemented by bioprocessing, including controlled photobioreactors or flue gases containing high concentrations of carbon dioxide, thereby optimizing the maximum rates of photosynthetic carbon fixation (Stephens et al., 2021). It should be understood that the above approaches aimed at photosynthetic improvement involve the use of genetic, synthetic, or other bioprocessing tools to achieve maximum carbon fixation rates and sustainable bioproduction of various products (Table 2).

Table 2. case studies demonstrating increased carbon assimilation in algae via genetic modification

Algal Species	Genetic Modification	Mechanism Strategy	Observed Outcomes
Chlamydomonas reinhardtii	Overexpression of Rubisco activase	Enhances Rubisco activation and CO_2 fixation	Increased photosynthetic rate by ~20–30%, higher biomass accumulation
Chlamydomonas reinhardtii	CRISPR-mediated knockdown of glycolate dehydrogenase	Reduces photorespiration and recycles carbon internally	15–25% increase in net carbon assimilation
Synechococcus elongates	Introduction of heterologous carbon-concentrating carboxysomes	Elevated CO_2 at Rubisco active sites	Up to 40% increase in CO_2 fixation efficiency
Nannochloropsis gaditana	Overexpression of key Calvin cycle enzymes (FBPase, SBPase)	Accelerates carbon fixation flux	Enhanced growth rate and lipid productivity
Chlamydomonas reinhardtii	Synthetic CETCH cycle integration	Engineered alternative CO_2 fixation pathway	Experimental increase in carbon assimilation under lab conditions
Dunaliella salina	Overexpression of bicarbonate transporters	Improved inorganic carbon uptake	Enhanced biomass accumulation and carotenoid production
Chlamydomonas reinhardtii	Chloroplast-targeted expression of pyrenoid proteins	Formation of synthetic CCM like compartments	Higher carbon fixation under low CO_2 conditions
Phaeodactylum tricornutum	CRISPR editing of carbonic anhydrase genes	Improved CO_2 conversion to bicarbonate for Rubisco	Increased photosynthetic efficiency and growth

SYNTHETIC BIOLOGY APPROACHES FOR ENHANCED NUTRIENT UPTAKE AND ASSIMILATION

Synthetic biology provides transformative approaches to increase nutrient uptake and assimilation in microalgae and other microorganisms owing to nitrogen, phosphorus, and other elemental limitations that limit growth and bioproduct synthesis (Mathew et al., 2022). One strategy is the overexpression or engineering of nutrient transporters, including high affinity nitrate, ammonium, and phosphate transport proteins, which are critical for increasing the uptake efficiency of limiting or variable nutrient sources, including wastewater streams. Complementing this are the rewiring of regulatory circuits and transcriptional networks to optimally express assimilation enzymes, and hence rapidly convert inorganic nitrogen and phosphorus into biomass or metabolites with minimized energy wastage. Synthetic metabolic pathways have been introduced to allow the utilization of alternative nitrogen sources, organic phosphorus compounds, or even waste derived nutrients that wild-type strains cannot process efficiently (Krysenko, 2023). Modular gene circuits and inducible promoters introduce dynamic control, enabling cells to adjust nutrient assimilation in response to environmental cues and to maintain metabolic balance under fluctuating conditions. Chloroplast targeted engineering further enhances assimilation by localizing enzymes, where reducing power and ATP are abundant, which improves metabolic efficiency. Moreover, synthetic biology enables the design of microbial consortia that cooperate in nutrient capture and transformation, creating synergistic systems that couple wastewater remediation with biomass production (Jaiswal et al., 2023). These strategies help increase nutrient uptake, reduce dependence on synthetic fertilizers, support circular bioeconomy approaches, and facilitate sustainable cultivation in industrial or environmental settings. Integration of transporter engineering, pathway optimization, design of regulatory circuits, and engineering of microbial communities thus makes synthetic biology a robust framework for maximizing nutrient assimilation and enhancing productivity and environmental performance in algal systems (Table 3).

Table 3. Importance of nitrogen, phosphorus, and sulfur metabolism in microalgal growth and bioproduct formation

Nutrient	Metabolic Role in Microalgae	Impact on Growth	Impact on Bioproduct Formation	Examples
Nitrogen (N)	Incorporated into amino acids, nucleotides, chlorophyll, and proteins	Essential for biomass accumulation and photosynthetic efficiency	Influences synthesis of proteins, pigments (e.g., phycobiliproteins, carotenoids), and secondary metabolites; nitrogen limitation can trigger lipid accumulation	Nitrate, ammonium, or urea are primary sources; engineering N-assimilation enzymes enhances uptake and storage
Phosphorus (P)	Component of nucleic acids, ATP, phospholipids, and regulatory molecules	Critical for energy transfer, cell division, and membrane synthesis	Supports ATP-dependent biosynthesis, lipid metabolism, and polyphosphate accumulation; P limitation may induce lipid or starch storage	Phosphate transporters and phosphatases regulate uptake; microbial consortia or engineered transporters can enhance utilization
Sulfur (S)	Incorporated into sulfur-containing amino acids (cysteine, methionine), coenzymes, and vitamins	Essential for protein synthesis, redox balance, and enzymatic activity	Affects production of glutathione, sulfolipids, and certain pigments; sulfur limitation can alter metabolic flux toward protective metabolites	Sulfate uptake and assimilation pathways can be engineered to improve growth and stress tolerance

Engineering of Transporters for Efficient Nutrient Uptake From Wastewater

Engineering nutrient transporters in microalgae represents a promising strategy for enhancing the efficiency of nutrient uptake from wastewater, allowing both bioremediation and biomass production (Ranjbar and Malcata, 2022). Nitrogen, phosphorus, and trace elements in wastewater are often present in chemical forms that are suboptimal or variable for natural uptake. Through synthetic biology and genetic engineering, the overexpression or optimization of high affinity transporters for nitrate, ammonium, phosphate, and other ions can be performed to increase the rates and capacity of nutrient acquisition. In addition, the broadening of transporter specificity can be aimed at utilizing alternative or organic nutrient sources, whereas designed regulatory circuits may induce transporter expression under nutrient-limited conditions. Chloroplast or plasma membrane targeted transporters improve assimilation efficiency by localizing nutrient entry points into the cell where metabolic demand is the highest (The et al., 2021). These modifications enhance growth rates and nutrient removal efficiencies in wastewater based cultivation systems, while simultaneously producing biomass rich in proteins, lipids, or pigments. Engineered transporters can also be integrated into microbial consortia, which have other complementary nutrient transformation pathways, creating synergistic systems for wastewater valorisation. The improving both uptake kinetics and metabolic integration, transporter engineering provides a scalable and sustainable approach to couple environmental remediation with biofuel and bioproduct generation, supporting the circular bioeconomy model and reducing reliance on synthetic fertilizers (Ebadi et al., 2025).

Modulation of Metabolic Pathways

Metabolic pathway modulation in microalgae is thus a potent means of optimizing carbon fixation, nutrient uptake, and biomaterial production, thus enabling microalgae to overcome their natural metabolic limitations and productivity constraints. Metabolic pathway modulation in microalgae is thus essentially aimed at engineered modifications of enzymes, regulatory elements, and flux-controlling steps in primary and secondary metabolism (Kumar et al., 2020). For example, overexpression of enzymes involved in Calvin-Benson-Bassham cycle or alternate carbon fixation pathways can lead to higher CO_2 fixation rates, while knockdown of competing pathways like photorespiration or glycolytic by-passes can prevent wasteful turnover of reducing equivalents. Similarly, nitrogen, phosphorus, and sulfur assimilation pathways can be engineered to optimize their uptake, storage, and utilization for proteins, lipids, and secondary compounds of value (Zenda et al., 2021). Modular gene circuits, inducible promoters, and feedback regulation systems in synthetic biology would thus enable microalgae to modulate their metabolic fluxes in response to environmental inputs or nutrient availability. Metabolic flux analysis, genome-scale models, and omics-assisted genetic engineering would thus help in identifying bottlenecks and pathways for efficient and predictable interventions. Moreover, pathway modulation can be combined with targeted gene expression in chloroplasts or mitochondria to boost cofactor synthesis and metabolic efficiency (Kumar et al., 2023). All these approaches enable microalgae to become programmable factories for optimized carbon and nutrient use, stress resilience, and increased lipids, carbohydrates, pigments, and secondary compound production thus finding applications in biotechnology and biomanufacturing processes with positive benefits in terms of sustainable biomanufacturing, waste bioremediation, and carbon negativity.

Synthetic Pathway Design for Coupling Nutrient Capture With Value-Added Product Synthesis

Synthetic pathway design integrates the capture of nutrients with the biosynthesis of valuable products in microalgae to create a favorable system for environmental bioremediation and biomanufacturing of valuable products. Utilizing the power of synthetic biology, it is possible to redesign metabolic pathways that can assimilate nutrients, such as nitrogen, phosphorus, and carbon, while using fixed nutrients for the biosynthesis of valuable products such as oils, bioplastics, pigments, and bioactive compounds (Roell and Zurbriggen, 2020). This process involves the application of a genetic circuit design for the coordinated sensing and assimilation of nutrients with downstream biosynthetic enzymes to ensure the effortless use of nutrients for the formation of biomass and value-added products. For example, the overexpression of high affinity transporters for nitrogen and phosphorus can improve the use of wastewater nutrients simultaneously utilized for the fixed carbon and lipid biosynthetic pathways to produce biofuels or biopolymers for different industrial applications. Feedback and feed forward regulatory components enable the continuous control of fluxes to maximize the use of nutrients while avoiding the formation of undesired byproducts for different biotechnological applications (Abinandan et al., 2018). Computer simulations and metabolic reconstruction algorithms facilitate the integration of metabolic pathways to identify bottlenecks that predict the effectiveness of multigene approaches for biotechnological applications. The application of chloroplast targeted genetic expression approaches improves the availability of cofactors for enhanced bioreaction rates and efficiencies for different biotechnological applications (Singhal et al., 2023). The integration of nutrient capture and synthetic biosynthesis for biotechnological applications makes algal systems multifunctional to clean the environment, reduce the use of chemical fertilizers, and produce valuable products to establish new biorevolution for the creation of a bioeconomy platform for biotechnological applications in the future.

METABOLIC REWIRING FOR COUPLED CARBON AND NUTRIENT UTILIZATION

Systems Biology Approaches to Identify Metabolic Bottlenecks

Using genome scale metabolic reconstructions, flux balance analysis, and constraint based modeling, simulations of metabolic flux distributions under different environmental or genetic conditions can be realized, indicating steps where substrate conversion is incomplete or energy might be lost (Qiu et al., 2023). Moreover, transcriptomics, proteomics, and metabolomics datasets provide more insight into the regulatory and post-translational constraints on pathways, including enzymes, transporters, and cofactor dependent reactions that impede holistic pathway performance. These approaches allow the identification of competing pathways, energy drains, and feedback loops, thereby reducing the efficacy of carbon and nutrient use. Based on these findings, synthetic biology and metabolic engineering strategies can be developed to overexpress rate-limiting enzymes, knockdown competing pathways, or introduce alternative synthetic routes to bypass the respective bottlenecks (Choi et al., 2019). Systems biology also sustains dynamic pathway optimization as a function of designs in regulatory circuits that lead to changing metabolic fluxes in response to environmental fluctuations or nutrient availability. Overall, the application of systems biology allows for a data-driven and rational strategy toward maximizing productivity, enhancing carbon and nutrient utilization, and designing and engineering high performance algal strains for biofuel production, bio manufacturing, and environmental remediation (Nneoma et al., 2025).

Integration of Carbon and Nitrogen Metabolism for Optimal Productivity

Coordinating these two metabolic networks ensures the balance of carbon and nitrogen supply, and importantly, minimizes the accumulation of unused intermediates and energy dissipation (Zhang et al., 2018). Synthetic biology and metabolic engineering allow the integration of these two metabolic networks by modulating regulatory enzymes, transporters, and signaling pathways, enabling microalgae to dynamically adjust the carbon flux in response to nitrogen availability and vice versa. For instance, the overexpression of nitrate or ammonium transporters increases nitrogen assimilation, which can then drive amino acid biosynthesis and optimize carbon flux toward biomass or lipid biosynthesis (Yang et al., 2025). Computational modeling and systems biology tools help map the interconnections between carbon and nitrogen pathways, identify bottlenecks, and enable the design of gene circuits that coordinate fluxes for maximal productivity. Moreover, coupling of these integrated metabolic networks with environmental or nutrient responsive promoters enables adaptive regulation under fluctuating light, CO_2, or nitrogen conditions. Altogether, the harmonization of carbon and nitrogen metabolism provides a basic strategy to enhance photosynthetic efficiency, nutrient use, and the sustainable production of biofuels, bioplastics, pigments, and other value-added compounds in engineered algal systems (Sarker and Kaparaju, 2025).

Flux Balance Analysis and Genome-Scale Metabolic Models

FBA and GEMs are powerful bioinformatics tools for the investigation, prediction, and optimization of metabolic fluxes in microalgae, with respect to enhanced carbon fixation, nutrient assimilation, and biosynthesis of value-added bio products (Tamburro and Boyle, 2025). GEMs represent a complete metabolic network, integrating knowledge of genes, enzymes, reactions, and metabolites encoding the full metabolic capacity of the cell. FBA utilizes these models with the application of stoichiometric constraints and linear programming to predict flux distributions across metabolic pathways under defined environmental and genetic conditions, identify rate-limiting steps, pathway bottlenecks, and trade-offs between growth and metabolite production. Simulation of gene knockouts, overexpression, or incorporation of alternative pathways enables rational design strategies through metabolic engineering or synthetic biological interventions (Ramakrishnan et al., 2025). Examples of GEM-based FBA in *Chlamydomonas reinhardtii* and *Nannochloropsis*, among other algal species, have targeted the optimization of lipid biosynthesis, carbon flux reallocation from storage carbohydrates, and improvement of nitrogen and phosphorus assimilation under nutrient limiting conditions. These models allow integration with omics experimental data, such as transcriptomics, proteomics, and metabolomics, which can provide dynamic insights into metabolic regulation and resource allocation (Satrio et al., 2024). When merged with approaches in synthetic biology, GEM-based FBA guides the rational rewiring of metabolic networks, enabling the design of algal strains with enhanced photosynthetic efficiencies, nutrient recovery, and the production of advanced biofuels, bioplastics, pigments, and other high value compounds. Thus, FBA and GEM form essential bioinformatic frameworks for the predictive system level optimization of algal metabolism.

Engineering Regulatory Networks and Signaling Cascades for Resource-Efficient Metabolism

Engineering regulatory networks and signaling cascades in microalgae represents a versatile approach for the optimization of resource usage, promotion of metabolic efficiency, and augmentation of bioproduct synthesis under dynamic environments and nutritional conditions. Cellular metabolism is carefully regulated by intricate networks involving transcription factors, kinases, small RNAs, and signaling molecules, all of which are involved in managing carbon fixation, nutrient acquisition, and stress responses in cells (Zhu and Thompson, 2019). Synthetic biology tools can be utilized for the strategic design of these regulatory networks, dynamically redirecting metabolic flux in response to internal and external signals, which ensures the optimized distribution of energy and resources towards growth, as well as bioproduct synthesis. For example, inducible promoters, synthetic transcription factors, and riboswitches can be utilized to promote the expression of genes involved in carbon fixation and nutrient acquisition under low CO_2 concentrations and low nutrient availability while repressing unnecessary pathways to maintain minimal metabolic burden. Feedback and feedforward elements can be introduced to maintain homeostasis, which helps to prevent the build-up of toxic intermediates, while modular genetic circuits can be designed for the coordinated regulation of multiple pathways concurrently (Karoui et al., 2022). Signaling pathways controlling nitrogen, phosphorus, and sulfur acquisition can also be rewired to work in coordination with carbon acquisition, which helps link bioproduct synthesis with biomass synthesis. Computer simulations and tools related to systems biology can help identify critical regulatory nodes in these networks, which can then predict the effect of modifications in these networks. The integrating these regulatory pathways with metabolic pathways, it is possible to achieve resource optimized metabolism in microalgae, which can maintain high product synthesis under dynamically changing light, CO_2 concentration, and nutritional conditions in the environment (Sun et al., 2018). These strategies can aid in sustainable biotechnological applications, including bioproduct synthesis, wastewater remediation, and biofuel synthesis in bioreactors.

SYNTHETIC CONSORTIA AND ALGAL-BACTERIAL SYNERGIES

Concept of Synthetic Microbial Consortia for Enhanced Carbon and Nutrient Cycling

Synthetic microbial consortia are defined as the combination of two or more different microbial species with complementary metabolic processes for augmented carbon and nutrient cycling that go beyond individual microbe capabilities. These engineered microbial populations have different specializations, in which photoautotrophic microalgae perform CO_2 fixation from the atmosphere to organic carbon, whereas heterotrophic bacteria perform functions such as nutrient remineralization, nitrogen fixation, phosphorus solubilization, and the removal of metabolic byproducts (Shilky, Patra et al., 2023). The harnessing these different functions, synthetic consortia form a closed loop in which carbon and nutrients are incessantly recycled, thereby promoting the utmost biomass and minimal waste formation. Synthetic Biology offers approaches for optimizing these different functions, such as optimizing interspecies interaction in microbe communities by engineered approaches based on quorum sensing, metabolic exchange networks, and inducible gene operating networks, which interact in dynamic networks controlling microbe utilization and metabolic transition in microbe consortia. Systems Biology and metabolic networks help predict microbe consortium behavior and pinpoint microbe consortium bottlenecks, which in turn aids in optimizing microbe ratio composition to form efficient microbe populations with optimal microbe ratios essential for microbe consortium stability and productivity (Semple and Pandhal, 2025). Rather, this method is particularly efficient in algae and bacterial microbe consortium based wastewater farming, where algae use their capability for carbon sequestration and removal of excess nutrients, whereas heterotrophic bacteria augment nutrient processing and augment algal growth and harvesting for bioproduct applications and uses. Furthermore, by strategically designing these microbe consortia, culture fixed carbon and processed nutrients can specifically produce high demand bioproducts such as lipids, bioproducts, pigments, and biopolymers.

Genetic Engineering of Symbiotic Interactions Between Algae and Bacteria

Genetic engineering of symbiotic interactions between algae and bacterial species has emerged as a promising area for improving nutrient assimilation, carbon fixation, and overall efficiency in algal-bacterial co-cultivation and synthetic consortium communities (Viswanaathan et al., 2022). In both natural and synthetic ecosystems, symbiotic interactions exist between algae and bacterial species, where algae provide the following benefits to the bacterial species: (i) oxygen, (ii) carbon, and (iii) organic metabolites, thereby promoting bacterial growth and reproduction. In contrast, bacterial species provide essential nutrients, vitamins, and enzymes, thereby promoting algal growth and metabolism. Genetic engineering of these interactions could make them optimal and predictable. For example, bacterial species can be made to overproduce growth promoting materials such as vitamins B_{12}, siderophores, or signaling molecules, which could potentially promote algal growth and metabolism (Balabanova et al., 2021). In contrast, algae can be genetically modified to secrete specific metabolites or signaling molecules, such as those involved in promoting beneficial bacterial species, thereby enabling efficient nitrogen and phosphorus cycling within the algal bacterial community. The latest tools in synthetic biology have made it possible to design circuits for interspecies communication, quorum sensing, and regulation of metabolic fluxes between species, such that efficient exchange of resources can be achieved. Systems biology and modeling have enabled the identification of key regulatory and bottleneck interactions between species for rational engineering (Balabanova et al., 2021). Genetically modified algal-bacterial systems have successfully demonstrated improvements in the efficiency of nutrient recovery from wastewater, CO_2 sequestration, and high-value biofuel production such as lipids, pigments, and biopolymers. Thus, the genetic engineering of symbiotic interactions between algae and bacterial species has emerged as a promising area for improving the overall efficiency of these algal-bacterial systems, with vast application prospects for bioindustrial and environment-related applications (Kumar and Verma, 2025).

Synthetic Communication Systems

These quorum sensing networks use signaling molecules, such as acyl-homoserine lactone or peptides, to provide individual cells with information about population density and coordinate gene expression in response (Fuqua et al., 2001). The genetically programming such quorum sensing networks, it is possible to dynamically control algal and bacterial metabolism to activate nutrient assimilation, carbon fixation, or metabolite secretion only when the algal-bacterial consortium attains optimum density, thus optimizing efficiency and preventing energy wastage. In addition to quorum sensing, metabolic cross feeding improves algal-bacterial symbiosis and reinforces cooperation between algae and bacteria. This involves enabling metabolite transfer in both algae to bacteria (carbon) and bacteria to algae (vitamins, amino acids, and algal growth factors). The applying synthetic biology to algal-bacterial consortia, programmable cross feeding components can be designed to ensure perfectly balanced metabolite exchange and prevent the excessive generation of inhibitory compounds (Balouch et al., 2025). An integrated quorum sensing and cross feeding system is capable of generating self-regulating algal-bacterial consortia, where both carbon and nutrient conversion rates are optimized to improve algal biomass productivity and algal bioproduct (lipids, pigments, and biopolymers) yield. Such algal-bacterial consortia with synthetic communication networks hold significant promise for designing and developing adaptive and multifaceted algal-bacterial biosystems for sustainable bioproduction, wastewater bioremediation, and carbon negative technologies.

Applications in Biofilm Reactors and Wastewater Systems

Algal-bacterial associations have been extensively harnessed in biofilm reactors and wastewater treatment bioreactors owing to their complementary interactions that promote carbon sequestration, nutrient removal, and bioproduct synthesis. In biofilm bioreactors, algae and bacteria coexist in stable, attached biofilms, maintaining high cellular concentrations and efficient biomass retention as opposed to suspended cultures (Manoharan et al., 2025). Algae and bacteria in biofilms co-occur in close physical proximity, enabling efficient oxygen, carbon, and nutrient transfer between them, during which algae can fix carbon dioxide and bacteria can convert nitrogen, phosphorus, and other micronutrient elements in the wastewater. In addition to their use as biological support systems in continuous wastewater treatment, biofilms can be specifically designed to accumulate lipids, pigments, or biopolymers. In wastewater treatment bioreactors, designed microbial communities efficiently remove excess nitrogen and phosphorus, mitigating eutrophication, whereas the organic carbon in wastewater supports bacterial growth, particularly heterotrophic bacteria, during the concurrent growth of algae in wastewater (Mathew et al., 2022). The incorporating the latest technological approaches in the field of synthetic biology, including the application of quorum sensing and metabolic cross feeding, nutrient uptake and bioproduct synthesis in algae/bacterial associations can be controlled in a dynamic, optimized, and wastewater composition dependent manner. Additionally, the growth of algae/bacteria in biofilms and wastewater treatment bioreactors can decrease the high costs of algae/bacteria separation in various wastewater treatment approaches and promote a closed loop bioeconomy by using wastewater treatment to produce high value compounds, in addition to other benefits (Mathur et al., 2024).

Advantages and Challenges of Engineered Consortia in Large-Scale Applications

Engineered algal bacterial consortia offer great advantages for large scale applications in biofuel production, wastewater treatment, and bioproduct synthesis, although some challenges must be overcome for industrial deployment. Among the main advantages, these consortia combine complementary metabolic capabilities, enabling simultaneous carbon fixation, nutrient recovery, and high-value metabolites production that improves the system's efficiency compared to the single monocultures (Rosero-Chasoy et al., 2021). Dynamic regulation through synthetic circuits and quorum sensing confers consortia flexibility against fluctuating environmental conditions and stability to productivity under light, CO_2, or nutrient fluctuations. High biomass retention within biofilm reactors and integrated wastewater systems reduces harvesting costs and improves resource utilization, whereas coupling waste remediation and metabolite synthesis supports models for a circular bioeconomy. However, scaling engineered consortia from lab or pilot scale systems to industrial reactors presents considerable challenges. The stable coexistence of population ratios and metabolic balance over time is challenging owing to competition, genetic drift, and environmental heterogeneity (van den Berg et al., 2022). Productivities may be constrained in large reactors owing to limitations in light, nutrients, and mass transfer, while engineered organisms must be well contained to avoid their release into natural ecosystems. In addition, some requirements such as regulatory compliance, biosafety considerations, and long-term genetic stability require continuous control. The economic feasibility depends on the optimization of the reactor design, operation parameters, and consortia composition. The strategic design and management of engineered microbial consortia allows for a very promising approach to realizing high efficiency, multifunctional, and environmentally sustainable biotechnological applications on a large scale by combining resource recovery, carbon sequestration, and bioproduct generation (Bilal et al., 2025).

TOOLS, TECHNIQUES, AND COMPUTATIONAL PLATFORMS

CRISPR/Cas-Based Genome Editing and Precision Engineering

CRISPR/Cas-mediated genome editing has been increasingly recognized as a revolutionary technology for precision manipulation of microalgal genomes to enhance carbon fixation rates, nutrient uptake rates, and production rates of valuable bio-products. This technology involves the use of a guide RNA molecule to guide Cas nuclease to a targeted genomic location to produce double-stranded breaks that are precisely repaired through natural cellular pathways to achieve insertions, deletions, and substitutions (Aledhari and Rahouti, 2024). This technology presents unprecedented accuracy and flexibility relative to traditional mutagenesis technologies to enable the editing of crucial enzymes involved in carbon fixation through the Calvin-Benson-Bassham cycle, transport proteins for nutrients, and regulatory proteins responsible for managing flux through these pathways. CRISPR/Cas systems have been engineered to enable precise gene regulation through dead Cas proteins (dCas) to enhance the dynamic control of these pathways by activating or repressing gene expression without changing the genetic information (Kazi and Biswas, 2021). Simultaneous editing of multiple genomes is achievable through multiplex editing to modify the multiple pathways involved in carbon and nutrient utilization. The integration of high-throughput screening technologies and precision genome editing through omics and modeling tools provides a deeper combinatorial enablement of precision manipulation of algal genomes to develop high performance algal species at a faster pace to meet the diverse needs of sustainable bio-based technologies such as production of biofuels, wastewater purifiers, and value-added bio-based products (Table 4).

Table 4. Omics-Driven Design in Microalgae

Omics Approach	Data Generated	Applications in Microalgal Engineering	Advantages Insights	Examples
Genomics	Complete genetic blueprint; gene identification; regulatory elements	Identifying targets for CRISPR/Cas editing, synthetic pathway integration, and metabolic engineering	Provides foundational information on potential genes and pathways; enables genome-scale metabolic reconstructions	Whole-genome sequencing of *Chlamydomonas reinhardtii*, *Nannochloropsis* spp.
Transcriptomics	Gene expression profiles under varying conditions; mRNA abundance	Revealing active metabolic pathways, stress response mechanisms, and regulatory nodes for dynamic control	Highlights condition-specific gene regulation; identifies bottlenecks and inducible pathways	RNA-Seq analysis for nutrient stress responses in microalgae
Proteomics	Protein abundance, modifications, enzyme activity	Determining functional enzyme levels, metabolic flux capacity, and pathway efficiency	Provides insight into post-transcriptional regulation; links gene expression to functional metabolism	Mass spectrometry-based profiling of Calvin cycle enzymes
Metabolomics	Intracellular and extracellular metabolite concentrations; pathway flux	Assessing metabolic activity, cofactor availability, and product formation	Direct readout of pathway output; informs flux optimization and pathway rewiring	LC-MS analysis of lipid precursors under nitrogen limitation

Bioinformatics and AI-Assisted Metabolic Optimization

Bioinformatics and artificial intelligence assisted metabolic optimization have emerged as crucial tools for the rational design of high-performance microalgae, facilitating the integration of multi-omics data, genome scale models, and environmental conditions for metabolic optimization (Yusof et al., 2025; Kumar et al., 2026). Bioinformatic tools offer functionalities in genome Annotation, Pathway Reconstruction, Promoter Discovery, and Codon Optimization, which serve as the starting points for genetic and synthetic approaches. The leveraging Artificial Intelligence and machine learning algorithms, it is possible to analyze such data to identify meaningful patterns, model metabolic bottlenecks, and compute optimized approaches in metabolic rewiring, nutrient management, and flux maximization. AI algorithms can assess multiple genetic, environmental, or nutrient stimulation conditions, indicating maximum efficiency gain, carbon fixation, nutrient uptake, and production of bioproducts (Kushwaha et al., 2023). Using deep learning techniques, it is relatively easy to make accurate models of enzyme kinetics, interaction, and network regulation, ensuring rapid analyses of complex metabolic networks, often incommensurate with experimental interpretations. Furthermore, Artificial Intelligence-assisted optimization approaches can help in strategy formulation to optimize the performance of the designed synthetic genetic regulatory elements, circuits, and microbial communities to achieve dynamic and adaptive metabolic control (Nguyen et al., 2025). In addition, implementation in high throughput algorithms will help gain multiple cycles of predictive, experimental, or implemented validation, accelerating the rapid development of strains. The integrating the potential of bioinformatics, Artificial Intelligence, and systems biology, it is hopefully feasible to create highly efficient microalgal platforms to gain sustainable biotechnological applications in the production of biofuels, bioremediation of wastewater resources, carbon sequestration, and production of high-value biological compounds in various industries (Nneoma et al., 2025).

Synthetic Promoter and Gene Circuit Design for Expression

The design of synthetic promoters and gene circuits is a major approach for the precise regulation of gene expression in microalgae to enable the regulated overexpression of carbon fixation, nutrient uptake, and metabolite production. The use of synthetic promoters involves the design of engineered DNA sequences that can be programmed for strength, inducibility, and sensitivity to external and internal cues, thus ensuring the temporal and conditional regulation of target genes (Artemyev et al., 2024). The strategic use of synthetic promoters and transcription regulators, riboswitches, or small RNAs for designing gene circuits enables the development of modular gene designs capable of integrating and coordinating several metabolic pathways to achieve optimal flux and alleviate metabolic burden. The use of gene circuits provides the benefits of logic-based controls, feedback regulation, and oscillatory controls for the dynamic regulation of enzyme expression based on external and internal cues of nutrient availability, CO_2 concentration, and stress responses. The design of gene circuits enables optimal coordination of carbon fixation and nutrient assimilation, including phosphorus and nitrogen, to ensure optimal allocation and biosynthesis of metabolites (Chubukov et al., 2014). Additionally, gene circuit designs enable the coordinated regulation of several pathways without interference owing to their modularity and orthogonality, thus offering optimal biotechnological development for biofuel, bioplastics, pigments, and bioremediation.

High-Throughput Screening, Automation, and Synthetic Biology Toolkits for Algal Systems

High-throughput screening, automation, and synthetic biology toolkits have revolutionized engineering in algal systems, allowing the rapid and systematic optimization of metabolic pathways, gene circuits, and performance at scale (Zammit et al., 2023). Microfluidic devices, droplet-based assays, and automated plate readers facilitate the evaluation of hundreds to thousands of engineered strains under a wide variety of environmental and nutrient conditions, thus enabling the identification of superior variants with improved carbon fixation, nutrient assimilation, and the synthesis of bioproducts. Automation of cultivation, sampling, and analytical procedures reduces labor intensity, minimizes human error, and enables the precise control of experimental parameters for reproducibility and scalability. Synthetic biology toolkits encompass modular promoters and ribosome-binding sites, CRISPR/Cas components, and standardized genetic parts that offer a versatile framework for the construction and testing of complex metabolic pathways and regulatory circuits in a predictable way (Müller et al., 2025). Integration with computational design, machine learning, and omics data enables rational design build test learn cycles to accelerate strain development and optimization. Thus, the integration of these techniques allows for the systematic assessment of metabolic rewiring, pathway flux optimization, and engineered interspecies interactions in consortia. This will enable the construction of robust algal platforms for the production of biofuels, synthesis of pigments, accumulation of lipids, and valorization of wastewater. The combination of high-throughput methodologies, automation, and modular synthetic biology tools can reduce the time and cost of algal strain engineering to an order of magnitude lower and enable scalable, efficient, and sustainable biotechnological applications that manage carbon and nutrients (Fabris et al., 2020).

CASE STUDIES AND RECENT ADVANCES

Engineered *Chlamydomonas reinhardtii* for High CO_2 Fixation Efficiency

Chlamydomonas reinhardtii has been recognized as a model organism for genetic engineering of high CO_2 fixation because of its well-understood genome, facile genetic modification, and strong photosynthetic ability (Vavitsas et al., 2019). Research in genetic and synthetic biology has aimed at the optimization of the Calvin-Benson-Bassham cycle, the main carbon fixation route in organisms. Genes involved in this cycle, including ribulose-1,5-bisphosphate carboxylase/oxygenase (Rubisco), sedoheptulose-1,7-bisphosphatase, and phosphoribulokinase, have been shown to improve the rate of carbon fixation and biomass production at various levels of CO_2 (De Porcellinis et al., 2018). CRISPR/Cas deletion of genes involved in photorespiration improves carbon fixation efficiency by reducing carbon loss. Furthermore, genetic modification of CCM, specifically the overexpression of genes involved in the transport of bicarbonate and carbonic anhydrase, improves the cellular concentration of CO_2, especially under conditions where the concentration of exogenous CO_2 is low. Synthetic promoters and dynamic gene switches facilitate the temporal regulation of these biological processes in organisms in relation to environmental signals, ensuring the efficient use of metabolic flux and energy (Ream and Prather, 2024). Synthetic integration of omics modeling and metabolic flux analyses in organisms allows the elucidation of limiting factors in biological systems, including the genetic engineering of high-performing strains. Engineered *C. reinhardtii* strains show high levels of improved CO_2 fixation, paving the way for the sustainable production of biofuels, carbon sequestration, and production of valuable metabolic products in various applications (Priyadharsini et al., 2022).

Nannochloropsis Mutants for Enhanced Nitrogen Utilization

Nannochloropsis species have emerged as promising microalgal resources for the production of biofuels and other biomaterials because of their fast growth rate, high lipid content, and efficient use of nutrients. Recently, there has been significant interest in the use of genetic modification and synthetic biology to generate Nannochloropsis mutants with improved nitrogen use to facilitate efficient biomass production, especially in areas where nitrogen is abundant, such as wastewater (Poliner et al., 2018). To this end, high-affinity nitrate and ammonium transporters have been overexpressed to facilitate efficient use of nitrogen, even at low concentrations. Enzymes such as nitrate reductase and glutamine synthetase, which participate in the assimilation of nitrogen, have also been upregulated to facilitate faster utilization of ingested nitrogen to produce amino acids, proteins, and other nitrogenous compounds. Knocking out other pathways using the CRISPR/Cas system has also contributed to redirecting the nitrogen used towards the production of biomass and other valuable products (Shi et al., 2022). Cooperation with pathways responsible for lipid synthesis ensures that ingested nitrogen is utilized towards the production of biomass and other products. Omics analyses have also enabled the rational selection of genes to be targeted in Nannochloropsis to produce superior strains that can be utilized to produce biofuel from wastewater, thereby eliminating the problem of nitrogen waste, which poses significant hazards to the environment (Sarwan et al., 2024).

Cyanobacterial Synthetic Circuits for CO_2 Capture and Biofuel Co-Production

Cyanobacteria have been found to be useful microbial chassis tools for CO_2 fixation and biofuel production because of their natural photosynthetic capacity, short doubling time, and easy genetic expertise. Cyanobacteria have been engineered to improve carbon fixation, manipulate metabolic pathways, and coordinate the biosynthesis of biofuels, such as ethanol, isobutanol, and fatty acids (Khan et al., 2019). The constructs utilize specific synthetic promoters, riboswitches, and transcriptional regulators to precisely control the gene expression of key enzymes involved in the Calvin-Benson-Bassham cycle, CO_2 concentrating mechanisms, and downstream biofuel biosynthesis. For instance, inducible or photoregulated promoters have been able to temporally regulate carbon metabolism, thereby increasing CO_2 fixation or assimilation, while reducing metabolic costs, especially under suboptimal conditions. Metabolic interactions and feedback controls can then be integrated to strategically coordinate the energy allocation between biomass and biofuels (Briat and Khammash, 2018). Recent expert systems have been able to incorporate sensors to monitor intracellular carbon and redox levels, thereby adaptively controlling biofuel metabolism according to the metabolic conditions. The integration of advanced engineered circuits and genome-scale metabolic reconstructions, combined with omic-based information, can help identify metabolic bottlenecks and optimize pathway performance. Engineered cyanobacterial systems exhibit improved CO_2 capture capabilities and simultaneous biofuel production, making them a viable, scalable, and sustainable tool for mitigating CO_2 pollution levels while producing valuable energy products (Sadvakasova et al., 2025).

ENVIRONMENTAL AND REGULATORY CONSIDERATIONS

The use of genetically modified microalgae for enhanced carbon sequestration, nutrient uptake, and bioproducts has critical environmental and regulatory aspects that require thorough consideration to ensure environmental safety and sustainability. Among the most direct concerns is the environmental risk potential for genetically modified strains released into environmental compartments, which poses the hazard of horizontal gene transfer, unexpected interactions with naturally occurring organisms, uncontrolled growth in ecological environments, and disruption of local biogeochemical cycles (Verma et al., 2025). To the environmental containment, such as physical barriers, closed bioreactors, and carefully managed growth, is critical for ensuring environmental safety. Moreover, biocontainment circuits, such as synthetic auxotrophy, inducible lethality, and metabolic dependencies on synthetic nutrients, are critical genetic controls that ensure failure in environmental growth, thus lowering the potential for accidental release. The regulatory policies covering genetically modified microalgae are quite different in various international platforms, including guidelines from international regulatory bodies such as the United States Environmental Protection Agency. The European Food Safety Authority, and biosafety clearances from various biosafety regulatory bodies in different countries, which require strict environmental risk assessments, environmental monitoring, and strict biosafety guideline compliance (Hundleby et al., 2018). Finally, focusing on regulatory guidelines, there is an increasing focus on the application of life cycle analysis (LCA) tools to determine the post-production environmental impacts associated with genetically modified microalgal technologies, including energy consumption, greenhouse gas production, water usage, nutrient recycling, and bioproduct management practices. The practice is directly helpful in systematically comparing genetically modified organisms and conventional technologies, enabling the verification process for potential optimization in sustainable technology applications, among other environmental factors (Thakur et al., 2023). Through the comprehensive integration of environmental risk management, strictly modified microalgae designs, regulatory guidelines, and life cycle analysis, it is ensured that genetically modified organisms can be successfully exploited in environmental applications, regardless of potential environmental disturbances, through a sustainable green economy conceptual approach.

FUTURE PERSPECTIVES AND RESEARCH DIRECTIONS

The future of microalgal biotechnology has emerged at the nexus of synthetic biology, machine learning, and digital biofoundries, where future algal strains can be rapidly designed, analyzed, and optimized to unprecedented levels of accuracy and speed (Ali et al., 2024). The synergistic implementation of computational modeling, artificial intelligence predictive modeling, and advanced automated platforms enables scientists to discern optimal gene modification, regulatory, and metabolic designs, rapidly developing strains for improved carbon fixation, nutrient utilization, and high-value bioproduction. The development of these strains provides a foundation for a biologically based closed-loop bioeconomy, where wastewater treatment, CO_2 capture, and biomass utilization can be directly integrated with bioproduction, waste reduction, and resource utilization (Kuppan et al., 2023). There is an increasing potential for carbon negative bioproduction, where CO_2 can be actively captured from the atmosphere while producing commercial quantities of renewable fuels, bioplastics, and other renewable products. However, there are hurdles to overcome regarding scaled commercial implementation, stable genetic expression when cultured for prolonged periods, and ensuring public acceptance and risk assessment. Looking to the future, the vision of synthetic ecosystems and self-optimizing biorefineries entails programmable, modular microbial communities, capable of adaptively reacting to external and internal environmental conditions, projecting carbon and nutrient balances to maximal productivity (Mitra et al., 2024). Through the integrated implementation of biological design strategies, including biological design of biological systems, machine learning process algorithms, and process control, such future visions could potentially implement autonomous, high-efficiency bioproduction continuously. In summary, future developments in synthetic biology, biological design, and biological system integration appear to hold promise for microalgae-based platforms for a variety of biological applications, carbon fixation, carbon utilization, climate-positive carbon strategies, and the future of biologically based, fully functional biorefineries to overcome both the biological and economic challenges of the 21st century.

CONCLUSION

Genetic engineering and synthetic biology approaches to microalgae have transformed a promising area for carbon and nutrient utilization. The power of combined state-of-the-art molecular biology, CRISPR/Cas genome engineering, synthetic promoters, gene circuits, and omics knowledge allows scientists to genetically engineer microalgae to optimize carbon fixation, nutrient uptake, and bioproduct synthesis. Microalgae-bacteria engineered composite systems and microbial ecosystems have further advanced this area, promoting microbe-microbe interactions and guiding efforts to optimize resource utilization, wastewater treatment, and CO_2 capture. Simulation tools, metabolic pathway optimization, and AI-driven predictive modeling of metabolism help rapidly design super performers, whereas automated, high-throughput throughput systems promote scalable validation. Notwithstanding the hurdles of scalability, genetic instability, biosafety, social acceptability, technological innovations, and leaps in microalgal containment, regulatory alignment, and environmental assessment now support responsible and aligned development. The future horizon of machine learning/synthetic biology combinations now opens promising directions toward self-optimized closed-loop bioeconomy, carbon-negative biorefineries, and a plethora of other applications. Taken together, these breakthrough innovations now position modified microalgae as multimodal biological platforms to address crucial climate, environmental problems of carbon capture, nutrient pollution, and biological production of carbon-based fuels and high-value specialty chemicals, thereby providing a foundation for the development of a next-generation climate-resilient, highly resource-efficient biological technology.

REFERENCES

Abinandan, S., Subashchandrabose, S. R., Venkateswarlu, K., & Megharaj, M. (2018). Nutrient removal and biomass production: Advances in microalgal biotechnology for wastewater treatment. *Critical Reviews in Biotechnology*, *38*(8), 1244–1260. DOI: 10.1080/07388551.2018.1472066 PMID: 29768936

Ahmad Kamal, A. H., Mohd Hamidi, N. F., Zakaria, M. F., Ahmad, A., Harun, M. R., Chandra Segaran, T., & Jusoh, M. (2024). Genetically engineered microalgae for enhanced bioactive compounds. *Discover Applied Sciences*, *6*(9), 482. DOI: 10.1007/s42452-024-06116-5

Akash, S., Sivaprakash, B., Rajamohan, N., & Vo, D. V. N. (2023). Biotechnology to convert carbon dioxide into biogas, bioethanol, bioplastic and succinic acid using algae, bacteria and yeast: A review. *Environmental Chemistry Letters*, *21*(3), 1477–1497. DOI: 10.1007/s10311-023-01569-3

Aledhari, M., & Rahouti, M. (2024). Gene and RNA editing: methods, enabling technologies, applications, and future directions. *arXiv preprint arXiv:2409.09057*.

Ali, M., Shabbir, K., Ali, S., Mohsin, M., Kumar, A., Aziz, M., Zubair, M., & Sultan, H. M. (2024). A New Era of Discovery: How Artificial Intelligence has Revolutionized the Biotechnology. *Nepal Journal of Biotechnology*, *12*(1), 1–11. DOI: 10.54796/njb.v12i1.312

Amin, M., Tahir, F., Akbar, I., Alessa, A. H., Alsaigh, A. A., Liu, C. G., Xin, F., Chi, Z., Syafiuddin, A., Mehmood, M. A., & Boopathy, R. (2025). Towards Environmental Sustainability: Employing Adaptive Laboratory Evolution to Develop Elite Algae Strains for Industrial and Environmental Applications. *Current Pollution Reports*, *11*(1), 14. DOI: 10.1007/s40726-025-00346-w

Artemyev, V., Gubaeva, A., Paremskaia, A. I., Dzhioeva, A. A., Deviatkin, A., Feoktistova, S. G., Mityaeva, O., & Volchkov, P. Y. (2024). Synthetic promoters in gene therapy: Design approaches, features and applications. *Cells*, *13*(23), 1963. DOI: 10.3390/cells13231963 PMID: 39682712

Balabanova, L., Averianova, L., Marchenok, M., Son, O., & Tekutyeva, L. (2021). Microbial and genetic resources for cobalamin (vitamin B12) biosynthesis: From ecosystems to industrial biotechnology. *International Journal of Molecular Sciences*, *22*(9), 4522. DOI: 10.3390/ijms22094522 PMID: 33926061

Balouch, H., Sadvakasova, A. K., Kossalbayev, B. D., Bauenova, M. O., Zaletova, D. E., Kumarbekuly, S., & Kirbayeva, D. K. (2025). Phyto-Algal Consortia as a Complementary System for Wastewater Treatment and Biorefinery. *Plants*, *14*(19), 3069. DOI: 10.3390/plants14193069 PMID: 41095212

Bilal, M., Niu, D., & Wang, Z. (2025). Biotechnological Strategies and Perspectives for Food Waste Treatment: The Role of Lactic Acid and Microbial Biomass. *Waste and Biomass Valorization*, *16*(2), 547–569. DOI: 10.1007/s12649-024-02705-y

Briat, C., & Khammash, M. (2018). Perfect adaptation and optimal equilibrium productivity in a simple microbial biofuel metabolic pathway using dynamic integral control. *ACS Synthetic Biology*, *7*(2), 419–431. DOI: 10.1021/acssynbio.7b00188 PMID: 29343065

Busch, F. A. (2020). Photorespiration in the context of Rubisco biochemistry, CO2 diffusion and metabolism. *The Plant Journal : For Cell and Molecular Biology*, *101*(4), 919–939. DOI: 10.1111/tpj.14674 PMID: 31910295

Cao, K., Cui, Y., Sun, F., Zhang, H., Fan, J., Ge, B., Cao, Y., Wang, X., Zhu, X., Wei, Z., Yao, Q., Ma, J., Wang, Y., Meng, C., & Gao, Z. (2023). Metabolic engineering and synthetic biology strategies for producing high-value natural pigments in Microalgae. *Biotechnology Advances*, *68*, 108236. DOI: 10.1016/j.biotechadv.2023.108236 PMID: 37586543

Cardiff, R. A., Carothers, J. M., Zalatan, J. G., & Sauro, H. M. (2024). Systems-level modeling for CRISPR-based metabolic engineering. *ACS Synthetic Biology*, *13*(9), 2643–2652. DOI: 10.1021/acssynbio.4c00053 PMID: 39119666

Chen, J., Liu, D., Wang, Y., Wang, S., & Huang, R. (2023). Effects of selective RNA processing and stabilization enzymes on carbon sequestration by photosynthesis of Synechococcus sp. PCC7002. *Frontiers in Marine Science*, *10*, 1223060. DOI: 10.3389/fmars.2023.1223060

Choi, K. R., Jang, W. D., Yang, D., Cho, J. S., Park, D., & Lee, S. Y. (2019). Systems metabolic engineering strategies: Integrating systems and synthetic biology with metabolic engineering. *Trends in Biotechnology*, *37*(8), 817–837. DOI: 10.1016/j.tibtech.2019.01.003 PMID: 30737009

Chubukov, V., Gerosa, L., Kochanowski, K., & Sauer, U. (2014). Coordination of microbial metabolism. *Nature Reviews. Microbiology*, *12*(5), 327–340. DOI: 10.1038/nrmicro3238 PMID: 24658329

De Porcellinis, A. J., Nørgaard, H., Brey, L. M. F., Erstad, S. M., Jones, P. R., Heazlewood, J. L., & Sakuragi, Y. (2018). Overexpression of bifunctional fructose-1, 6-bisphosphatase/sedoheptulose-1, 7-bisphosphatase leads to enhanced photosynthesis and global reprogramming of carbon metabolism in Synechococcus sp. PCC 7002. *Metabolic Engineering*, *47*, 170–183. DOI: 10.1016/j.ymben.2018.03.001 PMID: 29510212

Ebadi, A. G., Naeem, M. Y., & Selamoglu, Z. (2025). Biohybrid Systems for Waste Valorization: Synergizing Synthetic Biology and Green Engineering to Enable Circular Resource Recovery. *International Journal of Environmental Science and Green Technology*, *1*(2), 25–37.

Fabris, M., Abbriano, R. M., Pernice, M., Sutherland, D. L., Commault, A. S., Hall, C. C., Labeeuw, L., McCauley, J. I., Kuzhiuparambil, U., Ray, P., Kahlke, T., & Ralph, P. J. (2020). Emerging technologies in algal biotechnology: Toward the establishment of a sustainable, algae-based bioeconomy. *Frontiers in Plant Science*, *11*, 279. DOI: 10.3389/fpls.2020.00279 PMID: 32256509

Feng, J., Ma, D., Gao, S., Liao, Y., Feng, J., Xu, S., Wang, X., & Chen, K. (2023). Recent advances in engineering heterotrophic microorganisms for reinforcing CO2 fixation based on Calvin–Benson–Bassham cycle. *ACS Sustainable Chemistry & Engineering*, *11*(26), 9509–9522. DOI: 10.1021/acssuschemeng.2c06627

Forero-Pineda, C. (2006). The impact of stronger intellectual property rights on science and technology in developing countries. *Research Policy*, *35*(6), 808–824. DOI: 10.1016/j.respol.2006.04.003

Fuqua, C., Parsek, M. R., & Greenberg, E. P. (2001). Regulation of gene expression by cell-to-cell communication: Acyl-homoserine lactone quorum sensing. *Annual Review of Genetics*, *35*(1), 439–468. DOI: 10.1146/annurev.genet.35.102401.090913 PMID: 11700290

Goold, H. D., Wright, P., & Hailstones, D. (2018). Emerging opportunities for synthetic biology in agriculture. *Genes*, *9*(7), 341. DOI: 10.3390/genes9070341 PMID: 29986428

Grama, S. B., Liu, Z., & Li, J. (2022). Emerging trends in genetic engineering of microalgae for commercial applications. *Marine Drugs*, *20*(5), 285. DOI: 10.3390/md20050285 PMID: 35621936

Hundleby, P. A., Sack, M., & Twyman, R. M. (2018). Biosafety, risk assessment, and regulation of molecular farming. *Molecular Pharming: Applications, Challenges, and Emerging Areas*, 327-351.

Jaiswal, D. K., Verma, J. P., Belwal, T., Pereira, A. P. D. A., & Ade, A. B. (2023). Microbial co-cultures: A new era of synthetic biology and metabolic engineering. *Frontiers in Microbiology*, *14*, 1235565. DOI: 10.3389/fmicb.2023.1235565 PMID: 37426012

Jin, K., Chen, G., Yang, Y., Zhang, Z., & Lu, T. (2023). Strategies for manipulating Rubisco and creating photorespiratory bypass to boost C3 photosynthesis: Prospects on modern crop improvement. *Plant, Cell & Environment*, *46*(2), 363–378. DOI: 10.1111/pce.14500 PMID: 36444099

Karoui, H., Patwal, P. S., Pavan Kumar, B. V. V. S., & Martin, N. (2022). Chemical communication in artificial cells: Basic concepts, design and challenges. *Frontiers in Molecular Biosciences*, *9*, 880525. DOI: 10.3389/fmolb.2022.880525 PMID: 35720123

Kazi, T. A., & Biswas, S. R. (2021). CRISPR/dCas system as the modulator of gene expression. *Progress in Molecular Biology and Translational Science*, *178*, 99–122. DOI: 10.1016/bs.pmbts.2020.12.002 PMID: 33685602

Khan, A. Z., Bilal, M., Mehmood, S., Sharma, A., & Iqbal, H. M. (2019). State-of-the-art genetic modalities to engineer cyanobacteria for sustainable biosynthesis of biofuel and fine-chemicals to meet bio–economy challenges. *Life (Basel, Switzerland)*, *9*(3), 54. DOI: 10.3390/life9030054 PMID: 31252652

Kim, H., Hillson, N. J., Cho, B. K., Sung, B. H., Lee, D. H., Kim, D. M., Oh, M.-K., Chang, M. W., Jin, Y.-S., Rosser, S. J., Vegh, P., Fragkoudis, R., Le Feuvre, R., Scrutton, N. S., Storch, M., Seong, W., Freemont, P. S., & Lee, S. G. (2025). Abstraction hierarchy to define biofoundry workflows and operations for interoperable synthetic biology research and applications. *Nature Communications*, *16*(1), 6056. DOI: 10.1038/s41467-025-61263-6 PMID: 40593727

Krysenko, S. (2023). Impact of nitrogen-containing compounds on secondary metabolism in Streptomyces spp.—A source of metabolic engineering strategies. *SynBio*, *1*(3), 204–225. DOI: 10.3390/synbio1030015

Kumar, A., Pandey, S. S., Kumar, D., & Tripathi, B. N. (2023). Genetic manipulation of photosynthesis to enhance crop productivity under changing environmental conditions. *Photosynthesis Research*, *155*(1), 1–21. DOI: 10.1007/s11120-022-00977-w PMID: 36319887

Kumar, G., Shekh, A., Jakhu, S., Sharma, Y., Kapoor, R., & Sharma, T. R. (2020). Bioengineering of microalgae: Recent advances, perspectives, and regulatory challenges for industrial application. *Frontiers in Bioengineering and Biotechnology*, *8*, 914. DOI: 10.3389/fbioe.2020.00914 PMID: 33014997

Kumar, R., Karmakar, R., Sierra, B. G., & Andjelković, S. (2026). The Future of Respiratory Bioinformatics. In *Next-Generation Bioinformatics for Pulmonary Disease Research* (pp. 373–412). IGI Global Scientific Publishing.

Kumar, V., & Verma, P. (2025). Advances in microbial biotechnology for sustainable wastewater reclamation: Recent trends and future prospects. *World Journal of Microbiology & Biotechnology*, *41*(12), 478. DOI: 10.1007/s11274-025-04674-3 PMID: 41307588

Kuppan, P., Sudharsanam, A., Venkateswarlu, K., & Megharaj, M. (2023). Solar technology–closed loop synergy facilitates low-carbon circular bioeconomy in microalgal wastewater treatment. *NPJ Clean Water*, *6*(1), 43. DOI: 10.1038/s41545-023-00256-8

Kushwaha, O. S., Uthayakumar, H., & Kumaresan, K. (2023). Modeling of carbon dioxide fixation by microalgae using hybrid artificial intelligence (AI) and fuzzy logic (FL) methods and optimization by genetic algorithm (GA). *Environmental Science and Pollution Research International*, *30*(10), 24927–24948. DOI: 10.1007/s11356-022-19683-0 PMID: 35349067

Lanigan, T. M., Kopera, H. C., & Saunders, T. L. (2020). Principles of genetic engineering. *Genes*, *11*(3), 291. DOI: 10.3390/genes11030291 PMID: 32164255

Liu, J., Wu, X., Yao, M., Xiao, W., & Zha, J. (2020). Chassis engineering for microbial production of chemicals: From natural microbes to synthetic organisms. *Current Opinion in Biotechnology*, *66*, 105–112. DOI: 10.1016/j.copbio.2020.06.013 PMID: 32738762

Manoharan, L., Suparmaniam, U., Lam, M. K., Chai, Y. H., Loy, A. C. M., Tan, I. S., Foo, H. C. Y., & Lim, S. (2025). Advances in attached growth microalgae cultivation for third-generation biofuel production: Opportunities, challenges, and future prospect. *Brazilian Journal of Chemical Engineering*, *43*(2), 1–36. DOI: 10.1007/s43153-025-00567-y

Mathew, M. M., Khatana, K., Vats, V., Dhanker, R., Kumar, R., Dahms, H. U., & Hwang, J. S. (2022). Biological approaches integrating algae and bacteria for the degradation of wastewater contaminants—A review. *Frontiers in Microbiology*, *12*, 801051. DOI: 10.3389/fmicb.2021.801051 PMID: 35185825

Mathur, M., Choudhary, P., Sujathan, S., Naaz, F., Trenado-Yuste, C., & Malik, A. (2024). A biologically driven model for rural wastewater management: Feasibility and efficiency of algal-bacterial biofilm reactors for combined treatment and algae farming. *Frontiers in Water*, *6*, 1430900. DOI: 10.3389/frwa.2024.1430900

Matzko, R., & Konur, S. (2024). Technologies for design-build-test-learn automation and computational modelling across the synthetic biology workflow: A review. *Network Modeling and Analysis in Health Informatics and Bioinformatics, 13*(1), 22. DOI: 10.1007/s13721-024-00455-4

Meloni, M., Gurrieri, L., Fermani, S., Velie, L., Sparla, F., Crozet, P., Henri, J., & Zaffagnini, M. (2023). Ribulose-1, 5-bisphosphate regeneration in the Calvin-Benson-Bassham cycle: Focus on the last three enzymatic steps that allow the formation of Rubisco substrate. *Frontiers in Plant Science, 14*, 1130430. DOI: 10.3389/fpls.2023.1130430 PMID: 36875598

Mipeshwaree Devi, A., Khedashwori Devi, K., Premi Devi, P., Lakshmipriyari Devi, M., & Das, S. (2023). Metabolic engineering of plant secondary metabolites: Prospects and its technological challenges. *Frontiers in Plant Science, 14*, 1171154. DOI: 10.3389/fpls.2023.1171154 PMID: 37251773

Mitra, S., Paliya, S., & Mandpe, A. (2024). Microbial engineering in biofuel production—A global outlook, advances, and roadmap. In *Emerging sustainable technologies for biofuel production* (pp. 547–593). Springer Nature Switzerland. DOI: 10.1007/978-3-031-52167-6_22

Montaño López, J., Duran, L., & Avalos, J. L. (2022). Physiological limitations and opportunities in microbial metabolic engineering. *Nature Reviews. Microbiology, 20*(1), 35–48. DOI: 10.1038/s41579-021-00600-0 PMID: 34341566

Müller, M. M., Arndt, K. M., & Hoffmann, S. A. (2025). Genetic circuits in synthetic biology: Broadening the toolbox of regulatory devices. *Frontiers in Synthetic Biology, 3*, 1548572. DOI: 10.3389/fsybi.2025.1548572

Nguyen, D. D., Shin, M. C., Han, J. H., & Kim, G. M. (2025). Artificial Intelligence–Assisted Optimization of Microfluidic and Microbial Fuel Cells: An Updated Review. *International Journal of Precision Engineering and Manufacturing, 26*(9), 2389–2409. DOI: 10.1007/s12541-025-01340-4

Nneoma, U. C., Chukwudi, O. F., Nnenna, U. J., & Paul-Chima, U. O. (2025). Hybrid biofactories: Integrating microalgae and engineered microbiomes for enhanced biofuel production in circular carbon systems. *Frontiers in Energy Research, 13*, 1654079. DOI: 10.3389/fenrg.2025.1654079

Nneoma, U. C., Chukwudi, O. F., Nnenna, U. J., & Paul-Chima, U. O. (2025). Hybrid biofactories: Integrating microalgae and engineered microbiomes for enhanced biofuel production in circular carbon systems. *Frontiers in Energy Research, 13*, 1654079. DOI: 10.3389/fenrg.2025.1654079

Pan, M., Lyu, T., Zhan, L., Matamoros, V., Angelidaki, I., Cooper, M., & Pan, G. (2021). Mitigating antibiotic pollution using cyanobacteria: Removal efficiency, pathways and metabolism. *Water Research, 190*, 116735. DOI: 10.1016/j.watres.2020.116735 PMID: 33352526

Poliner, E., Farré, E. M., & Benning, C. (2018). Advanced genetic tools enable synthetic biology in the oleaginous microalgae Nannochloropsis sp. *Plant Cell Reports, 37*(10), 1383–1399. DOI: 10.1007/s00299-018-2270-0 PMID: 29511798

Prasad, R., Gupta, S. K., Shabnam, N., Oliveira, C. Y. B., Nema, A. K., Ansari, F. A., & Bux, F. (2021). Role of microalgae in global CO2 sequestration: Physiological mechanism, recent development, challenges, and future prospective. *Sustainability (Basel), 13*(23), 13061. DOI: 10.3390/su132313061

Priyadharsini, P., Nirmala, N., Dawn, S. S., Baskaran, A., SundarRajan, P., Gopinath, K. P., & Arun, J. (2022). Genetic improvement of microalgae for enhanced carbon dioxide sequestration and enriched biomass productivity: Review on CO2 bio-fixation pathways modifications. *Algal Research, 66*, 102810. DOI: 10.1016/j algal.2022.102810

Qiu, S., Yang, A., & Zeng, H. (2023). Flux balance analysis-based metabolic modeling of microbial secondary metabolism: Current status and outlook. *PLoS Computational Biology, 19*(8), e1011391. DOI: 10.1371/journal.pcbi.1011391 PMID: 37619239

Ramakrishnan, P., Sundaram, T., Lahiri, D., Nag, M., & Bhattacharya, D. (2025). Genetic Engineering and Modulation of Metabolic Pathways. In *Introduction to Metabolic Engineering andApplication* (pp. 295–330). Springer Nature Switzerland. DOI: 10.1007/978-3-031-93189-5_12

Ranjbar, S., & Malcata, F. X. (2022). Is genetic engineering a route to enhance microalgae-mediated bioremediation of heavy metal-containing effluents? *Molecules (Basel, Switzerland), 27*(5), 1473. DOI: 10.3390/molecules27051473 PMID: 35268582

Ravichandran, M., Kumar, T. T. A., & Dineshkumar, R. (2024). Carbon dioxide capture, sequestration, and utilization models for carbon management and transformation. *Environmental Science and Pollution Research International, 31*(44), 55895–55916. DOI: 10.1007/s11356-024-34861-y PMID: 39256334

Ray, S., Abraham, J., Jordan, N., Lindsay, M., & Chauhan, N. (2022). Synthetic, photosynthetic, and chemical strategies to enhance carbon dioxide fixation. *C, 8*(1), 18.

Ream, M., & Prather, K. L. (2024). Engineered autonomous dynamic regulation of metabolic flux. *Nature Reviews Bioengineering*, *2*(3), 233–243. DOI: 10.1038/s44222-023-00140-7

Roell, M. S., & Zurbriggen, M. D. (2020). The impact of synthetic biology for future agriculture and nutrition. *Current Opinion in Biotechnology*, *61*, 102–109. DOI: 10.1016/j.copbio.2019.10.004 PMID: 31812911

Rosero-Chasoy, G., Rodríguez-Jasso, R. M., Aguilar, C. N., Buitrón, G., Chairez, I., & Ruiz, H. A. (2021). Microbial co-culturing strategies for the production high value compounds, a reliable framework towards sustainable biorefinery implementation–an overview. *Bioresource Technology*, *321*, 124458. DOI: 10.1016/j.biortech.2020.124458 PMID: 33338739

Rubin-Blum, M., Dubilier, N., & Kleiner, M. (2019). Genetic evidence for two carbon fixation pathways (the Calvin-Benson-Bassham cycle and the reverse tricarboxylic acid cycle) in symbiotic and free-living bacteria. *MSphere*, *4*(1), 10–1128. DOI: 10.1128/mSphere.00394-18 PMID: 30602523

Sadvakasova, A. K., Balouch, H., Bauenova, M. O., Kossalbayev, B. D., Allakhverdiev, S. I., & Shabala, S. (2025). Strategies to intensify CO2 capture by microalgae for the circular bioeconomy. *Trends in Plant Science*. PMID: 41168026

Sarker, N. K., & Kaparaju, P. (2025). Advances in algae-based bioplastics: From strain engineering and fermentation to commercialization and sustainability. *Fermentation (Basel, Switzerland)*, *11*(10), 574. DOI: 10.3390/fermentation11100574

Sarwan, J., Nair, M. V., Sharma, S., Uddin, N., & Bose, K. J. C. (2024). Omics technology approaches for the generation of biofuels. In *Biomass Energy for Sustainable Development* (pp. 165–191). CRC Press. DOI: 10.1201/9781003406501-9

Satrio, R. D., Fendiyanto, M. H., & Miftahudin, M. (2024). Tools and techniques used at global scale through genomics, transcriptomics, proteomics, and metabolomics to investigate plant stress responses at the molecular level. In *Molecular dynamics of plant stress and its management* (pp. 555–607). Springer Nature Singapore. DOI: 10.1007/978-981-97-1699-9_25

Semple, A., & Pandhal, J. (2025). Engineering the phycosphere: Fundamental concepts and tools for the bottom-up design of microalgal-bacterial consortia. *Applied Phycology*, *6*(1), 21–51. DOI: 10.1080/26388081.2024.2435040

Sharma, P. K., Goud, V. V., Yamamoto, Y., & Sahoo, L. (2021). Efficient Agrobacterium tumefaciens-mediated stable genetic transformation of green microalgae, Chlorella sorokiniana. *3 Biotech, 11*(4), 196.

Shi, S., Qi, N., & Nielsen, J. (2022). Microbial production of chemicals driven by CRISPR-Cas systems. *Current Opinion in Biotechnology, 73*, 34–42. DOI: 10.1016/j.copbio.2021.07.002 PMID: 34303184

Shilky, Patra, S., Harshvardhan, A., Kumar, A., & Saikia, P. (2023). Role of microbes in controlling the geochemical composition of aquatic ecosystems. *Hydrogeochemistry of Aquatic Ecosystems*, 265-281.

Singhal, R., Pal, R., & Dutta, S. (2023). Chloroplast engineering: Fundamental insights and its application in amelioration of environmental stress. *Applied Biochemistry and Biotechnology, 195*(4), 2463–2482. DOI: 10.1007/s12010-022-03930-8 PMID: 35484466

Singla, P. (2021). Plant synthetic biology: A paradigm shift targeting stress mitigation, reduction of ecological footprints and sustainable transformation in agriculture. In *Plant Stress Biology: Strategies and Trends* (pp. 435–489). Springer Singapore.

Stephens, S., Mahadevan, R., & Allen, D. G. (2021). Engineering photosynthetic bioprocesses for sustainable chemical production: A review. *Frontiers in Bioengineering and Biotechnology, 8*, 610723. DOI: 10.3389/fbioe.2020.610723 PMID: 33490053

Sun, H., Zhao, W., Mao, X., Li, Y., Wu, T., & Chen, F. (2018). High-value biomass from microalgae production platforms: Strategies and progress based on carbon metabolism and energy conversion. *Biotechnology for Biofuels, 11*(1), 227. DOI: 10.1186/s13068-018-1225-6 PMID: 30151055

Tamburro, J., & Boyle, N. R. (2025). Charting the state of GEMs in microalgae: Progress, challenges, and innovations. *Frontiers in Plant Science, 16*, 1614397. DOI: 10.3389/fpls.2025.1614397 PMID: 40584862

Thakur, N., Nigam, M., Mann, N. A., Gupta, S., Hussain, C. M., Shukla, S. K., Shah, A. A., Casini, R., Elansary, H. O., & Khan, S. A. (2023). Host-mediated gene engineering and microbiome-based technology optimization for sustainable agriculture and environment. *Functional & Integrative Genomics, 23*(1), 57. DOI: 10.1007/s10142-023-00982-9 PMID: 36752963

The, S. V., Snyder, R., & Tegeder, M. (2021). Targeting nitrogen metabolism and transport processes to improve plant nitrogen use efficiency. *Frontiers in Plant Science, 11*, 628366. DOI: 10.3389/fpls.2020.628366 PMID: 33732269

Ullah, F., Ali, S., Siraj, M., Akhtar, M. S., & Zaman, W. (2025). Plant Microbiomes Alleviate Abiotic Stress-Associated Damage in Crops and Enhance Climate-Resilient Agriculture. *Plants, 14*(12), 1890. DOI: 10.3390/plants14121890 PMID: 40573878

Ummalyma, S. B., Maharajan, T., Sreelekshmy, B. R., & Eswaran, K. (2025). Carbon Sequestration by Microalgae and Engineering of Microalgal Cells by CRISPR Technology for Efficient CO2 Capture for Carbon Neutral Bioproducts. *ACS ES&T Water*, *5*(9), 4969–4984. DOI: 10.1021/acsestwater.5c00517

van den Berg, N. I., Machado, D., Santos, S., Rocha, I., Chacón, J., Harcombe, W., Mitri, S., & Patil, K. R. (2022). Ecological modelling approaches for predicting emergent properties in microbial communities. *Nature Ecology & Evolution*, *6*(7), 855–865. DOI: 10.1038/s41559-022-01746-7 PMID: 35577982

Vavitsas, K., Crozet, P., Vinde, M. H., Davies, F., Lemaire, S. D., & Vickers, C. E. (2019). The synthetic biology toolkit for photosynthetic microorganisms. *Plant Physiology*, *181*(1), 14–27. DOI: 10.1104/pp.19.00345 PMID: 31262955

Verma, K., Dogra, R., & Rathour, N. S. (2025). Ecosystem dynamics: Exploring types, components and the forces shaping their transformation. *Journal of Geography. Environment and Earth Science International*, *29*(4), 177–191.

Viswanaathan, S., Perumal, P. K., & Sundaram, S. (2022). Integrated approach for carbon sequestration and wastewater treatment using algal–bacterial consortia: Opportunities and challenges. *Sustainability (Basel)*, *14*(3), 1075. DOI: 10.3390/su14031075

Yang, Z., Yan, H., Liu, H., Yang, L., Mi, G., & Wang, P. (2025). Enhancing Crop Nitrogen Efficiency: The Role of Mixed Nitrate and Ammonium Supply in Plant Growth and Development. *Biology (Basel)*, *14*(5), 546. DOI: 10.3390/biology14050546 PMID: 40427735

Yusof, Z., Tong, Y. W., Selvarajoo, K., Parakh, S. K., & Foo, S. C. (2025). Overcoming challenges in microalgal bioprocessing through data-driven and computational approaches. *Current Opinion in Food Science*, *61*, 101253. DOI: 10.1016/j.cofs.2024.101253

Zammit, G., Zammit, M. G., & Buttigieg, K. G. (2023). Emerging technologies for the discovery of novel diversity in cyanobacteria and algae and the elucidation of their valuable metabolites. *Diversity*, *15*(11), 1142. DOI: 10.3390/d15111142

Zenda, T., Liu, S., Dong, A., & Duan, H. (2021). Revisiting sulphur—The once neglected nutrient: It's roles in plant growth, metabolism, stress tolerance and crop production. *Agriculture*, *11*(7), 626. DOI: 10.3390/agriculture11070626

Zhang, C. C., Zhou, C. Z., Burnap, R. L., & Peng, L. (2018). Carbon/nitrogen metabolic balance: Lessons from cyanobacteria. *Trends in Plant Science*, *23*(12), 1116–1130. DOI: 10.1016/j.tplants.2018.09.008 PMID: 30292707

Zhang, S., & Liu, Z. (2021). Advances in the biological fixation of carbon dioxide by microalgae. *Journal of Chemical Technology and Biotechnology*, *96*(6), 1475–1495. DOI: 10.1002/jctb.6714

Zhang, Y. T., Jiang, J. Y., Shi, T. Q., Sun, X. M., Zhao, Q. Y., Huang, H., & Ren, L. J. (2019). Application of the CRISPR/Cas system for genome editing in microalgae. *Applied Microbiology and Biotechnology*, *103*(8), 3239–3248. DOI: 10.1007/s00253-019-09726-x PMID: 30877356

Zhu, J., & Thompson, C. B. (2019). Metabolic regulation of cell growth and proliferation. *Nature Reviews. Molecular Cell Biology*, *20*(7), 436–450. DOI: 10.1038/s41580-019-0123-5 PMID: 30976106

Chapter 5

Multiomics Insights Into Advanced Carbon and Nutrient Capture: Omics-Driven Strategies for Microalgae Photosynthetic Efficiency and Resource Utilization

Amrutha Shivaraju
Sri Sathya Sai University for Human Excellence, India

Harika Vaddi
Sri Sathya Sai University for Human Excellence, India

Amaramma B. Yaraladdi
Sri Sathya Sai University for Human Excellence, India

Ramya Raghavan
http://orcid.org/0000-0002-9953-543X
Sri Sathya Sai University for Human Excellence, India

ABSTRACT

Multiomics technologies are transforming photosynthetic microorganisms engineered for coupled carbon and nutrient capture. Here, we synthesize recent advances in omics, and fluxomics to elucidate regulatory networks for photosynthetic efficiency, and nutrient utilization in microalgae and algae–bacteria consortia. The systems-level studies, highlight antenna size minimization, and optimization of carbon concentrating pyrenoid architectures sequestration. Metabolic engineering redirect carbon flux toward biomass, biofuels, and high-value co-products from wastewater

DOI: 10.4018/979-8-3373-7630-1.ch005

and flue-gas coupled processes. Algae workflows integrate strain and consortium design, stress and nutrient modelling for predicting process conditions. Finally, we advocate for rural offgrid omics-driven chassis that connect molecular traits to reactor-scale deployment. Algae systems for advanced carbon sequestration, nutrient recovery, promote circular bioeconomy applications aligned with the Sustainable Development Goals.

INTRODUCTION

Anthropogenic emissions of carbon dioxide and the excessive release of nitrogen and phosphorus into aquatic systems have intensified global climate change, eutrophication, and ecosystem degradation. Conventional physicochemical approaches for carbon capture and nutrient remediation remain energy-intensive and often economically unsustainable at scale. In this context, photosynthetic microalgae have emerged as promising biological platforms capable of simultaneously sequestering atmospheric CO_2 and assimilating inorganic nutrients while generating valuable biomass. Microalgae possess rapid growth rates, high photosynthetic efficiencies, and metabolic plasticity that allows them to convert inorganic carbon and nutrients into lipids, carbohydrates, pigments, and other bioactive compounds. Their cultivation can therefore integrate environmental remediation with biofuel production, nutraceutical development, and circular bioeconomy strategies.

Microalgae, due to their rapid growth rates, high adaptability to extreme environments, have applications across industries such as food, feed, biofuels, pharmaceuticals, and environmental remediation. Microalgae are important commercial sources of carotenoids. The zeaxanthin, fucoxanthin, violaxanthin, astaxanthin, and β-carotene are health-promoting antioxidants and colorants in food, feed, and cosmetics. Microalgae play a critical role in achieving numerous UN Sustainable Development Goals by sequestering carbon, purifying wastewater, and producing clean energy. (Veale et al., 2025) The carbon capture potential of microalgal cultivation suggests that energy consumption and nutrient inputs can significantly reduce carbon emissions through carbon sequestration. Algae are highly promising third-generation feedstocks for sustainable biofuel production. Unlike traditional agricultural crops, algae offer exceptionally high growth rates, superior carbon dioxide sequestration, and high lipid (oil) content. Alage can be cultivated on non-arable land using wastewater or saline water. Microalgae using renewable energy and inexpensive nutrient sources could mitigate carbon emissions. Further literature advocates the use of produced microalgae as biofertilizers or aquafeeds to promote nutrient recycling. (Sharma et al., 2026) Microalgae are promising organisms for sustainable biofuel production and environmental remediation. Biological carbon sequestration (microalgae, cyanobacteria, soil microbiomes) and nutrient removal (wastewater treatment plants, activated sludge) are enabled by their rapid growth and carbon storage capabilities Microalgae demonstrate effective removal of contaminants and nutrients from wastewater, indicating their potential for environmentally friendly wastewater treatment and addressing global climate change and resource scarcity. (Hou et al., 2024) The central thesis of this work is that omics-guided bioengineering represents a critical frontier for unlocking the full potential of microalgae as platforms for carbon capture and nutrient recovery. Through comparative analysis of existing studies, we examine how advances in multi-omics technologies, synthetic biology, and metabolic modeling are reshaping the understanding of algal metabolism and enabling next-generation strategies for sustainable biomanufacturing. By synthesizing evidence across experimental systems and engineering approaches, this review aims to clarify emerging trends, identify unresolved controversies, and outline a roadmap for omics-driven optimization of algal carbon and nutrient capture systems.

HISTORY

Over the past two decades, several landmark studies have demonstrated the potential of algal systems for carbon and nutrient capture. Early work on model microalgae such as Chlamydomonas reinhardtii revealed that nutrient deprivation, particularly nitrogen limitation, redirects cellular carbon flux from protein synthesis toward storage compounds such as starch and triacylglycerols (TAGs). Subsequent studies using oleaginous species, including *Chromochloris zofingiensis and Microchloropsis gaditana*, confirmed that environmental cues such as elevated CO_2 or nutrient stress can significantly increase biomass productivity and lipid accumulation through co-ordinated metabolic reprogramming. More recent investigations have incorporated multi-omics approaches to elucidate the regulatory networks underlying carbon allocation and nutrient assimilation. For example, transcriptome-resolved analyses of nitrogen deprivation in microalgae revealed synchronized down-regulation of photosynthetic machinery with activation of lipid biosynthetic pathways, while metabolomic profiling under elevated CO_2 conditions has identified shifts in central carbon metabolism and increased production of long-chain polyunsaturated fatty acids. Parallel advances in genome editing and synthetic biology have enabled targeted manipulation of metabolic pathways, including modifications in phosphatase signaling and lipid biosynthetic enzymes that redirect cellular carbon toward energy-dense metabolites. Algal biofuels are identified as a sustainable alternative to fossil fuels, which are limited in availability and pose environmental risks, underscoring the need for greener energy sources. (Veale et al., 2025) The production of algal biofuels is influenced by various factors, including algal strain, cultivation system (open or closed photobioreactors), cultivation methods, and environmental conditions such as temperature, light, CO2, and pH. Microalgae demonstrate superior carbon fixation capabilities compared to terrestrial plants, with the ability to fix 1.8 kg of CO_2 per kilogram of biomass and tolerate elevated CO_2 concentrations. Microalgae are potent biological platforms for carbon dioxide sequestration and renewable biomass production. (Prajapati et al., 2026)

Table 1. Photosynthesis & Carbon Fixation aims to improve the efficiency of converting CO_2 and light energy into chemical energy (biomass). (Prajapati et al., 2026) (Sharma et al., 2026)

	Function in Microalgae	Carbon Sequestration Role
RuBisCO (Large Subunit)	Carbon Fixation incorporating atmospheric CO_2 into the Calvin-Benson-Bassham cycle.	The rate-limiting step due to its slow catalytic rate and oxygenase activity. Synthetic biology hyper-active variants from cyanobacteria boost carbon fixation rates.
Sedoheptulose-1,7-bisphosphatase (SBPase)	Calvin Cycle Regeneration dephosphorylation of sedoheptulose-1,7-bisphosphate to sedoheptulose-7-phosphate.	Calvin Cycle Flux bottleneck in the regeneration of RuBP (the CO_2 acceptor). Overexpression of SBPase has been shown to significantly increase biomass and photosynthetic efficiency by ensuring RuBisCO has enough substrate.
Photosynthetic Antenna Proteins (LHC, FCP)	Light Harvest light energy and transfers it to the reaction centers (PSI and PSII).	Antenna proteins (size/density) in dense cultures allow light to penetrate deeper into the bioreactor, increasing productivity and carbon capture per volume.
D1 Protein (PSII Core)	The reaction center of Photosystem II binds chlorophyll and facilitates electron transport.	Photo-damage Repair photoinhibition. Engineering faster D1 turnover variants can sustain high carbon fixation rates under intense sunlight.

OMICS TECHNOLOGIES FOR ALGAE-BASED ENERGY AND NUTRIENT SEQUESTRATION

Innovations in sustainable algal biofuel production offer a greener alternative to fossil fuels, thanks to their high lipid content. Microalgae can grow in environments, including wastewater. The biofuel yield depends on innovative extraction methods. Artificial Intelligence and genetic engineering enhance sustainability in the production process. (Mohapatra et al., 2025) The industrial flue gas and direct air capture, with advancements in CRISPR/Cas and synthetic biology, can enhance the efficiency of microalgal systems for carbon assimilation and biomass production. Optimized microalgal systems can achieve CO_2 removal efficiencies exceeding 70% and can valorize algal biomass into various bioproducts, supporting a circular bioeconomy. Microalgae are a primary source for third-generation biofuels, offering high lipid yields and carbon-neutral potential. The *Chlorella* genus is highly resilient and economical for biofuel production. Mixotrophic growth modes have been found to yield the highest lipid accumulation (45–70 mg L^{-1}) compared to photoautotrophic or heterotrophic modes. Algal biodiesel with a high cetane index and calorific value, showing improved ignition quality and lower emissions. Nanoparticles can be added directly to algal cultivation media to boost productivity. For example, supplementing *Chlorella vulgaris* with iron-based nanoparticles redirects the organism's metabolic flux toward triacylglycerol synthesis, resulting in a massive increase in lipid yield and improved biodiesel quality. Similarly, nickel-doped ceria nanoparticles have been shown to increase microalgal biomass production, protein and lipid content, and CO2 fixation rates. Furthermore, "fourth-generation" (4G) biofuels are emerging by combining these advanced cultivation techniques with genetic engineering to maximize the accumulation of energy storage compounds. (Prajapati et al., 2026)

Despite these advances, translating algal carbon capture systems from laboratory studies to scalable biotechnological platforms remains constrained by several unresolved challenges. First, substantial variability exists in reported carbon fixation rates and nutrient removal efficiencies across species, cultivation systems, and environmental conditions. While certain studies report a two-fold increase in biomass productivity under high CO_2 conditions, others demonstrate limited improvements due to metabolic bottlenecks, photo-inhibition, or nutrient imbalances. Second, the molecular mechanisms governing carbon partitioning between growth, storage compounds, and stress responses remain incompletely understood. Biolipids, such as palmitic and oleic acids, enhance sustainability by improving resource utilization and efficiency in biofuel production. *Graesiella emersonii*, a green microalga known for its high adaptability to extreme environmental conditions, is relevant for biotechnological applications. The integration of microalgae growth with biophotovoltaic systems, which can simultaneously capture carbon dioxide and generate electricity. *Alcaligenes faecalis* ZS-1 and *Shewanella oneidensis* as biocathode bioanode partners to drive extracellular electron transfer (EET), boosting CO_2 reduction and nutrient transformations in a microbial fuel cell. Optimization of cultivation conditions, electrode materials, and system configurations has led to significant improvements in both power generation and carbon storage rates. In the USA, the FDA's GRAS framework has been applied to multiple microalgal ingredients, including Chlamydomonas reinhardtii biomass (Triton's TAI114 strain). *Chlorella spp., Dunaliella spp., Haematococcus spp., Arthrospira (Spirulina), Schizochytrium spp., Porphyridium cruentum and Crypthecodinium cohnii* facilitate bioremediation. Microalgae work through accumulation, adsorption, and metabolism of refractory pollutants, including heavy metals, pesticides, and pharmaceuticals. Alginate-based immobilization matrices and hybrid sorbent–biological systems to link fast chemical CO_2 capture with slower biological assimilation. Algae also help produce reactive oxygen species to tolerate heavy metal stress and can even degrade plastic structures in aquatic environments. (Singh et al., 2024)

A major translational hurdle is the "energy return on energy invested" (EROEI). Current LCA models often reveal that the energy required for reactor maintenance, artificial lighting, and mixing in closed photobioreactors (PBRs) can exceed the energy recovered from the algal biomass. Furthermore, if the cultivation power source is fossil-derived, the system's overall carbon footprint may be net-positive, negating the biological CO_2 fixation.(Veale et al., 2025)

Table 2. Genomic and Transcriptomic Insights (Jain & Kumar, 2026)

Capture Type	Key Gene / Operon	Functional Insight
Carbon Fixation	rbcL, rbcS (Rubisco subunits)	Carbon fixation when P is limiting suggests Carbon capture is strictly coupled to N/P availability.
Carbon Fixation	ccm operon	Efficiency in low-carbon environments critical target for genetic engineering.
P-Removal	pstS, pstABC (Phosphate transport)	Expression spikes >50-fold during starvation; Phormidium sp. dominates P-uptake, starving heterotrophs (niche partitioning).
N-Removal	amoA (Ammonia monooxygenase)	Bacterial nitrifiers expression patterns compared to archaeal nitrifiers on oxygen/carbon ratios.
Lipid Storage	DGAT (Diacylglycerol acyltransferase)	Carbon into triacylglycerols (TAGs) rather than biomass, enhancing biofuel energy density.

CARBON FIXATION AND FLUX REDIRECTION

Chlamydomonas is a significant model organism for research in photosynthesis, metabolism, and cellular processes, and is important for green bioengineering applications. The genome of Chlamydomonas comprises 17 haploid chromosomes that encode 16,883 protein-coding genes, making it a target for functional genomics and biotechnological advances. Early work relied on UV mutagenesis and classical genetics. Random mutagenesis, plus crossing, has been used to manipulate the genome for decades and is still used. Key milestones include the establishment of the Chlamydomonas Resource Center, stable nuclear transformation (1982), and chloroplast transformation. The electroporation and glass-bead transformation are practical routes for nuclear transformation. These knowledge gaps highlight the need for integrative frameworks that couple systems biology with bioengineering strategies to optimize algal performance. Omics-driven approaches provide a powerful means to bridge this gap by enabling comprehensive characterization of metabolic networks and regulatory pathways controlling carbon fixation and nutrient assimilation. High-throughput transcriptomic and metabolomic analyses can identify candidate genes and metabolic fluxes associated with enhanced carbon sequestration, while genome-scale metabolic models and machine learning tools can predict system-level responses to environmental and genetic perturbations. By integrating these datasets, it becomes possible to rationally design algal strains with improved photosynthetic efficiency, enhanced nutrient uptake capacity, and optimized carbon partitioning toward desired bioproducts. Microbial interactions within the phycosphere, including bacterial symbioses that influence nutrient cycling and algal growth, introduce an additional layer of complexity that is rarely incorporated into engineering frameworks. Finally, there is an ongoing debate about the scalability and economic feasibility of genetically engineered microalgae, particularly in open cultivation systems, where ecological and regulatory constraints may limit deployment. *Nannochloropsis oceanica* exhibits strong CO2 tolerance and superior biomass productivity. CRISPR-Cas technology for genome engineering in microalgae and carbon-based engineering result in the production of valuable metabolites.

Photosynthesis (KEGG) based genome editing in the green alga Chlamydomonas is significant as a model organism for various biological studies and biotechnological applications. *Haematococcus pluvialis* (astaxanthin) and *Dunaliella salina* (β-carotene) are classic platforms for pigment production, with ongoing efforts to engineer additional strains to tailor carotenoid profiles. A multi-stage deep learning framework designed to predict transcriptional regulators using large-scale epigenomic data.(Nievergelt, 2025) *Chlorella* sp. and Isochrysis galbana are valued as feedstocks for biofuels and nutraceuticals. *Nannochloropsis oceanica* advances genome engineering strategies, sgRNA libraries, and synthetic biology applications. A cytidine base editor with an episomal vector enables cytidine-to-thymidine base substitutions without causing toxic large deletions in host cells. Successful base substitutions were achieved at six target sites across five endogenous genes, with substitution rates ranging from 29.2% to 47.6%. The removal of the base editor plasmids post-editing was confirmed, which is crucial for developing transgene-free algal strains, enhancing the potential for industrial applications. (Moroi et al., 2025)

Figure 1. Multiomics Insights into Carbon Sequestration by Microalgae

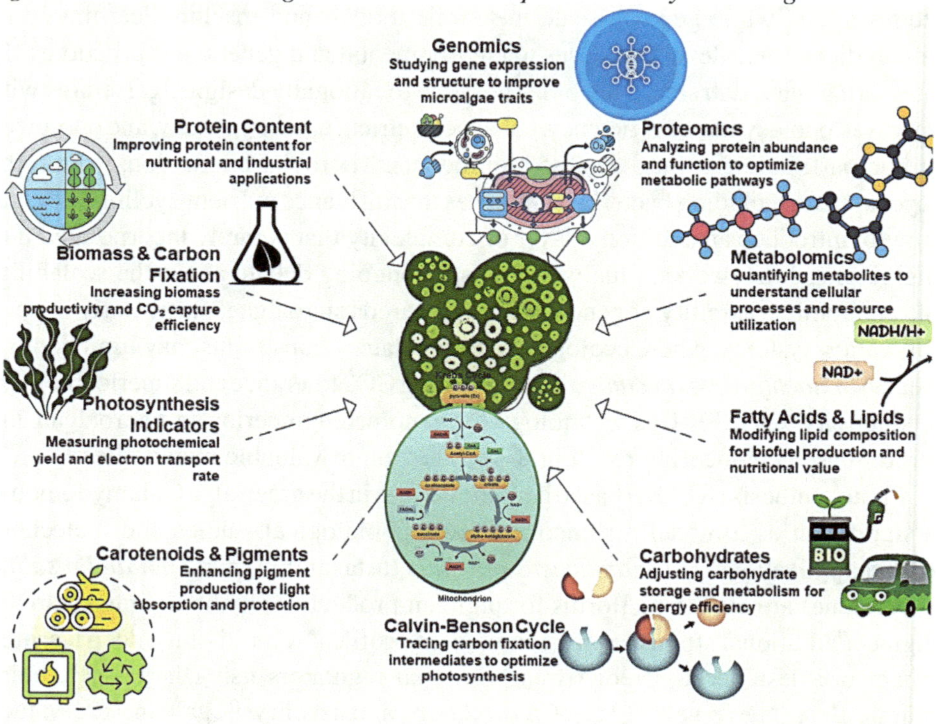

CARBON FIXATION (CALVIN CYCLE)

The modes of nutrition for microalgae are photoautotrophic and heterotrophic. Many species maintain photosynthesis and growth under very different atmospheres. Adaptation to varying CO_2, O_2, or mixed gas streams is via photoheterotrophic and mixotrophic pathways. Microalgal strains are highlighted as nutrient-dense biofactories. Microalgae accumulate high levels of carbohydrates, lipids, proteins, pigments, minerals, and vitamins. Astaxanthin production is achieved through the modulation of the microalgae metabolic pathway and nanobiotechnologies. Strategies to enhance astaxanthin production include precise genetic modifications of the carotenoid biosynthetic pathway and downregulation of competing metabolic routes. Targeted engineering of acyltransferases and lipid metabolism regulators is reported using RNAi and artificial microRNA knockdowns, transformation of nuclear, chloroplast, and mitochondrial genomes, and genome-wide mutant libraries enabled by random insertional mutagenesis. (Saranya et al., 2024)

The integration of nanoparticles into microalgae cultivation systems can improve nutrient delivery and light management, further boosting astaxanthin production. Nanotechnology offers a highly efficient alternative through nanoparticle-assisted harvesting. By using magnetic nanoparticles functionalized with polymers, the algae were separated from the water using an external magnetic field. (Jain & Kumar, 2026) Another innovative approach uses fungi-nanoparticle conjugates incorporating NiO and ZrO as highly effective bio-flocculants to quickly gather the microalgae. Researchers have successfully incorporated Nickel oxide (NiO) and Zirconium oxide (ZrO) nanoparticles onto fungal pellets to harvest *Chlorella vulgaris*. The fungi-nanoparticle conjugates achieved harvesting efficiencies of 94.25% (NiO) and 85.67% (ZrO) within just two hours. Furthermore, the water medium recovered showed minimal residual metals and successfully supported the regrowth of new microalgae.

Magnetic nanocatalysts combine the high reactivity and efficiency of homogeneous catalysts with easy magnetic recoverability, preventing catalyst loss and complex purification steps when converting algal lipids to biodiesel. Nanocatalysts can harvest biomass, disrupt algal cell walls, and enhance lipid extraction (oils), streamlining the downstream processing of photosynthetic organisms to produce biofuel. (Ananthakumar et al., 2025)

Nanotechnology overcomes the *Chlamydomonas* associated with low transformation efficiencies. The nickel-doped ceria nanoparticles have been shown to significantly increase microalgal biomass, lipid and protein content, and CO2 fixation rates. *Chlamydomonas* elicits strong defense and repair responses to exogenous DNA and to random nuclear integration, rather than to efficient homology-directed repair. Improved precision editing with more reliable HDR, base-prime editing analogs, cyclical marker systems for iterative edits, and sophisticated regulatory control of nuclear and plastid genes.(Nievergelt, 2025)

Table 3. Proteomic and Metabolomic Insights

Protein Metabolite	Change Flux	Physiological Outcome
Photosystem I & II complexes	Declining abundance (Succession phase)	Energy capture machinery suggests proteomic shift to maximize C-capture.
Polyphosphate Kinase (PPK)	High Abundance	Confirms active P-accumulation in *Accumulibacter* Biomarker EBPR (Enhanced Bio-P Removal).
13C-labeled Amino Acids	Rapid Turnover	C-sources (glucose) are mineralized (CO2 release), microbial biomass (sequestration).

continued on following page

Table 3. Continued

Protein Metabolite	Change Flux	Physiological Outcome
Triacylglycerols (TAGs)	Accumulation	N-starvation sink for captured Carbon, algal metabolome shifts C-flux from protein synthesis to lipid storage
Extracellular Polymeric Substances (EPS)	Enriched	Biofilms secrete C-rich EPS to protect against desiccation, stable Carbon sink in wastewater systems.

FATTY ACID OR LIPID SYNTHESIS PATHWAYS

Triacylglyceride assembly, triacylglycerol (TAG) productivity, and wild microalgae are essential for the cost-competitive microalgal biodiesel industry.

Diacylglycerol acyltransferase (DGAT) is identified as a critical enzyme in TAG synthesis, specifically mediating the last and rate-limiting step. DGAT is crucial for genetic engineering efforts. The microalga *Nannochloropsis oceanica* IMET1 harbors 13 DGAT homologs (NoDGAT), which exhibit distinct temporal and spatial transcription patterns, suggesting potential functional diversity. The TAG synthetic activities of NoDGAT in a yeast model elucidate their physiological roles and the functional mechanism underlying the generation of transgenic strains with improved TAG traits. Metagenomic analysis indicated a metabolic shift from denitrification-driven pathways to assimilation-dominated nitrogen metabolism, as well as beneficial interactions within algal–bacterial consortia. (Tazon et al., 2026)

Glycerol-3-phosphate dehydrogenase (GPDH) is a vital enzyme in microalgal lipid synthesis, converting dihydroxyacetone phosphate to glycerol-3-phosphate. The initial substrate for the Kennedy pathway is responsible for the production of triacylglycerol. The characterization of glycerol-3-phosphate:acyl-CoA acyltransferase (GPAT) in the model green alga *Chlamydomonas reinhardtii*, the endoplasmic reticulum-associated GPAT (CrGPATer), plays a crucial role in glycerolipid biosynthesis. Enzymatic assays demonstrated that CrGPATer functions as an sn-1 acyltransferase, utilizing various acyl-CoA as acyl donors, and is localized to ER membranes and lipid droplets. Overexpression (OE) of CrGPATer in transgenic C. reinhardtii lines led to an increase in galactolipid content, particularly monogalactosyldiacylglycerol, under nitrogen deficiency, while knockdown (KD) lines showed reduced galactolipid levels. The study indicates that CrGPATer is involved in the eukaryotic Kennedy pathway for galactolipid biosynthesis, thereby increasing the production of high-value storage lipids. Industrial applications of triacylglycerol and 1,3-olein-2-palmitin range from infant formula to space exploration. (Shelake et al., 2025) Monogalactosyldiacylglycerol (MGDG) is a predominant photosynthetic membrane lipid derived from plants and microalgae. Heterologous production of monogalactosyldiacylglycerol was achieved by overexpression of codon-optimized monogalactosyldiacylglycerol synthase gene MGD1 and Kennedy pathway genes (GAT1, ICT1, and PAH1), through the multi-copy integration of the MGD1 expression cassette. (Gu et al., 2024)

System-level analysis of cell growth and metabolism would be beneficial for understanding the regulatory networks governing metabolic engineering. The microalga *Chlorella vulgaris* is encapsulated in a specialized zinc-based metal-organic framework encapsulated in alginate beads. Accelerated photosynthesis achieved a total carbon capture of 4941.15 mg/L with growth, stress resilience, and lipid accumulation in microalgae. An increased glycerol production and mono-unsaturated fatty acids were shown in the overexpression of chloroplastic NAD(P)+-dependent G3PDH (NsGPDH1) in *Nannochloropsis salina*. Overexpression of GPAT, AGPAT1, DGAT, and LPAAT in Kennedy Pathway genes can enhance the resilience and productivity of microalgae in contaminated waters, addressing significant environmental pollution issues. By studying the interactions between microalgae and bacteria under chemical stress, consortia for improved bioremediation and commodity production in saline wastewater treatments. The characterization of the plastidial enzyme GPAT1 in the green microalga *Chlamydomonas reinhardtii,* which is a model organism for lipid metabolism research. GPAT1 exhibits not only glycerol-3-phosphate acyltransferase activity but also lysophosphatidic acid acyltransferase (LPAAT) activity, with a preference for C18:1 and C16:0 substrates. Knockdown of GPAT1 reduces triacylglycerol content, especially C16 species, during mixotrophic growth and nitrogen deprivation, suggesting a role in lipid biosynthesis. Knocking down GPAT1 results in upregulation of GPAT2, which significantly increases the production of 1,3-olein-2-palmitin (OPO), a valuable lipid for infant formula, suggesting GPAT1's potential as a target for biotechnological applications to enhance lipid production in microalgae. (Chen et al., 2025) Blocking Lipid Degradation in PLA2 and ACO knockouts increases productivity by 64%. The development of a high-oleic palm oil alternative through the fermentation of the microalgae *Prototheca moriformis*, achieving fatty acid profiles similar to conventional palm oil. (Parker et al., 2025)

CARBON FLUX REDIRECTION

Natural strains of microalgae face limitations in lipid yield, stress tolerance, and metabolic engineering for industrial applications. For example, the growth of *Synechocystis* sp. PCC 6803 is inhibited due to acidification from ammonium consumption. When algal biodiesel is combusted in an engine, blending it with nano-additives heavily optimizes its performance. For example, adding barium titanate (BaTiO3) or graphene oxide (GO) nanoparticles to *biodiesel blends from Spirulina or marine macroalgae* improves fuel atomization and thermal conductivity. This results in increased brake thermal efficiency, increased engine power, and significant reductions in harmful emissions like carbon monoxide, unburned hydrocarbons, and smoke.

Transcriptomic analysis shows that proteases and antioxidant enzymes are up-regulated, while photosynthesis and nitrogen assimilation are repressed. The small RNA HsrR1, encoded by the plasmid of *Synechocystis* sp. PCC 6803 plays a role in the heat shock response of photosynthetic organisms under high-temperature stress. Furthermore, green algae *Chloropicon primus* and *Chloropicon roscoffensis* reveal interspecific variation in chromosome number, gene content, gene arrangement, and ploidy state. *Chloropicon roscoffensis* has a unique C4-like inorganic carbon-concentrating mechanism that recycles CO2 for carbon fixation and improved resilience to abiotic stresses compared to C. primus. A novel eukaryotic recycling pathway for the micronutrient queuosine is crucial for tRNA post-transcriptional modification.(Turmel et al., 2025) Genetic engineering and synthetic biology applications increased bioethanol production in yeast and microalgae consortia. Improved lipid content in microalgae or reduced lignin in plants facilitates the release of carbon and sugars for biofuel production. CRISPR improves microalgal strains by enhancing lipid accumulation and stress tolerance. CRISPR/Cas9 technology, particularly the Cas9-driven Type II system, is being used for directed genome modification in microalgae by precisely altering genetic and metabolic pathways. CrPEPC1 knockdown via successful CRISPRi targeting to regulate PEPC1 expression in *Chlamydomonas reinhardtii*, achieving 94% efficiency and stability over seven generations. The down-regulation of PEPC1 led to lower chlorophyll levels but increased biomass concentration and lipid accumulation. Specific examples include the miRNA-silencing of the OEE2 gene in *Chlamydomonas reinhardtii*, which doubled H2 production, and CRISPR interference in *Phaeodactylum tricornutum*, resulting in 7- and 20-fold increases in lipid productivity. (Kirbayeva et al., 2025)

Table 4. Lipid Biosynthesis (Biofuel Production) enzymes are key targets for diverting sequestered carbon into energy-dense lipids (triacylglycerols) for biofuel applications.

Enzyme / Protein	Function in Microalgae	Carbon Sequestration Role
ACCase (Acetyl-CoA Carboxylase)	Fatty Acid Initiation irreversible conversion of Acetyl-CoA to Malonyl-CoA.	Overexpressing ACCase carbon flux away from starch/protein and toward lipid production.
Fatty Acid Synthase (FAS)	Chain Elongation of multi-enzyme complex that elongates the carbon chain of fatty acids.	FAS components can allow for the production of specific chain lengths (medium-chain fatty acids) desirable for drop-in jet fuels.
Hydroxyacyl-ACP dehydrogenase (HAD)	FAS Component in the dehydration step in the fatty acid elongation cycle.	Saturation Engineering HAD activity can influence the saturation level of the fatty acids, affecting fuel quality (oxidative stability).
Enoyl-ACP reductase (ENR)	FAS catalyzes the final reduction step in the elongation cycle.	Rate-determining enzyme in the elongation cycle, ENR overexpression can enhance the overall rate of lipid accumulation.
DGAT (TAG Synthesis)	TAG Assembly final step of Triacylglycerol (TAG) formation from diacylglycerol.	DGAT fatty acids into storage oil droplets, preventing feedback inhibition and maximizing total carbon storage as oil.

The engineering of synthetic microbial consortia through rational design, bottom-up and top-down approaches, highlights the roles of quorum sensing. (Mahalakshmi et al., 2026) The metabolite cross-feeding stabilizes consortia in nutrient and heavy metal removal. Nanoparticle separation works through electrostatic attraction, bridging flocculation, and Lewis acid–base interactions. Microalgal biomass, such as *Chlorella sorokiniana* thin films, can remove 100% of cadmium from industrial wastewater while remaining viable for biofuel conversion. The Fe_3O_4-based systems surface-functionalized with polyethyleneimine (PEI), chitosan, or urea can bind to microalgae. Once attached, the algae-nanoparticle clusters can be rapidly and effortlessly pulled out of the water using an external magnetic field, allowing for high recyclability. Additionally, the residual "deoiled" algal biomass left over after lipid extraction can be pyrolyzed into biochar. The biochar can then be utilized as a green catalyst support itself, or as a highly effective adsorbent to remove synthetic dyes and heavy metals from wastewater.

In the dairy industry, microalgae are highly effective at capturing nitrogen and phosphorus from whey wastewater for carbon neutrality through symbiosis. Bacteria-algae interactions amplify CO_2 fixation through molecular quorum sensing and recalcitrant dissolved organic carbon, thereby enabling long-term storage. These synergistic consortia can increase algal photosynthesis by 20–40%. The presence of essential ions like Ca(II) can trigger microalgae-induced carbonate precipitation, which immobilizes microplastics and reduces the toxic effects of contaminants on the algae's photosynthetic machinery. (Gupta & Philip, 2026)

The overall removal efficiencies across different systems reached 88-92% for amoxicillin, 63-76% for ciprofloxacin, and 23-44% for sulfamethoxazole, while nutrient and organic matter removal remained high, indicating stability despite antibiotic exposure. The presence of antibiotic-resistant genes and integrons confirmed microbial genetic adaptation, and the study highlighted the importance of enhanced biodegradation and extracellular polymeric substances in maintaining cell integrity and resilience under antibiotic stress. (Gupta & Philip, 2026) Advanced techniques such as CRISPR-Cas9 for genome editing and Machine Learning for predictive modeling are emphasized as tools to enhance SMC capabilities, improving their effectiveness in addressing both wastewater management and plastic pollution. (Thajuddin et al., 2026)

Figure 2. Multiomics Understanding of Nitrogen Deprivation In Microalgae for Biofuel Production

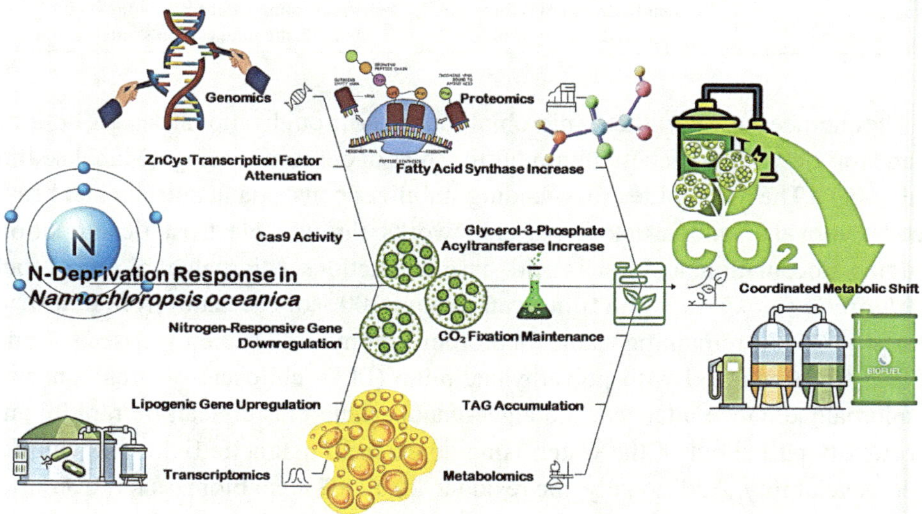

CARBOHYDRATE REDIRECTION

AGP (starch synthesis blocking) increases oil productivity by completely suppressing starch synthesis. Overexpression of the DpAGP gene, which encodes the small subunit of ADP-glucose pyrophosphorylase (AGPase), enhances its metabolic capabilities for bioethanol production. self-flocculation as a sustainable, cost-effective method for harvesting microalgae in biorefineries. (Anabtawi et al.,

2025) Microalgae can be manipulated to produce various isoprenoids, especially under challenging cultivation conditions, by adjusting their metabolic pathways to generate compounds with significant technological potential. The effects of nitrogen, sulfur, and combined nitrogen and sulfur deprivations on the growth and lipid body accumulation of the starchless mutant of Chlamydomonas reinhardtii. The gene expression in the starchless mutant of Chlamydomonas reinhardtii (cc5373-sta6) deprived of sulfur (TAP-S) exhibited significantly higher total chlorophyll, cell density, and protein content after six days. Nitrogen deprivation for 6 days led to a significant increase in lipid body size, and higher expression of acetyl-CoA carboxylase (ACC) was observed after 8 and 24 hours of nitrogen and sulfur deprivation. The optimal cultivation condition for cc5373-sta6 cells is suggested to be nutrient-deprived (-N, -S, or -NS) for four days, resulting in enhanced cell growth and increased lipid body production. (Gonzalez & Ynalvez, 2023) Synthetic biology toolkits, transcriptional modulators, epigenome editors, and biosensor-integrated logic gates can be used to modulate BGS/TGS. Bioengineering thereby decreases soluble carb synthesis and improves fatty acid production. CRISPR-derived tools have facilitated significant advancements in microalgal engineering, enabling precise control of gene expression and stable epigenetic modifications. (Pallavi et al., 2026) The ability of CmGLG1 or GLG2 to rewire complex metabolic networks in response to environmental changes affects glycogenin genes, starch initiation, and starch accumulation. (Lim, 2025)

Table 5. Scalability and Economic Constraints

Microalgal Strain Consortium	Genetic Modification Strategy	Key Limitations
Nannochloropsis sp.	CRISPR-Cas9 knockout of ZnCys (Transcription factor)	Severe growth penalty under nutrient stress; confined to closed PBRs.
Chlamydomonas reinhardtii	Overexpression of RuBisCO activase	Strictly a model organism; poor performance in raw wastewater.
Chlorella vulgaris & Azospirillum (Consortium)	Co-cultivation (metabolic cross-feeding)	Complex population dynamics; susceptible to invasive rotifers/ciliates.
Phaeodactylum tricornutum	Overexpression of Malic Enzyme (ME)	Requires silicate supplementation; marine species limits freshwater applications.
Synechocystis sp. PCC 6803 (Cyanobacteria)	Synthetic carbon concentrating mechanism (CCM)	High risk of environmental horizontal gene transfer; stringent regulatory hurdles.

CHALLENGES TO SCALABILITY

TEA consistently identifies capital expenditure (CAPEX) for PBRs and operational expenditure (OPEX) for nutrient inputs and water management as prohibitive. While utilizing algal-bacterial consortia in open raceway ponds reduces CAPEX and allows for wastewater integration (lowering nutrient OPEX), it introduces significant vulnerabilities to contamination and environmental fluctuations, destabilizing predictable lipid and biomass yields. Under nutrient limitation or nitrogen deprivation, carbon flux shifts from protein synthesis toward lipid accumulation. In *Chlamydomonas reinhardtii*, coordinated down-regulation of photosynthetic flux precedes high TAG accumulation In *Chromochloris zofingiensis,* nitrogen starvation stimulates coordinated carbon reallocation toward TAG accumulation. The large-scale microalgae production faces significant hurdles. (Jain & Kumar, 2026) There is a lack of cohesive, international regulatory guidelines specifically addressing the outdoor cultivation of transgenic microalgae. Consequently, achieving biosafety approval requires exhaustive, localized environmental impact assessments, creating a massive barrier to market entry for engineered strains.

Economic Viability: High production costs and inconsistent biomass characterization remain major obstacles to large-scale competitiveness. Traditional centrifugation is energy-prohibitive at scale. Alternative methods like chemical flocculation introduce metal salts (e.g., alum, ferric chloride) or synthetic polymers, which can contaminate the biomass and limit its downstream utility in high-value bio-products or feed. Membrane separation (ultrafiltration/microfiltration) offers high recovery rates but is severely limited by rapid membrane fouling caused by extracellular polymeric substances (EPS) secreted by the algae and bacterial consortia.

Biological Barriers: Many species, such as *Nannochloropsis gaditana*, have rigid cell walls that limit the extraction of internal nutrients and lipids. To bypass the energy-intensive drying phase, modern downstream processing must pivot toward wet lipid extraction. Techniques such as hydrothermal liquefaction (HTL), supercritical fluid extraction, and cell-wall disruption via enzymatic or ultrasonic methods enable direct processing of wet slurries. However, these technologies remain largely confined to pilot-scale and require substantial energy and financial inputs, which challenge their immediate commercialization.

Contamination: Biological contaminants can cause crashes in outdoor ponds. Innovations like using 1,8-cineole (from eucalyptus oil) have shown promise in reducing predator abundance without harming the microalgae. Cultivating GE strains in open raceway ponds poses a significant risk of accidental release into natural aquatic ecosystems. The ecological consequences of introducing highly competitive, hyper-fixating synthetic strains include potential disruptions to native food webs, localized eutrophication, and horizontal gene transfer (HGT) of antibiotic resistance markers (often used in laboratory selection) to environmental microbiota.

Figure 3. Multiomics For Algae techno-economic and life cycle assessment perspective, shows translational gaps between laboratory productivity gains and pilot-scale performance.

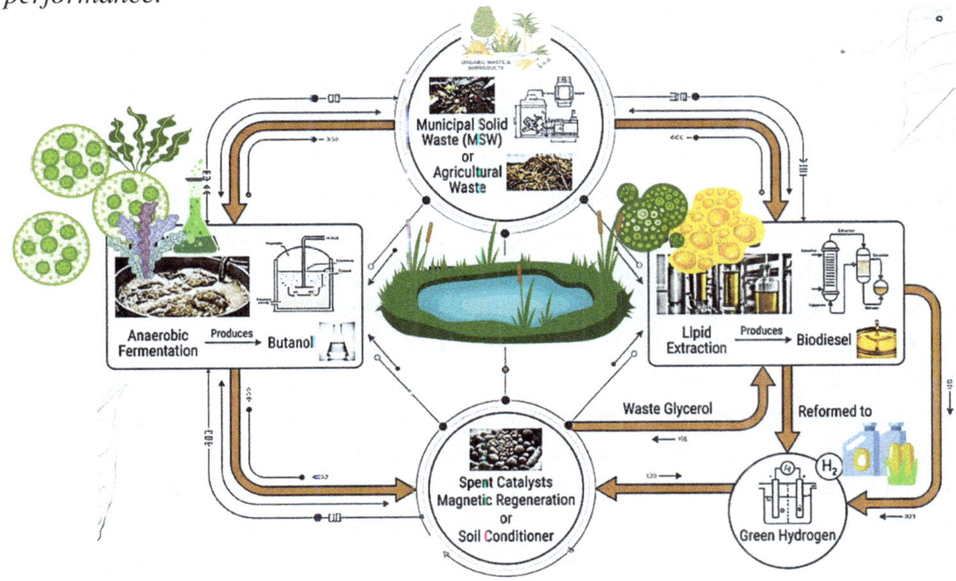

FUTURE DIRECTIONS

Multi-omics Integration with phosphoproteomics offers promise of a bio-hybrid system. The microalga *Chlorella* vulgaris with carbonic anhydrase-functionalized Zn-based metal-organic framework-5 to enhance CO2 hydration and carbonate precipitation, thereby improving CO2 capture and ocean alkalinity. Experiments showed significant biomass and CaCO3 production at optimal CO2 levels, demonstrating the potential of this approach for scalable ocean-based CO2 sequestration and alkalinity enhancement. (Fazal et al., 2026) Metabolic Engineering systems-level approaches to identify transcription factors that regulate triacylglycerol (TAG) accumulation. Zeolitic imidazolate framework-based nanozymes improved microalgal growth, Rubisco activity, and carbonic anhydrase activity, enhancing CO_2 affinity and photosynthetic efficiency. Engineering these pathways can rewire metabolism to enhance lipid production under nutrient stress. Carbon fixation by engineered bacteria-algae consortia can increase CO2 fixation efficiency by 22–35% compared to axenic algal systems. High-value metabolites, including esterified astaxanthin, eicosapentaenoic acid (EPA), and docosahexaenoic acid (DHA), are manufactured via genetic engineering and nanoformulation to enhance their accumulation and stability. (Chakraborty et al., 2025)

Chlorella vulgaris, genome-scale metabolic modeling quantify plastid–mitochondria carbon transport and subsystem activity changes under different trophic modes. Using similar models for *P. tricornutum and C. reinhardtii* allows computational allocation of carbon among growth, storage lipids, and excreted organics that can feed heterotrophs. The ZS-1, genome sequencing plus metabolomics, has identified rTCA and C_4-associated genes upregulated during EET-driven CO_2 fixation and lipid production. Transcriptomics proteomics under engineered EET and light–carbon regimes is giving concrete engineering targets and constraints for bioelectrochemical operation. In ZS-1 and Shewanella, omics under varying electrode potentials and flavin carbon-dot supplementation can define regulons controlling EET components, riboflavin shuttles, and central metabolism, enabling rational rewiring to favor CO_2 reduction vs. respiration. In *C. reinhardtii*, high-light and carbon acclimation omics has already revealed major remodeling of pigments, antenna size, and carbon metabolism, which can be leveraged to design dynamic promoters and light-responsive circuits.

PHOTOSYNTHESIS ENHANCEMENT

Traditional methods for improving microalgae have reached their limits, underscoring the need for genetic engineering to enhance photosynthesis and productivity. Understanding the mechanisms that control photosynthesis is essential for identifying genetic engineering targets to increase biomass yield and reduce cultivation costs. RuBisCO activase overexpression increases biomass efficiency. The key enzyme, ribulose-1,5-bisphosphate carboxylase/oxygenase (Rubisco), catalyzes carbon fixation in the Calvin–Benson cycle. Many microalgae possess a carbon concentrating mechanism (CCM) that increases CO_2 availability near Rubisco, enhancing fixation efficiency. (Jain & Kumar, 2026)

Elevated CO_2 supplementation (e.g., 3% v/v) in Microchloropsis gaditana significantly increased biomass productivity (up to 2-fold) and redirected carbon flux toward long-chain polyunsaturated fatty acids (LC-PUFAs). In plant responses, microRNAs (miRNAs) function as post-transcriptional regulators of genes involved in antioxidant defense, growth, and hormonal signaling. The Calvin-Benson enzymes (FBA, SBPase) and chlorophyll antenna tuning (CAO, CpSRP) increase photosynthesis rate. Heterologous carbonic anhydrase benefits biomass or lipid productivity.

Additionally, inhibiting photosystem II, photosystem I, and the respiratory complexes I and III in *Chlorella pyrenoidosa* SHOU-1002 under nitrogen limitation enhances lipid biosynthesis by increasing reactive oxygen species (ROS). Redirected carbon flux towards lipid production, yields biodiesel-suitable lipids, and provides insights for advancing microalgal biofuel development. (Wang et al., 2026) Transcription Factors ZnCys (bZIP) improve 2× lipid productivity and maintain CO_2 fixation. NobZIP77 (bZIP) affects TAG under N-deprivation, while CryP (cryptochrome) optimizes light sensing in fucoxanthin. The bidirectional promoter (BDP) intergenic regions in the chloroplast genome of *Chlamydomonas reinhardtii*, revealing diverse evolutionary patterns and potential for multigene engineering in microalgae. Three candidate bidirectional promoters (BDP1, BDP2, and BDP3) were selected for functional analysis; BDP1 supported the expression of two transgenes in opposing orientations, while BDP2 and BDP3 showed lower and minimal activity, respectively. (Tazon et al., 2026)

Zeolitic imidazolate framework-8 (ZIF-8)- based nanozymes (ZNZs) increase microalgae's affinity for CO2 by providing dual functionality: they directly adsorb CO2 and mimic the catalytic activity of carbonic anhydrase (CA). When these nanozymes are added to microalgae cultures, they significantly boost the activity of essential photosynthetic enzymes. Research shows that introducing ZNZs increases both intra- and extracellular CA activity, as well as the activity of ribulose-1,5-bisphosphate carboxylase oxygenase (Rubisco). Photosynthetic efficiency gains show baseline microalgae 5-7% light-to-biomass conversion (vs. C_3 plants: 3.5%, C_4 plants: 4.3%). Genetic engineering targets truncated light-harvesting antenna, RuBisCO enhancement, and NPQ tuning. (Torres-Momber et al., 2026)

Methyl jasmonate can selectively enhance expression of the tdTomato transgene from BDP1, providing a chemical method for regulating chloroplast transgene expression. The chloroplast transcriptional mechanisms underscore the functional potential of natural BDPs, particularly BDP1, as tools for multigene engineering. Nickel-doped ceria nanoparticles in growth media improve the physiology of the microalgae's photosystem for chemical modulation. (Tazon et al., 2026) The outdoor photobioreactors using single-gene knockouts (AGP, ZnCys attenuation) or multi-gene engineering (AGP + ZEP double knockout) report 450.09 mg/g DW TAG accumulation and 81% higher oil productivity. (Nievergelt, 2025)

Cofactor and energy-balancing multi-omics investigation of *Pseudomonas putida* in tandem with algae. Lignin valorization reveals how cells coordinate phenolic carbon processing with the generation of essential cofactors, such as NADPH and NADH. Bio-Hybrid biomimetic systems for accelerated nutrient capture enable the prediction of metabolic bottlenecks. Supplementing *Chlorella vulgaris* cultures with Fe_2O_3 nanoparticles (up to 75 mg/L) significantly increases NADPH generation derived from photosynthetic electron transport resource utilization. Mutation of

PhoD-type alkaline phosphatase in *Phaeodactylum tricornutum* altered lipid composition and redirected carbon toward lipid synthesis. Regulatory mechanisms of photosynthesis integrated proteomic profiling has elucidated how photothermally responsive nanoparticles (T-NPs) regulate the microenvironmental temperature in *Chlorella pyrenoidosa*. Encapsulating *Chlorella vulgaris* within a carbonic anhydrase-functionalized zinc metal-organic framework in alginate beads mimics natural enzymatic CO_2 hydration. The availability of bicarbonate (HCO_3^-) promotes photosynthesis and microalgae-induced carbonate precipitation with carbon capture capacities. The upregulation of Rubisco activity increased ATP and NADPH levels, thereby enhancing both light-dependent and light-independent photosynthetic reactions.

PHOTOPERIOD: TEMPORAL RESOLUTION

Most studies use a 12h:12h light-dark cycle; diurnal seasonal variation is underexplored. Light color was identified as the primary factor influencing both the growth of *Chlorella sorokiniana* and the production of extracellular enzymes, followed by CO2 concentration and light intensity. *A Tetradesmus obliquus*-based algae-bacteria symbiotic system treats wastewater with a low carbon-to-nitrogen ratio. At 95.85% nitrogen and 97.96% phosphorus removal, it demonstrated significant biomass production and enhanced nutrient assimilation, for sustainable wastewater treatment and bioresource recovery. The findings suggest that controlling light and carbon supplies can effectively modulate microalgal growth and enzyme activity, which is beneficial for developing sustainable microalgal bioprocesses. (Torres-Momber et al., 2026)

NANOPARTICLES

Integrating nanoparticles such as boron oxide (B2O3), cerium oxide (CeO2), or magnetized iron oxide (Fe3O4) into algal biodiesel blends improves engine efficiency (BTE) and significantly reduces harmful emissions like CO, HC, and smoke. Microalgae are also used in microbial fuel cells to degrade pollutants, such as antibiotics, while simultaneously producing biohydrogen, a bioenergy. Photothermal conjugated polymer nanoparticles (T-NPs) can accurately regulate the microenvironment temperature of *Chlorella pyrenoidosa,* increasing ATP and oxygen evolution. Low concentrations of zinc oxide (ZnO) nanoparticles can enhance growth and alleviate salt stress in microalgae like *Lobosphaera* sp. Nano-Sensors from microalgae such as *Haematococcus pluvialis* can be upcycled into high-value nitrogen-doped carbon dots for ultrasensitive detection of environmental toxins, such as copper ions. (Cozma et al., 2025) Another advanced strategy for capturing industrial carbon emissions involves a cascade cultivation of *C. vulgaris* using exogenous plant hormones indole-3 acetic acid and varying nanoparticle concentrations. This approach effectively promotes both carbon fixation and high lipid accumulation for biofuels.

METABOLITE ANNOTATION

Microalgae serve as a powerful biotechnological tool for detoxifying environmental pollutants without generating additional waste sludge. Heavy Metal Removal by *Chlorella sorokiniana* has been successfully used in thin-film platforms to remove up to 100% of cadmium (Cd^{2+}) from aqueous solutions and industrial wastewater. Microalgae are highly effective at removing nitrogen and phosphorus from dairy and whey wastewater. Integration with membrane technologies enables simultaneous cultivation and biomass separation, thereby maximizing cost-effectiveness. Microalgae-microbe interactions can help mitigate the toxic effects of microplastics. For example, Microcystis aeruginosa can use calcium ions to initiate carbonate precipitation, thereby immobilizing microplastics and reducing cellular exposure. (Anabtawi et al., 2025)

GENETIC STABILITY

Long-term maintenance of transgene expression and epigenetic silencing in engineered strains. The METE promoter in *Phaeodactylum tricornutum* enables high transgene expression and can be regulated by vitamin B12, allowing for optimized production levels while minimizing potential negative effects on the host organism. Transgenic lines utilizing the METE promoter produced approximately 2 mg/L of the diterpenoid enzyme casbene synthase, with production levels adjustable by B12 concentration, highlighting the potential of *Phaeodactylum* METE for industrial biotechnology applications.(Hickland et al., 2025) Transcriptomic data from species such as Synechocystis sp. and Geminocystis sp. have identified hub genes involved in salt and nitrate stress responses, providing targets for engineering strains that can thrive in marginal or saline environments. Microalgae exhibit flexible carbon partitioning in response to light, nutrient availability, and genetic modifications. Phototrophic conditions favor starch accumulation. Nitrogen-deprived limitation and engineered knockdowns redirect flux toward lipid enhancement. (Figure 2) The application of genetic engineering and nanotechnology in microalgae cultivation raises biosafety concerns, regulatory approval requirements, and ecological impacts.

ECOLOGICAL CONTEXT

Environmental metagenomics and metabolomics are now used for real-time ecological surveillance. The microbial identity and in situ activity across spatial and temporal scales, which reduces uncertainty in carbon flux predictions. The integration of artificial intelligence (AI) into algal systems to enhance wastewater treatment and carbon sequestration. It highlights how AI can address traditional challenges in algal biotechnology, such as high harvesting costs and scalability, by optimizing growth conditions in photobioreactors through real-time monitoring and machine learning. The integration of synthetic biology with advances in photobioreactor design and automated cultivation is paving the way for scalable, economically viable microalgal biomanufacturing, contributing to a low-carbon, bio-based economy. (Thajuddin et al., 2026) A household-scale photobioreactor (PBR) concept was proposed specifically for poor communities in developing countries, using small transparent reactors to capture CO_2 from indoor air and produce algal biomass that can be used locally as feed, fertilizer, or for bioproducts. Several real-world rural and off-grid projects already show how algae and bioelectrochemical systems can advance multiple SDGs, especially SDG 6, 7, 11, 12 and 13. (Jain & Kumar, 2026) (Veale et al., 2025)

Nano-Bio Synergistic Fermentation recovers nutrients from agro-residues like rice husk and sugarcane bagasse. Researchers developed a process that integrates zinc oxide (ZnO) nanoparticles with the fungus Aspergillus oryzae. This combination significantly promotes nutrient recovery, increasing total soluble nutrients enriching protein content while reducing environmental organic loads.

Microalgae-Sludge Hybrid Systems in wastewater treatment, combine microalgae and bacterial consortia are used to degrade persistent antibiotic contaminants. The integrated systems simultaneously enhance nutrient recovery, supply oxygen, and generate biomass that can be valorized for other uses.

Slaughterhouse Waste Valorization use biochemical processes and waste-to-energy technologies to manage slaughterhouse waste, demonstrating significant success in nutrient recovery and high-efficiency biogas production.(Sharma et al., 2026)

NUTRIENT CAPTURE AND WASTEWATER APPLICATIONS

Microalgae deploy multiple DNA repair pathways. *Phaeodactylum tricornutum* is recognized as a versatile photosynthetic chassis for sustainable bioproduction. *P. tricornutum* is capable of producing bioactive metabolites and engineered heterologous compounds through synthetic biology. The modular cloning systems have improved metabolic pathway control and strain scalability, paving the way for more efficient microalgal biomanufacturing. Strategies to enhance *P. tricornutum* chloroplast gene expression and DNA site-specific integration could scale up production. (Shelake et al., 2025) *Phaeodactylum* upregulates enzymatic and non-enzymatic antioxidant systems under stress. *Phaeodactylum* synthesizes UV-screening molecules like mycosporine-like amino acids and sporopollenin-like wall components to reduce radiation damage and control reactive oxygen species. The photoreactivation, UV-damage endonuclease-dependent excision, and nucleotide excision repair efficiently remove UV-induced lesions, including cyclobutane pyrimidine dimers. The synthesis of algal nanoparticles provides high adsorptive efficiency for waste removal. Nanomaterial-based biosensors and quantum-dot light-emitting diodes for intelligent edge image detection enable real-time monitoring of microalgal density and health.(Moroi et al., 2025)

The METE promoter in the diatom *Phaeodactylum tricornutum*, enables high expression and B12-tuneable regulation of transgenes, for metabolic engineering. The METE promoter exhibits significantly higher expression levels than previously characterized promoters and enables tight regulation of transgene expression in response to B12. (Hickland et al., 2025) *Platymonas helgolandica* in mariculture wastewater had the highest removal efficiencies of total nitrogen and COD at a 15L:9D photoperiod. Significant factors affecting microalgal growth were identified as light intensity and illumination time. Photoperiod affects the removal efficiencies of chemical oxygen demand (COD) and ammonium nitrogen (NH4+-N). Different photoperiods influence enzymatic activities and gene expression related to nitrogen metabolism and photosynthesis. It was found that the circadian rhythm significantly enhances starch accumulation and increases amylose content by regulating key enzymes involved in starch synthesis. (Priyanka et al., 2026) Transcriptomic analysis indicated that gene expression related to nitrogen metabolism and photosynthesis pathways was significantly up-regulated, highlighting the impact of photoperiod on enzymatic activities involved in nitrogen transformation. (Torres-Momber et al., 2026)

The algal-bacterial symbiosis system in a rotating biological contactor can treat mariculture wastewater under varying light/dark cycles. In the conversion of anaerobic digestion sludge and wine cellar sludge into valuable caproate, metagenomic analysis identified the methanogen *Methanothrix* as a key participant in the chain elongation process. Furthermore, metatranscriptomic analysis revealed cooperative interactions among microbial species during this chain-elongation reaction. The biomass growth, with dry weight increasing from 0.05 to 0.37 g/L, produced substantial amounts of protein, polysaccharides, and lipids, indicating strong potential for bioresource recovery. Metagenomic and transcriptomic analyses revealed microbial succession and upregulation of genes involved in nitrogen and phosphorus metabolism. (Byalahunasishetter et al., 2026) In bioethanol production, transcriptomic analysis was used to study *Saccharomyces cerevisiae* (yeast) cells treated with magnetic nanoparticles. The analysis confirmed the upregulation of key xylose-utilizing genes XKS1 and YJR096W, unlocking a breakthrough in producing second-generation biofuels from complex sugars. (Prajapati et al., 2026) (Zhao et al., 2026) In unicellular red algae species, *Porphyridium purpureum* and *Porphyridium aerugineum,* short photoperiods enhanced phycobiliprotein accumulation. The long photoperiods increased biomass and exopolysaccharide production in *P. purpureum*, suggesting enhanced nutrient assimilation under low C/N stress. The GS-GOGAT cycle, Glutamine Synthetase, or Glutamate Synthase combine wastewater treatment with biosynthetic productivity. TOR kinase effects of CO2 availability, nitrogen levels, light color, and light intensity on the growth and extracellular enzyme activities of *Chlorella sorokiniana* using a Taguchi L4 experimental design. Cell growth and lipase activity were optimized under purple light and moderate light intensity with low CO2 supplementation, while amylolytic activity was enhanced under purple light with increased CO2 availability and lower light intensity. (Torres-Momber et al., 2026) Hydrothermal Liquefaction (HTL) allows whole-algae conversion bypass the need to extract lipids first. Catalysts like Mg-modified red mud or cerium oxide are used in HTL processes to convert wet algal biomass directly into energy-dense biocrude or bio-oil. Trimetallic nanocatalysts (Co-Mo-Ni) are used to deoxygenate and break bonds in microalgae oil, upgrading it into high-density hydrocarbons suitable for aviation bio-jet fuel. (Ananthakumar et al., 2025)

Table 6. Candidate Chassis In Sequestration for Biofuel and Bioproduct Engineering

Species	Primary role in system	Key native traits for capture
Alcaligenes faecalis ZS-1	Electrotrophic CO_2-fixing biocathode bacterium	In biocathodic MFC, ZS-1 simultaneously produces current and fixes CO_2; genome and metabolomics show rTCA and C_4-like pathways with ppc upregulation under inward EET, accompanied by lipid accumulation.
Alcaligenes faecalis (other strains)	Nutrient removal, xenobiotic/metal bioremediation	Genomes harbor broad xenobiotic and metal tolerance/degradation genes; widely used in hydrocarbon removal and nutrient-aided bioremediation.
Shewanella oneidensis	EET hub in bioanode / redox shuttle partner	Exhibits strong flavin-mediated extracellular electron transfer; flavin shuttles dominate EET to insoluble acceptors, and metabolic rate and current output can be boosted by carbon dots feeding.
Chlorella vulgaris	Workhorse CO_2 and nutrient sequestration alga	High-efficiency green alga; genome-scale models show elevated carbon metabolism fluxes and increased plastid–mitochondrial CO_2 transport under mixotrophy/heterotrophy, with large boosts in carbon metabolism subsystems.
Chlamydomonas reinhardtii	Advanced model for Calvin cycle/CCM engineering	Detailed omics and physiology; overexpression of chloroplast GAPDH enhanced Calvin cycle capacity and carbon conversion, improving energy storage compound accumulation; acclimation studies show flexible high-light and carbon responses.
Chromochloris zofingiensis	High-value carbon sink (astaxanthin, lipids)	Known as a robust green alga and strong triacylglycerol/astaxanthin producer under stress
Chlorella ohadii	Extremophile for high-light CO_2 capture	Desert alga with exceptional high-light tolerance and very efficient light utilization

continued on following page

Table 6. Continued

Species	Primary role in system	Key native traits for capture
Tetradesmus obliquus (Scenedesmus)	Wastewater-compatible CO_2 and nutrient scavenger	Commonly used in wastewater, tolerant to variable nutrients; can deliver high biomass and lipid content
Phaeodactylum tricornutum	Diatom chassis for alternative carbon routes	Model diatom with distinct carbon metabolism and silica frustules, suited to different light spectra and possibly flue gas compositions
Alginates Alginate hydrogels	Immobilization matrix, chemical CO_2 pre-capture	Alginate is an algal cell-wall biopolymer crosslinking with divalent cations into hydrogels; calcium alginate gels can entrap monoethanolamine and other amines, enabling rapid CO_2 absorption and delayed amine diffusion while remaining biocompatible.

MULTI-OMICS FOR NUTRIENT CAPTURE IN RURAL REGIONS

Microbial fuel cells and related BES are being evaluated as decentralized power sources for off-grid communities, converting local organic waste and wastewater into electricity while treating the effluents. A micro-wastewater treatment and bio-energy system based on biological electrochemical systems (BES) was piloted in the Dadaab refugee camp, Kenya, to treat human waste and generate energy in a remote, infrastructure-poor setting. (Chakraborty et al., 2025) For wastewater-oriented strains (*Chlorella vulgaris, Tetradesmus obliquus, A. faecalis*), combining transcriptomics and metabolomics in real effluents can map regulatory and transport bottlenecks for ammonium, nitrate, phosphate, and organics. (Veale et al., 2025) Materials–biology integration via omics-informed immobilization inform synthetic up-regulation of transporters or storage polymers. Alginate hydrogels can immobilize both chemical sorbents (MEA) and microalgae The tuning polymer composition and layering controls CO_2 and nutrient gradients experienced by cells, can be detected via spatial transcriptomics and proteomics. Omics data from immobilized vs. planktonic cultures of *C. vulgaris or C. reinhardtii* can guide redesign of cell walls, CCM, and stress responses to thrive in high-CO_2, diffusion-limited microenvironments. Microalgae like Chlorella pyrenoidosa can be cultivated directly in photofermentation effluents, which are otherwise highly loaded with organic pollutants and extreme pH levels. (Anabtawi et al., 2025)By optimizing light intensity (7000 lux) and initial pH (8) to mitigate inhibitory effects. The microalgae achieve high protein yields while removing over 74% of the chemical oxygen demand (COD) and stripping nitrogen and phosphorus from the waste. Co-producing biohydrogen and microalgae from agricultural waste like corn stover improves overall carbon conversion efficiency. (Hou et al., 2024)

Omics technologies provide a systems-level understanding of microalgae, enabling researchers to engineer more efficient pathways for carbon sequestration, nutrient capture, and high-value metabolite production within a sustainable bioeconomy. The transition from a linear to a circular bioeconomy relies on the high-value utilization of microalgal residues and hybrid coatings and nanoproducts. (Jangama, 2026) Despite these advancements, the sources emphasize that economic viability, large-scale scalability, and regulatory approval remain the primary challenges to moving microalgal innovations from the laboratory to industrial implementation. Omics-driven strategies are used to boost the production of esterified astaxanthin, beta-carotene, and omega-3 fatty acids (EPA and DHA), which are essential for human and animal nutrition. (Nanjundeshwara et al., 2025) Microalgae-microbial fuel cells represent a sustainable platform for simultaneously removing emerging contaminants, such as antibiotics, and producing bioenergy. AI and Machine Learning integration in microbial production is increasingly required to manage large-scale cultivation systems in the bioeconomy.

CONCLUSION

Advanced omics-driven strategies are transforming microalgal biotechnology by providing systems-level insights into metabolic pathways, allowing for more efficient carbon sequestration and nutrient valorization. Microalgae omics data reveal coordinated upregulation of genes, proteins, and metabolites involved in carbon fixation and nutrient assimilation, as well as in technological applications. Genomics show genes in photosynthesis, carbon fixation, and related metabolic pathways are significantly upregulated under carbon/nutrient capture conditions, with robust statistical validation. Proteomics reflect that enzymes in fatty acid, glycolysis, and amino acid biosynthesis pathways are upregulated during carbon sequestration. Metabolomics demonstrate lipids, carbohydrates, and amino acids accumulate under nutrient limitation.

REFERENCES

Anabtawi, H. M., Ikhlaq, A., Kumar, S., Rafique, S., & Hassan, A. A. (2025). Addressing Challenges for Eco-Friendly and Sustainable Wastewater Treatment Solutions Using Extremophile Microorganisms. *Sustainability (Basel), 17*(6), 2339. DOI: 10.3390/su17062339

Ananthakumar, B., Nanjundeshwara, N., Lakshmaiah, A., & Raghavan, R. (2025). Nanocarriers for Delivering Nucleic Acids in Plants: Mechanisms and Applications. *Nanocarriers in Plant Science and Agriculture*, 207–234. DOI: 10.4018/979-8-3693-9869-2.ch008

Byalahunasishetter, B. I., Chowdareddy, S., & Raghavan, R. (2026). Cancer Risk Prediction Based on Gut Microbiota Analysis. *IGI Global Scientific Publishing*, 307–342. DOI: 10.4018/979-8-3373-8172-5.ch009

Chakraborty, S., Talukdar, A., Dey, S., & Bhattacharya, S. (2025). Role of fungi, bacteria and microalgae in bioremediation of emerging pollutants with special reference to pesticides, heavy metals and pharmaceuticals. *Discover Environment, 3*(1), 91. DOI: 10.1007/s44274-025-00217-7

Chen, H., Ma, H., Yu, L., Zou, S., Li, Y., Lu, Y., Yang, M., Ge, F., Yuan, L., Zeng, X., Zhang, C.-C., Yoon, K , Han, D., & Hu, Q. (2025). Functional characterization of the plastidial enzyme GFAT in microalgae and its potential role in lipid biosynthesis. *Plant Physiology, 199*(1), kiaf395. Advance online publication. DOI: 10.1093/plphys/kiaf395 PMID: 40892446

Cozma, P., Ro ca, M., Minu , M., & Gavrilescu, M. (2025). Phytoremediation: A sustainable and promising bio-based approach to heavy metal pollution management. *The Science of the Total Environment, 1001*, 180458. DOI: 10.1016/j.scitotenv.2025.180458 PMID: 40992187

Fazal, T., Wang, Y., Zhang, Y., Mushtaq, A., Hou, S., & Zhang, C. (2026). Enhancing ocean alkalinity and CO2 sequestration via microalgae-driven carbonate precipitation and biomimetic catalysis. *Journal of Environmental Management, 398*, 128457. DOI: 10.1016/j.jenvman.2025.128457 PMID: 41494387

Gonzalez, D. I., & Ynalvez, R. A. (2023). Comparison of the effects of nitrogen-, sulfur- and combined nitrogen- and sulfur-deprivations on cell growth, lipid bodies and gene expressions in Chlamydomonas reinhardtii cc5373-sta6. *BMC Biotechnology, 23*(1), 35. DOI: 10.1186/s12896-023-00808-3 PMID: 37684579

Gu, X., Shi, Y., Luo, C., & Cheng, J. (2024). Establishment of Saccharomyces cerevisiae as a cell factory for efficient de novo production of monogalactosyldiacylglycerol. *Biotechnology for Biofuels and Bioproducts*, *17*(1), 111. DOI: 10.1186/s13068-024-02560-y PMID: 39129014

Gupta, Y., & Philip, L. (2026). Mechanistic insights and resistance profiling of chlorophyta–proteobacteria–cyanobacteria dominated consortia in removing antibiotic mixtures and nutrients from synthetic and real municipal wastewater. *Journal of Environmental Chemical Engineering*, *14*(1), 121004. DOI: 10.1016/j.jece.2025.121004

Hickland, P. R., Geisler, K., Newsad, S., Pasquina, M. L., Faessler, A. C., Mendoza-Ochoa, G. I., Holzer, A., Mehrshahi, P., & Smith, A. G. (2025). A new tool for engineering Phaeodactylum tricornutum: The METE promoter drives both high expression and B12-tuneable regulation of transgenes. *The Plant Journal : For Cell and Molecular Biology*, *124*(2), e70210. DOI: 10.1111/tpj.70210 PMID: 41116758

Hou, C., Zhao, J., Huang, B., Zhou, X., & Zhang, Y. (2024). Microalgae-based technologies for carbon neutralization and pollutant remediation: A comprehensive and systematic review. *Resources, Conservation and Recycling*, *202*, 107323. DOI: 10.1016/j.resconrec.2023.107323

Jain, S., & Kumar, S. (2026). Advancing bioethanol: Exploring feedstock diversity, production pathways, and environmental implications. *RSC Sustainability*, *4*(3), 1129–1159. Advance online publication. DOI: 10.1039/D5SU00152H

Jangama, R. (2026). *Hybrid Coating Systems: Bioactive Molecules and Nanoparticles for Cancer Therapy*. IGI Global Scientific Publishing., DOI: 10.4018/979-8-3373-4104-0.ch003

Kirbayeva, D. K., Shayakhmetova, A. Y., Kossalbayev, B. D., Sadvakasova, A. K., & Bauenova, M. O. (2025). Harnessing MicroRNAs and CRISPR to enhance biofuel production in microalgae. *International Journal of Hydrogen Energy*, *157*, 150399. DOI: 10.1016/j.ijhydene.2025.150399

Lim, L. W. K. (2025). CRISPR-Cas applications in extremophiles: Thermophiles, psychrophiles, halophiles, acidophiles, alkaliphiles, piezophiles, xerophiles, radiophiles and metallophiles. *Total Environment Microbiology*, *1*(3), 100029. DOI: 10.1016/j.temicr.2025.100029

Mahalakshmi, K., P, D. A., J, S. P., Moger, S. S., & Raghavan, R. (2026). Spatial Multi-Omics in Cancers: Integrating Molecular Landscapes for Precision Oncology. *Spatially Resolved Single-Cell and Multiple Omics in Human Diseases*, 217–248. DOI: 10.4018/979-8-3373-4849-0.ch007

Mohapatra, A., Guliani, E., Nandal, V., Solanki, M., Pathak, V. V., Banga, S., & Sharma, D. K. (2025). Recent innovations in sustainable algal biofuel production: A comprehensive review. *Brazilian Journal of Chemical Engineering*, *43*(2), 549–577. Advance online publication. DOI: 10.1007/s43153-025-00598-5

Moroi, K., Yamamoto, T., & Kurita, T. (2025). Double-strand break-free and transgene-free genome editing in the microalga Nannochloropsis oceanica using removable vectors containing the CRISPR base editing system. *Scientific Reports*, *15*(1), 42431. DOI: 10.1038/s41598-025-26657-y PMID: 41309901

Nanjundeshwara, N., Lakshmaiah, A., & Raghavan, R. (2025). Aflatoxigenic Fungi. *Aflatoxigenic Fungi*, 225–239. DOI: 10.1201/9781003487043-29

Nievergelt, A. P. (2025). Genome editing in the green alga Chlamydomonas: Past, present practice and future prospects. *The Plant Journal : For Cell and Molecular Biology*, *122*(1), e70140. DOI: 10.1111/tpj.70140 PMID: 40186543

Pallavi, V. N., Chowdareddy, S., Byalahunasishetter, B. I., & Raghavan, R. (2026) CRISPR-Cas9 Technology for Precision Cancer Medicine: Target Discovery, Disease Modeling, and Therapeutics. *CRISPR-Cas9 Technology for Precision Medicine*, 35–72. DOI: 10.4018/979-8-3373-4857-5.ch002

Parker, L., Ward, K., Pilarski, T., Price, J., Derkach, P., Correa, M., Miller, R., Benites, V., Athanasiadis, D., Doherty, B., Edy, L., Alnozaili, G., Reyes, N., Wittenberg, J., Elares, G., Destaillats, F., Rakitsky, W., & Franklin, S. (2025). Development and Production of High-Oleic Palm Oil Alternative by Fermentation of Microalgae. *Fermentation (Basel, Switzerland)*, *11*(4), 207. DOI: 10.3390/fermentation11040207

Prajapati, A. K., Mahajan, A., Jadhav, S. M., & Kumar, K. (2026). Fourth-generation (4G) biodiesel: Paving the way for a greener and sustainable energy future in emerging economies. *Renewable & Sustainable Energy Reviews*, *225*, 116103. DOI: 10.1016/j.rser.2025.116103

Priyanka, S. J., Moger, S. S., Byalahunasishetter, B. I., & Raghavan, R. (2026). The One Health Framework and Antimicrobial Resistance: Role of Herbal Medicine. *Herbal Medicine Approaches to Antimicrobial Resistance and One Health*, 1–36. DOI: 10.4018/979-8-3373-7810-7.ch001

Saranya, S., Thamanna, L., & Chellapandi, P. (2024). Unveiling the potential of systems biology in biotechnology and biomedical research. *Systems Microbiology and Biomanufacturing*, *4*(4), 1217–1238. DOI: 10.1007/s43393-024-00286-4

Sharma, A., Hussain Mehdi, S. E., Shahzad, S., Hussain, F., Pandey, S., Kang, W., & Oh, S.-E. (2026). Microalgal biorefinery: Innovations in sustainable biofuel production. *Green Chemistry*, *28*(4), 1852–1871. DOI: 10.1039/D5GC02940F

Shelake, R. M., Khalid, M. A., & Kim, J.-Y. (2025). Microalgal bioengineering for futuristic applications in synthetic and space biology. *Biotechnology Advances*, *84*, 108665. DOI: 10.1016/j.biotechadv.2025.108665 PMID: 40738416

Tazon, A. W., Mérindol, N., Fantino, E., Bouhadada, A., Awwad, F., Meddeb-Mouelhi, F., & Desgagné-Penix, I. (2026). Engineering bidirectional chloroplast promoters for tunable co-expression of multiple genes in microalgae (Chlamydomonas reinhardtii). *Communications Biology*, *9*(1), 202. Advance online publication. DOI: 10.1038/s42003-025-09478-7 PMID: 41484447

Thajuddin, F., Rasheed, A. S., Elumalai, A., Palanivel, P., Nooruddin, T., & Dharumadurai, D. (2026). Harnessing microalgae and cyanobacteria for sustainable bioenergy Production: Integration of molecular engineering and AI-driven approaches. *Renewable Energy*, *256*, 123961. DOI: 10.1016/j.renene.2025.123961

Torres-Momber, S., Morales-Rivera, F. N., Jacuinde-Ruíz, J. C., & González-Hernández, J. C. (2026). Extracellular Enzyme Activity and Growth of Chlorella sorokiniana in Response to CO2, Light Color, and Light Intensity in a Photobioreactor. *Current Microbiology*, *83*(2), 116. DOI: 10.1007/s00284-025-04710-3 PMID: 41483157

Turmel, M., Pombert, J.-F., Kosanovic, C., Julian, A. T., Otis, C., & Lemieux, C. (2025). Comparative Genomics of Chloropicon primus and Chloropicon roscoffensis Provide Insights into the Evolutionary Dynamics and Ecological Success of These Tiny Green Algae in Marine Environments. *Genome Biology and Evolution*, *17*(7), evaf140. Advance online publication. DOI: 10.1093/gbe/evaf140 PMID: 40644408

Veale, C. G. L., Woodland, J. G., Wicht, K. J., & Chibale, K. (2025). Chemistry as a Catalyst for Transforming the Health and Wealth of South Africa. *Angewandte Chemie*, *137*(12), e202419681. Advance online publication. DOI: 10.1002/ange.202419681 PMID: 39887802

Wang, L., Ma, X., Huang, X., Wen, H., Jin, W., Chen, W., Wang, Y., Xu, P., Yang, H., & Zhang, Y. (2026). Inhibition of photosystems and respiratory complexes enhances lipid biosynthesis via ROS in Chlorella pyrenoidosa SHOU-1002 under nitrogen limitation. *Planta*, *263*(2), 37. DOI: 10.1007/s00425-025-04909-5 PMID: 41483224

Zhao, L., Xu, Y., Zhang, Y., Huang, Y., Huang, G., Hu, Z., Wu, C., & Tang, J. (2026). Performance and mechanisms of an algae-bacteria symbiotic system for treating low carbon-to-nitrogen ratio wastewater. *Biochemical Engineering Journal*, *226*, 109995. DOI: 10.1016/j.bej.2025.109995

Chapter 6
AI, IoT, and Digital Twins in Optimizing Algal Bioprocesses

Abhinandan Choudhury
http://orcid.org/0009-0005-5505-6284
Cotton University, India

ABSTRACT

The integration of Artificial Intelligence (AI), Internet of Things (IoT), and Digital Twin (DT) technologies represents a paradigm shift in microalgal bioprocessing for sustainable production, environmental remediation, and carbon capture. Traditional cultivation faces variable productivity, inconsistent monitoring, and scalability challenges. This chapter examines how digital transformation addresses these through intelligent sensing, data analytics, and virtual modeling. AI, IoT, and digital twin frameworks are explored, including their functionality in monitoring and control. Data-driven approaches for optimizing carbon fixation, biomass productivity, and metabolite synthesis are presented with case studies in CO_2 sequestration, wastewater treatment, and biofuel generation. AI enables pattern recognition, IoT provides real-time monitoring, and digital twins serve as virtual replicas for optimization. Critical challenges including data standardization, costs, and cybersecurity are addressed. Future prospects include autonomous biorefineries.

1. INTRODUCTION

In the present era of escalating global challenges, humanity faces the closely linked problems of ensuring sustainable energy supplies and achieving nutritional security. According to the United Nations, the global population is expected to reach

DOI: 10.4018/979-8-3373-7630-1.ch006

9.7 billion by 2050, reflecting an approximate 19% rise compared to current levels (United Nations, 2022). At present, global energy demand is largely met through fossil fuels, which contribute significantly to environmental pollution and are characterized by finite availability and rising extraction costs. This rapid population expansion necessitates innovative strategies to mitigate increasing stress on food production systems and energy resources (Michael et al., 2026). These converging challenges underscore the urgent need for sustainable, multifunctional biological solutions that can simultaneously address energy production and nutritional requirements, while minimizing environmental impact and ensuring long-term resource security. Phycoremediation represents a sustainable and environmentally benign technology that harnesses the pollutant-removal capacity of algae (Parthiban et al., 2026). The growing global imperative to adopt sustainable resources has renewed interest in microalgae as a viable platform for the production of both renewable energy and nutritionally valuable products. Microalgae have emerged as versatile biological platforms capable of addressing some of humanity's most pressing environmental and energy challenges. These microscopic photosynthetic organisms possess the unique ability to convert carbon dioxide (CO_2) and nutrients into valuable biomass and metabolites while simultaneously contributing to climate change mitigation and environmental remediation (Sivaramakrishnan et al., 2026). With their rapid growth rates, high photosynthetic efficiency, and ability to accumulate lipids, proteins, and other high-value compounds, microalgae offer tremendous potential for producing biofuels, pharmaceuticals, nutraceuticals, and bioplastics (Chaudry et al., 2025). Furthermore, their capacity for wastewater treatment and CO_2 sequestration positions them as key players in the transition toward a circular bioeconomy.

Despite their promising attributes, the commercial-scale cultivation and processing of microalgae face numerous technical and economic challenges. Traditional algal bioprocesses often suffer from variable productivity due to fluctuating environmental conditions, inadequate monitoring capabilities, suboptimal resource utilization, and high operational costs. The biological complexity of algal systems, combined with their sensitivity to parameters such as light intensity, temperature, pH, nutrient availability, and CO_2 concentration, makes process optimization highly challenging (Balamurugan et al., 2025). Conventional control strategies, which rely on periodic manual sampling and offline analysis, are insufficient for managing the dynamic nature of algal cultivation systems and often result in delayed responses to process disturbances.

The advent of Industry technologies characterized by the integration of cyber-physical systems, artificial intelligence, and interconnected digital networks has introduced transformative possibilities for bioprocess management and optimization (Malik et al., 2026). Among these technologies, Artificial Intelligence (AI), the Internet of Things (IoT), and Digital Twin (DT) systems have emerged as particularly powerful tools for revolutionizing microalgal biotechnology. AI encompasses a range of computational techniques including machine learning, deep learning, and neural networks that enable systems to learn from data, recognize patterns, and make intelligent decisions (An et al., 2026). IoT refers to networks of interconnected sensors, actuators, and devices that collect, transmit, and exchange data in real-time, enabling continuous monitoring and control of bioprocesses (Sharma et al., 2022). Digital twins are virtual replicas of physical systems that integrate real-time data, simulation models, and predictive analytics to optimize performance, predict outcomes, and support decision-making (Tao et al., 2022).

The convergence of these digital technologies offers unprecedented opportunities to address the limitations of conventional algal bioprocessing. By enabling real-time process monitoring, predictive optimization, adaptive control, and intelligent decision-making, AI, IoT, and digital twin systems collectively enhance system stability, reduce operational costs, maximize yield, and improve the overall sustainability of microalgal biotechnology. Recent advances have demonstrated the potential of these technologies to optimize critical process parameters, predict biomass productivity, detect system anomalies, and enable autonomous operation of algal cultivation facilities. However, the successful implementation of digital technologies in algal bioprocessing requires careful consideration of various technical, economic, and organizational challenges (Venkatrayalu et al., 2026).

This chapter provides a comprehensive examination of the integration of AI, IoT, and digital twin technologies in microalgal bioprocesses. Fundamental principles and architectures underlying these technologies are first presented, followed by detailed discussions of their applications in smart monitoring, process control and optimization. Through analysis of case studies and real-world implementations, we explore how these digital tools are being deployed in carbon sequestration, wastewater treatment, and bioproduct generation. Finally, the challenges, ethical considerations, and future prospects associated with digital transformation in algal biotechnology are addressed, providing readers with a balanced and forward-looking perspective on this rapidly evolving field.

2. FUNDAMENTALS OF AI, IoT, AND DIGITAL TWIN TECHNOLOGIES

2.1 Artificial Intelligence in Bioprocessing

Artificial Intelligence represents a collection of computational methodologies designed to enable machines to perform tasks that traditionally require human intelligence. In the context of bioprocessing, AI techniques are employed to analyze complex datasets, identify patterns in process behavior, predict future system states, and optimize operational parameters (Rayamajhi et al., 2025). The foundation of AI in bioprocessing rests on several key technologies: machine learning, which enables systems to learn from historical data and improve performance without explicit programming; deep learning, which uses multi-layered neural networks to extract hierarchical features from raw data; and reinforcement learning, which allows systems to learn optimal control strategies through trial-and-error interactions with their environment (Butean et al., 2025).

2.2 Internet of Things Architecture

The Internet of Things represents a paradigm shift in bioprocess monitoring and control, enabling the creation of intelligent, interconnected systems that can sense, communicate, and respond to changing conditions in real-time (Gotovtsev et al., 2025). An IoT ecosystem for algal bioprocessing typically consists of four fundamental layers: the perception layer, network layer, processing layer, and application layer. The perception layer comprises various sensors and actuators deployed in the cultivation system, including optical sensors for cell density measurement, electrochemical sensors for pH and dissolved oxygen monitoring, temperature sensors, light intensity sensors, and CO_2 analyzers. These devices collect real-time data on critical process parameters and can also include actuators for controlling pumps, valves, lighting systems, and gas injection mechanisms (Bin et al., 2024).

The network layer facilitates data transmission between sensors, controllers, and cloud platforms using various communication protocols. Common technologies include Wi-Fi for high-bandwidth local area networks, Bluetooth and Zigbee for low-power short-range communication, LoRaWAN for long-range wide-area networks, and cellular networks (4G/5G) for mobile connectivity. The choice of communication technology depends on factors such as data transmission rate requirements, power consumption constraints, coverage area, and deployment costs (Thiviyanathan et al., 2022).

The processing layer encompasses cloud computing platforms and data analytics engines that store, process, and analyze the vast amounts of data generated by IoT sensors. This layer typically includes databases for time-series data storage, data preprocessing pipelines for cleaning and normalizing sensor measurements, analytics engines for real-time and batch processing, and machine learning platforms for model training and deployment. The application layer presents processed information to users through web-based dashboards, mobile applications, and visualization tools, enabling remote monitoring, process control, and decision support.

2.3 Digital Twin Concepts and Implementation

Digital twins represent one of the most sophisticated applications of digitalization in bioprocessing, serving as virtual replicas that mirror the behavior and state of physical algal cultivation systems in real-time. A digital twin integrates three fundamental components: a physical system (the actual photobioreactor or cultivation facility), a virtual model (computational representations of system behavior), and bidirectional data flows that connect the physical and virtual domains. The virtual model incorporates multiple types of computational representations, including mechanistic models based on first principles (such as mass balance equations and kinetic models of algal growth), data-driven models derived from machine learning algorithms, and hybrid models that combine mechanistic and empirical approaches (Canzoneri et al., 2024).

The operational cycle of a digital twin involves continuous synchronization between the physical and virtual systems. Real-time sensor data from the physical algal cultivation system is fed into the digital twin, updating the virtual model's state and ensuring that it accurately reflects current conditions. The digital twin then uses this updated state, combined with historical data and predictive models, to simulate future scenarios, optimize process parameters, and identify potential issues before they manifest in the physical system (Sheik et al., 2024). This predictive capability enables proactive rather than reactive management of algal bioprocesses, reducing the risk of process failures and improving overall productivity.

Implementation of digital twins for algal bioprocesses requires careful consideration of model fidelity, computational efficiency, and data integration. High-fidelity models that capture detailed biochemical and physical phenomena may provide more accurate predictions but require substantial computational resources and extensive calibration data.

3. SMART MONITORING AND CONTROL SYSTEMS IN ALGAL CULTIVATION

3.1 IoT-Enabled Sensor Networks

Modern algal cultivation systems increasingly rely on sophisticated sensor networks that provide continuous, real-time monitoring of critical process parameters. These IoT-enabled sensing systems represent a fundamental departure from traditional manual sampling approaches, offering unprecedented temporal resolution and spatial coverage (Soon et al., 2025). Key parameters monitored in smart algal cultivation systems include biomass concentration (typically measured using optical density sensors or turbidity meters), pH (monitored using electrochemical sensors that are crucial for maintaining optimal growth conditions and preventing culture crashes), dissolved oxygen (DO) concentration (measured with Clark-type electrodes or optical DO sensors to track photosynthetic activity and prevent hyperoxic conditions), temperature (monitored using thermistors or resistance temperature detectors to ensure optimal enzymatic activity), light intensity (measured with quantum sensors or photosynthetically active radiation meters to optimize photon delivery), and CO_2 concentration in both the gas and liquid phases (monitored using infrared sensors or dissolved CO_2 probes).

Advanced sensor systems incorporate multiple measurement modalities to provide comprehensive process characterization. Spectroscopic sensors, including fluorescence-based probes, enable non-invasive monitoring of cellular physiological states, photosynthetic efficiency, and intracellular metabolite accumulation. These optical methods can detect stress conditions, nutrient limitations, and photoinhibition before they significantly impact productivity (Alaka et al., 2025). Image-based monitoring systems, utilizing digital cameras and computer vision algorithms, provide additional information on cell morphology, aggregation patterns, and contamination by predators or competing microorganisms (Reyes et al., 2024). The integration of multiple sensor types through sensor fusion techniques enables more robust and reliable process monitoring by cross-validating measurements and compensating for individual sensor limitations.

Sensor network design for algal cultivation systems must address several practical considerations including sensor placement optimization, calibration and maintenance schedules, data quality assurance, and redundancy requirements. Strategic sensor placement ensures adequate spatial coverage while minimizing costs and avoiding interference with fluid dynamics or light distribution. Automated calibration procedures, triggered by digital twin-based anomaly detection algorithms, help maintain measurement accuracy over extended operational periods. Data quality assurance mechanisms, including outlier detection and sensor drift compensation algorithms, ensure the reliability of information used for process control and decision-making. Redundant sensor deployment for critical parameters provides fault tolerance and enables continuous operation even when individual sensors fail or require maintenance.

3.2 AI-Driven Feedback Control

Artificial intelligence-driven control systems represent a significant advancement over conventional PID (Proportional-Integral-Derivative) controllers in managing the complex dynamics of algal bioprocesses. While traditional control approaches rely on linear approximations and fixed parameter values, AI-based controllers can adapt to nonlinear system behavior, handle multiple interacting variables, and learn optimal control strategies from operational data (Guo et al., 2025). Model Predictive Control (MPC) enhanced with machine learning algorithms has emerged as a particularly powerful approach for algal cultivation, combining the systematic optimization framework of MPC with the adaptive capabilities of AI. These systems use predictive models—either mechanistic, data-driven, or hybrid—to forecast future process states and compute optimal control actions that maximize a specified objective function while respecting system constraints.

Reinforcement learning-based controllers have demonstrated impressive performance in optimizing complex bioprocesses by learning control policies through trial-and-error interactions with either real cultivation systems or digital twin simulations. These controllers frame the control problem as a Markov Decision Process, where the system state is determined by current process measurements, actions correspond to manipulated variables (such as nutrient feed rates, light intensity, or mixing speed), and rewards reflect the achievement of process objectives (such as maximizing biomass productivity or minimizing energy consumption) (Pataro et al., 2023). Deep reinforcement learning algorithms, combining deep neural networks with reinforcement learning principles, can handle high-dimensional state and action spaces, making them suitable for controlling large-scale algal cultivation facilities with numerous sensors and actuators.

Adaptive control strategies that continuously update controller parameters based on observed system behavior offer robustness to changes in cultivation conditions, strain characteristics, or equipment performance. These approaches are particularly valuable for managing seasonal variations in outdoor cultivation systems, accommodating strain adaptation or mutation during long-term continuous culture, and maintaining performance as photobioreactors age or foul. Neural adaptive controllers, which combine neural networks for system identification with adaptive control laws, can automatically adjust to changing process dynamics without requiring manual retuning (Hu et al., 2008). The integration of AI-driven controllers with digital twin systems enables offline testing and validation of control strategies before deployment, reducing the risk of process disruptions during controller commissioning or modification.

3.3 Predictive Analytics and Fault Detection

Predictive analytics systems leverage historical and real-time data to anticipate future process behavior, identify potential problems before they escalate, and enable proactive intervention strategies. In algal cultivation systems, predictive models can forecast biomass concentration trajectories, estimate the timing of nutrient depletion, predict the onset of photoinhibition under high light conditions, and anticipate the need for harvesting or culture dilution. Time-series forecasting algorithms, including ARIMA (AutoRegressive Integrated Moving Average) models, Prophet (a procedure for forecasting time series data developed by Facebook), and LSTM neural networks, analyze historical patterns in process data to generate short-term and medium-term predictions of key performance indicators (Greene et al., 2021). These forecasts inform scheduling decisions, resource allocation, and preventive maintenance activities, improving operational efficiency and reducing unexpected downtime. Anomaly detection algorithms continuously monitor process measurements and computed features to identify deviations from normal operating conditions. Statistical methods such as control charts and hypothesis tests provide baseline anomaly detection capabilities, while more sophisticated approaches including isolation forests, one-class support vector machines, and autoencoders can detect subtle anomalies in high-dimensional data streams. Early detection of anomalies enables timely corrective actions, preventing minor issues from developing into major process failures or product quality problems.

4. DATA-DRIVEN MODELING AND PROCESS OPTIMIZATION

4.1 Machine Learning for Biomass Prediction

Machine learning approaches have revolutionized the modeling and prediction of microalgal biomass production by enabling the development of accurate, data-driven models that capture complex nonlinear relationships between environmental conditions and growth outcomes (Coşgun et al., 2023). Traditional mechanistic models, while providing valuable insights into the underlying biological and physical processes, often require extensive parametrization and may struggle to capture the full complexity of real-world cultivation systems influenced by numerous interacting factors. In contrast, machine learning models can automatically learn patterns from experimental and operational data without requiring explicit formulation of underlying mechanisms. Regression techniques, including linear regression, polynomial regression, and regularized regression methods (such as Ridge and Lasso regression), provide interpretable models for biomass prediction and can identify the most influential process parameters through feature importance analysis (Islam et al., 2022).

Ensemble learning methods, which combine multiple base learners to improve prediction accuracy and robustness, have demonstrated exceptional performance in modeling algal growth dynamics. Random Forest algorithms, which construct multiple decision trees using bootstrapped datasets and random feature subsets, provide robust predictions while offering resistance to overfitting and the ability to handle nonlinear relationships and feature interactions. Gradient Boosting methods, including XGBoost and LightGBM, build ensembles of weak learners (typically shallow decision trees) in a sequential manner, with each new learner focusing on correcting the errors of previous learners (Li et al., 2025). These algorithms can achieve state-of-the-art prediction accuracy for biomass concentration and have been successfully applied to optimize cultivation parameters in both laboratory and industrial settings.

Deep learning architectures offer powerful capabilities for modeling the temporal dynamics of algal growth and metabolite production. Recurrent neural networks, particularly LSTM and GRU (Gated Recurrent Unit) variants, can capture long-term dependencies in time-series data and learn the sequential patterns that characterize bioprocess trajectories. These networks excel at multi-step ahead prediction, enabling anticipation of future biomass concentrations based on current and historical states (VP et al., 2024). Hybrid modeling approaches that combine mechanistic models with machine learning components offer a promising direction for balancing physical interpretability with predictive accuracy. These models typically use mechanistic formulations to represent well-understood processes (such as mass balances and basic kinetics) while employing machine learning to model complex or poorly characterized phenomena (such as photoadaptation dynamics or species interactions in mixed cultures).

4.2 Optimization Algorithms and Strategies

Process optimization in algal bioprocessing involves identifying the optimal values of controllable parameters (such as light intensity, temperature, pH, nutrient concentrations, and CO_2 supply rates) that maximize desired objectives (such as biomass productivity, lipid content, or specific metabolite yields) while satisfying operational constraints and resource limitations. Classical optimization methods, including gradient-based algorithms and derivative-free methods, can effectively solve well-defined optimization problems with smooth objective functions (Xia et al., 2021). However, the complex, nonlinear, and often multi-modal nature of algal bioprocesses frequently requires more sophisticated optimization approaches. Evolutionary algorithms, inspired by biological evolution principles, have proven particularly effective for bioprocess optimization. Genetic Algorithms (GA) maintain a population of candidate solutions and iteratively improve them through selection, crossover, and mutation operations, exploring the solution space efficiently while avoiding premature convergence to local optima. Particle Swarm Optimization (PSO) simulates the social behavior of bird flocking or fish schooling, with candidate solutions (particles) moving through the search space influenced by their own best positions and the global best position found by the swarm. Multi-objective optimization techniques, such as the Non-dominated Sorting Genetic Algorithm

(NSGA-II), enable simultaneous optimization of multiple conflicting objectives, generating Pareto-optimal solutions that represent optimal trade-offs between competing goals such as maximizing productivity while minimizing energy consumption or maximizing both biomass and lipid yields (Kushwaha et al., 2023).

Table 1. Comparison of Optimization Algorithms for Algal Bioprocess Control

Algorithm	Strengths	Limitations	Best Applications
Genetic Algorithm	Global search, handles multi-modal functions	Computationally intensive, requires parameter tuning	Multi-parameter optimization, complex landscapes
Particle Swarm	Fast convergence, simple implementation	May converge prematurely	Continuous optimization, real-time control
Bayesian Optimization	Sample efficient, handles uncertainty	Scales poorly with dimensions	Expensive experiments, hyperparameter tuning
Reinforcement Learning	Adaptive, learns from interaction	Requires extensive training data	Dynamic control, autonomous systems

Bayesian optimization represents another powerful approach for bioprocess optimization, particularly when experimental evaluations are expensive or time-consuming. This method builds a probabilistic surrogate model (typically a Gaussian Process) of the objective function and uses acquisition functions to balance exploration of uncertain regions with exploitation of promising areas. The sequential nature of Bayesian optimization makes it well-suited for optimizing cultivation parameters through systematic experimentation, minimizing the number of trials required to identify near-optimal conditions (Lapierre et al., 2025). Integration of optimization algorithms with digital twin systems enables rapid evaluation of candidate solutions through simulation, dramatically accelerating the optimization process compared to purely experimental approaches.

5. APPLICATIONS IN CARBON SEQUESTRATION AND WASTEWATER TREATMENT

5.1 AI-Enabled CO_2 Capture Monitoring

Microalgae-based carbon capture systems represent a biologically-mediated approach to mitigating greenhouse gas emissions, with the potential to sequester significant quantities of CO_2 from industrial flue gases, ambient air, or concentrated CO_2 streams. The effectiveness of these systems depends critically on optimizing CO_2 transfer rates, maintaining appropriate culture conditions, and maximizing carbon fixation efficiency. AI-enabled monitoring systems provide comprehensive oversight of carbon capture performance through integration of multiple measurement modalities including direct CO_2 concentration measurements in gas inlet and outlet streams, pH monitoring as an indicator of dissolved inorganic carbon availability, biomass productivity tracking as a measure of carbon fixation rates, and isotopic analysis of carbon sources when applicable (Obidi, et al., 2025).

Machine learning models trained on historical carbon capture data can predict CO_2 fixation efficiency under varying operational conditions, enabling proactive adjustment of process parameters to maintain high capture rates. These predictive models account for complex interactions between light availability, temperature, culture density, and CO_2 supply, optimizing the balance between mass transfer limitations and biological utilization capacity. Real-time optimization algorithms adjust gas flow rates, mixing intensity, and light supply to maximize carbon fixation while minimizing energy consumption and preventing carbon limitation or inhibition. Digital twin systems simulate the coupled processes of gas-liquid mass transfer, carbonate chemistry, and photosynthetic carbon fixation, providing insights into bottlenecks and opportunities for process improvement.

5.2 Smart Wastewater Treatment Systems

Integration of microalgae into wastewater treatment processes offers a sustainable approach to nutrient recovery, pollutant removal, and simultaneous biomass production, converting waste streams into valuable resources. IoT-enabled monitoring systems for algal wastewater treatment continuously track influent characteristics (including nitrogen and phosphorus concentrations, organic matter content, heavy metal levels, and pathogen loads), culture performance (biomass growth, nutrient uptake rates, and metabolic activity), and effluent quality (residual nutrient concentrations, suspended solids, and compliance with discharge standards) (Sheik et al., 2024). These comprehensive monitoring capabilities enable adaptive management strategies that optimize treatment performance while producing high-quality biomass suitable for downstream applications.

AI algorithms analyze patterns in wastewater composition and treatment performance to predict optimal operating conditions for different influent characteristics. Machine learning models identify relationships between hydraulic retention time, light intensity, culture density, and nutrient removal efficiency, enabling dynamic adjustment of process parameters in response to variations in wastewater properties. Predictive models forecast treatment outcomes and biomass quality, supporting decisions about biomass harvesting schedules and downstream processing routes. Digital twin representations of algal wastewater treatment systems facilitate scenario analysis and process optimization, enabling evaluation of different operational strategies, assessment of scale-up options, and identification of opportunities for process intensification. Integration with automated control systems enables implementation of optimized strategies in real-time, improving treatment efficiency, reducing operational costs and enhancing the sustainability of wastewater management.

6. INDUSTRIAL SCALE-UP AND SMART BIOREFINERIES

6.1 Digital Integration for Process Automation

The transition from laboratory-scale algal cultivation to industrial-scale production requires sophisticated process automation and control systems capable of managing the increased complexity, variability, and economic constraints of commercial operations. Digital technologies enable comprehensive automation of cultivation, harvesting, and downstream processing operations, reducing labor requirements, improving consistency, and enhancing operational efficiency. Industrial automation architectures typically employ hierarchical control structures with multiple layers including field devices (sensors and actuators), programmable logic controllers (PLCs) for local control and safety interlocks, supervisory control and data acquisition (SCADA) systems for facility-wide monitoring and coordination and manufacturing execution systems (MES) for production management and optimization (Serhane et al., 2019).

Cloud computing platforms provide the computational infrastructure necessary for advanced analytics, digital twin simulations, and AI model training at industrial scales. Edge computing devices deployed at cultivation facilities perform local data processing and immediate control actions, reducing latency and bandwidth requirements while maintaining functionality during network disruptions. Hybrid cloud-edge architectures balance the advantages of centralized computing resources with the responsiveness and resilience of distributed computing, enabling sophisticated analytics and control while maintaining reliable local operations. Integration with enterprise resource planning (ERP) systems connects bioprocess operations with broader business functions including supply chain management, inventory control, quality assurance, and financial planning.

6.2 Case Studies of Smart Biorefinery Implementation

Several pioneering facilities have demonstrated the potential of digital technologies for commercial-scale algal production. One notable example involves the integration of IoT sensors, AI-driven control systems, and digital twin models in a large-scale photobioreactor facility producing high-value carotenoids for nutraceutical applications. The facility employs over 200 sensors monitoring critical parameters across multiple cultivation units, with data aggregated in real-time through industrial IoT gateways and transmitted to cloud-based analytics platforms. Machine learning models predict optimal harvest timing based on product accumulation profiles, while reinforcement learning algorithms dynamically adjust light intensity, temperature, and nutrient feeding to maximize product quality and minimize production costs. The digital twin system simulates alternative operating strategies and evaluates the impact of equipment modifications before physical implementation, reducing experimental costs and accelerating process development (Dębowski et al., 2025). Results from this facility demonstrate significant improvements in productivity, product quality, and resource efficiency compared to conventional control approaches, validating the commercial viability of digitally-enabled algal production systems.

Table 2. Key Performance Indicators for Digital Algal Biorefinery Systems

Performance Metric	Conventional System	Digital Smart System
Biomass Productivity (g/L/day)	0.15-0.25	0.35-0.50
CO_2 Fixation Efficiency (%)	45-60	75-90
Energy Consumption (kWh/kg biomass)	8-12	5-7
Process Downtime (%)	15-25	3-8
Product Quality Consistency (CV%)	20-35	5-12
Labor Requirements (hours/day)	12-16	2-4

7. CHALLENGES AND FUTURE PERSPECTIVES

7.1 Technical and Implementation Challenges

Despite the promising potential of AI, IoT, and digital twin technologies in algal bioprocessing, several significant challenges must be addressed to enable widespread adoption and maximize the benefits of digital transformation. Data quality and availability represent fundamental challenges, as machine learning models require large quantities of high-quality training data to achieve reliable performance. Many algal cultivation facilities lack historical datasets with sufficient coverage of operating conditions, process disturbances, and outcomes to support robust model development. Sensor drift, calibration errors, missing data, and measurement noise further complicate the development and deployment of data-driven models. Standardization of data formats, protocols, and interfaces is essential for enabling interoperability between different systems and facilitating data sharing across the research and industrial communities.

Integration costs and complexity pose significant barriers to adoption, particularly for small and medium-sized enterprises. Implementation of comprehensive IoT sensor networks, cloud computing infrastructure, and AI analytics platforms requires substantial capital investment and technical expertise. Legacy equipment may lack digital connectivity or standardized communication protocols, necessitating costly retrofitting or replacement. Cybersecurity concerns become increasingly important as bioprocessing facilities become more connected and reliant on digital systems. Protecting sensitive operational data, intellectual property, and control systems from cyber threats requires robust security measures including network segmentation, encryption, access controls, and continuous monitoring for suspicious activities.

7.2 Emerging Trends and Future Directions

The future of algal biotechnology will be shaped by continued advances in digital technologies and their integration with biological systems. Hybrid digital-biological twins that combine mechanistic bioprocess models with genomic, transcriptomic, and metabolomic data will enable unprecedented understanding and control of cellular metabolism and physiology. These systems will incorporate multi-scale models spanning molecular, cellular, and reactor scales, providing insights into how genetic modifications, environmental conditions, and process operations collectively determine production outcomes. Integration with synthetic biology and metabolic engineering will enable design and optimization of algal strains specifically tailored for digitally-controlled cultivation systems.

Autonomous biorefinery systems leveraging advances in robotics, computer vision, and artificial intelligence will enable fully automated operation with minimal human intervention. These facilities will continuously optimize their own performance, adapt to changing conditions, detect and diagnose problems, and execute corrective actions without requiring operator input. Advanced human-machine interfaces utilizing augmented reality, natural language processing, and intelligent decision support will empower operators to interact more effectively with complex bioprocessing systems and make informed decisions rapidly.

Integration with emerging technologies including blockchain for supply chain traceability and carbon credit verification, 5G networks for ultra-reliable low-latency communication, and quantum computing for solving complex optimization problems will further enhance the capabilities of digital algal bioprocessing systems. Development of standardized platforms, open-source tools, and collaborative frameworks will accelerate innovation and democratize access to advanced digital technologies, enabling broader participation in the algal biotechnology sector and accelerating progress toward sustainable bioeconomy goals.

8. CONCLUSION

The integration of Artificial Intelligence, Internet of Things, and Digital Twin technologies represents a transformative paradigm shift in microalgal bioprocessing, offering powerful solutions to long-standing challenges in process monitoring, control, and optimization. These digital technologies collectively enable real-time, intelligent, and predictive management of algal cultivation systems, enhancing productivity, efficiency, and sustainability while reducing operational costs and environmental impacts. AI algorithms provide sophisticated capabilities for pattern recognition, predictive modeling, and adaptive control, enabling bioprocesses to learn from experience and continuously improve performance. IoT sensor networks deliver comprehensive real-time visibility into process states and dynamics, supporting data-driven decision-making and rapid response to disturbances. Digital twins serve as powerful platforms for simulation, optimization, and scenario analysis, enabling virtual testing and refinement of process strategies before physical implementation.

The applications of these technologies span the full spectrum of algal biotechnology, from fundamental research and process development to industrial-scale production and environmental remediation. In carbon capture applications, AI-enabled monitoring and optimization maximize CO_2 fixation efficiency while minimizing energy consumption. In wastewater treatment, intelligent control systems optimize nutrient recovery and pollutant removal while producing valuable biomass. In bioproduct manufacturing, digital technologies enable consistent, high-quality production with improved economics and sustainability metrics. Case studies from pioneering implementations demonstrate significant improvements in key performance indicators including productivity, product quality, resource efficiency, and operational reliability.

However, realizing the full potential of digital transformation in algal bioprocessing requires addressing several important challenges. Data quality, availability, and standardization must be improved to support robust development and deployment of AI models. Integration costs and technical complexity must be reduced to enable broader adoption, particularly among smaller enterprises. Cybersecurity measures must be strengthened to protect critical bioprocessing systems and sensitive data. Workforce development and training programs must be expanded to build the interdisciplinary expertise required to implement and operate digitally-enabled bioprocessing facilities. Policy frameworks and industry standards must be developed to guide responsible implementation of autonomous and AI-driven systems.

Looking forward, the convergence of digital technologies with advances in synthetic biology, metabolic engineering, and process intensification promises to unlock unprecedented capabilities for sustainable bioproduction. Hybrid digital-biological twins incorporating genomic and metabolomic data will enable rational design and optimization of both organisms and processes. Autonomous biorefinery systems will achieve levels of performance, reliability, and efficiency unattainable with conventional approaches. Integration with blockchain, 5G, and emerging computational paradigms will expand the frontiers of what is possible in bioprocessing. As these technologies mature and costs decline, digital transformation will become increasingly accessible, driving widespread adoption and accelerating progress toward a sustainable bioeconomy. The future of algal biotechnology is inherently digital, and the continued development and deployment of AI, IoT, and digital twin systems will play a central role in realizing the full potential of these remarkable organisms to address pressing global challenges in climate change, resource scarcity, and environmental sustainability.

REFERENCES

Alaka, S. (2025). Integrating machine learning and IoT technologies for smart water quality monitoring: Methods, challenges, and future directions.

An, H., Hu, W., Huang, S., Huang, S., Li, R., Liang, Y., Shao, J., Song, Y., Wang, Z., Yuan, C., Zhang, C., Zhang, H., Zhuang, W., & Li, X. (2026). Ai flow: Perspectives, scenarios, and approaches. *Vicinagearth, 3*(1), 1. DOI: 10.1007/s44336-025-00031-y

Balamurugan, S., Li, D. W., Wang, X., & Li, H. Y. (2025). Unleashing the potential of biotechnological strategies for the sustainable production of microalgal polysaccharides. *Critical Reviews in Food Science and Nutrition, 65*(31), 1–18. DOI: 10.1080/10408398.2025.2475240 PMID: 40084718

Bin Abu Sofian, A. D. A., Lim, H. R., Manickam, S., Ang, W. L., & Show, P. L. (2024). Towards a sustainable circular economy: Algae-based bioplastics and the role of internet-of-things and machine learning. *ChemBioEng Reviews, 11*(1), 39–59 DOI: 10.1002/cben.202300028

Butean, A., Cutean, I., Barbero, R., Enriquez, J., & Matei, A. (2025). A review of artificial intelligence applications for biorefineries and bioprocessing: From data-driven processes to optimization strategies and real-time control. *Processes (Basel, Switzerland), 13*(8), 2544. DOI: 10.3390/pr13082544

Canzoneri, M., Horner, M., & Banerjee, P. (2024). Digital twins in biomanufacturing. In *Bioconvergence Revolution: Challenges and Opportunities in Research Productivity* (pp. 59–81). Springer Nature Switzerland. DOI: 10.1007/978-3-031-61593-1_4

Chaudry, S., Hurtado-McCormick, V., Cheng, K. Y., Willis, A., Speight, R., & Kaksonen, A. H. (2025). Microalgae to bioplastics–Routes and challenges. *Cleaner Engineering and Technology, 25*, 100922. DOI: 10.1016/j.clet.2025.100922

Coşgun, A., Günay, M. E., & Yıldırım, R. (2023). Machine learning for algal biofuels: A critical review and perspective for the future. *Green Chemistry, 25*(9), 3354–3373. DOI: 10.1039/D3GC00389B

D, R. G., v P, H., K P, R. A. H., & Bhide, A.VP. (2024). Attention-driven LSTM and GRU deep learning techniques for precise water quality prediction in smart aquaculture. *Aquaculture International, 32*(6), 8455–8478. DOI: 10.1007/s10499-024-01574-5

Dębowski, M., Kazimierowicz, J., & Zieliński, M. (2025). Multi-Sensing Monitoring of the Microalgae Biomass Cultivation Systems for Biofuels and Added Value Products Synthesis—Challenges and Opportunities. *Applied Sciences (Basel, Switzerland), 15*(13), 7324. DOI: 10.3390/app15137324

Gotovtsev, P., Sergeeva, Y., Sukhinov, D., Petrova, M., Vishnevskaya, M., Shestakova, A., & Vasilov, R. (2025). Internet of things (IoT) use for remote monitoring of algal bioreactors. In *Algal Bioreactors* (pp. 205–214). Elsevier Science Ltd. DOI: 10.1016/B978-0-443-14058-7.00044-0

Greene, J. M., Quiroz, D., Compton, S., Lammers, P. J., & Quinn, J. C. (2021). A validated thermal and biological model for predicting algal productivity in large scale outdoor cultivation systems. *Algal Research, 54*, 102224. DOI: 10.1016/j.algal.2021.102224

Guo, B., Lu, X., Jiang, X., Shen, X. L., Wei, Z., & Zhang, Y. (2025). Artificial Intelligence in Advancing Algal Bioactive Ingredients: Production, Characterization, and Application. *Foods, 14*(10), 1783. DOI: 10.3390/foods14101783 PMID: 40428562

Hu, D., Liu, H., Yang, C., & Hu, E. (2008). The design and optimization for light-algae bioreactor controller based on Artificial Neural Network-Model Predictive Control. *Acta Astronautica, 63*(7-10), 1067–1075. DOI: 10.1016/j.actaastro.2008.02.008

Islam, M. S., Senaha, I., Rahman, M. M., Yoda, Y., & Saha, B. B. (2022). Mathematical modelling and statistical optimization of fast cultivation of Agardhiella subulata: Response surface methodology. *Energy Nexus, 7*, 100115. DOI: 10.1016/j.nexus.2022.100115

Knox, S. W. (2026). *Machine learning: a concise introduction*. John Wiley & Sons. DOI: 10.1002/9781394325313

Kushwaha, O. S., Uthayakumar, H., & Kumaresan, K. (2023). Modeling of carbon dioxide fixation by microalgae using hybrid artificial intelligence (AI) and fuzzy logic (FL) methods and optimization by genetic algorithm (GA). *Environmental Science and Pollution Research International, 30*(10), 24927–24948. DOI: 10.1007/s11356-022-19683-0 PMID: 35349067

Lapierre, F. M., Mattaliano, P., Raith, D., Castillo-Cota, M., Bermeitinger, J., & Huber, R. (2025). Multi-cycle high-throughput growth media optimization using batch Bayesian optimization. *Journal of Chemical Technology and Biotechnology, 100*(8), 1571–1583. DOI: 10.1002/jctb.7860

Li, H., Chen, L., Zhang, F., & Cai, Z. (2025). Graph-learning-based machine learning improves prediction and cultivation of commercial-grade marine microalgae Porphyridium. *Bioresource Technology*, *416*, 131728. DOI: 10.1016/j.biortech.2024.131728 PMID: 39521188

Li, J. (2026). Policy Gradient Algorithms. In *Artificial Intelligence for Drug Design* (pp. 169–177). Springer Nature Singapore. DOI: 10.1007/978-981-95-2525-6_16

Malik, H. A.. (2026). *"Development of algal industry 4.0 for efficient cultivation, resource recovery, and bioproducts production." Algae Biotechnology*. Elsevier.

Michael, M., Vanness, V., Henry, H., Tjandra, T. M., Kristian, K., Nasution, J. A., Masytha, M., Sidabutar, R., Trisakti, B., Irvan, I., Dalimunthe, N. F., Al Fath, M. T., Nabilah, Y., Kobayashi, T., Daimon, H., & Takriff, M. S. (2026). Microalgae for sustainable biodiesel and omega-3: A comprehensive review of production, processing, and implementation. *Renewable & Sustainable Energy Reviews*, *226*, 116327. DOI: 10.1016/j.rser.2025.116327

Obidi, P. O., & Bayless, D. J. (2025). The synergy of artificial intelligence and algal systems for sustainable wastewater treatment and carbon sequestration. *Environmental Technology Reviews*, *14*(1), 1040–1067. DOI: 10.1080/21622515.2025.2581330

Parthiban, N., & Thiagarajan, K. (2026). Biodegradation of phthalate esters by microalgae: A comprehensive review. *The Science of the Total Environment*, *1012*, 181248. DOI: 10.1016/j.scitotenv.2025.181248 PMID: 41447808

Pataro, I. M., Gil, J. D., Guzman, J. L., Berenguel, M., & Lemos, J. M. (2023). A learning-based model predictive strategy for pH control in raceway photobioreactors with freshwater and wastewater cultivation media. *Control Engineering Practice*, *138*, 105619. DOI: 10.1016/j.conengprac.2023.105619

Porras Reyes, L., Havlik, I., & Beutel, S. (2024). Software sensors in the monitoring of microalgae cultivations. *Reviews in Environmental Science and Biotechnology*, *23*(1), 67–92. DOI: 10.1007/s11157-023-09679-8

Rayamajhi, V., Hussain, M., Shin, H., & Jung, S. (2025). Recent advances in the application of artificial intelligence in microalgal cultivation. *Processes (Basel, Switzerland)*, *13*(12), 3764. DOI: 10.3390/pr13123764

Serhane, A., Raad, M., Raad, R., & Susilo, W. (2019). Programmable logic controllers based systems (PLC-BS): Vulnerabilities and threats. *SN Applied Sciences*, *1*(8), 924. DOI: 10.1007/s42452-019-0860-2

Sharma, A., & Bhushan, K. (2026). A comprehensive survey on IoT security: Challenges, security issues, and countermeasures. *Computer Science Review, 59*, 100839. DOI: 10.1016/j.cosrev.2025.100839

Sheik, A. G., Kumar, A., Ansari, F. A., Raj, V., Peleato, N. M., Patan, A. K., Kumari, S., & Bux, F. (2024). Reinvigorating algal cultivation for biomass production with digital twin technology-a smart sustainable infrastructure. *Algal Research, 84*, 103779. DOI: 10.1016/j.algal.2024.103779

Sheik, A. G., Kumar, A., Ansari, F. A., Raj, V., Peleato, N. M., Patan, A. K., Kumari, S., & Bux, F. (2024). Reinvigorating algal cultivation for biomass production with digital twin technology-a smart sustainable infrastructure. *Algal Research, 84*, 103779. DOI: 10.1016/j.algal.2024.103779

Sivaramakrishnan, R., Kumar, S., Hosseinzadeh, H., & Balan, V. (2026). Harnessing microalgae from wastewater for biocrude via hydrothermal Liquefaction: A sustainable pathway to biofuels and biochemicals. *Renewable & Sustainable Energy Reviews, 231*, 116759. DOI: 10.1016/j.rser.2026.116759

Soon, C. F., Jamia'an, M., Mohamed Sunar, N., Arifin, S. N. H., Tay, K. G., Heng, C., See, C. H., Mat Nayan, N. H., & Tee, K. S. (2025). Smart sensing and anomaly detection for microalgae culture based on LoRaWAN sensors and LSTM autoencoder. *Aquaculture International, 33*(6), 438. DOI: 10.1007/s10499-025-02104-7

Tao, F., Xiao, B., Qi, Q., Cheng, J., & Ji, P. (2022). Digital twin modeling. *Journal of Manufacturing Systems, 64*, 372–389. DOI: 10.1016/j.jmsy.2022.06.015

Thiviyanathan, V. A., Lim, H. R., Tham, P. E., & Show, P. L. (2022). How far has the development for industrial Internet of Things (IoT) in microalgae? In *Microalgae for Environmental Biotechnology* (pp. 145–174). CRC Press. DOI: 10.1201/9781003202196-4

United Nations Department of Economic and Social Affairs. Population Division. World population prospects 2022: summary of results. In: UN DESA/POP/2022/TR/NO. 3. New York, NY, USA: Un

Venkatrayalu, G., Choudhary, S., & Poluri, K. M. (2026). AI-Assisted Microalgal Bioprocessing: Innovations, Challenges, and Future Prospects in Sustainable Manufacturing. *Industrial Microbiology and Biotechnology: Trends in Microbial Biotechnology*, 269-307.

Xia, W., & Shoemaker, C. (2021). GOPS: Efficient RBF surrogate global optimization algorithm with high dimensions and many parallel processors including application to multimodal water quality PDE model calibration. *Optimization and Engineering*, *22*(4), 2741–2777. DOI: 10.1007/s11081-020-09556-1

Yusof, Z., Tong, Y. W., Selvarajoo, K., Parakh, S. K., & Foo, S. C. (2024). Overcoming challenges in microalgal bioprocessing through data-driven and computational approaches. *Current Opinion in Food Science*, •••, 101253.

Chapter 7
Stress-Induced Biorefinery Approaches for Enhanced Microalgal Carbon Sequestration and Wastewater Valorization

Salim Kichouh Aiadi
http://orcid.org/0000-0002-5438-3962
University of Mons, Belgium

ABSTRACT

Global environmental challenges, necessitate innovative resource recovery strategies. This chapter explores the "stress-induced biorefinery" paradigm, which leverages the heterogeneous nature of wastewater as a tunable metabolic signal for microalgae. By manipulating environmental stressors such as nutrient limitation, light intensity, and salinity, microalgal metabolism can be "programmed" to accumulate high-value compounds, including triacylglycerols for biofuels, carotenoids for nutraceuticals, and carbohydrates for biopolymers. Integrated systems utilizing algal-bacterial consortia offer energy-efficient treatment by replacing mechanical aeration with photosynthetic oxygenation. Furthermore, the coupling of nutrient recovery with carbon sequestration provides a pathway for industrial carbon management. This chapter synthesizes biological mechanisms, engineering strategies for diverse effluent streams, and cascading valorization routes to transform wastewater treatment from a cost-intensive disposal process into a sustainable, value-generating manufacturing platform.

DOI: 10.4018/979-8-3373-7630-1.ch007

INTRODUCTION

Global Context and Challenges

The contemporary world is confronting overlapping environmental problems that demand creative and comprehensive responses. Atmospheric carbon dioxide has risen beyond 400 ppm, well above pre-industrial levels, and continues to drive climate change (Global Monitoring Laboratory 2025; Hashimoto 2019). In this context, Carbon Capture and Utilization (CCU) technologies have transitioned from a theoretical concept to a practical necessity (Hepburn et al. 2019). Yet the destabilization of the atmosphere cannot be separated from the way society manages water, nutrients, and the wider Earth system (Rockstrom et al. 2009).

Global wastewater generation has escalated to an estimated 380 billion cubic meters annually (Qadir et al. 2020), creating a dual challenge: untreated discharges promote eutrophication and pathogen spread, while the same streams contain recoverable nutrients and chemical energy (Hasan et al. 2021; Qadir et al. 2020). In emerging economies and rapidly urbanizing regions, the discharge of untreated or partially treated effluents into water bodies creates "dead zones" driven by nitrogen and phosphorus saturation (Diaz and Rosenberg 2008; Jones et al. 2021).

At the same time, the agricultural sector faces a looming crisis of resource scarcity, particularly regarding phosphorus (Alewell et al. 2020). Conventional wastewater treatment plants (WWTPs) reduce organic loads and health risks, but they largely operate as linear, energy-intensive disposal systems and can be notable sources of greenhouse gas emissions (Derco et al. 2024). The activated sludge process remains the dominant paradigm and requires substantial electricity for aeration to sustain microbial oxidation (Acién Fernández et al. 2016; C. Wang et al. 2023). This process creates a paradox: the very nitrogen and phosphorus compounds that act as pollutants in aquatic ecosystems are the rate-limiting fertilizers required to sustain global agriculture.

The challenge extends beyond municipal wastewater. Each sector produces wastewater with a distinct chemical signature. Industrial effluents are highly specialized, agricultural operations generate nitrogen-rich streams from livestock waste, and aquaculture facilities discharge nutrient-laden water with varying salinity. Food processing plants contribute organic-rich effluents with fermentable sugars, while textile and mining industries add heavy metal contamination and extreme pH fluctuations (Acién Fernández et al. 2016; Liu et al. 2024; Sciuto et al. 2025; Vaishnav et al. 2023). Importantly, this heterogeneity is not merely a limitation; it can also be viewed as a set of tunable "stress profiles" that steer microalgal metabolism toward targeted products. Capitalizing on this diversity, however, requires moving beyond standardized treatment toward tailored, biorefinery-oriented strategies.

Microalgae as a Sustainable Solution

Meeting the dual challenge of energy-intensive treatment and nutrient pollution requires reframing wastewater management from remediation to resource recovery. Microalgae are uniquely positioned to drive this transition by establishing a symbiotic link between water treatment and biomass production.

A key advantage is the substitution of mechanical energy with solar energy. In conventional systems, electricity is consumed to aerate bacteria; in algal-bacterial consortia, photosynthesis supplies the dissolved oxygen needed for bacterial degradation of organic matter (Acién Fernández et al. 2016). Bacterial respiration, in turn, provides CO_2 for algal growth, closing a local carbon loop within the treatment process (Acién Fernández et al. 2016)(Figure 1). By replacing energy-intensive aeration with biological oxygenation, microalgal systems can reduce both operating costs and the carbon footprint of treatment.

Figure 1. Algae–bacteria coupling in wastewater treatment. Photosynthetic O_2 produced by microalgae supports bacterial oxidation of organic matter, while bacterial respiration provides CO_2 for algal growth, enabling simultaneous nutrient recovery into harvestable biomass.

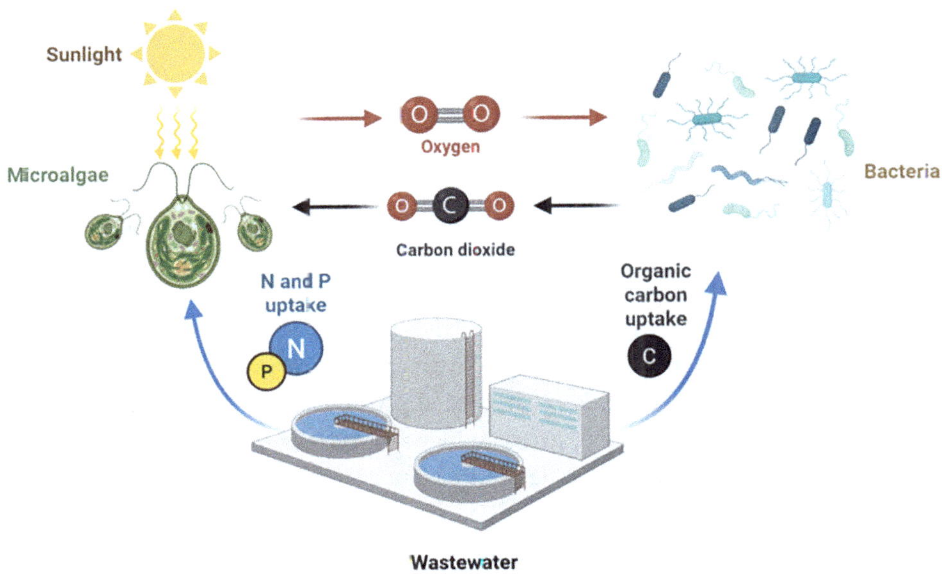

Microalgae also change the logic of nutrient removal. Instead of converting nitrogen to gaseous emissions or immobilizing nutrients in low-value sludge, algal cells assimilate nitrogen and phosphorus into harvestable biomass. Their high photosynthetic productivity, often exceeding that of terrestrial crops on an area basis, enables rapid nutrient uptake without competing for arable land (Prasad et al. 2021). The result is a pathway to resolve the "pollution paradox" by converting eutrophication-driving nutrients into a feedstock for fuels, chemicals, and functional ingredients.

Crucially, wastewater is not a uniform growth medium. Different streams impose different constraints (e.g., nutrient limitation, salinity, metals, light attenuation), each of which can be interpreted as a controllable signal for metabolic programming. However, high biomass productivity alone rarely guarantees economic viability. Value must be extracted from multiple fractions of the biomass, motivating the biorefinery paradigm and its cascading approach to product recovery.

The Biorefinery Paradigm

A microalgal biorefinery extends beyond a single-product model by maximizing value through sequential or parallel recovery of multiple compounds (Chew et al. 2017). Analogous to fractionation in the petrochemical industry, algal biomass can be separated into high-value ingredients (e.g., pigments, polyunsaturated fatty acids) and bulk fractions (e.g., proteins, carbohydrates) (Chew et al. 2017). Diversifying the product slate improves economic resilience and is particularly attractive when cultivation is coupled to wastewater treatment, where the primary service (pollution control) can be complemented by revenue from bioproducts (Chew et al. 2017; Wang et al. 2022).

Stress as a Strategic Tool

Microalgae have developed advanced mechanisms to handle stress, enabling them to thrive in changing natural environments (Shetty et al. 2019). These responses often involve predictable reallocations of carbon and reducing power, leading to the accumulation of target compounds: triacylglycerols under nitrogen limitation, carotenoids under high light or oxidative stress, or compatible solutes and specific amino acids under salinity stress (Chen et al. 2017; Kichouh-Aiadi et al. 2024; Shetty et al. 2019). The concept of stress-induced biorefinery essentially "programs" metabolic processes toward these desired products (Chen et al. 2017), combining physiological insights with process engineering to develop economically feasible production systems. This method requires a balance between growth and product accumulation, as too much stress can harm cell viability, while too little stress might not trigger the desired responses (Chen et al. 2017).

Objectives and Structure of the Chapter

This chapter offers an in-depth analysis of stress-induced microalgal biorefineries, connecting core biological concepts with applied biotechnology. Particular focus is placed on integrating stress methods with wastewater valorization, showing how environmental challenges can be transformed into feedstock opportunities.

MICROALGAL BIOLOGY AND STRESS RESPONSE MECHANISMS

Microalgal Cellular Architecture and Metabolism

Microalgae are highly adaptable photosynthetic microorganisms whose cellular organization supports rapid metabolic switching. The chloroplast serves as both the primary site for carbon fixation and an early stress sensor (Gorelova et al. 2018; Naduthodi et al. 2021). Structurally, the chloroplast envelope regulates transport of precursors for lipid biosynthesis, while thylakoid membranes host the photophosphorylation machinery that generates ATP and NADPH (Gorelova et al. 2018; Launay et al. 2023). When stress intensifies, microalgae actively remodel these structures to protect the photosynthetic apparatus: thylakoid unstacking (grana disruption) reduces the effective light-absorbing cross-section and helps prevent oxidative damage (Gorelova et al. 2018). In parallel, many species form specialized lipid bodies that function as dynamic storage organelles, accumulating triacylglycerols (TAGs) during nitrogen or phosphorus scarcity (Gorelova et al. 2018; Launay et al. 2023).

At the biochemical level, carbon fixation in microalgae is driven primarily by the Calvin–Benson–Bassham (CBB) cycle, in which RuBisCO fixes CO_2 into 3-phosphoglycerate (3-PGA) (Naduthodi et al. 2021). However, RuBisCO's inefficiency and oxygen sensitivity create strong selection for mechanisms that elevate CO_2 near the enzyme. Many species therefore rely on carbon concentrating mechanisms (CCMs) that actively transport bicarbonate into the cell and convert it to CO_2 within the pyrenoid, a specialized chloroplast sub-compartment (Naduthodi et al. 2021; Prasad et al. 2021). Other taxa employ C4-like processes that concentrate CO_2 around RuBisCO and improve performance under carbon-limited conditions (Ashour et al. 2024; Kichouh-Aiadi, López-Rosales, et al. 2025; Singh et al. 2016). These adaptations are especially relevant for cultivation strategies involving industrial flue gas or atmospheric carbon capture because they stabilize photosynthetic performance under fluctuating dissolved inorganic carbon (DIC) availability (Ashour et al. 2024).

Crucially, stress responses are coordinated across organelles. Under balanced conditions, fixed carbon supports growth, especially protein synthesis, whereas nitrogen scarcity triggers rapid metabolic reprogramming: stress-response genes are activated and carbon is redirected toward storage pathways such as lipids or carbohydrates (Msanne et al. 2012; Tan et al. 2016). Throughout this shift, mitochondria and peroxisomes play stabilizing roles by supporting energy homeostasis and managing oxidative balance during the transition from growth to survival metabolism (Gomez-Casati, Barchiesi, and Busi 2022; He, Dean, and Lodhi 2021; Tang et al. 2025).

Biochemistry of Stress-Induced Biosynthesis Pathways

Designing stress-induced biorefineries requires understanding which biochemical pathways dominate under specific stresses. Microalgae use diverse strategies to cope with environmental change, but three biosynthetic pathways are most relevant for stress-induced biorefineries and are detailed below: lipid accumulation, carotenoid synthesis, and carbohydrate polymerization.. Each reflects a distinct survival strategy that can be strategically exploited through controlled cultivation conditions.

Lipid Accumulation Pathways

A hallmark response to nitrogen stress is the accumulation of neutral lipids, particularly TAGs, largely governed by the Kennedy pathway (Li et al. 2014; Miller et al. 2010). Nitrogen starvation can inhibit isocitrate dehydrogenase in the TCA cycle, promoting citrate accumulation within mitochondria (Ratledge and Wynn 2002). Citrate is then exported to the cytosol and cleaved by ATP-citrate lyase to generate acetyl-CoA, the key precursor for fatty-acid synthesis (Ratledge and Wynn 2002) Acetyl-CoA carboxylase (ACCase) then converts acetyl-CoA to malonyl-CoA, a widely recognized rate-limiting step (Huerlimann and Heimann 2013). Fatty acids are assembled onto a glycerol-3-phosphate backbone in the endoplasmic reticulum, with the final committed step catalyzed by diacylglycerol acyltransferase (DGAT); in oleaginous species such as *Nannochloropsis*, DGAT2 upregulation correlates strongly with cytosolic lipid body formation (Li et al. 2014, 2016; Wei et al. 2017). This pathway is summarized in Figure 2.

Figure 2. Metabolic pathway of nitrogen-stress-induced TAG accumulation. Under nitrogen starvation, inhibition of isocitrate dehydrogenase promotes citrate export and acetyl-CoA generation. Key control enzymes (ACCase, DGAT2) channel carbon flux toward triacylglycerol (TAG) synthesis and lipid droplet formation.

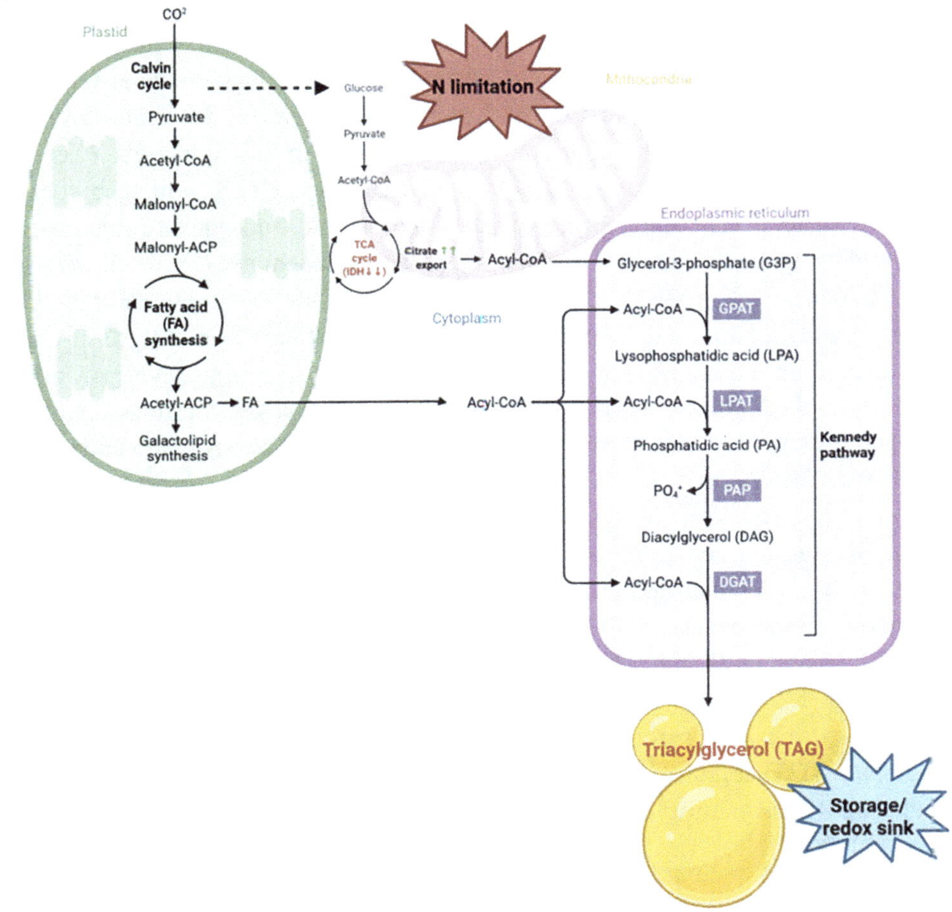

From a biorefinery perspective, this shift diverts photosynthetically derived ATP and NADPH away from cell division and into reduced-carbon storage, effectively "banking" energy as TAGs (Li et al. 2014), simultaneously 'banking' energy and alleviating over-reduction of the electron transport chain (L. Wang et al. 2023). In wastewater cultivation, progressive nitrogen depletion can therefore act as a natural induction phase for lipogenesis, enabling conversion of waste nutrients into biodiesel precursors (El-Sheekh et al. 2023).

Carotenoid and Antioxidant Biosynthesis Pathways

Carotenoids function as photoprotective pigments and valuable nutraceutical products, and their accumulation is commonly stimulated by stressors such as high light, heavy metals, and nutrient limitation (Chen et al. 2017; Xiao et al. 2023). Biosynthesis begins in the plastid through the MEP pathway, which produces the isoprenoid precursors IPP and DMAPP (Cazzonelli 2011; Paniagua-Michel, Olmos-Soto, and Ruiz 2012; Sun et al. 2010). Stress increases carbon flux through this pathway, particularly via upregulation of *dxs*, the first committed enzyme (Cen et al. 2022; Lv et al. 2016). IPP and DMAPP condense to form GGPP, which phytoene synthase (PSY) converts to phytoene, the first C40 carotenoid. Subsequent desaturation steps yield lycopene through a series of conjugated intermediates (Sun et al. 2010).

At lycopene, cyclization yields α- and β-carotenes, which feed into xanthophyll production and the xanthophyll cycle (Cazzonelli 2011). This cycle rapidly dissipates excess excitation energy as heat via non-photochemical quenching (NPQ). Under high light, violaxanthin de-epoxidase (VDE) converts violaxanthin to zeaxanthin within minutes, while zeaxanthin epoxidase (ZEP) reverses the conversion under low light (Fernández-Marín et al. 2021). Under more severe stress, β-carotene ketolase (BKT) and β-carotene hydroxylase (CHY) drive conversion toward downstream carotenoids through ketocarotenoid intermediates linked to astaxanthin synthesis (Mei et al. 2025; Vidhyavathi et al. 2008).

Reactive oxygen species (ROS), particularly H_2O_2, also function as signaling molecules that initiate carotenoid induction. One well-described route involves activation of MAPK cascades, which phosphorylate transcription factors (e.g., MYB, bZIP) that bind carotenogenic promoters and upregulate the pathway (L. Wang et al. 2023) (Figure 3). In wastewater, heavy metals can act as pro-oxidants that elevate ROS signaling, effectively converting pollutants into metabolic stimuli (Xiao et al. 2023).

Figure 3. Oxidative-stress signaling cascade linking wastewater-derived stressors to carotenoid biosynthesis and antioxidant defenses. Heavy metals and other pro-oxidants elevate ROS, triggering MAPK-mediated transcriptional activation of carotenogenic genes, xanthophyll cycle operation (violaxanthin–zeaxanthin interconversion), and enzymatic detoxification via superoxide dismutase (SOD), catalase (CAT), and ascorbate peroxidase (APX).

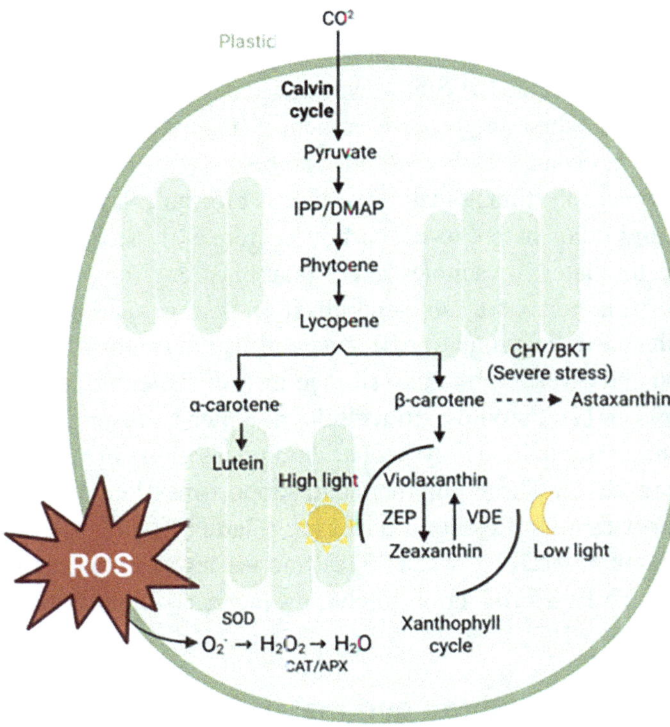

Carotenoid induction is typically coupled with upregulation of enzymatic antioxidants. Superoxide dismutase (SOD) converts superoxide to H_2O_2, which is subsequently detoxified by catalase (CAT) and ascorbate peroxidase (APX) (L. Wang et al. 2023). Together with carotenoids and the xanthophyll cycle, these enzymes help maintain redox balance and support productivity under variable heavy-metal stress that would otherwise be inhibitory or lethal (Niekerk et al. 2024; Xiao et al. 2023).

Carbohydrate Storage Pathways

Carbohydrate accumulation, particularly starch, which can reach ~50% dry weight in Chlorella under sulfur starvation (Brányiková et al. 2011), provides an alternative storage strategy that is often induced by phosphorus, sulfur, or silicon limitation (Hildebrand et al. 2017; Tan et al. 2016). Under phosphorus limitation, continued photosynthetic carbon fixation despite reduced ATP availability (Wykoff et al. 1998) redirects excess carbon into starch synthesis via ADP-glucose pyrophosphorylase (AGPase) (Li et al. 2013), whose upregulation correlates with declining phosphate availability (Bajhaiya et al. 2016). Starch synthase and branching enzymes then build the characteristic branched amylopectin structure (Ball et al. 1996). This route is biorefinery-relevant because carbohydrates can be converted into bioethanol, biogas, or long-term carbon storage materials such as biochar (Nagappan et al. 2020).

In diatoms and other heterokonts, carbon is frequently stored as chrysolaminarin, which accumulates in vacuoles rather than plastids (Yang et al. 2022). This storage polymer functions as a rapid-mobilization reserve, and silicon limitation can strongly enhance its accumulation because cells cannot complete frustule (silica shell) formation and divert carbon into storage instead (Hildebrand et al. 2017).

Another stress-linked output is extracellular polymeric substances (EPS). Many microalgae secrete extracellular polysaccharides under salinity or heavy-metal stress (Xiao and Zheng 2016). These polymers can support metal chelation outside the cell, biofilm formation, and osmotic buffering (García-Abad et al. 2022; Nagar, Antony, and Thamban 2021; Nowicka 2022), and wastewater conditions have been reported to enhance EPS production, including increases under moderate heavy-metal stress (Nowicka 2022).

Cross-pathway competition and metabolic steering

The pathways described above do not operate in isolation. A practical design constraint in stress-induced biorefineries is that lipid and carbohydrate pathways compete for shared precursors, particularly acetyl-CoA, and are steered by different nutrient limitations (Nagappan et al. 2020). Nitrogen starvation tends to favor lipid accumulation, whereas phosphorus or sulfur limitation tends to favor carbohydrate storage (Bajhaiya et al. 2016; Miller et al. 2010). In wastewater systems with variable N:P ratios, these shifting limitations can produce dynamic changes in the product spectrum (Li et al. 2011). Because of this, the chapter highlights metabolic "steering" as a process opportunity: real-time nutrient monitoring combined with pulsed dosing can push cultures toward a desired product profile rather than accepting whatever pathway dominates by chance (Bajhaiya et al. 2016; Nagappan et al. 2020).

Table 1. Stress-induced biosynthesis outcomes relevant to microalgal biorefineries

Trigger (stress)	Dominant pathway	Key control points	Main product	Location	Biorefinery value
N limitation	TAG synthesis	ACCase; DGAT2	TAG (neutral lipids)	Lipid droplets	Biofuel precursors
High light / ROS / metals	Carotenoid + NPQ	VDE/ZEP; BKT/CHY	Xanthophylls, ketocarotenoids	Chloroplast	Pigments/nutraceuticals
P or S limitation	Starch storage	AGPase	Starch	Chloroplast granules	Bioethanol/biogas feedstock
Si limitation (diatoms)	Chrysolaminarin storage	β-1 3-glucan synthesis	Chrysolaminarin	Vacuoles	Sugar stream for energy/ chemicals
Salinity / metals	EPS secretion	Polysaccharide assembly	Exopolysaccharides	Extracellular matrix	Bioflocculant/chelation aid

This multi-pathway framework demonstrates that wastewater's complex, stochastic stress profile is not a limitation but an opportunity. By understanding the molecular switches that control each biosynthetic route, biorefinery operators can manipulate cultivation conditions to "program" the metabolic output, transforming wastewater treatment from a cost center into a value-generating manufacturing platform (Table 1)

WASTEWATER VALORIZATION: ENGINEERING FOR SPECIFIC STREAMS

The "one-size-fits-all" approach to cultivating microalgae in wastewater is increasingly impractical because each effluent stream presents a distinct chemical and biological context. A more productive perspective treats wastewater not merely as a low-cost nutrient source but as *engineerable stress environments* that can be leveraged to steer metabolism toward desirable products (e.g., lipids, pigments, antioxidants), while simultaneously improving effluent quality. In practice, this means matching strain selection and cultivation mode (batch, semi-continuous, sequencing batch) to the specific stress signature of the wastewater being treated. Operational controls such as light, pH, salinity, mixing, and hydraulic retention time must be adjusted accordingly. Reviews of microalgae-based wastewater treatment emphasize that performance (both remediation and biomass composition) is strongly shaped by wastewater type, co-culturing conditions, and the dominant removal mechanisms (biosorption, bioaccumulation, biodegradation/biotransformation, and photodegradation) (Abdelfattah et al. 2023; Gupta et al. 2024; Srimongkol et al. 2022).

Wastewater Matrices as Complex Stress Environments

Unlike synthetic media, real wastewaters are chemically heterogeneous and biologically "alive," containing variable ratios of inorganic nutrients (N, P), organic carbon (COD/BOD), dissolved salts, metals, suspended solids, and diverse microbial communities (Table 2). This complexity can support robust growth when conditions are favorable, but it also creates layered stress (osmotic, oxidative, nutrient-imbalance, and xenobiotic exposure) that differs not only by wastewater category, but also by season, upstream practices, and pretreatment history. Consequently, wastewater should be treated as a *dynamic cultivation substrate* whose composition needs to be profiled and managed to balance growth stability with targeted metabolite induction (Abdelfattah et al. 2023; Gupta et al. 2024).

Municipal Wastewater: Municipal effluents often contain moderate to high nitrogen and phosphorus with comparatively lower concentrations of extreme toxicants, making them broadly compatible with common wastewater-tolerant green microalgae (Acién Fernández et al. 2016). However, fluctuations in nutrient loading, intermittent disinfectants, and microbial competition can shift community composition and productivity. Municipal streams are therefore frequently positioned as "baseline" wastewaters for polishing and biomass generation, but they still require operational control to stabilize pH, light exposure, and residence time (Abdelfattah et al. 2023; Srimongkol et al. 2022).

Agricultural Effluents: Manure-derived wastewaters (e.g., swine, dairy) commonly impose strong selective pressure via high ammonium, elevated COD, high turbidity, and often higher salinity and micronutrient loads. These conditions can be advantageous for inducing storage compounds (when growth becomes nutrient-imbalanced) but can also suppress growth through ammonia toxicity and light limitation. The presence of veterinary antibiotics and co-occurring metals can add additional stress layers and complicate downstream biomass use, reinforcing the need for careful strain choice and risk-aware product targeting (Gupta et al. 2024; Vaishnav et al. 2023).

Aquaculture Discharge: Aquaculture effluents are distinctive because they combine dissolved nutrients with a strong biological component: suspended bacteria, fish-associated microbiota, and organic particulates. In marine or brackish systems, salinity acts as a chronic osmotic stressor, often pushing cells toward osmoprotective and storage responses. This "biotic + salinity" profile can be useful for selecting halotolerant strains and for designing consortia where algal–bacterial interactions contribute to nutrient removal and biomass structuring (Abdelfattah et al. 2023; Gupta et al. 2024; Liu et al. 2024).

Industrial Wastewaters: Industrial streams are the most compositionally diverse, ranging from nutrient-rich (e.g., fermentation, some food industries) to highly inhibitory (e.g., metal-containing, dye-containing, solvent-containing, or high-salt streams). Metals are especially important because they can generate oxidative stress and damage membranes, proteins, and nucleic acids; yet tolerant strains can sometimes use these same environments for selective enrichment and for triggering antioxidant defenses. Industrial wastewater valorization typically requires either pretreatment (to reduce acute toxicity) or robust strains/consortia designed specifically for that industrial signature (Abdelfattah et al. 2023; Chu and Phang 2019).

Food-Processing Wastewaters: Dairy, meat-processing, and brewery wastewaters often contain high organic loads (high COD/BOD), fats and proteins, and variable nutrient ratios. High organic carbon can promote bacterial overgrowth and shading, but it can also be exploited in algal–bacterial systems where bacteria mineralize organics and recycle CO_2 for algal photosynthesis. These streams can therefore be promising for coupled remediation and biomass production, provided that process design prevents oxygen depletion and uncontrolled microbial dominance (Abdelfattah et al. 2023; Gupta et al. 2024; Li et al. 2019; Sciuto et al. 2025).

Table 2. Wastewater comparison

Wastewater Type	Dominant Stress Factors	Primary Metabolic Target
Municipal	Low/Moderate N & P, Seasonal variations	Baseline biomass
Agricultural	High NH_4^+, Turbidity, Salinity	Lipids (storage compounds)
Industrial	Heavy metals, Extreme pH, Xenobiotics	Antioxidants, Carotenoids
Food Processing	High COD/BOD, Sugars, Fats	Protein concentrates, Biogas

Wastewater-Induced Stress Responses and Metabolite Enhancement

Because real wastewater simultaneously imposes multiple stresses, microalgae often respond through coordinated physiological and metabolic reprogramming. A central pattern is the shift from growth-associated synthesis (proteins, nucleic acids) toward storage and protection (neutral lipids, carbohydrates, and antioxidant pigments) when conditions become nutrient-imbalanced or otherwise inhibitory. Multi-factor stress strategies are widely discussed as a route to enhance lipid accumulation and alter lipid profiles, including combinations of nutrient limitation, salinity, and light stress, although trade-offs with biomass productivity remain a consistent constraint (Suparmaniam et al. 2023).

Operationally, wastewater systems can be designed to intentionally deliver controlled stress. For example, sequencing batch and feast–famine feeding regimes create alternating nutrient availability that can reshape community structure and biochemical composition; however, outcomes depend strongly on pH, community composition, and cycle design, and do not universally increase lipid content. This highlights that "stress engineering" must be tuned to the biological context rather than assumed to be generically beneficial (Loria et al. 2023).

Oxidative stress is another key steering mechanism, particularly in wastewaters containing metals or other redox-active toxicants. Heavy metals commonly elevate reactive oxygen species (ROS), which can trigger antioxidant enzyme systems and non-enzymatic defenses; if uncontrolled, ROS can drive lipid peroxidation and broad cellular damage. In wastewater valorization, this implies a narrow but exploitable window: conditions must be strong enough to induce protective metabolite pathways (e.g., carotenoid/antioxidant responses) without collapsing growth (Danouche et al. 2022).

Finally, the convergence of stressors in real wastewater frequently produces non-additive outcomes. Nutrient limitation combined with salinity or high light can amplify storage compound accumulation (Ahmad et al. 2025; Nzayisenga et al. 2020; Osman et al. 2023), while simultaneous exposure to antibiotics, metals, and organic toxicants may reduce productivity and raise downstream biomass-safety concerns (Danouche et al. 2022; Manaia et al. 2018; Nowicka 2022). For this reason, wastewater valorization is best approached as a constrained optimization problem: maximize product-relevant pathways while maintaining enough growth and remediation performance to keep the process stable and scalable (Abdelfattah et al. 2023; Suparmaniam et al. 2023)

Emerging Contaminants and Degradation Pathways

Beyond nutrients and conventional pollutants, many wastewaters carry "emerging contaminants" such as pharmaceuticals, personal care products, disinfectant by-products, and other bioactive micropollutants. Microalgae-based systems can contribute to mitigation through multiple routes: biosorption onto cell surfaces/EPS, bioaccumulation, and transformation/photodegradation (Chu et al. 2022; Chu and Phang 2019; Kariyawasam et al. 2024; Xiong, Kurade, and Jeon 2017), often enhanced when algae are integrated with bacteria or combined with complementary treatment technologies. Recent reviews synthesize removal mechanisms and report that performance is compound- and system-dependent, but can be substantial under optimized conditions (Narayanan et al. 2025).

Antibiotics and antibiotic resistance genes (ARGs) are particularly important because wastewater treatment plants can act as reservoirs and dissemination points for ARGs, shaped by mixed urban/agricultural inputs and by selection pressures within treatment processes (Gao et al. 2022). Microalgae-based wastewater treatment has attracted attention as a potential contributor to resistome control, with reviews emphasizing mechanisms such as biosorption, biodegradation, and photodegradation for antibiotics, alongside system-level effects on microbial community structure and gene propagation (Inuwa, Pervez, and Nazir 2023; Zachee et al. 2025).

However, the valorization framing adds an additional constraint: if biomass is intended for downstream use (feed, fertilizer, bioproducts), contaminant fate and biosafety must be explicitly managed. This motivates process designs that (i) separate "high-risk" streams into product pathways compatible with safety requirements, (ii) incorporate polishing/validation steps when needed, and (iii) select product targets aligned with the contaminant profile of the source wastewater. In this way, emerging contaminants are not only a treatment challenge but also a design variable that shapes which valorization routes are technically and socially acceptable (Gao et al. 2022; Narayanan et al. 2025).

INTEGRATED CARBON SEQUESTRATION

Microalgae can convert inorganic carbon into biomass while removing nutrients from wastewater, but the climate value of this process depends on where the carbon comes from (point source vs. atmosphere) and where it ends up (short-lived products vs. durable storage).

Carbon Capture vs. Carbon Dioxide Removal (CDR)

Carbon capture, utilization and storage (CCUS) typically refers to capturing CO_2 from point sources (e.g., industrial facilities and power plants) and then using it or storing it (e.g., geological injection) (McLaughlin et al. 2023). Carbon dioxide removal (CDR) refers to deliberate removal of CO_2 from the atmosphere with durable storage in reservoirs or long-lived products (IPCC 2022). The distinction matters for microalgae: most algal cultivation supplied with flue gas or concentrated CO_2 is best framed as carbon utilization, and only qualifies as CDR when atmospheric CO_2 is the source and storage is durable with credible accounting (IPCC 2022). Carbon-negative claims therefore require explicit attention to permanence and leakage risk.

Stoichiometric Coupling of Nutrient Removal and Carbon Fixation

Wastewater-fed microalgal cultivation links nutrient recovery directly to carbon fixation through biomass synthesis. Under favorable growth conditions, microalgal biomass maintains relatively consistent elemental requirements, with a representative macronutrient stoichiometry of $C_{124}N_{16}S_{1.3}P_1$. However, this composition can shift under stress and induction regimes (Gerotto, Norici, and Giordano 2020). A useful heuristic is that microalgal biomass contains ~50% carbon by dry mass; thus producing 1 kg of dry biomass corresponds to ~1.83 kg CO_2 fixed (Chisti 2007). This does not automatically imply durable removal, but it clarifies why high-rate algal treatment can simultaneously support nutrient removal and carbon uptake. It also links directly to earlier sections: when stress engineering shifts carbon partitioning toward lipids or carbohydrates (see Biochemistry of Stress-Induced Biosynthesis Pathways), it changes biomass composition and downstream carbon fate, so carbon accounting must track the product pathway rather than assuming that fixation equals storage.

CO_2 Supply Strategies for Algal Wastewater Systems

Despite being nutrient-rich, many wastewaters are inorganic-carbon limited relative to microalgal growth demands. Municipal wastewater commonly lacks sufficient dissolved inorganic carbon to match the C:N:P ratios needed for sustained high productivity, which can constrain nutrient removal and biomass generation (Posadas et al. 2015). Engineered CO_2 supply is therefore often required, and it must be designed as both a productivity input and an operational control lever because CO_2 dosing affects pH and carbon speciation.

Three practical CO_2 supply routes are most relevant:

Direct CO_2 sparging. Gaseous CO_2 can be sparged into ponds or photobioreactors to increase dissolved inorganic carbon. While straightforward, sparging is constrained by gas–liquid mass-transfer limits and can cause rapid pH swings if not controlled.

Bicarbonate/alkalinity management. Carbon can be supplied as bicarbonate or via alkalinity control to stabilize carbon availability and buffer pH, but this adds chemical inputs and may increase salinity depending on the source.

Flue gas and biogas integration (industrial symbiosis). Co-locating algae systems with CO_2-emitting infrastructure can address carbon limitation while supporting integrated resource recovery. Flue gas streams can contain elevated CO_2 (often ~15–30%), and, when compatible with process control, may improve productivity and help stabilize pH (Kumar, Banerjee, and Das 2014). Microalgae can tolerate variable CO_2 conditions because carbon concentrating mechanisms (CCMs) support fixation across a range of dissolved inorganic carbon levels (Morales, Sánchez, and Revah 2018; Zhao and Su 2014). However, real gas streams introduce impurities (NO_X, SO_X and trace contaminants) and thermal variability that can push cultures beyond a productive stress window, reducing stability. This makes CO_2 integration a control problem that must be evaluated together with reactor design, mixing, pH management, and strain robustness.

Carbon Balance and Permanence: When is Algae Carbon-Negative?

Even when fixation is substantial, climate benefit depends on permanence. Many algal pathways return carbon rapidly: fuels and other short-lived products typically re-emit CO_2 within months. By contrast, routing biomass into durable sinks can extend storage from decades to centuries and potentially meet CDR criteria (IPCC 2022). A practical way to frame this is a carbon fate hierarchy:

Short-term (displacement, not removal): biofuels (biodiesel, biogas). These can reduce emissions by displacing fossil fuels, but they are not durable removal; net benefits depend strongly on harvesting/dewatering/extraction energy, and LCA results vary with process design (Lardon et al. 2009; Xu et al. 2011).

Medium-to long-term storage: include nutraceuticals and food/feed ingredients, which may improve economics but return carbon through respiration and waste processing, and durable biomaterials (e.g., selected polymers/composites), whose storage duration depends on product lifetime and end-of-life fate (Koller and Braunegg 2018).

Durable removal candidates: biochar/hydrochar, mineralisation, or engineered burial routes. Pyrolysis can convert biomass carbon into more recalcitrant forms, with stability inferred from compositional indicators and storage conditions (Yang et al. 2019). Engineered burial concepts aim to suppress decomposition under controlled conditions, but feasibility depends on energy requirements and robust monitoring.

Across these options, credible carbon-negative claims require life-cycle boundaries that include energy inputs for cultivation and processing, and they require measurement, reporting and verification (MRV) consistent with CDR definitions (IPCC 2022). Without LCA/MRV, systems may appear beneficial based on fixation alone while failing to deliver durable storage in practice.

Microalgae in Industrial Carbon Management

Microalgae can contribute to industrial carbon management by linking point-source CO_2 utilization with wastewater nutrient recovery and co-product generation. When biomass is routed into durable sinks, algal systems can move toward CDR-aligned outcomes, provided that permanence and system-wide accounting support net removal. However, practical constraints remain significant, including land footprint (especially for open systems), seasonal productivity variation, and the energy burden of harvesting and dewatering. For these reasons, microalgae are best evaluated as integrated environmental infrastructure. Techno-economic analysis (TEA) and life-cycle assessment (LCA) are essential prerequisites for identifying scalable, credible designs and for avoiding overstated carbon claims.

BIOREFINERY ENGINEERING: CASCADING PRODUCT VALORIZATION

Having established how different wastewater streams and stress conditions shape biomass composition, the focus now shifts to downstream processing. The economic viability of wastewater-fed microalgal systems depends on maximizing value extraction from every kilogram of biomass produced. In practice, this means implementing a biorefinery cascade that sequentially recovers products in order of decreasing value and increasing robustness, while minimizing energy-intensive steps such as drying. An additional design constraint is biomass eligibility: depending on the wastewater source, algae can co-accumulate metals, xenobiotics, or microbial contaminants. These may restrict use in food, feed, or high-purity markets, pushing the process toward fuels, materials, or controlled disposal endpoints.

Cascading Extraction Logic: from High-Value to Bulk Commodities

Cascading valorization mirrors the refinery concept, where one feedstock is fractionated into multiple outputs and residues are still routed to useful endpoints (Chew et al. 2017). For microalgae, the ordering is dictated by compound stability (Figure 4):

Tier 1: High-value, labile compounds (recover first). Pigments (e.g., astaxanthin, β-carotene, fucoxanthin), PUFAs, and other bioactives are sensitive to heat, oxidation, and harsh solvents; they should be extracted before aggressive processing to preserve yield and quality (Cuellar-Bermudez et al. 2015; Kichouh-Aiadi, Kawakita, et al. 2025; Molino et al. 2020). Although typically a small fraction of dry mass, these compounds can dominate revenue in biorefinery scenarios (Monlau et al. 2021; Shah et al. 2016). In wastewater systems, Tier 1 is only realistic when contaminant profiles, QA/QC capacity, and regulatory pathways support high-purity markets.

Tier 2: Lipids for fuels/oleochemicals. After Tier 1 recovery, remaining neutral lipids can be converted into biodiesel-range products or oleochemical intermediates, often via transesterification or related routes (Malik et al. 2022; Monlau et al. 2021). In wastewater biorefineries, lipid-to-fuel frequently serves as a reliable outlet when premium markets are constrained.

Tier 3: Protein-rich fraction. De-oiled biomass often remains protein-rich and can be processed into protein concentrates or hydrolysates via aqueous extraction and enzymatic treatment (Chew et al. 2017). These pathways can improve overall resource efficiency, but their feasibility depends heavily on contaminant limits for the intended use.

Tier 4: Carbohydrates and biopolymers. Residual carbohydrates can be fermented to ethanol, anaerobically digested to biomethane, or converted to higher-value materials via bacterial PHA production using algal hydrolysates as feedstock (Chew et al. 2017; Malik et al. 2022; Monlau et al. 2021). PHAs are attractive as a "bridge" product because they can upgrade low-value carbon streams into a durable material pathway (Koller and Braunegg 2018).

Tier 5: Energy recovery and nutrient recycling. Remaining residues can be routed to anaerobic digestion or thermochemical conversion (e.g., HTL, pyrolysis) to recover energy and concentrate nutrients for potential recycling, converting disposal into a controlled endpoint (Monlau et al. 2021; Xu et al. 2011; Yang et al. 2019).

Design rule: choose a cascade where early steps are mild and selective, and later steps are robust and tolerant, so that product quality is protected and residues still have a guaranteed outlet.

Figure 4. Cascading biorefinery extraction sequence for wastewater-derived microalgal biomass. Products are recovered in order of decreasing value and increasing process robustness, from high-value pigments and PUFAs (Tier 1) through lipids (Tier 2), proteins (Tier 3), and carbohydrates (Tier 4) to residual energy recovery and nutrient recycling (Tier 5).

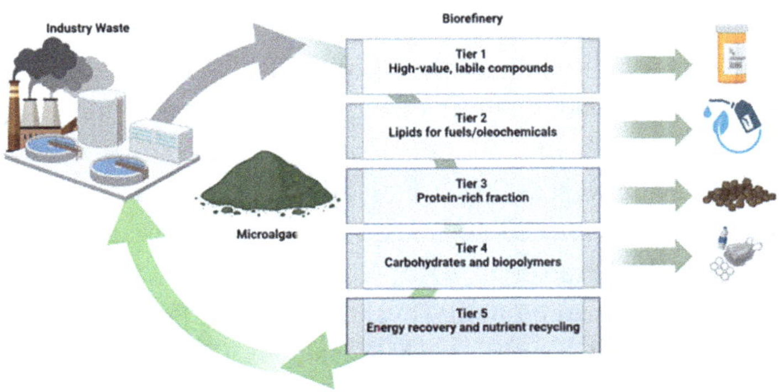

Process Engineering: Wet Extraction Routes and Energy Efficiency

A major barrier in microalgal biorefineries is the energy penalty of drying. Because cultures are dilute, removing water can dominate operating energy; classic analyses highlight that "dry route" processing can undermine sustainability and economics (Xu et al. 2011). For this reason, modern designs prioritize wet processing, keeping biomass as slurry/paste and extracting/fractionating in aqueous or low-temperature conditions (Papachristou et al. 2021; Wetterwald et al. 2023).

Key wet-route unit operations include:

Harvesting + thickening without full drying. Harvesting can represent a major share of costs, so processes aim for low-energy thickening (settling, flocculation, filtration) before any extraction (Hattab et al. 2015). Importantly, stress regimes used for induction can also improve harvestability (e.g., EPS-mediated aggregation), enabling process intensification (Boonchai, Kaewsuk, and Seo 2015; Lv et al. 2019).

Cell disruption in aqueous suspension. Technologies such as pulsed electric fields (PEF), homogenization, and enzymatic lysis can release intracellular compounds while limiting heat exposure, supporting Tier 1 recovery and reducing drying demand (Papachristou et al. 2021).

Selective extraction aligned with the cascade. Supercritical CO_2 (SFE) is often discussed for lipophilic compounds because it can preserve thermolabile pigments and reduce solvent residues, though it may increase CAPEX (Molino et al. 2020; Wetterwald et al. 2023). Hydrophilic fractions (proteins/polysaccharides) can often be recovered via aqueous extraction with pH shifts and enzymatic fractionation (Papachristou et al. 2021).

Techno-Economic Framing: What Makes Systems Viable

TEA results repeatedly indicate that profitability is rarely achieved by a single commodity stream; instead, viability emerges from a combination of (i) sufficient productivity and scale, (ii) product diversification, and (iii) monetizing services (Chew et al. 2017; Monlau et al. 2021). Cost drivers tend to cluster around cultivation productivity, harvesting/thickening, and downstream processing energy, especially when drying is used (Hattab et al. 2015; Xu et al. 2011).

Fuel-only pathways are commonly evaluated via minimum selling price (MSP). Representative TEA work has reported MSP values for integrated wastewater-algae-to-fuel schemes in the range where competitiveness depends strongly on system integration and assumptions (Ranganathan and Savithri 2019; Xin et al. 2016). Across studies, the consistent strategic conclusion is that fuels often function best as the stable "sink pathway", while higher-value fractions (when allowed by quality constraints) and/or service revenues improve resilience (Ferreira et al. 2024; Monlau et al. 2021).

Environmental services can meaningfully affect the business case because wastewater algae systems provide treatment and nutrient recovery in addition to products. Depending on local policy and contracting structures, treatment value may be captured as avoided cost or service payments; integrated circular models also show benefits when energy recovery (e.g., anaerobic digestion) and fertilizer displacement are included (Nobre et al. 2024; Vázquez-Romero et al. 2024).

Regulation and Product Eligibility: Wastewater Source and Biomass Applications

A wastewater biorefinery must treat regulatory compliance as a design input, not an afterthought, because the wastewater source often determines which product markets are feasible. Rules vary by country and end-use, but the underlying regulatory logic is similar: the higher the human/animal exposure, the stricter the requirements for contaminants, pathogens, and trace organics.

A practical engineering approach is to classify wastewater sources by risk category and map them to allowable product classes, based on contaminant profiles and regulatory thresholds.

Lower-risk sources: Some municipal tertiary effluents, controlled aquaculture streams, or food stream wastewaters may be candidates for higher-value products only if monitoring demonstrates compliance (metals, pathogens, micropollutants) and the facility can implement GMP-like controls and traceability (Abdelfattah et al. 2023; Sciuto et al. 2025; Srimongkol et al. 2022). Even so, many projects prefer non-food bioproducts due to regulatory complexity and market restrictions (Gupta et al. 2024).

Medium-risk sources: Typical municipal wastewaters, mixed urban effluents, and food-processing wastewaters can often support products such as biofertilizers, biogas, industrial materials, or low-risk chemical precursors, provided that purification and verification steps are in place (Li et al. 2019; Monlau et al. 2021). However, food or feed applications remain challenging, as even small amounts of residual heavy metals, organic contaminants, or antibiotic residues can breach safety standards.

Higher-risk sources: Industrial wastewaters containing metals, solvents, dyes, or persistent pharmaceuticals pose the highest biosafety challenges. These contaminants can accumulate in biomass or interfere with metabolic pathways, creating risks of toxic compound transfer or secondary pollution if products enter open cycles (Chu and Phang 2019; Danouche et al. 2022; Nowicka 2022). In such cases, valorization is typically limited to energy recovery (e.g., anaerobic digestion, HTL, pyrolysis) or durable material pathways designed for controlled use and traceability. Although these streams can still yield economic value, they demand strict product eligibility assessment and contaminant management, consistent with circular bioeconomy and safety principles (Abdelfattah et al. 2023; Gao et al. 2022).

CONCLUSION

Transitioning from linear wastewater disposal to circular resource recovery is essential for addressing global climate and nutrient crises. Microalgae serve as a versatile biological platform capable of synchronizing wastewater remediation with carbon sequestration. The stress-induced biorefinery model demonstrates that environmental constraints, once viewed as limitations, can be engineered to steer metabolic pathways toward specific, high-value bioproducts. However, achieving economic viability requires a shift toward cascading extraction and energy-efficient wet-route processing to minimize the energy penalty of biomass drying.

Several research priorities emerge from this analysis. First, pilot-scale validation is critically needed: most stress-induced biorefinery data originate from laboratory or bench-scale experiments, and the transition to outdoor, continuously operated systems introduces variables, such as diurnal and seasonal light fluctuations, temperature swings, biological contamination, and predator–prey dynamics, that are difficult to replicate at small scale. Demonstrating stable, long-term operation of stress-engineered microalgal systems at pilot scale, ideally co-located with real wastewater sources and CO_2-emitting infrastructure, would provide the operational data required to validate techno-economic models and attract investment. Second, comprehensive life-cycle assessment (LCA) must become standard practice rather than an afterthought. Current LCA studies vary widely in system boundaries, functional units, and allocation methods, making cross-study comparisons unreliable. Harmonized LCA frameworks that explicitly account for the multi-product nature of cascading biorefineries, the variable carbon permanence of different product pathways, and the avoided environmental burden of wastewater treatment are essential for credible sustainability claims and for distinguishing carbon utilization from genuine carbon dioxide removal.

Third, regulatory frameworks remain a major barrier to commercialization. As discussed in this chapter, the wastewater source largely determines which product markets are accessible, yet regulations governing biomass derived from wastewater vary considerably across jurisdictions and are often not designed with biorefinery outputs in mind. Developing clear, science-based guidelines for contaminant thresholds in microalgal products, differentiated by end-use (food, feed, fertilizer, fuel, materials), would reduce market uncertainty and enable rational process design from the outset. Finally, advances in real-time monitoring and adaptive process control, including integration of omics-based diagnostics with automated nutrient dosing and harvesting, could close the gap between the precise metabolic steering described in this chapter and the operational reality of large-scale, variable-input systems. Addressing these priorities in concert will determine whether stress-induced microalgal biorefineries remain a promising laboratory concept or mature into a scalable component of circular bioeconomy infrastructure.

REFERENCES

Abdelfattah, A., Ali, S. S., Ramadan, H., El-Aswar, E. I., Eltawab, R., Ho, S. H., Elsamahy, T., Li, S., El-Sheekh, M. M., Schagerl, M., Kornaros, M., & Sun, J. (2023). Microalgae-Based Wastewater Treatment: Mechanisms, Challenges, Recent Advances, and Future Prospects. *Environmental Science and Ecotechnology, 13*, 100205. DOI: 10.1016/j.ese.2022.100205 PMID: 36247722

Acién Fernández, F. G. (2016). Wastewater Treatment Using Microalgae: How Realistic a Contribution Might It Be to Significant Urban Wastewater Treatment? *Applied Microbiology and Biotechnology, 100*(21), 9013–9022. DOI: 10.1007/s00253-016-7835-7 PMID: 27645298

Ahmad, Maqbool, David Plischke, Petar Zuanović, and Zoran Zorić. 2025. "Interactive Effects of Salinity Variation and Nitrogen Deficiency on Fatty Acids Composition in Nannochloropsis." *Journal of Applied Phycology 2025 37:5* 37(5):3487–99. .DOI: 10.1007/s10811-025-03624-8

Alewell, Christine, Bruno Ringeval, Cristiano Ballabio, David A. Robinson, Panos Panagos, and Pasquale Borrelli. 2020. "Global Phosphorus Shortage Will Be Aggravated by Soil Erosion." *Nature Communications 2020 11:1* 11(1):4546-. .DOI: 10.1038/s41467-020-18326-7

Ashour, M., Mansour, A. T., Alkhamis, Y. A., & Elshobary, M. (2024). Usage of Chlorella and Diverse Microalgae for CO2 Capture - towards a Bioenergy Revolution. *Frontiers in Bioengineering and Biotechnology, 12*, 1387519. DOI: 10.3389/fbioe.2024.1387519 PMID: 39229458

Bajhaiya, A. K., Dean, A. P., Zeef, L. A. H., Webster, R. E., & Pittman, J. K. (2016). PSR1 Is a Global Transcriptional Regulator of Phosphorus Deficiency Responses and Carbon Storage Metabolism in Chlamydomonas Reinhardtii. *Plant Physiology, 170*(3), 1216–1234. DOI: 10.1104/pp.15.01907 PMID: 26704642

Ball, S., Han, P. G., James, M., Myers, A., Keeling, P., Mouille, G., Buléon, A., Colonna, P., & Preiss, J. (1996). From Glycogen to Amylopectin: A Model for the Biogenesis of the Plant Starch Granule. *Cell, 86*(3), 349–352. DOI: 10.1016/S0092-8674(00)80107-5 PMID: 8756717

Boonchai, R., Kaewsuk, J., & Seo, G. (2015). Effect of Nutrient Starvation on Nutrient Uptake and Extracellular Polymeric Substance for Microalgae Cultivation and Separation. *Desalination and Water Treatment, 55*(2), 360–367. DOI: 10.1080/19443994.2014.939501

Brányiková, I., Maršálková, B., Doucha, J., Brányik, T., Bišová, K., Zachleder, V., & Vítová, M. (2011). Microalgae-Novel Highly Efficient Starch Producers. *Biotechnology and Bioengineering*, *108*(4), 766–776. DOI: 10.1002/bit.23016 PMID: 21404251

Cazzonelli, C. I. (2011). Carotenoids in Nature: Insights from Plants and Beyond. *Functional Plant Biology : FPB*, *38*(11), 833–847. DOI: 10.1071/FP11192 PMID: 32480941

Cen, S. Y., Li, D. W., Xue, L. H., Huang, D., Balamurugan, S., Wan, J. L., Jian, W. Z., Wei, D. Y., & Hong, Y. L. (2022). Crucial Carotenogenic Genes Elevate Hyperaccumulation of Both Fucoxanthin and β-Carotene in Phaeodactylum Tricornutum. *Algal Research*, *64*, 102691. DOI: 10.1016/j.algal.2022.102691

Chen, B., Wan, C., Mehmood, M. A., Chang, J. S., Bai, F., & Zhao, X. (2017). Manipulating Environmental Stresses and Stress Tolerance of Microalgae for Enhanced Production of Lipids and Value-Added Products–A Review. *Bioresource Technology*, *244*, 1198–1206. DOI: 10.1016/j.biortech.2017.05.170 PMID: 28601395

Chew, K. W., Yap, J. Y., Show, P. L., Suan, N. H., Juan, J. C., Ling, T. C., Lee, D. J., & Chang, J. S. (2017). Microalgae Biorefinery: High Value Products Perspectives. *Bioresource Technology*, *229*, 53–62. DOI: 10.1016/j.biortech.2017.01.006 PMID: 28107722

Chisti, Y. (2007). Biodiesel from Microalgae. *Biotechnology Advances*, *25*(3), 294–306. DOI: 10.1016/j.biotechadv.2007.02.001 PMID: 17350212

Chu, Wan Loy, and Siew Moi Phang. 2019. "Biosorption of Heavy Metals and Dyes from Industrial Effluents by Microalgae." *Microalgae Biotechnology for Development of Biofuel and Wastewater Treatment* 599–634. doi:.DOI: 10.1007/978-981-13-2264-8_23/TABLES/2

Chu, Y., Zhang, C., Wang, R., Chen, X., Ren, N., & Ho, S. H. (2022). Biotransformation of Sulfamethoxazole by Microalgae: Removal Efficiency, Pathways, and Mechanisms. *Water Research*, *221*, 118834. DOI: 10.1016/j.watres.2022.118834 PMID: 35839594

Cuellar-Bermudez, S. P., Aguilar-Hernandez, I., Cardenas-Chavez, D. L., Ornelas-Soto, N., Romero-Ogawa, M. A., & Parra-Saldivar, R. (2015). Extraction and Purification of High-Value Metabolites from Microalgae: Essential Lipids, Astaxanthin and Phycobiliproteins. *Microbial Biotechnology*, *8*(2), 190–209. DOI: 10.1111/1751-7915.12167 PMID: 25223877

Danouche, Mohammed, Naïma El Ghatchouli, and Hicham Arroussi. 2022. "Overview of the Management of Heavy Metals Toxicity by Microalgae." *Journal of Applied Phycology 2021 34:1* 34(1):475–88. .DOI: 10.1007/s10811-021-02668-w

Derco, Ján, Patrícia Guľašová, Maša Legan, Ronald Zakhar, and Andreja Žgajnar Gotvajn. 2024. "Sustainability Strategies in Municipal Wastewater Treatment." *Sustainability 2024, Vol. 16, Page 9038* 16(20):9038. .DOI: 10.3390/su16209038

Diaz, R. J., & Rosenberg, R. (2008). Spreading Dead Zones and Consequences for Marine Ecosystems. *Science*, *321*(5891), 926–929. DOI: 10.1126/science.1156401 PMID: 18703733

El-Sheekh, Mostafa M., Hamdy R. Galal, Amal Sh H. Mousa, and Abla A. M. Farghl. 2023. "Coupling Wastewater Treatment, Biomass, Lipids, and Biodiesel Production of Some Green Microalgae." *Environmental Science and Pollution Research 2023 30:12* 30(12):35492–504. .DOI: 10.1007/s11356-023-25628-y

Fernández-Marín, B., Roach, T., Verhoeven, A., & García-Plazaola, J. I. (2021). Shedding Light on the Dark Side of Xanthophyll Cycles. *The New Phytologist*, *230*(4), 1336–1344. DOI: 10.1111/nph.17191 PMID: 33452715

Ferreira, J., Quintão Braga, M., Carvalho Nogueira da Gama, R., Barbosa Magalhães, I., Barros Marangon, B., de Siqueira Castro, J., Ferreira Lorentz, J., Silva Henriques, B., Saleme Aona de Paula Pereira, A., Peixoto Assemany, P., & Lúcia Calijuri, M. (2024). Carotenoids from Wastewater-Grown Microalgae Biomass: Life Cycle Assessment and Techno-Economical Analysis. *Journal of Cleaner Production*, *434*, 140526. Advance online publication. DOI: 10.1016/j.jclepro.2023.140526

Gao, Y. X., Li, X., Xiao, Y. F., Jun, R. Z., & Zhong, X. Z. (2022). Wastewater Treatment Plants as Reservoirs and Sources for Antibiotic Resistance Genes: A Review on Occurrence, Transmission and Removal. *Journal of Water Process Engineering*, *46*, 102539. DOI: 10.1016/j.jwpe.2021.102539

García-Abad, L., López-Rosales, L., María del Carmen, C.-G., Fernández-García, M., García-Camacho, F., & Molina-Grima, E. (2022). Influence of Abiotic Conditions on the Biofouling Formation of Flagellated Microalgae Culture. *Biofouling*, *38*(5), 507–520. DOI: 10.1080/08927014.2022.2089564 PMID: 35729852

Gerotto, C., Norici, A., & Giordano, M. (2020). Toward Enhanced Fixation of CO2 in Aquatic Biomass: Focus on Microalgae. *Frontiers in Energy Research*, *8*, 550930. DOI: 10.3389/fenrg.2020.00213

Global Monitoring Laboratory. 2025. https://gml.noaa.gov/ccgg/trends/global.html

Gomez-Casati, D. F., Barchiesi, J., & Busi, M. V. (2022). Mitochondria and Chloroplasts Function in Microalgae Energy Production. *PeerJ*, *10*, e14576. DOI: 10.7717/peerj.14576 PMID: 36545385

Gorelova, Olga, Olga Baulina, Tatiana Ismagulova, Kamilya Kokabi, Elena Lobakova, Irina Selyakh, Larisa Semenova, Olga Chivkunova, Olga Karpova, Pavel Scherbakov, Inna Khozin-Goldberg, and Alexei Solovchenko. 2018. "Stress-Induced Changes in the Ultrastructure of the Photosynthetic Apparatus of Green Microalgae." *Protoplasma 2018 256:1* 256(1):261–77. .DOI: 10.1007/s00709-018-1294-1

Gupta, Riya, Niranjan Mishra, Gayatri Singh, Sandhya Mishra, Neelu Lodhiyal, and Blue Biotechnology. 2024. "Microalgae Cultivation and Value-Based Products from Wastewater: Insights and Applications." *Blue Biotechnology 2024 1:1* 1(1):20-. .DOI: 10.1186/s44315-024-00019-1

Hasan, M. N., Altaf, M. M., Khan, N. A., Khan, A. H., Khan, A. A., Ahmed, S., Kumar, P. S., Naushad, M., Rajapaksha, A. U., Iqbal, J., Tirth, V., & Islam, S. (2021). Recent Technologies for Nutrient Removal and Recovery from Wastewaters: A Review. *Chemosphere*, *277*, 130328. DOI: 10.1016/j.chemosphere.2021.130328 PMID: 33794428

Hashimoto, Koji. 2019. "Global Temperature and Atmospheric Carbon Dioxide Concentration." 5–17. .DOI: 10.1007/978-981-13-8584-1_3

Hattab, M. A., & Ghaly, A. (2015). Microalgae Harvesting Methods for Industrial Production of Biodiesel:Critical Review and Comparative Analysis. *Journal of Fundamentals of Renewable Energy and Applications*, *05*(02). Advance online publication. DOI: 10.4172/2090-4541.1000154

He, A., Dean, J. M., & Lodhi, I. J. (2021). Peroxisomes as Cellular Adaptors to Metabolic and Environmental Stress. *Trends in Cell Biology*, *31*(8), 656–670. DOI: 10.1016/j.tcb.2021.02.005 PMID: 33674166

Hepburn, Cameron, Ella Adlen, John Beddington, Emily A. Carter, Sabine Fuss, Niall Mac Dowell, Jan C. Minx, Pete Smith, and Charlotte K. Williams. 2019. "The Technological and Economic Prospects for CO2 Utilization and Removal." *Nature 2019 575:7781* 575(7781):87–97. .DOI: 10.1038/s41586-019-1681-6

Hildebrand, M., Manandhar-Shrestha, K., & Abbriano, R. (2017). Effects of Chrysolaminarin Synthase Knockdown in the Diatom Thalassiosira Pseudonana: Implications of Reduced Carbohydrate Storage Relative to Green Algae. *Algal Research*, *23*, 66–77. DOI: 10.1016/j.algal.2017.01.010

Huerlimann, R., & Heimann, K. (2013). Comprehensive Guide to Acetyl-Carboxylases in Algae. *Critical Reviews in Biotechnology*, *33*(1), 49–65. DOI: 10.3109/07388551.2012.668671 PMID: 22524446

Inuwa, A. B., A. Pervez, and R. Nazir. 2023. "Microalgae-Based Wastewater Treatment System: Current State, Antibiotic Resistant Bacteria and Antibiotic Resistance Genes Reduction Potentials." *International Journal of Environmental Science and Technology 2023 20:12* 20(12):14053–72. .DOI: 10.1007/s13762-023-05069-3

IPCC. 2022. "IPCC AR6 WGIII: CDR Factsheet."

Jones, E. R., Van Vliet, M. T. H., Qadir, M., & Bierkens, M. F. P. (2021). Country-Level and Gridded Estimates of Wastewater Production, Collection, Treatment and Reuse. *Earth System Science Data*, *13*(2), 237–254. DOI: 10.5194/essd-13-237-2021

Kariyawasam, T., Helvig, C., Petkovich, M., & Vriens, B. (2024). Pharmaceutical Removal from Wastewater by Introducing Cytochrome P450s into Microalgae. *Microbial Biotechnology*, *17*(6), e14515. Advance online publication. DOI: 10.1111/1751-7915.14515 PMID: 38925623

Kichouh-Aiadi, S., Gallardo-Rodríguez, J. J., Cerón-García, M. C., López-Rosales, L., García-Camacho, F., & Sánchez-Mirón, A. (2024). Exploring the Potential of Epigenetic Chemicals to Increase Metabolite Production in the Dinoflagellate Microalga Amphidinium Carterae. *Journal of Applied Phycology*, *36*(3), 1169–1179. Advance online publication. DOI: 10.1007/s10811-024-03189-y

Kichouh-Aiadi, S., H. Kawakita, J. J. Gallardo-Rodríguez, A. Sánchez-Mirón, and F. García-Camacho. 2025. "Sequential Recovery of Phycobiliproteins and Polysaccharides from Spirulina Using a Novel Two-Step Process Integrating Adsorption and Cross-Flow Dialysis." *Journal of Applied Phycology 2025 37:5* 37(5):3581–93. .DOI: 10.1007/s10811-025-03606-w

Kichouh-Aiadi, S., López-Rosales, L., Gallardo-Rodríguez, J. J., Cerón-García, M. C., Sánchez-Mirón, A., & García-Camacho, F. (2025). Effects of Hormones on the Growth and Metabolite Production of Amphidinium Carterae under Carbon Sufficient and Carbon Limited Conditions. *Algal Research*, *85*, 103810. DOI: 10.1016/j.algal.2024.103810

Koller, M., & Braunegg, G. (2018). Advanced Approaches to Produce Polyhydroxyalkanoate (PHA) Biopolyesters in a Sustainable and Economic Fashion. *The EuroBiotech Journal*, *2*(2), 89–103. DOI: 10.2478/ebtj-2018-0013

Kumar, K., Banerjee, D., & Das, D. (2014). Carbon Dioxide Sequestration from Industrial Flue Gas by Chlorella Sorokiniana. *Bioresource Technology, 152*, 225–233. DOI: 10.1016/j.biortech.2013.10.098 PMID: 24292202

Lardon, L., Hélias, A., Sialve, B., Steyer, J. P., & Bernard, O. (2009). Life-Cycle Assessment of Biodiesel Production from Microalgae. *Environmental Science & Technology, 43*(17), 6475–6481. DOI: 10.1021/es900705j PMID: 19764204

Launay, H., Avilan, L., Gérard, C., Parsiegla, G., Receveur-Brechot, V., Gontero, B., & Carriere, F. (2023). Location of the Photosynthetic Carbon Metabolism in Microcompartments and Separated Phases in Microalgal Cells. *FEBS Letters, 597*(23), 2853–2878. DOI: 10.1002/1873-3468.14754 PMID: 37827572

Li, D. W., Shi, Y. C., Yu, H. L., Balamurugan, S., Xin, Y. Z., Alimujiang, A., Wei, D. Y., Jie, S. L., & Hong, Y. L. (2016). A Type 2 Diacylglycerol Acyltransferase Accelerates the Triacylglycerol Biosynthesis in Heterokont Oleaginous Microalga Nannochloropsis Oceanica. *Journal of Biotechnology, 229*, 65–71. DOI: 10.1016/j.jbiotec.2016.05.005 PMID: 27164260

Li, J., Han, D., Wang, D., Ning, K., Jia, J., Wei, L., Jing, X., Huang, S., Chen, J., Li, Y., Hu, Q., & Xu, J. (2014). Choreography of Transcriptomes and Lipidomes of Nannochloropsis Reveals the Mechanisms of Oil Synthesis in Microalgae. *The Plant Cell, 26*(4), 1645–1665. DOI: 10.1105/tpc.113.121418 PMID: 24692423

Li, S., Zhao, S., Yan, S., Qiu, Y., Song, C., Li, Y., & Kitamura, Y. (2019). Food Processing Wastewater Purification by Microalgae Cultivation Associated with High Value-Added Compounds Production — A Review. *Chinese Journal of Chemical Engineering, 27*(12), 2845–2856. DOI: 10.1016/j.cjche.2019.03.028

Li, Y., Han, D., Yoon, K., Zhu, S., Sommerfeld, M., & Hu, Q. 2013. "Molecular and Cellular Mechanisms for Lipid Synthesis and Accumulation in Microalgae: Biotechnological Implications." *Handbook of Microalgal Culture: Applied Phycology and Biotechnology: Second Edition* 545–65. doi:DOI: 10.1002/9781118567166.CH28;JOURNAL:JOURNAL:BOOKS;WGROUP:STRING:PUBLICATION

Li, Y., Yi, F. C., Chen, P., Min, M., Zhou, W., Martinez, B., Zhu, J., & Ruan, R. (2011). Characterization of a Microalga Chlorella Sp. Well Adapted to Highly Concentrated Municipal Wastewater for Nutrient Removal and Biodiesel Production. *Bioresource Technology, 102*(8), 5138–5144. DOI: 10.1016/j.biortech.2011.01.091 PMID: 21353532

Liu, X., Wang, Y., Liu, H., Zhang, Y., Zhou, Q., Wen, X., Guo, W., & Zhang, Z. (2024). A Systematic Review on Aquaculture Wastewater: Pollutants, Impacts, and Treatment Technology. *Environmental Research, 262*(Pt 1), 119793. Advance online publication. DOI: 10.1016/j.envres.2024.119793 PMID: 39147181

Loria, M. H., Griffin, J. S., Wells, G. F., & Rhoads, K. R. (2023). Effects of Feast-Famine Nutrient Regimes on Wastewater Algal Biofuel Communities. *PLoS One, 18*(1), e0279943. Advance online publication. DOI: 10.1371/journal.pone.0279943 PMID: 36598899

Lv, H., Cui, X., Wahid, F., Xia, F., Zhong, C., & Jia, S. (2016). Analysis of the Physiological and Molecular Responses of Dunaliella Salina to Macronutrient Deprivation. *PLoS One, 11*(3), e0152226. DOI: 10.1371/journal.pone.0152226 PMID: 27023397

Lv, J., Guo, B., Feng, J., Liu, Q., Nan, F., Liu, X., & Xie, S. (2019). Integration of Wastewater Treatment and Flocculation for Harvesting Biomass for Lipid Production by a Newly Isolated Self-Flocculating Microalga Scenedesmus Rubescens SX. *Journal of Cleaner Production, 240*, 118211. DOI: 10.1016/j.jclepro.2019.118211

Malik, S., Shahid, A., Betenbaugh, M. J., Chen, G. L., & Mehmood, M. A. (2022). A Novel Wastewater-Derived Cascading Algal Biorefinery Route for Complete Valorization of the Biomass to Biodiesel and Value-Added Bioproducts. *Energy Conversion and Management, 256*, 115360. DOI: 10.1016/j.enconman.2022.115360

Manaia, C. M., Rocha, J., Scaccia, N., Marano, R., Radu, E., Biancullo, F., Cerqueira, F., Fortunato, G., Iakovides, I. C., Zammit, I., Kampouris, I., Vaz-Moreira, I., & Nunes, O. C. (2018). Antibiotic Resistance in Wastewater Treatment Plants: Tackling the Black Box. *Environment International, 115*, 312–324. DOI: 10.1016/j.envint.2018.03.044 PMID: 29626693

McLaughlin, H., Littlefield, A. A., Menefee, M., Kinzer, A., Hull, T., Sovacool, B. K., Bazilian, M. D., Kim, J., & Griffiths, S. (2023). Carbon Capture Utilization and Storage in Review: Sociotechnical Implications for a Carbon Reliant World. *Renewable & Sustainable Energy Reviews, 177*, 113215. DOI: 10.1016/j.rser.2023.113215

Mei, R., Yang, H., Guo, C., Hong, Z., Hu, Z., Wu, Y., Huang, D., & Wang, C. (2025). Effects of High Light Exposure and Heterologous Expression of β-Carotene Ketolase on the Metabolism of Carotenoids in Chlamydomonas Reinhardtii. *Frontiers in Bioengineering and Biotechnology, 13*, 1533661. Advance online publication. DOI: 10.3389/fbioe.2025.1533661 PMID: 40144396

Miller, R., Wu, G., Deshpande, R. R., Vieler, A., Gartner, K., Li, X., Moellering, E. R., Zauner, S., Cornish, A. J., Liu, B., Bullard, B., Sears, B. B., Min, H. K., Hegg, E. L., Shachar-Hill, Y., Shiu, S. H., & Benning, C. (2010). Changes in Transcript Abundance in Chlamydomonas Reinhardtii Following Nitrogen Deprivation Predict Diversion of Metabolism. *Plant Physiology*, *154*(4), 1737–1752. DOI: 10.1104/pp.110.165159 PMID: 20935180

Molino, Antonio, Sanjeet Mehariya, Giuseppe Di Sanzo, Vincenzo Larocca, Maria Martino, Gian Paolo Leone, Tiziana Marino, Simeone Chianese, Roberto Balducchi, and Dino Musmarra. 2020. "Recent Developments in Supercritical Fluid Extraction of Bioactive Compounds from Microalgae: Role of Key Parameters, Technological Achievements and Challenges." *Journal of CO2 Utilization* 36:196–209. .DOI: 10.1016/j.jcou.2019.11.014

Monlau, F., Suarez-Alvarez, S., Lallement, A., Vaca-Medina, G., Giacinti, G., Munarriz, M., Urreta, I., Raynaud, C., Ferrer, C., & Castañón, S. (2021). A Cascade Biorefinery for the Valorization of Microalgal Biomass: Biodiesel, Biogas, Fertilizers and High Valuable Compounds. *Algal Research*, *59*, 102433. DOI: 10.1016/j.algal.2021.102433

Morales, M., Sánchez, L., & Revah, S. (2018). The Impact of Environmental Factors on Carbon Dioxide Fixation by Microalgae. *FEMS Microbiology Letters*, *365*(3) Advance online publication. DOI: 10.1093/femsle/fnx262 PMID: 29228188

Msanne, J., Di Xu, A. R. K., Casas-Mollano, J. A., Awada, T., Cahoon, E. B., & Cerutti, H. (2012). Metabolic and Gene Expression Changes Triggered by Nitrogen Deprivation in the Photoautotrophically Grown Microalgae Chlamydomonas Reinhardtii and Coccomyxa Sp. C-169. *Phytochemistry*, *75*, 50–59. DOI: 10.1016/j.phytochem.2011.12.007 PMID: 22226037

Naduthodi, M. I. S., Claassens, N. J., D'Adamo, S., van der Oost, J., & Barbosa, M. J. (2021). Synthetic Biology Approaches To Enhance Microalgal Productivity. *Trends in Biotechnology*, *39*(10), 1019–1036. DOI: 10.1016/j.tibtech.2020.12.010 PMID: 33541719

Nagappan, S., Devendran, S., Tsai, P. C., Jayaraman, H., Alagarsamy, V., Pugazhendhi, A., & Ponnusamy, V. K. (2020). Metabolomics Integrated with Transcriptomics and Proteomics: Evaluation of Systems Reaction to Nitrogen Deficiency Stress in Microalgae. *Process Biochemistry (Barking, London, England)*, *91*, 1–14. DOI: 10.1016/j.procbio.2019.11.027

Nagar, S., Antony, R., & Thamban, M. (2021). Extracellular Polymeric Substances in Antarctic Environments: A Review of Their Ecological Roles and Impact on Glacier Biogeochemical Cycles. *Polar Science*, *30*, 100686. DOI: 10.1016/j.polar.2021.100686

Narayanan, Ishvarya, Ricky Rajamanickam, Rishu Kumari Singh, Pardeep Singh, Ashish A. Prabhu, Ashutosh Pandey, Nirakar Pradhan, Y. V. Nancharaiah, Vikram Kumar, Rangabhashiyam Selvasembian, and Joyabrata Mal. 2025. "Microalgae-Based Remediation of Pharmaceutical Contaminants: Emerging Strategies and Technological Synergies." *Discover Applied Sciences 2025 7:11* 7(11):1288-. .DOI: 10.1007/s42452-025-07226-4

Niekerk, L. A., Gokul, A., Basson, G., Badiwe, M., Nkomo, M., Klein, A., & Keyster, M. (2024). Heavy Metal Stress and Mitogen Activated Kinase Transcription Factors in Plants: Exploring Heavy Metal-ROS Influences on Plant Signalling Pathways. *Plant, Cell & Environment*, *47*(8), 2793–2810. DOI: 10.1111/pce.14926 PMID: 38650576

Nobre, M. L. F., Tavares, D., Fraga, C., Oliveira, B., Dias, M., Mesquita, S., Oliveira, C. M., & Pires, J. C. M. (2024). Techno-Economic Analysis of a Circular Microalgal Approach for Enhanced Wastewater Treatment and Resource Recovery in Northern Portugal. *Journal of Cleaner Production*, *434*, 140389. Advance online publication. DOI: 10.1016/j.jclepro.2023.140389

Nowicka, B. (2022). Heavy Metal-Induced Stress in Eukaryotic Algae-Mechanisms of Heavy Metal Toxicity and Tolerance with Particular Emphasis on Oxidative Stress in Exposed Cells and the Role of Antioxidant Response. *Environmental Science and Pollution Research International*, *29*(12), 16860–16911. DOI: 10.1007/s11356-021-18419-w PMID: 35006558

Nzayisenga, Jean Claude, Xavier Farge, Sophia Leticia Groll, and Anita Sellstedt. 2020. "Effects of Light Intensity on Growth and Lipid Production in Microalgae Grown in Wastewater." *Biotechnology for Biofuels 2020 13:1* 13(1):4-. .DOI: 10.1186/s13068-019-1646-x

Osman, Mohamed E. H., Atef M. Abo-Shady, Saly F. Gheda, Samy M. Desoki, and Mostafa E. Elshobary. 2023. "Unlocking the Potential of Microalgae Cultivated on Wastewater Combined with Salinity Stress to Improve Biodiesel Production." *Environmental Science and Pollution Research 2023 30:53* 30(53):114610–24. .DOI: 10.1007/s11356-023-30370-6

Paniagua-Michel, J., Olmos-Soto, J., & Ruiz, M. A. (2012). Pathways of Carotenoid Biosynthesis in Bacteria and Microalgae. *Methods in Molecular Biology (Clifton, N.J.)*, *892*, 1–12. DOI: 10.1007/978-1-61779-879-5_1 PMID: 22623294

Papachristou, I., S. Akaberi, A. Silve, E. Navarro-López, R. Wüstner, K. Leber, N. Nazarova, G. Müller, and W. Frey. 2021. "Analysis of the Lipid Extraction Performance in a Cascade Process for Scenedesmus Almeriensis Biorefinery." *Biotechnology for Biofuels 2021 14:1* 14(1):20-. .DOI: 10.1186/s13068-020-01870-1

Posadas, E., del Mar Morales, M., Gomez, C., Acién, F. G., & Muñoz, R. (2015). Influence of PH and CO2 Source on the Performance of Microalgae-Based Secondary Domestic Wastewater Treatment in Outdoors Pilot Raceways. *Chemical Engineering Journal*, *265*, 239–248. DOI: 10.1016/j.cej.2014.12.059

Prasad, Ravindra, Sanjay Kumar Gupta, Nisha Shabnam, Carlos Yure B. Oliveira, Arvind Kumar Nema, Faiz Ahmad Ansari, and Faizal Bux. 2021. "Role of Microalgae in Global CO2 Sequestration: Physiological Mechanism, Recent Development, Challenges, and Future Prospective." *Sustainability 2021, Vol. 13, Page 13061* 13(23):13061. .DOI: 10.3390/su132313061

Qadir, M., Drechsel, P., Cisneros, B. J., Kim, Y., Pramanik, A., Mehta, P., & Olaniyan, O. (2020). Global and Regional Potential of Wastewater as a Water, Nutrient and Energy Source. *Natural Resources Forum*, *44*(1), 40–51. DOI: 10.1111/1477-8947.12187

Ranganathan, P., & Savithri, S. (2019). Techno-Economic Analysis of Microalgae-Based Liquid Fuels Production from Wastewater via Hydrothermal Liquefaction and Hydroprocessing. *Bioresource Technology*, *284*, 256–265. DOI: 10.1016/j.biortech.2019.03.087 PMID: 30947140

Ratledge, C., & Wynn, J. P. (2002). The Biochemistry and Molecular Biology of Lipid Accumulation in Oleaginous Microorganisms. *Advances in Applied Microbiology*, *51*, 1–52. DOI: 10.1016/S0065-2164(02)51000-5 PMID: 12236054

Rockström, J., Steffen, W., Noone, K., Persson, Å., Chapin III, F. S., Lambin, E., ... & Foley, J. (2009). Planetary boundaries: exploring the safe operating space for humanity. Ecology and society, 14(2).

Sciuto, G., Russo, N., Randazzo, C. L., & Caggia, C. (2025). Valorization of Second Cheese Whey Through Microalgae-Based Treatments: Advantages, Limits, and Opportunities. BioTech, 14(4), 79.

Shah, M. M. R., Liang, Y., Cheng, J. J., & Daroch, M. (2016). Astaxanthin-producing green microalga Haematococcus pluvialis: from single cell to high value commercial products. Frontiers in plant science, 7, 531

Shetty, P., Gitau, M. M., & Maróti, G. (2019). Salinity stress responses and adaptation mechanisms in eukaryotic green microalgae. Cells, 8(12), 1657

Singh, S. K., Sundaram, S., Sinha, S., Rahman, M. A., & Kapur, S. (2016). Recent Advances in CO2 Uptake and Fixation Mechanism of Cyanobacteria and Microalgae. *Critical Reviews in Environmental Science and Technology, 46*(16), 1297–1323. DOI: 10.1080/10643389.2016.1217911

Srimongkol, P., Sangtanoo, P., Songserm, P., Watsuntorn, W., & Karnchanatat, A. (2022). Microalgae-Based Wastewater Treatment for Developing Economic and Environmental Sustainability: Current Status and Future Prospects. *Frontiers in Bioengineering and Biotechnology, 10*, 904046. DOI: 10.3389/fbioe.2022.904046 PMID: 36159694

Sun, T. H., Cheng, Q. L., Yuan, Y. H., Wen, K. W., Zhi, G. Z., & Lu, S. (2010). Coordinated Regulation of Gene Expression for Carotenoid Metabolism in Chlamydomonas Reinhardtii. *Journal of Integrative Plant Biology, 52*(10), 868–878. DOI: 10.1111/j.1744-7909.2010.00993.x PMID: 20883439

Suparmaniam, U., Lam, M. K., Lim, J. W., Yusup, S., Tan, I. S., Lau, S. Y., Kodgire, P., & Kachhwaha, S. S. (2023). Influence of Environmental Stress on Microalgae Growth and Lipid Profile: A Systematic Review. *Phytochemistry Reviews, 22*(4), 879–901. DOI: 10.1007/s11101-022-09810-7

Tan, Kenneth Wei Min, Huixin Lin, Hui Shen, and Yuan Kun Lee. 2016. "Nitrogen-Induced Metabolic Changes and Molecular Determinants of Carbon Allocation in Dunaliella Tertiolecta." *Scientific Reports 2016 6:1* 6(1):1–13. .DOI: 10.1038/srep37235

Tang, D., Li, X., Zhang, L., Xiao, P., Nie, Y., Qiu, F., Cheng, Z., Li, W., & Zhao, Y. (2025). Reactive Oxygen Species-Mediated Signal Transduction and Utilization Strategies in Microalgae. *Bioresource Technology, 418*, 132004. DOI: 10.1016/j.biortech.2024.132004 PMID: 39710205

Vaishnav, S., Saini, T., Chauhan, A., Gaur, G. K., Tiwari, R., Dutt, T., & Tarafdar, A. (2023). Livestock and Poultry Farm Wastewater Treatment and Its Valorization for Generating Value-Added Products: Recent Updates and Way Forward. *Bioresource Technology, 382*, 129170. DOI: 10.1016/j.biortech.2023.129170 PMID: 37196748

Vázquez-Romero, B., Villar-Navarro, E., Perales, J. A., Garrido-Pérez, C., & Ruiz, J. (2024). Techno-Economic Analysis of Using Microalgae to Treat Streams from Fish RAS Farming and Replace Fish Meal: A Case Study. *Journal of Water Process Engineering*, *59*, 104904. Advance online publication. DOI: 10.1016/j.jwpe.2024.104904

Vidhyavathi, R., Venkatachalam, L., Sarada, R., & Ravishankar, G. A. (2008). Regulation of Carotenoid Biosynthetic Genes Expression and Carotenoid Accumulation in the Green Alga Haematococcus Pluvialis under Nutrient Stress Conditions. *Journal of Experimental Botany*, *59*(6), 1409–1418. DOI: 10.1093/jxb/ern048 PMID: 18343887

Wang, C., Shi, H. D., You, N., Bai, Y., Jin, P., & Han, J. (2023). Pathways of Wastewater Treatment for Resource Recovery and Energy Minimization towards Carbon Neutrality and Circular Economy: Technological Opinions. *Frontiers in Environmental Chemistry*, *4*, 4. DOI: 10.3389/fenvc.2023.1255092

Wang, L., Yang, T., Pan, Y., Shi, L., Jin, Y., & Huang, X. (2023). The Metabolism of Reactive Oxygen Species and Their Effects on Lipid Biosynthesis of Microalgae. *International Journal of Molecular Sciences*, *24*(13), 11041. DOI: 10.3390/ijms241311041 PMID: 37446218

Wang, S., Mukhambet, Y., Esakkimuthu, S., & Abomohra, A. E. L.-F. (2022). Integrated Microalgal Biorefinery – Routes, Energy, Economic and Environmental Perspectives. *Journal of Cleaner Production*, *348*, 131245. DOI: 10.1016/j.jclepro.2022.131245

Wei, H., Shi, Y., Ma, X., Pan, Y., Hu, H., Li, Y., Luo, M., Gerken, H., & Liu, J. (2017). A Type-I Diacylglycerol Acyltransferase Modulates Triacylglycerol Biosynthesis and Fatty Acid Composition in the Oleaginous Microalga, Nannochloropsis Oceanica. *Biotechnology for Biofuels*, *10*(1), 174. Advance online publication. DOI: 10.1186/s13068-017-0858-1 PMID: 28694845

Wetterwald, L., Leybros, A., Fleury, G., Delrue, F., Dimitriades-Lemaire, A., Chambonniere, P., & Hertz, A. (2023). Supercritical CO2 Extraction of Neutral Lipids from Dry and Wet Chlorella Vulgaris NIES 227 Microalgae for Biodiesel Production. *Journal of Environmental Chemical Engineering*, *11*(5), 110628. DOI: 10.1016/j.jece.2023.110628

Wykoff, D. D., Davies, J. P., Melis, A., & Grossman, A. R. (1998). The Regulation of Photosynthetic Electron Transport during Nutrient Deprivation in Chlamydomonas Reinhardtii. *Plant Physiology*, *117*(1), 129–139. DOI: 10.1104/pp.117.1.129 PMID: 9576782

Xiao, R., & Zheng, Y. (2016). Overview of Microalgal Extracellular Polymeric Substances (EPS) and Their Applications. *Biotechnology Advances, 34*(7), 1225–1244. DOI: 10.1016/j.biotechadv.2016.08.004 PMID: 27576096

Xiao, X., Li, W., Jin, M., Zhang, L., Qin, L., & Geng, W. (2023). Responses and Tolerance Mechanisms of Microalgae to Heavy Metal Stress: A Review. *Marine Environmental Research, 183*, 105805. DOI: 10.1016/j.marenvres.2022.105805 PMID: 36375224

Xin, C., Addy, M. M., Zhao, J., Cheng, Y., Cheng, S., Mu, D., Liu, Y., Ding, R., Chen, P., & Ruan, R. (2016). Comprehensive Techno-Economic Analysis of Wastewater-Based Algal Biofuel Production: A Case Study. *Bioresource Technology, 211*, 584–593. DOI: 10.1016/j.biortech.2016.03.102 PMID: 27039331

Xiong, J. Q., Kurade, M. B., & Jeon, B. H. (2017). Ecotoxicological Effects of Enrofloxacin and Its Removal by Monoculture of Microalgal Species and Their Consortium. *Environmental Pollution, 226*, 486–493. DOI: 10.1016/j.envpol.2017.04.044 PMID: 28449968

Xu, L., Wim Brilman, D. W. F., Withag, J. A. M., Brem, G., & Kersten, S. (2011). Assessment of a Dry and a Wet Route for the Production of Biofuels from Microalgae: Energy Balance Analysis. *Bioresource Technology, 102*(8), 5113–5122. DOI: 10.1016/j.biortech.2011.01.066 PMID: 21334889

Yang, C., Li, R., Zhang, B., Qiu, Q., Wang, B., Yang, H., Ding, Y., & Wang, C. (2019). Pyrolysis of Microalgae: A Critical Review. *Fuel Processing Technology, 186*, 53–72. DOI: 10.1016/j.fuproc.2018.12.012

Yang, Y. F., Li, D. W., Balamurugan, S., Wang, X., Wei, D. Y., & Hong, Y. L. (2022). Chrysolaminarin Biosynthesis in the Diatom Is Enhanced by Overexpression of 1,6-β-Transglycosylase. *Algal Research, 66*, 102817. DOI: 10.1016/j.algal.2022.102817

Zachee, G., Kayiranga, A., Nizeyimana, J. C., Tian, S., Rugema, J., You, L., Huang, X., & Jian, Q. S. (2025). Removal of Antibiotics and Antibiotic Resistance Genes Using Microalgae-Based Wastewater Treatment System: A Bibliometric Review and Mechanism Analysis. *Journal of Water Process Engineering, 72*, 107496. DOI: 10.1016/j.jwpe.2025.107496

Zhao, B., & Su, Y. (2014). Process Effect of Microalgal-Carbon Dioxide Fixation and Biomass Production: A Review. *Renewable & Sustainable Energy Reviews, 31*, 121–132. DOI: 10.1016/j.rser.2013.11.054

Chapter 8
Anaerobic Reactor Technologies for Algal Biomass and Sludge Valorization

S. Dayana
Sacred Heart College, India

S. Harini
http://orcid.org/0009-0008-0300-5081
Sacred Heart College, India

P. Mythili
Sacred Heart College, India

K. Kesavardhini
http://orcid.org/0009-0005-8235-042X
Sacred Heart College, India

M. Manigandan
http://orcid.org/0000-0002-8115-2713
Sacred Heart College, India

ABSTRACT

Anaerobic reactors and algal sludge conversion have become potential technologies of sustainable bioenergy generation and waste management. Anaerobic digestion can be defined as a biological process where the microorganisms decompose the organic matter to produce biogas with high levels of methane and carbon- dioxide. The algal sludge is formed as a result of wastewater treatment, as well as the

DOI: 10.4018/979-8-3373-7630-1.ch008

harvesting of algal biomass, which is a useful substrate because it contains a large amount of organic and nutrient matters. The process is improved and stabilized by the next generation reactor designs, such as and anaerobic sludge blanket and anaerobic membrane bioreactors. Performance is among the factors that are greatly affected by operational factors such as temperature, retention time and substrate composition. The hybrid solution ensures a decrease of environmental pollution, greenhouse gas emissions, and circular bioeconomy objectives.

1. INTRODUCTION

Bioenergy is an important part of the process of switching to the use of petroleum-derived fuels with more environmentally friendly ones. In the long run, it can help reduce CO_2 emissions, especially if environmental and economic factors are balanced. As the world's population grows and more people seek higher living standards, energy consumption is increasing. This tendency and increased awareness of the climate impact of CO_2 have led to a renewed interest in plant-based biofuels, which are fuels used in automobiles and are mostly generated using bioenergy. These include both gases and liquid ones such Fischer-Tropsch diesel, methanol, ethanol, and biofuels like biogas and biohydrogen. The generation of algal biogas is better than that of hydrogen, green fuel, biodiesel, and bioethanol (Kumar et al., 2008; Parakh et al., 2023).

Anaerobic digestion can also be used in conjunction with other conversion methods. The Biogas is produced when organic material decomposes without oxygen even found in practically any substance, including sewage, livestock and waste from industries, plant life, bacteria, and algal. The first methane digester was built in Bombay, India, in 1859. This process involves microorganisms that break down organic substances in the absence of oxygen. Anaerobic digestion is gaining importance in the field of renewable energy on a global scale. Biogas is produced when bacteria break down organic molecules without oxygen. It is mainly made up of carbon dioxide and methane with a few other gases. Important byproducts like liquid or solid digestate are also produced by this process (Li et al., 2011).

Bacteria break down organic stuff in anaerobic digestion when oxygen is not present. In both industrial and household settings, it is used to generate food, drinks, and fuels as well as manage waste. It is a reliable and practical technology that helps with waste management. The market depends on expanding the use of microalgae biomass and developing larger technologies for commercial use (Saranraj et al., 2025). Anaerobic digestion, which produces Methane restores the amount of bioenergy and biofuel derived from micro algae. Given that algae generate energy while picking up micronutrients through waste, the method's effluent can be used to grow them again (Liang et al., 2011).

2. GROWTH CHARACTERISTICS AND IMPORTANCE OF ALGAE

Algae are the principal producers in aquatic environments. Green, red, and brown algae are among the many types of algae that belong to the categories *Chlorophyta*, *Rhodophyta* and *Phaeophyta*. Algae can grow in a variety of settings, including fresh water, oceans, deserts, hot springs, open and closed ponds, photobioreactors, sewage, and CO_2-producing businesses. They can be found in moistened places, on land, and in water. Algae include *Cyanobacteria*, Eukaryotic Microalgae, and Macroalgae. This diverse group of mostly aquatic photosynthetic organisms is essential to the food chain. Algae are helpful for water purification and nutrient removal (Oswald, 2003).

In 1957, Oswald and Gotaas first suggested using algae for bioremediation. This process creates methane, an essential energy source. The aquatic food chain, bio-geochemical processes, and the carbon cycle are all supported by algae. Most algae require sunshine to generate their own food, much like plants do. Algae can be used in a variety of ways to provide biomass for energy and energy carriers. Anaerobic fermentation and digestion, gasification, direct burning, pyrolysis, and photosynthesis for hydrogen are examples of biological processes. Fatty acids are esterified in order to create biodiesel. Algae can fix carbon dioxide, according to numerous studies. According to Driver et al. (2014) algae are a good source of gaseous and liquid fuels that do not compete with food. This is because of advancements in genomic technology that allow for genetic and metabolic alteration, as well as better techniques for finding suitable strains (Saranraj et al., 2026). Figure 1 show that the Algae can be transformed to give several valuable products such as biofuels such as biodiesel and bioethanol. They also supply food, pharmaceuticals and cosmetics with proteins, carbohydrates as well as bioactive compounds. Besides, algae may be utilized in food to animals, fertilizers, and bioplastics.

Figure 1. Potential products from algae

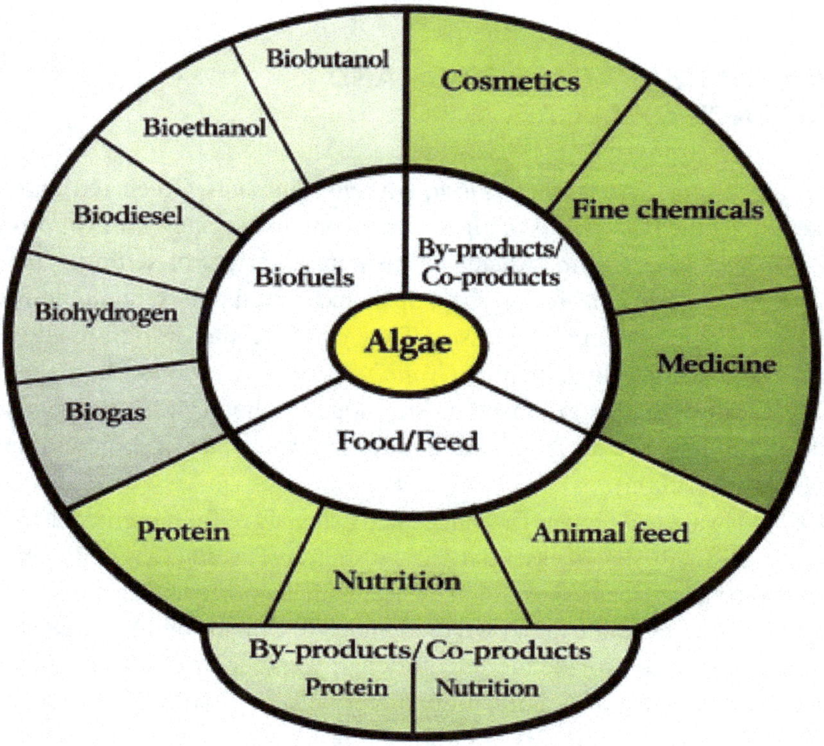

3. ANAEROBIC DECOMPOSITION FEEDSTOCK IS MICROALGAE

The algae is an massive and diverse species of living organisms, including small single-celled organisms to massive seaweeds. Seaweeds and macroalgae are incredibly different in their chemical composition, although they have similar structures to plants found on the land. Starch and fructosan are the primary carbon compounds held by plants on the earth that can be easily converted into biogas. Terrestrial plants on the other hand have cell walls that consist of cellulose and hemicellulose fibres which are interlaced with lignin, and coated with a coat of pectin. Nature has problems digesting this mixture, which is called lignocellulose (Subhalakshmi et al., 2026). It implies that the materials have to undergo chemical treatments such as acid hydrolysis or Thermal treatments such as steam bursting or warm water, ammonium fiber development or alkaline moist oxidation and then can be transformed into useful products. Macroalgae on the other hand consists of polysaccharide. One of the characteristics that make them different to terrestrial plants is the absence of lignin in their cell walls. Xylan and Ulvan are the major components of the cell walls of green algae (Thajuddin and Subramanian, 2005).

Brown algae include Fucoidan and Alginate, whereas the red algae contain carrageen, agar particles, and xylose. The quantities of cellulose in green algae are only seldom comparable to those in terrestrial plants, despite the fact that it forms a structural component of the walls of many algae species. Mannitol and laminarin are the main storage polysaccharides present in brown macroalgae, while chlorophycean is present in green macroalgae and floridean starch is present in red algae. Microalgae and cyanobacteria differ in composition from macroalgae. Carbohydrates frequently make up only a small percentage of the dry weight of microalgae, with proteins and lipids making up the majority (Balan et al., 2009).

3.1 *Cyanophyta* (Blue-Green Algae)

The *Cyanophyta* are a unique class of prokaryotic microorganisms that fall under the larger category of photosynthetic organisms. Some filamentous blue-green algae can produce heterocysts, which help fix atmospheric nitrogen and contain the nitrogenase enzyme. Unlike green and purple bacteria, cyanobacteria generate oxygen through photosynthesis. Among the components of *cyanobacteria* that capture light are phycobiliproteins and chlorophyll a. However, they lack the membrane-bound organelles found in eukaryotes, including the mitochondria, chloroplasts, and nucleus.

Cyanobacteria inhabit a wide variety of environments, including freshwater, brackish, marine, and terrestrial ones. They could be unicellular, filamentous, or colonial. Because of their high protein and vitamin content, these organisms are used in a wide range of goods, such as feed and food additives. Algae are high in fibre and easily digested. Pharmaceuticals, wastewater remediation, biofertilizers, and enzyme production are further uses. One of the well-known traits of cyanobacteria is their tremendous production (Thajuddin and Subramanian, 2005).

3.2 *Rhodophyta* (Red Algae)

The *Rhodophyta* is a well-defined group of algae that includes more than 6,000 species. They differ from other algae in that they have additional phycobilin pigments and do not have centrioles or flagella. Red algae are the predominant type of seaweed, multicellular, and marine-dwelling. These species are found in coastal and offshore zones as low as 40 to 60 meters in both tropical and temperate climates. Because of their distinctive hues, they can grow up to 200–250 meters in deeper waters. Certain algae with cemented cell walls are necessary for the development of coral reefs. Additionally, freshwater, brackish water, and soil all contain red algae. Some species, including Porphyra, are essential food sources in Asia. Some of the main sources of both agar and agarose are Rhodophyta organisms like *Gelidium* and *Gracilaria* which are used in labs to cultivate cultures and separate DNA. Carrageenan, another material made from red algae, is used in the food industry as a thickening and stabilising agent (Rasmussen and Morrissey, 2007).

3.3 *Chlorophyta* (Green Algae)

The primary pigments of *Chlorophyta*, sometimes referred to as "Green Algae" are chlorophyll a and b. Green algae get their yellow or reddish-green colour from secondary pigments called carotenoids, which include Beta-Carotene, Prasinoxanthin, Siphonaxanthin, and Astaxanthin. Green algae are primarily found in freshwater, though they can also be found in soil and brackish or saltwater. Green algae can produce filamentous, coccoid, parenchymatous, or siphonous structures and can be unicellular, colonial, motile, or non-motile. *Dunaliella* is a halophilic microalga that is cultivated to provide dietary supplements such as beta-carotene. Numerous *Ulvophyceae* species, or marine macroalgae, are utilised in coastal regions for both food and nitrogen extraction from effluent (Hsiao and Blanch, 2006).

3.4 *Heterokonts* (Brown, Yellow-Green Algae, Golden Algae, Diatoms)

Heterokonts, a large and diverse collection of algae with comparable ultrastructure, structure, pigments, and genetic makeup, were recently identified. Two examples of *Heterokont* groups that are related to formerly separate phyla are *Chrysophyta* and *Phaeophyta*. These algae can be found in terrestrial, marine, brackish, and freshwater habitats. Fucoxanthin or vaucheriaxanthin are the main types of carotenoids, whereas chlorophyll a and c. both constitute the primary pigments used in photosynthesis.

The farming industry makes use of some heterokont microalgae. In addition to species like *Isochrysisgalbana* and *Chaetocerosmuelleri*, *molluscs* are fed diatoms such as *Skeletonemacostatum, Nitzschia*, and *Navicula*. Nannochloropsis is fed to rotifers. The use of diatoms in the synthesis of biomolecules, nanomaterials, medications, health goods, and bioremediators has lately been studied. Numerous diatoms and Pinguiophyceae species exhibit significant biotech potential because of their high eicosapentenoic acid concentration. Phaeophyceae, or brown algae, are among the most prevalent types of algae. Some brown algae may function as the anaerobic breakdown material (Vandermeulen and Gordin, 1990).

4. CULTIVATION OF ALGAL BIOMASS

Algal biodiesel production is influenced by the lipid productivity and growth rate of the algae. It has been found that over 40 strains of microalgae can store lipids in amounts ranging from 2 to 75% of their mass (Mata et al., 2010). Certain species of *Scenedesmus* and *Chlorella* contain lipid levels that can surpass 50% of their entire cell mass, according to some research. However, elevated lipid content is usually the result of nutritional stress, such as nitrate deficiency. Triacylglycerols (TAGs), lipids that store carbon and energy, are produced by algae at these times. However, severe growing conditions might cause algae's cell division to slow down, which lowers their overall lipid output. Numerous cultivation methods have been used to produce algae feedstocks for biofuel. They involve various kinds of photo bioreactors, raceway ponds, and treatment of wastewater ponds (Lardon et al., 2009).

4.1 Wastewater Treatment Ponds

Aerobic wastewater treatment ponds help to remove harmful germs and reduce Biochemical Oxygen Demand (BOD). Because algae absorb nutrients and produce oxygen through photosynthesis, which bacteria may immediately use to break down waste, algae are essential to these systems. However, mechanical, Energy-intensive agitating systems are necessary for wastewater management techniques like the process of activated sludge. Shilton et al., (2008) state that up to 85% of the electrical energy used by these systems may be used for aeration. Wastewater ponds therefore offer a less expensive and effective treatment alternative with little energy and mechanical needs. Therefore, wastewater ponds are a suitable choice for societies with restricted material and financial assets (Kivaisi, 2001).

4.2 Raceway Ponds

A raceway pond is one type of apparatus used to grow algae. It is a closed loop with water continuously flowing around it in a channel and is often composed of one or more ponds connected in a line. These systems are usually built from concrete or compacted earth. A paddlewheel is used to keep the water moving and the algae mixing. Controlling the types of algae in these systems is difficult since other species can easily enter them. This can be reduced in two ways, by either cultivating algae which organically replaces the pond or by creating environments that will foster a particular kind of algae. The second problem of using raceway ponds is the loss of water over the pond through evaporation bearing in mind that the surface area of the pond is large. This could be a very grave issue in case the water is specifically utilized to generate algae with beneficial properties, such as rich in lipids. In wastewater systems utilized in such a way, however, these problems are not so much of an issue. The reason behind this consists of the fact that the growing medium contains high nitrogen levels and this restricts the chances of algae flourishing (Sosnowski et al., 2003).

Usually dominant in such environments are *Scenedesmus* and *Chlorella*. Evaporative losses are not critical in these systems because the medium is a waste product and has less applications. The benefits of raceway ponds include the fact that they are cheap to construct and run, compared to other systems, and they are not difficult to upgrade. Compared to previous wastewater-based systems, they are more efficient and have fewer mechanical and electrical components. These features make alternatives to open pond designs, especially considering the low cost of producing fuel, according to a USDOE Aquatic Species Program report (Shen et al., 2009).

4.3 Photobioreactors

Airtight containers such as tubes, bags, or transparent materials are used in photobioreactors, also referred to as PBR systems. Because these systems are confined, concerns regarding pollution and water loss are lower than with open ponds. Algae in PBR systems can grow denser because they receive more sunlight. Similar to raceway ponds, PBRs require nutrient addition unless wastewater is used in the system. Tubular PBRs are widely used because they can be built in a variety of ways and have a large surface area relative to their size. These include Helical Tubes, Fence-Like Designs, and Horizontal configurations. Each of these systems needs a certain quantity of water flow to prevent algae from forming in the tubes. However, Chisti (2007) pointed out that concentrated algae cultures create enormous amounts of oxygen during photosynthesis, they can exceed the oxygen saturation limitations of water. This high oxygen content could damage and finally kill the algae. To prevent this, degassing vessels are recommended to remove excess oxygen. Heat exchangers or external cooling are also needed to keep the system's temperature stable (Muga et al., 2008).

5. ALGAE BIOGAS PRODUCTION PROCESS AND TECHNOLOGY

Anaerobic digestion is one technique for decomposing various organic materials and creating biogas. Methane can be produced from both macro and microalgae through anaerobic decomposition. Recently, microalgae have gained interest as a possible source of process for creating biogas is anaerobic fermentation. Algal digestion is an environmentally friendly way to generate renewable energy for use in homes and businesses. Algal anaerobic digestion is an essential step that provides benefits and efficiency in the process of turning algae into biofuels. Anaerobic digestion is possible for both macro- and micro-algae. The production of biogas from algal is an alternative technology that produces greater electricity than renewable Diesel, Biodiesel, Bioethanol, and Hydrogen is gas from biomass produced by algae generation. Additionally, anaerobic digestion can be combined with other methods to improve sustainability and energy balance. Unlike biohydrogen, bioethanol, or biodiesel, which exclusively use specific components of the algae (such lipids and carbohydrates), anaerobic digestion generates biogas by turning all of the algae's macromolecules into methane (Lastella et al., 2002).

In the biological process known as anaerobic digestion, methanogens decompose organic materials, when oxygen is not present. In a variety of settings, such as Particles, Flooding Soils, Intestines from Livestock, and Landfills, numerous varieties of bacteria are capable of carrying out this task. It is a complex process that calls for specific environmental conditions as well as a range of bacterial communities. These bacteria break down organic materials to produce biogas, which is a mixture of gases mainly made up of carbon dioxide and methane. Methane fermentation occurs in four stages: Hydrolysis, Acidogenesis, Acetogenesis/Dehydrogenation, and Methanation. These four stages take place when organic matter breaks down and releases methane (Milledge et al., 2019).

Hydrolysis breaks down complex compounds like Proteins, Lipids, and Carbohydrates into simpler forms like sugars, fatty acids, and amino acids. Carbon dioxide, hydrogen gas, acetic acid, and volatile fatty acids are the products of the fermentation of these basic materials. Hydrogen gas, carbon dioxide, those fundamental elements ferment to produce volatile lipids and acetic acid. The reaction produces acetic acid as well as carbon dioxide, hydrogen gas, and volatile lipids. The volatile fatty acids are additionally transformed by acetogenesis into hydrogen gas, carbon dioxide, and acetic acid. Methanogens are special bacteria that can convert carbon dioxide and hydrogen gas from earlier stages or acetic acid into methane (Kao et al., 2012). Figure 2 illustrate the Anaerobic digestion is a biological process where microorganisms break down organic waste in the absence of oxygen. This process produces biogas, mainly methane and carbon dioxide, which can be used as renewable energy. The leftover material, called digestate, can be used as nutrient-rich fertilizer.

Figure 2. Anaerobic digestion for Biogas Production

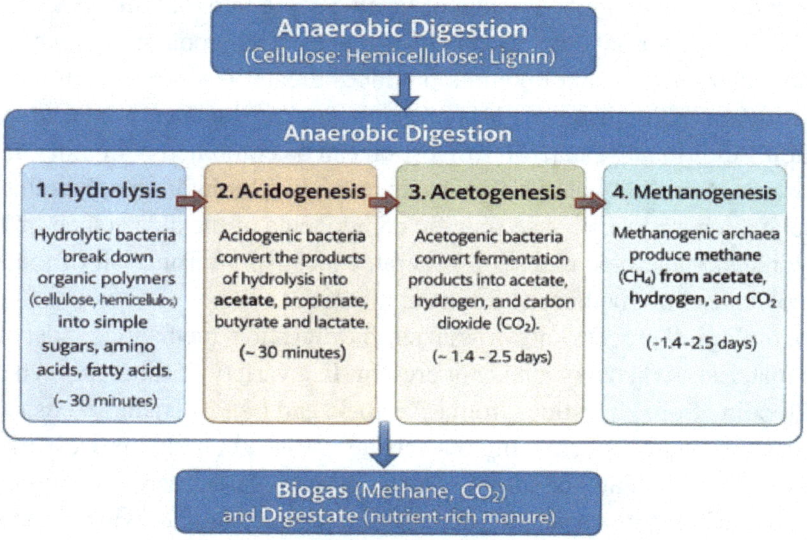

5.1 Anaerobic digestion of Macroalgae biomass

Macro algae are one type of aquatic biomass that may prove to be a practical renewable energy source. Compared to land-based plants, which have a photosynthetic efficiency of 1.8–2.2%, some reports suggests that macroalgae have an efficiency of about 6–8%. Some of these algae grow rapidly, reaching lengths of up to 60 meters. More primary material is produced each year by big marine macroalgae than by most land plants. The four groups of Red Algae or Rhodophyta, Brown Algae, or Phaeophyta, Green Algae, and Blue-Green Algae, or Cyanophyta, are macroalgae that occur naturally. Seaweed and kelp are examples of freshwater and marine macroalgae that are used to convert solar energy and create ethanol.

Macroalgae are becoming increasingly popular as a feedstock for biofuel due to their ability to grow in freshwater, along beaches, and along contaminated coastlines. They can be converted into biogas *via* Anaerobic Digestion (AD), which can produce about 60% methane. However, since then, AD technology has developed considerably and gained popularity. Macroalgae break down more quickly than terrestrial plants because they have very little cellulose and no lignin. AD has been used to handle this material and produce biogas. This approach could reduce Eutrophication and generate Renewable Energy. Because macroalgae are abundant, especially in coastal or polluted areas, their conversion may be highly beneficial for countries with long coastlines or eutrophic conditions (Abatzoglou et al., 2009).

5.2 Anaerobic digestion of microalgae biomass

The anaerobic digestion of biomass in a variety of algae species has been studied. Several cultures of microscopic algae have been successfully used to create methane. These include mixed cultures of *Scenedesmus* and *Chlorella* species, as well as combinations of *Scenedesmus, Chlorella, Euglena, Oscillatoria*, and *Synechocystis*. In certain studies, *Scenedesmus* was used either by itself or in conjunction with other species including *Spirulina, Euglena, Micractinium, Melosira*, or *Oscillatoria*. Anaerobic digestion has several benefits over conventional bioenergy methods for creating biogas. As AD breaks down organic carbon, biogas and organic acids are created. Biogas is mostly composed of Carbon Dioxide (about 35%) and Methane (about 65%). Other gases like Hydrogen, Ammonia, Nitrogen, and Hydrogen Sulphide are also present, albeit in lower amounts.

The amount of biogas that plants can create is usually limited by the presence of lignocellulose, which is difficult to digest. The efficiency of biogas production depends on the type of plant, the rate at which cells break down, and the presence or absence of particular chemicals. However, microalgae with low lignocellulose concentration, such as *Chlorella vulgaris*, *Phaeodactylum tricornutum*, and *Spirulina platensis*, allow for nearly complete utilisation of organic materials. Golueke et al., (1957) have shown that microalgae may pass through an anaerobic digester without breaking down. Therefore, techniques had been created in order to improve the absorption of biomass (Hughes et al., 2012).

6. ANAEROBIC CO-DIGESTION OF MICROALGAL BIOMASS

The breakdown of two or more feedstock types is known as anaerobic co-digestion. While reducing some of the negative effects of burning a single fuel, this method can boost the production of bioenergy. It's also financially viable. Microalgae can be mixed with other agricultural waste to boost biogas generation. For instance, co-digesting microalgae and wheat straw in 1:1 ratio resulted in 77% more methane than digesting microalgae alone. Compared to processing each of these materials independently, combining food waste leachate, algal biomass, and uncooked sludge can enhance digestion and raise methane emissions. This is because the combination allows methanogens to diffuse more quickly and reach a wider range of substrates (Siddique and Wahid, 2018; Peng & Colosi, 2016).

In a *Tetraselmis suecica* investigation, mesophilic bacteria produced twice as much methane as thermophilic bacteria. It was examined to what extent cellulose co-digests *Chlorella vulgaris* and *Dunaliella tertiolecta* using TC60. Co-digestion of *D. tertiolecta* with cellulose generated more hydrogen than the *C. vulgaris* culture alone, as *C. vulgaris* did not degrade efficiently in the presence of TC60. During a 20-day hydraulic retention period (HRT), microalgae and primary sludge were co-digested in continuous reactors at 1:3 ratio. Methane output was 65% higher because primary sludge is less ammonia hazardous and easier to break down than primary sludge alone (Kim and Kang, 2015).

However, microalgae digestion alone proved more effective in removing a number of contaminants, including galactoside, co-digestion using the main waste material, followed by triphenyl phosphate, methyl dihydrojasmonate, caffeine, tonalide, and triclosan. The inclusion of initial sewage significantly improved both the anaerobic breakdown of unprocessed phytoplankton material with the absorption of algal with robust cell walls was further enhanced by increasing HRT. Regardless of the nutritional value or biodegradability of the substrate, the energy output is not up to par. To increase the biomass's biodegradability and yield greater final product, the supporting materials must be pretreated prior to the AD method (Carver et al., 2011).

7. CURRENT ALGAE BIOREACTORS LIMITS FOR PRODUCING RENEWABLE ENERGY

Large biomass production, the high lipid content useful for biofuel, and the fact that they don't compete with food crops for land are only a few benefits of microalgae. Micro algae-based biofuel remains too expensive, though Compared with plant oils, the oil produced by micro algae is about between three and four times greater expensive. The largest barriers to making ethanol from microalgae are the procedures needed to grow and harvest the algae. The cultivation stage alone accounts for about 40% of the costs and energy needed to manufacture micro algae biofuel. Numerous factors affect manufacturing, such as the abundance of minerals, land and water, gas exchange and transfer, light delivery, and culture integrity. The harvesting phase accounts for 20 to 30 percent of the entire cost of producing microalgal biofuel (Molina Grima et al., 2003; Rai et al., 2025).

7.1 Types of Bioreactors

The majority of algae bioreactors today use suspended cultures, which come in both closed and open systems. These suspended cultures need a lot of energy to collect the algae and extract water throughout the processing stage, whether open or closed systems are used. Twenty to thirty percent of the entire cost of production can go towards harvesting. It is thought that the open pond method is the most straightforward and cost-effective way to raise algae. Open pond systems come in a variety of forms, including raceway ponds, slope systems, and circular ponds. The fact that open systems are less expensive to build and maintain than closed systems is one of its advantages (Ugwu et al., 2008).

Nevertheless, these systems have a number of disadvantages that may outweigh their advantages. Systems with open ponds have less however, there are a number of factors that limit biomass output. Some of problems include temperature swings, poor light penetration, restricted CO_2 transport, and contamination by other organisms, including Protozoa. Another problem is evaporation, which may alter the water's ionic strength. The drawbacks of open systems, on the other hand, have led to the development of closed culture systems, also known as customised photobioreactors. In addition to allowing for controlled CO_2, light, and temperature, these closed systems have fewer contamination problems (Hase et al., 2000).

7.2 Limitation of Nutrient Supply

The three main nutrients that microalgae need are Phosphate, Oxygen, Nitrogen, and Carbon. Although it is a component of the cell's architecture and provides energy that powers numerous biological processes, including the creation of metabolites and development, carbon is significant. Carbon comes in two varieties: inorganic (like CO_2) and organic (like other substrates). Nitrogen normally makes up 1–10% of the dry mass of microalgae, while carbon typically makes up 40–50%. A lack of nitrogen and phosphate can limit the growth of microalgae. Nitrogen is necessary for the creation of Proteins, Amino Acids, and Chlorophyll. Low quantities of nitrogen can slow down cell development, reduce the quantity of chlorophyll, and promote the synthesis of lipids (Mulbry et al., 2008).

Phosphate is necessary for Photosynthesis, Energy Transfer, and the creation of DNA and RNA. Micro nutrients like Sulphur, Calcium, Salt, Potassium, Magnesium, Iron, Manganese, Nickel, Zinc, and Cobalt are also necessary for phytoplankton in along with Nitrogen and Phosphorous. These nutrients are present in household and agricultural wastewater, which makes it a suitable substrate for the growth of microalgae. There are several ways to integrate wastewater treatment and microalgae biofuel production. Although strain-specific differences may exist, the ideal C: N:

P ratio for microalgae development is typically 106:16:1. Wastewater can be used as a source of fertilizer to boost biomass productivity after being treated to alter the nutrient levels (Niyogi, 1999).

7.3 Gas Transfer and PAR (Photosynthetically Active Radiation)

Efficient gas exchange, including O_2 removal and CO_2 supply, is essential. In aquatic systems, inorganic carbon exists mainly as carbon dioxide (CO_2), bicarbonate (HCO_3^-), and carbonate (CO_3^{2-}). Microalgae primarily utilize CO_2, can use bicarbonate to a lesser extent, and only rarely use carbonate. Continuous CO_2 injection is often required for biomass production and wastewater treatment due to the low solubility of CO_2 in water and the need to maintain an appropriate C:N:P ratio. High biomass productivity depends on efficient CO_2 transfer in various types of bioreactors, and algal biomass production also requires effective removal of dissolved O_2 (Blanken et al., 2014).

The creation of PBRs (photobioreactors) should be taken into consideration for O_2 removal since it can hinder photosynthesis. For example, CO_2 and O_2 can be efficiently transported and exchanged *via* a gas-liquid contactor such as the Rotating Biological Contactor (RBC). In algal culture, light delivery which takes into account light intensity, quality, and cycle is essential. Even though they can't consume all of the light, microalgae absorb the majority of it and expel any extra. Since light cannot penetrate far inside the bioreactor, it's probable that cells won't receive enough light to grow. The possibility that the thickness of the culture will keep the algae from receiving enough light for photosynthesis is known as light constraint. However, because of the photoinhibition effect, too much light may decrease cell growth and output. Reactor design, effective mixing, and quick light replenishment can all help to maximize light utilisation and reduce photoinhibition (Molina Grima et al., 2003).

7.4 Harvesting of Algal Cells

Separating the algae from the culture broth is one of the most difficult tasks in the production of algal biofuel. This is because algal cells, which range in size from 0.2 to 2 micrometres for *cyanobacteria* and 3 to 30 micrometres for eukaryotic algae, have a strong negative charge. These features make cell collection 20–30% of the overall cost of making algal fuel, which can be an inhibitor in the process of making other types of biofuels. Electrical, chemical, and mechanical techniques have all been used to study the collection and dried of algae. Centrifugation is the most popular method for collecting cells from algal plants which uses differences in density to separate cells. However, this method is expensive and could harm the cell structure due to the tremendous gravity forces involved. Filtration procedures may cause membrane clogging when cell density is high. Tangential flow filtration, on the other hand, is an efficient method of separating algae from water (Uduman et al., 2010).

This method however is costly to operate and also consumes a lot of energy due to the membrane fouling. Though the process of sedimentation involving gravity is cheap, the particles move gradually. Also, there are small populations of cells that are not well settled and are, therefore, hard to separate. Dissolved Air Flotation (DAF) is a commonly used technique in the treatment of wastewater used to remove floating particulates. Bubbles of air that the algae attach to air during injection can be captured using a skimming. Nevertheless, this approach cannot be suitable when harvesting on a large scale because it involves pre-treatment of some kind. Chemical flocculation of most strains can be done using organic or inorganic coagulants. Negative cell surface charge of microalgal cells enables them to flock with positively charged chemicals. Flocculants may be effective chemical flocculants like aluminium sulphate and ferric chloride. But, the method is prone to changes in pH and produces a high level of sludge. Algae may also be affected by substances such as iron salts and ammonium.

Natural coagulants, including cationic starch and chitosan, are also employed although they are large molecules, neutral-charged polymers in their case although too costly to be used widely. An electric field is another safe mechanism of separating negatively charged algae cells in the absence of the usage of pesticides though it consumes a lot of electricity and is not suitable in large-scale culture and harvesting (Danquah et al., 2009). The following table provides the key drawbacks of the existing algae bioreactors to generate renewable energy. It highlights factors such as high costs, low light penetration, inefficient CO_2 transfer, oxygen buildup, contamination, harvesting challenges, and nutrient demands. For each limitation, it also suggests causes and potential mitigation strategies to improve productivity and energy efficiency.

Table 1. Current Algae Bioreactors: Limits for Producing Renewable Energy

Limitation	Description	Impact on Energy Production	Main Causes	Possible Mitigation
High capital cost	Expensive reactor materials and setup	Increases overall production cost	Photobioreactor design, materials	Use low-cost materials, scale optimization
Operational cost	Energy needed for mixing, lighting, and pumping	Reduces net energy gain	Continuous operation demands	Energy-efficient equipment
Light limitation	Poor light penetration in dense cultures	Low photosynthetic efficiency	Self-shading, reactor depth	Improved reactor geometry
CO_2 transfer inefficiency	Low CO_2 solubility in water	Limits algal growth and yield	Gas–liquid transfer barriers	Enhanced aeration systems
Oxygen accumulation	O_2 buildup inhibits photosynthesis	Reduces biomass productivity	Insufficient degassing	Efficient O_2 removal systems
Contamination risk	Invasion by unwanted microbes	Culture crashes and yield loss	Open systems exposure	Closed reactors, monitoring
Harvesting difficulty	Small cell size and low density	High harvesting cost	Dilute cultures	Efficient flocculation/ filtration
Nutrient supply demand	Large nutrient requirements	Raises production cost	Need for N and P inputs	Wastewater integration

8. PRE-TREATMENT STRATEGIES BASED ON MICROALGAE CELL WALL COMPOSITION

8.1 Microalgae Cell Wall Composition and Degradability

The most important factor in anaerobic digestion is the characteristics of the algal cell wall, which is made up of intricate organic compounds and controls how substrate is turned into methane. About 11–37% of the dry material includes components of the micro algae wall of cells. The effectiveness of anaerobic breakdown is directly affected by the protecting outermost layers of some microalgae, which are less susceptible to destruction. These materials include Polysaccharides, Glycolipids, and Glycoproteins, Cellulose, Hemicellulose, and Pectin (Gayathri et al., 2025). Additionally, the microscopic algae cell wall is reinforced by a variety of biopolymers that are including Uronic Acid, Glucosamine, Hydroxyproline, Proline, Sporopollenin, Alganeans, and Carotenoids (Ward et al., 2014).

One of the biggest difficulties in micro algae anaerobic metabolism is thought to be the cell wall's resilience. The methane emissions were lower in microalgae with carbohydrate-based cell walls than in those without protein-based cell walls.

Another study demonstrated that the cellulose and stiff outer layer of algae present in the cell wall of *Chlorella vulgaris* directly affect anaerobic digesting ability. *Scenedesmus obliquus's* rigid cell wall, which is composed of Glucose, Mannose, and Galactose, limits the anaerobic digesting process. The literature states that several pretreatments have been studied to improve hydrolysis and anaerobic digestion of the stiff cell wall and break down the microalgal cell wall (Mussgnug et al., 2010).

8.2 Pre-treatment Approaches

Microalgal cells and biomass often exhibit reduced biodegradability due to their resistance to disruption. An important step that improves methane conversion and reduces the organic component that is difficult to break down is pretreatment. Pretreatment is necessary to increase the effectiveness of anaerobic digestion and make the substrate available to microorganisms. Numerous methods, such as Physical, Chemical, Sterilization, Microwave sonication, Lipid Separation, and Microbiological methods can be used to pretreat algae material. Steam explosion, thermo-mechanical, thermo-acidic, moderate thermal pretreatment, and heavy metal removal are further pretreatment techniques.

The effectiveness of any pretreatment method is impacted by several elements, such as kind of biomass, metabolic characteristics, economic viability, the equilibrium of energy, and the quantity of material produced throughout the procedure. Sometimes pretreatment is not necessary for better anaerobic digestion. Some researchers have demonstrated that desalting algal biomass lowers methane output, most likely due to the loss of salts and fermentable compounds. The rigid cell walls of some microalgal species, such *Scenedesmus obliquus*, are composed of the biopolymer sporopollenin, which leads to very low biogas emissions. Pretreatment methods can improve biodegradability by accelerating the bacterial hydrolytic reaction rate and breaking down the cell wall before the delipidized microalgal biomass is added to the anaerobic digester (Montgomery and Bochmann, 2014).

9. TYPES OF PRE-TREATMENT

The four main categories of microalgae pretreatments are Mechanical, Chemical, Biological, and Physical. Microalgal cells have been reported to be successfully dissolved by mechanical and physical pretreatments. In continuously digestive systems, the two most popular methods of pretreatment are thermally and microwaves. Maceration and cutting are examples of mechanical treatments which expand the area covered and release carbohydrates for the anaerobic digestion-related bacteria to feed. Mechanical pretreatments need more energy and are somewhat dependent on the type of microalgae compared to thermal, biological, and chemical pretreatments (Tamilarasan et al., 2017).

Chemical pretreatments are commonly employed in conjunction with thermal procedures and are also effective. However, the usage of chemicals may contaminate byproducts. A number of pretreatment with biological techniques, including the use of cellulose and other hydrolytic agents, altering the microorganism inoculum, or developing bacteria resistance to anaerobic fermenting byproducts of absorption, have been extensively studied to improve digestion efficacy. Biological pretreatment is suggested as a low-cost method for algae biomass preparation since it uses less power (Kavitha et al., 2017).

Figure 3. Pretreatments for improving algae biogas production

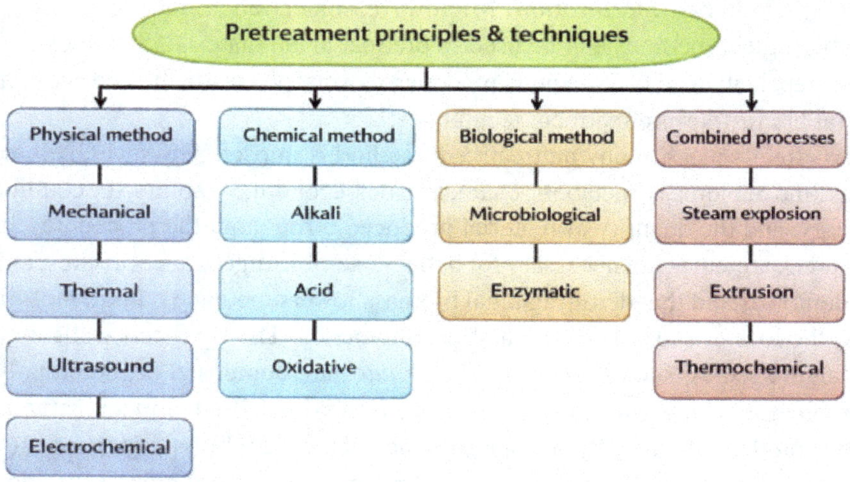

9.1 Thermal Pre-treatment

Thermal pretreatment is considered a physical method of cell disintegration since it uses heat to break down microalgal biomass. Heat energy during thermal pre-treatment disrupts the cell membrane and breaks chemical bonds within the biomass structure, leading to cell wall degradation and the release of intracellular components into the liquid. The temperature range for organic liquefaction is 50°C to 270°C. However, the optimal substrate's material properties determine its thermal range. For instance, it has been shown that digesting microalgal biomass is effective at temperature ranging from 55°C to 170°C enhanced the synthesis of methane. According to Passos et al. (2014), there are three main types of thermal pretreatments based on the temperature at which they are used: Explosions of steam prior to treatment, thermal pretreatment, and very low degree heat treatment. During anaerobic environments, temperature is essential in order to liquefy and breakdown biomass to generate more methane. Low temperature thermal pretreatments are carried out at temperatures lower than 100°C. The primary advantage of the method is that it consumes less energy (YukeshKannah et al., 2017).

9.2 Microwave Pre-treatment

Microwave radiation is electromagnetic radiation of frequencies between the order of 300 MHz and 300 GHz. The frequency normally used is approximately 2450 MHz. Microwaves offer a high-speed alternating electric field which influences polar or dielectric substances. The heat is generated during the friction of the molecules. This temperature increases the boiling of water. Although biomass has a high energy level, the treatment with microwaves is able to break hydrogen bonds, but not chemical bonds. Treatment time and output power are the two major parameters that influence microwave pretreatment. Research demonstrates that the quantity of produced methane and the level of liquefaction of microalgae depend on the energy required in the process of microwave pretreatment. Using microwaves, there is 8 per cent increase in liquefaction when microwaves are applied to microalgae and a 78 per cent increase in biomethane production (Frigon et al., 2013).

9.3 Ultrasonic Pre-treatment

Application of ultrasonic waves which undergo rapid compression and decompression cycles is called ultrasonic pretreatment. It is a mechanical process, which creates a Cavitation Effect. The consequence of the formation of microscopic sized liquid and vapor bubbles within the biomass leads to cavitation. These are referred to as microbubbles that are generated by the movement of liquid molecules in reaction to sound waves. Such bubbles might burst with time once they are very small depending on the intensity of ultrasound. This brings about shockwave, heat, hydroxyl radical and high pressure, which will lead to the dissolution of algal cell walls. The ultrasonic pretreatment is primarily affected by the energy input used. Nonetheless, a rise in the amount of energy intake to 100-200 MJ/kg TS could lead to a rise in the level of methane, to 80-90%. In ultrasonic pretreatment, H^+ and OH^- reactive radicals are produced, which contribute to the destruction of cells and liberation of intracellular contents. This accelerates the hydrolysis process and it creates volatile fatty acids, which are further transformed into methane (Alzate et al., 2012).

9.4 Chemical Pre-treatment

Another effective treatment involves chemical pretreatment which involves use of acids or alkalis. Alkaline agents are especially efficient at the disintegration of the biomass cell walls. Some of the processes that are facilitated by alkali include hydrolysis and saponification, which degrade the cell walls to enhance solubilization. Thermal methods can be used together with chemical and thermochemical treatments to enhance liquefaction and the production of methane. Methane production was increased by 30–40% with thermochemical processing of microalgal species including Chlorella and Nannochloropsis (Bohutskyi et al., 2014).

9.5 Biological Pre-treatment

Another process that does not produce any toxic compounds and that needs less energy consumption compared to other processes is biological pretreatment. Enzymatic and bacterial methods are both discussed. The cellulose and hemicellulose can be broken down using enzymes, thereby simplifying the process of anaerobic bacteria to feed on biomass. The rate of biological pretreatment depends on several factors such as the temperature, the pH, quantity of bacteria or enzymes, and the duration required to decompose the bacteria.

After the pretreated biomass was anaerobically digested, methane production increased from 254.73 mL for untreated biomass to 324.38 mL CH_4 per gram of volatile solids. The combination of calf manure and processed biomass increased the methane emission to 413.89 mL per gram of volatile solids. The usage of single and mixed enzymes for pretreatment and found that cellulose pretreatment produced 0.133 liters of methane per gram of dry biomass, while using mixed enzymes produced higher yields of about 0.145 liters per gram. These results show that pretreatment and co-digestion can significantly improve methane output and anaerobic digestion from algal biomass (Ehime et al., 2013).

10. MICROALGAE IN WASTEWATER TREATMENT

Microalgae can efficiently remove harmful Metals, Nitrogen, and Phosphorus from various types of waste water (Razaviarani et al., 2023). The cultivation of microalgae in a variety of waste streams, including Municipal, Agricultural, and Industrial Ones, has been the subject of numerous research. Microalgal development, the accumulation of biochemical components, and production are all impacted by the nutrients that are accessible. The composition of wastewater will differ according to the source. Growing urbanization has made municipal waste water an important source of water. Nitrogen and phosphorus are usually less in the municipal wastewater (Chinnasamy et al., 2010; Chu et al., 2025).

The use of microalgae to remove these nutrients has been the subject of numerous investigations. Excess nutrients from wastewater can be efficiently absorbed by microalgae to promote their development and output. Autotrophic microalgae can absorb carbon dioxide from industrial exhaust gases or the atmosphere through photosynthesis. Because they are heterotrophic, certain algae consume organic carbon. Nitrogen is a crucial element for microalgae growth. Microalgal biomass can effectively use wastewater with high ammonium content. It's important to realize that, unlike nitrogen removal, phosphorus removal from wastewater requires more than just the algae absorbing it into their cells. It is also affected by external factors like the pH and dissolved oxygen level of the water (Cai et al., 2013).

11. ENERGY PRODUCTION FROM ALGAE

Despite for over 50 years, the concept of utilising algae to produce energy has existed, as has the concept of utilizing algae oils to create that fluid is relatively recent. Research on using microalgae to produce liquid fuels started in the 1980s. Aquatic biomass may be a viable option because it develops faster compared to plants on land. Since microalgae may thrive throughout natural and salted water, they had been thoroughly investigated. They could be applied to the cost-effective and sustainable production of biofuels. Algae can be used to make biofuel, often known as third-generation biofuel, algal fuel, or algae fuel. Algal fuels are significantly more efficient than second-generation biofuels, producing 30 to 100 times more energy per hectare than arable crops. Microalgae have been used to study a variety of biofuels are substances such as Plant-Based Oils, Biodiesel Biological Oil, Bio-Syngas, Bio-Hydrogen, and Bio-Ethanol (Xu et al., 2006).

The manufacturing of such type of biofuels can also contribute to the treatment of waste water, high value compounds production and reduction of CO_2 emissions of exhaust gases. The relative innovation in post-growth processing and growth of microalgae has the potential to reduce the cost of this biofuel system to a degree. Microalgae can help meet high biofuel demands easily due to their rapid rate of growth and do not require much area or contribute to the biomass deficit. Microalgae require minimal water to cultivate compared to terrestrial plants. They are efficient in reducing carbon emission due to their capacity to cope with a high level of CO_2. Microalgae farming might be less costly than conventional farming (Mathimani et al., 2024). One challenge in using microalgae for biofuel is the low biomass in the culture, which is caused by the small size of the algae cells and limited light penetration. Harvesting the biomass is hence expensive. Commercializing this biofuel strategy may also be difficult because microalgae farms require more stringent oversight and greater initial costs than conventional farms (Li et al., 2008).

12. COMPARISON OF ENERGY CONVERSION PROCESSES

The pilot stage of fermentation-based hydrogen production is already underway, although MFCs (microbial fuel cells) are still in the research and development phase. Progress on MFCs is slow because of their low power output and high material and maintenance expenses. Research should focus on improving Design, Materials, and process scaling. With current technology, a large-scale commercial deployment of MFCs to produce electricity to the grid seems unlikely. However, MFCs may be useful in processing solid and liquid waste in circumstances when energy is otherwise wasted. These systems may also work well in biorefineries, which transform biomass into Fuels, Heat, Electricity, and other useful products. Methane (CH_4) and ethanol are already manufactured using a variety of feedstocks and used as energy sources. Butanol synthesis using ABE fermentation is being explored for industrial use, despite the fact that it has already been used in various countries. In the chemical industry, butanol is widely used as a solvent (Ni and Sun, 2009; Tiwari et al., 2026).

The most fruitful producers are *Clostridium* species. The method is less cost-effective on a large scale than methane and ethanol since high yields need the sterile cultivation of these bacteria. Butanol-producing microbes can tolerate up to 18% ethanol but only 1% to 3% butanol. Using lower heating values, the energy outputs of CH_4, H_2, ethanol, and butanol were compared, normalizing the highest conversion rates to kilojoules per gram of microalgal biomass. The highest energy yield from methanogenic digestion was obtained from untreated microalgal biomass and was nearly as high as that from ethanol fermentation. H_2 and butanol yields from microalgal biomass were the highest, accounting for around 10% and 50% of the ethanol yields, respectively. The energy levels were similar to those of hydrogen fermentation when butanol and electricity were generated concurrently in MFCs (Ras et al., 2011).

After producing Hydrogen, Methane, or Ethanol, MFCs might be a suitable final step because they perform best with low amounts of organic material that is easily biodegradable. Based on current technology and energy conversion rates, methanogenic digestion and ethanol fermentation are the most efficient ways to transform microalgal material into energy carriers. It's also exciting to see the substantial energy generation from lipid-extracted microalgal biomass leftover that has fermented to produce methane and hydrogen. Microalgal biomass is often dried and then re-suspended in water before being used. Furthermore, drying may also decrease the energy generation (Mussgnug et al., 2010).

13. CONCLUSION

Anaerobic reactors provide an excellent opportunity to transform algal sludge to renewable bioenergy and at the same time solve the waste disposal problem. This combination of microalgae growth and anaerobic digestion forms a circular solution whereby nutrients are recycled, the organic waste is stabilized and useful biogas is generated. Even though there are still technical obstacles like cell wall recalcitrance, optimization of processes and cost efficiency, pre-treatment techniques and reactor design are still being enhanced to further refine conversion performance. Altogether, the anaerobic conversion of algal sludge is one of the sustainable strategies that can help to recover the energy, to protect and to save resources. Further research and extensive deployment will be significant to achieve the potential in the future bioenergy systems.

REFERENCES

Abatzoglou, N., & Boivin, S. A. (2009). A review of biogas purification processes. *Biofuels, Bioproducts & Biorefining : Biofpr, 3*(1), 42–71. DOI: 10.1002/bbb.117

Alzate, M. E., Muñoz, R., Rogalla, F., Fdz-Polanco, F., & Perez-Elvira, S. I. (2012). Biochemical methane potential of microalgae: Influence of substrate to inoculum ratio, biomass concentration and pretreatment. *Bioresource Technology, 123*, 488–494. DOI: 10.1016/j.biortech.2012.06.113 PMID: 22940359

Balan, V., Bals, B., Chundawat, S. P., Marshall, D., & Dale, B. E. (2009). Lignocellulosic biomass pretreatment using AFEX. *Methods in Molecular Biology (Clifton, N.J.), 581*, 61–77. DOI: 10.1007/978-1-60761-214-8_5 PMID: 19768616

Blanken, W., Janssen, M., Cuaresma, M., Libor, Z., Bhaiji, T., & Wijffels, R. H. (2014). Biofilm growth of *Chlorella sorokiniana* in a rotating biological contactor-based photobioreactor. *Biotechnology and Bioengineering, 111*(12), 2436–2445. DOI: 10.1002/bit.25301 PMID: 24895246

Bohutskyi, P., & Bouwer, E. (2013). Biogas production from algae and cyanobacteria through anaerobic digestion: A review, analysis, and research needs. In Lee, J. W. (Ed.), *Advanced Biofuels and Bioproducts* (pp. 873–975). Springer. DOI: 10.1007/978-1-4614-3348-4_36

Cai, T., Park, S. Y., & Li, Y. (2013). Nutrient recovery from wastewater streams by microalgae: Status and prospects. *Renewable & Sustainable Energy Reviews, 19*, 360–369. DOI: 10.1016/j.rser.2012.11.030

Carver, S. M., Hulatt, C. J., Thomas, D. N., & Tuovinen, O. H. (2011). Thermophilic, anaerobic co-digestion of microalgal biomass and cellulose for H_2 production. *Biodegradation, 22*(4), 805–814. DOI: 10.1007/s10532-010-9419-z PMID: 20878208

Chinnasamy, S., Bhatnagar, A., Hunt, R. W., & Das, K. C. (2010). Microalgae cultivation in a wastewater dominated by carpet mill effluents for biofuel applications. *Bioresource Technology, 101*(9), 3097–3105. DOI: 10.1016/j.biortech.2009.12.026 PMID: 20053551

Chisti, Y. (2007). Biodiesel from microalgae. *Biotechnology Advances, 25*(3), 294–306. DOI: 10.1016/j.biotechadv.2007.02.001 PMID: 17350212

Chu, Y., Jiao, X., Li, D., Yu, J., Wang, C., & He, M. (2025). Integration of dairy manure wastewater treatment and biofuel production by microalgae-bacteria consortia in a biofilm reactor. *Algal Research, 86*, 103912. DOI: 10.1016/j.algal.2025.103912

Danquah, M. K., Ang, L., Uduman, N., Moheimani, N., & Forde, G. M. (2009). Dewatering of microalgal culture for biodiesel production: Exploring polymer flocculation and tangential flow filtration. *Journal of Chemical Technology and Biotechnology*, *84*(7), 1078–1083. DOI: 10.1002/jctb.2137

Driver, T., Bajhaiya, A., & Pittman, J. K. (2014). Potential of bioenergy production from microalgae. *Current Sustainable/Renewable Energy Reports*, *1*(3), 94–103. DOI: 10.1007/s40518-014-0011-8

Ehimen, E. A., Holm-Nielsen, J. B., Poulsen, M., & Boelsmand, J. E. (2013). Influence of different pretreatment routes on the anaerobic digestion of a filamentous algae. *Renewable Energy*, *50*, 476–480. DOI: 10.1016/j.renene.2012.06.064

Frigon, J. C., Matteau-Lebrun, F., Abdou, R. H., McGinn, P. J., O'Leary, S. J. B., & Guiot, S. R. (2013). Screening microalgae strains for their productivity in methane following anaerobic digestion. *Applied Energy*, *108*, 100–107. DOI: 10.1016/j.apenergy.2013.02.051

Gayathri, K., Saranraj, P., Manigandan, M., Lokeshwari, B., Yuvalakshmi, L., & Monisha, S. (2025). Applications of nanosciences and nanotechnology in life sciences. *American-Eurasian Journal of Scientific Research*, *20*(3), 91–97.

Golueke, C. G., Oswald, W. J., & Gotaas, H. B. (1957). Anaerobic digestion of algae. *Applied Microbiology*, *5*(1), 47–55. DOI: 10.1128/am.5.1.47-55.1957 PMID 13403639

Hase, R., Oikawa, H., Sasao, C., Morita, M., & Watanabe, Y. (2000). Photosynthetic production of microalgal biomass in a raceway system under greenhouse conditions in Sendai City. *Journal of Bioscience and Bioengineering*, *89*(2), 157–163. DOI: 10.1016/S1389-1723(00)88730-7 PMID: 16232719

Hsiao, T. Y., & Blanch, H. W. (2006). Physiological studies of eicosapentaenoic acid production in the marine microalga *Glossomastix chrysoplasta*. *Biotechnology and Bioengineering*, *93*(3), 465–475. DOI: 10.1002/bit.20761 PMID: 16304671

Hughes, A. D., Kelly, M. S., Black, K. D., & Stanley, M. S. (2012). Biogas from macroalgae: Is it time to revisit the idea? *Biotechnology for Biofuels*, *5*(1), 86. DOI: 10.1186/1754-6834-5-86 PMID: 23186536

Kao, C. Y., Chiu, S. Y., Huang, T. T., Dai, L., Wang, G. H., Tseng, C. P., Chen, C. H., & Lin, C. S. (2012). A mutant strain of microalga *Chlorella sp.* for the carbon dioxide capture from biogas. *Biomass and Bioenergy*, *36*, 132–140. DOI: 10.1016/j.biombioe.2011.10.046

Kavitha, S., Subbulakshmi, P., Rajesh Banu, J., Gobi, M., & Yeom, I. T. (2017). Enhancement of biogas production from microalgal biomass through cellulolytic bacterial pretreatment. *Bioresource Technology*, *233*, 34–43. DOI: 10.1016/j.biortech.2017.02.081 PMID: 28258994

Kim, J., & Kang, C. M. (2015). Increased anaerobic production of methane by co-digestion of sludge with microalgal biomass and food waste leachate. *Bioresource Technology*, *189*, 409–412. DOI: 10.1016/j.biortech.2015.04.028 PMID: 25911192

Kivaisi, A. K. (2001). The potential for constructed wetlands for wastewater treatment and reuse in developing countries: A review. *Ecological Engineering*, *16*(4), 545–560. DOI: 10.1016/S0925-8574(00)00113-0

Kumar, R., Singh, S., & Singh, O. V. (2008). Bioconversion of lignocellulosic biomass: Biochemical and molecular perspectives. *Journal of Industrial Microbiology & Biotechnology*, *35*(5), 377–391. DOI: 10.1007/s10295-008-0327-8 PMID: 18338189

Lardon, L., Hélias, A., Sialve, B., Steyer, J. P., & Bernard, O. (2009). Life-cycle assessment of biodiesel production from microalgae. *Environmental Science & Technology*, *43*(17), 6475–6481. DOI: 10.1021/es900705j PMID: 19764204

Lastella, G., Testa, C., Cornacchia, G., Notornicola, M., Voltasio, F., & Sharma, V. K. (2002). Anaerobic digestion of semi-solid organic waste: Biogas production and its purification. *Energy Conversion and Management*, *43*(1), 63–75. DOI: 10.1016/S0196-8904(01)00011-5

Li, Y., Horsman, M., Wu, N., Lan, C. Q., & Dubois-Calero, N. (2008). Biofuels from microalgae. *Biotechnology Progress*, *24*(4), 815–820. DOI: 10.1021/bp070371k PMID: 18335954

Li, Y., Park, S. Y., & Zhu, J. (2011). Solid-state anaerobic digestion for methane production from organic waste. *Renewable & Sustainable Energy Reviews*, *15*(1), 821–826. DOI: 10.1016/j.rser.2010.07.042

Liang, Y., Zheng, Z., Hua, R., & Luo, X. (2011). A preliminary study of simultaneous lime treatment and dry digestion of smooth cord grass for biogas production. *Chemical Engineering Journal*, *174*(1), 175–181. DOI: 10.1016/j.cej.2011.08.071

Mata, T. M., Martins, A. A., & Caetano, N. S. (2010). Microalgae for biodiesel production and other applications: A review. *Renewable & Sustainable Energy Reviews*, *14*(1), 217–232. DOI: 10.1016/j.rser.2009.07.020

Mathimani, T., Le, T. T. H., Salmen, S. H., Chinnathambi, A., & Jhanani, G. K. (2024). A first-of-its-kind study on founding concatenated trio fuels production platform.... *Fuel*, *361*, 130639. DOI: 10.1016/j.fuel.2023.130639

Milledge, J. J., Nielsen, B. V., Maneein, S., & Harvey, P. J. (2019). A brief review of anaerobic digestion of algae for bioenergy. *Energies*, *12*(6), 1166. DOI: 10.3390/en12061166

Molina Grima, E., Belarbi, E. H., Acién Fernández, F. G., Robles Medina, A., & Chisti, Y. (2003). Recovery of microalgal biomass and metabolites: Process options and economics. *Biotechnology Advances*, *20*(7-8), 491–515. DOI: 10.1016/S0734-9750(02)00050-2 PMID: 14550018

Montgomery, L. F. R., & Bochmann, G. (2014). *Pretreatment of feedstock for enhanced biogas production*. IEA Bioenergy.

Muga, H. E., & Mihelcic, J. R. (2008). Sustainability of wastewater treatment technologies. *Journal of Environmental Management*, *88*(3), 437–447. DOI: 10.1016/j.jenvman.2007.03.008 PMID: 17467148

Mulbry, W., Kondrad, S., Pizarro, C., & Kebede-Westhead, E. (2008). Treatment of dairy manure effluent using freshwater algae: Algal productivity and recovery of manure nutrients using pilot-scale algal turf scrubbers. *Bioresource Technology*, *99*(17), 8137–8142. DOI: 10.1016/j.biortech.2008.03.073 PMID: 18487042

Mussgnug, J. H., Klassen, V., Schlüter, A., & Kruse, O. (2010). Microalgae as substrates for fermentative biogas production in a combined biorefinery concept. *Journal of Biotechnology*, *150*(1), 51–56. DOI: 10.1016/j.jbiotec.2010.07.030 PMID: 20691224

Ni, Y., & Sun, Z. (2009). Recent progress on industrial fermentative production of acetone–butanol–ethanol by *Clostridium acetobutylicum* in China. *Applied Microbiology and Biotechnology*, *83*(3), 415–423. DOI: 10.1007/s00253-009-2003-y PMID: 19430776

Niyogi, K. K. (1999). Photoprotection revisited: Genetic and molecular approaches. *Annual Review of Plant Biology*, *50*(1), 333–359. DOI: 10.1146/annurev.arplant.50.1.333 PMID: 15012213

Oswald, W. J. (2003). My sixty years in applied algology. *Journal of Applied Phycology*, *15*(2-3), 99–106. DOI: 10.1023/A:1023871903434

Parakh, S. K., Tian, Z., Wong, J. Z. E., & Tong, Y. W. (2023). From microalgae to bioenergy: Recent advances in biochemical conversion processes. *Fermentation (Basel, Switzerland)*, *9*(6), 529. DOI: 10.3390/fermentation9060529

Passos, F., Hernandez-Marine, M., Garcia, J., & Ferrer, I. (2014). Long-term anaerobic digestion of microalgae grown in HRAP for wastewater treatment: Effect of microwave pretreatment. *Water Research*, *49*, 351–359. DOI: 10.1016/j.watres.2013.10.013 PMID: 24183557

Peng, S., & Colosi, L. M. (2016). Anaerobic digestion of algae biomass to produce energy during wastewater treatment. *Water Environment Research : A Research Publication of the Water Environment Federation*, *88*(1), 29–39. DOI: 10.2175/106143014X14062131179195 PMID: 26803024

Prajapati, S. K., Malik, A., & Vijay, V. K. (2014). Comparative evaluation of biomass production and bioenergy generation potential of Chlorella spp. through anaerobic digestion. *Applied Energy*, *114*, 790–797. DOI: 10.1016/j.apenergy.2013.08.021

Rai, S. K., Kim, G., & Song, H. (2025). Algae to biofuels: Catalytic strategies and sustainable technologies for green energy conversion. *Catalysts*, *15*(9), 806. DOI: 10.3390/catal15090806

Ras, M., Lardon, L., Bruno, S., Bernet, N., & Steyer, J.-P. (2011). Experimental study on a coupled process of production and anaerobic digestion of *Chlorella vulgaris*. *Bioresource Technology*, *102*(1), 200–206. DOI: 10.1016/j.biortech.2010.06.146 PMID: 20678925

Rasmussen, R. S., & Morrissey, M. T. (2007). Marine biotechnology for production of food ingredients. In Taylor, S. L. (Ed.), *Advances in Food and Nutrition Research, 52, 237–292*. Academic Press. DOI: 10.1016/S1043-4526(06)52005-4

Razaviarani, V., Arab, G., Lerdwanawattana, N., & Gadia, Y. (2023). Algal biomass dual roles in phycoremediation of wastewater and production of bioenergy and value-added products. *International Journal of Environmental Science and Technology*, *20*(7), 8199–8216. DOI: 10.1007/s13762-022-04696-6

Saranraj, P., Singh, S., Manigandan, M., Kesavardhini, K., Lokeshwari, B., & Gayathri, K. (2026). Gold Nanoparticles: Green Synthesis, Characterization, and Biomedical Applications. In Plant-Associated Medicinal Nano-Chemistry (pp. 37-49). CRC Press.

Saranraj, P., Singh, S., Manigandan, M., Nisha, R., Lokeshwari, B., & Gayathri, K. (2025). Pharmacological Effects of Medicinal Herbs: Recent Development and Progress. In Pharmacology of Plants and Plant Derived Biologically Active Molecules (pp. 30-45). CRC Press.

Shen, Y., Yuan, W., Pei, Z., Wu, Q., & Mao, E. (2009). Microalgae mass production methods. *Transactions of the ASABE*, *52*, 1275–1287. DOI: 10.13031/2013.27771

Shilton, A., Mara, D., Craggs, R., & Powell, N. (2008). Solar-powered aeration and disinfection, anaerobic co-digestion, biological CO_2 scrubbing and biofuel production: The energy and carbon management opportunities of waste stabilization ponds. *Water Science and Technology*, *58*(1), 253–258. DOI: 10.2166/wst.2008.666 PMID: 18653962

Siddique, M. N. I., & Wahid, Z. B. A. (2018). Enhanced methane yield by codigestion of sewage sludge with microalgae and catering waste leachate. *Water Environment Research : A Research Publication of the Water Environment Federation*, *90*(9), 835–839. DOI: 10.2175/106143017X15131012153031 PMID: 29618407

Sosnowski, P., Wieczorek, A., & Ledakowicz, S. (2003). Anaerobic co-digestion of sewage sludge and organic fraction of municipal solid wastes. *Advances in Environmental Research*, *7*(3), 609–616. DOI: 10.1016/S1093-0191(02)00049-7

Subhalakshmi, U., Saranraj, P., Kanshiya, S., Vithiya, G., Lokeshwari, B., & Manigandan, M. (2026). Diagnostic challenges of Parkinson disease in under-resourced settings. In *Evolution and innovation in Parkinson's disease therapies in under-resourced settings* (pp. 37–64). IGI Global Scientific Publishing. DOI: 10.4018/979-8-3373-7437-6.ch002

Tamilarasan, K., Kavitha, S., Banu, J. R., Arulazhagan, P., & Yeom, I. T. (2017). Energy-efficient methane production from macroalgal biomass through chemo-disperser liquefaction. *Bioresource Technology*, *228*, 156–163. DOI: 10.1016/j biortech.2016.12.102 PMID: 28064132

Thajuddin, N., & Subramanian, G. (2005). Cyanobacterial biodiversity and potential applications in biotechnology. *Current Science*, *89*(1), 11.

Tiwari, M. D., Dhengare, S. W., Yenurkar, G. K., Bhagat, R. M., Pande, P. B., & S, A. (2026). Machine learning and adaptive control for optimizing MSW-algal biomass anaerobic co-digestion. *Scientific Reports*, *16*(1), 17884. DOI: 10.1038/s41598-026-48288-7 PMID: 41998065

Uduman, N., Qi, Y., Danquah, M. K., Forde, G. M., & Hoadley, A. (2010). Dewatering of microalgal cultures: A major bottleneck to algae-based fuels. *Journal of Renewable and Sustainable Energy*, *2*(1), 012701. DOI: 10.1063/1.3294480

Ugwu, C. U., Aoyagi, H., & Uchiyama, H. (2008). Photobioreactors for mass cultivation of algae. *Bioresource Technology*, *99*(10), 4021–4028. DOI: 10.1016/j.biortech.2007.01.046 PMID: 17379512

Vandermeulen, H., & Gordin, H. (1990). Ammonium uptake using *Ulva* (Chlorophyta) in intensive fishpond systems: Mass culture and treatment of effluent. *Journal of Applied Phycology*, *2*(4), 363–374. DOI: 10.1007/BF02180927

Ward, A. J., Lewis, D. M., & Green, F. B. (2014). Anaerobic digestion of algae biomass: A review. *Algal Research*, *5*, 204–214. DOI: 10.1016/j.algal.2014.02.001

Xu, H., Miao, X., & Wu, Q. (2006). High quality biodiesel production from a microalga *Chlorella protothecoides* by heterotrophic growth in fermenters. *Journal of Biotechnology*, *126*(4), 499–507. DOI: 10.1016/j.jbiotec.2006.05.002 PMID: 16772097

Yukesh Kannah, R., Kavitha, S., Rajesh Banu, J., & Yeom, I. T. (2017). Synergetic effect of combined pretreatment for energy-efficient biogas production. *Bioresource Technology*, *232*, 235–246. DOI: 10.1016/j.biortech.2017.02.042 PMID: 28236758

Chapter 9
Coupling Biohydrogen Production With Carbon Capture:
Technologies, Pathways, and Future Prospects

Rehab Mostafa Mohamed
http://orcid.org/0000-0003-1541-1833
Independent Researcher, Egypt

Imran Ahmad
Institute for Water and Wastewater Technology, South Africa

Ahmed Hamdy Zabady
Independent Researcher, Egypt

Mofida Elsayed Makhlof
http://orcid.org/0000-0002-8636-7430
Independent Researcher, Egypt

ABSTRACT

Biohydrogen (bio-H_2) is a sustainable, carbon-neutral energy vector with strong potential to complement electrochemical hydrogen production. This chapter provides a critical overview of recent advances in biohydrogen generation Future developments offer avenues to reduce greenhouse gas emissions and promote a clean energy transition, as well as insights into the viability of CCS–hydrogen integrationThe chapter highlights emerging innovations in enzyme and strain engineering, immobilized photobioreactor design, and in-situ gas recovery. By combining biological insights

DOI: 10.4018/979-8-3373-7630-1.ch009

with process intensification and techno-economic considerations, biohydrogen is positioned as a practical and decentralized pathway for renewable hydrogen production, contributing significantly to the transition toward a circular, low-carbon energy economy.

1. INTRODUCTION

Fossil fuels drive the greenhouse effect, climate change, and particulate pollution, while their reserves are rapidly diminishing (Lak et al., 2024). Biofuels—including bioethanol, biodiesel, and biohydrogen (bio-H_2)—are attractive alternatives. Four generations of biofuels are typically distinguished: first-generation (food crops), second-generation (non-food crops and wastes), third-generation (microorganisms), and fourth-generation (genetically engineered systems) (Bhagat et al., 2025). Biohydrogen technologies produce H_2 from water, organic waste, and biomass via solar energy and microorganisms, offering an environmentally friendly option, particularly for transportation (Cha et al., 2025).

Key advantages driving interest in H_2 include: (1) abundance and clean energy potential (Malode et al., 2025); (2) zero direct CO_2 emissions at point of use (Pivac et al., 2024); (3) non-toxicity compared with coal, nuclear, and gasoline (Niescioruk et al., 2025); and (4) higher conversion efficiencies in fuel cells (up to ~65%) relative to typical thermal power plants (~33–35%) (Ma et al., 2024). Consequently, H_2 is poised to be a significant chemical energy carrier and high-energy storage medium for transport. Microorganisms producing hydrogen are commonly grouped into fermentative bacteria, photosynthetic bacteria, cyanobacteria, and green microalgae (Demirkaya & De la Hoz Siegler, 2024). Photosynthetic bacteria span Rhodospirillaceae (purple non-sulfur), Chromatiaceae (purple sulfur), and Chlorobiaceae (green sulfur) (Qian et al., 2024). Among algae, *Chlamydomonas reinhardtii* is the most intensively studied strain for H_2 evolution (Chandrasekhar et al., 2025). Anaerobic bacteria include strict and facultative anaerobes. This chapter surveys bio-H_2 production routes, associated microorganisms, and strategies to achieve commercially relevant yields.

Although biohydrogen (bio-H2) is a special and useful part of the larger hydrogen economy, its traditional production processes from fossil fuels, such as coal gasification or steam methane reforming, are linked to considerable carbon emissions. Technologies for carbon capture and storage (CCS) have become essential for cutting greenhouse gas emissions and slowing down climate change. In order to further improve the environmental sustainability of hydrogen as a clean energy carrier, there has recently been an increase in interest in combining CCS with hydrogen production processes. By incorporating CCS technologies into these processes, it is possible to capture and store CO2 emissions, thereby lowering the overall carbon footprint of hydrogen production (Shaban, et al., 2026). This thorough analysis attempts to investigate the potential, difficulties, and future prospects of combining CCS with hydrogen production. Additionally, the generation of biohydrogen (bio-H2) utilizing environmentally friendly and sustainable manufacturing techniques could lower CO2 emissions (J & Saritha, 2026). Biohydrogen is generated biologically using renewable biomass sources—such as organic waste, agricultural residues, and wastewater—through processes like dark fermentation and microbial electrolysis. This biological route results in a lower carbon footprint and often achieves carbon-negative emissions, making biohydrogen a distinctly green and low-emission form of hydrogen. Additionally, biohydrogen production can contribute to waste treatment and resource recovery, offering integrated environmental and economic benefits that conventional hydrogen lacks. These characteristics position biohydrogen as a crucial player in transitioning to a sustainable hydrogen economy focused on decarbonization and circular bioeconomy principles(Ganguly et al., 2025). The processes involved in biohydrogen production are illustrated in Figure 1.

Figure 1. Biohydrogen production pathways

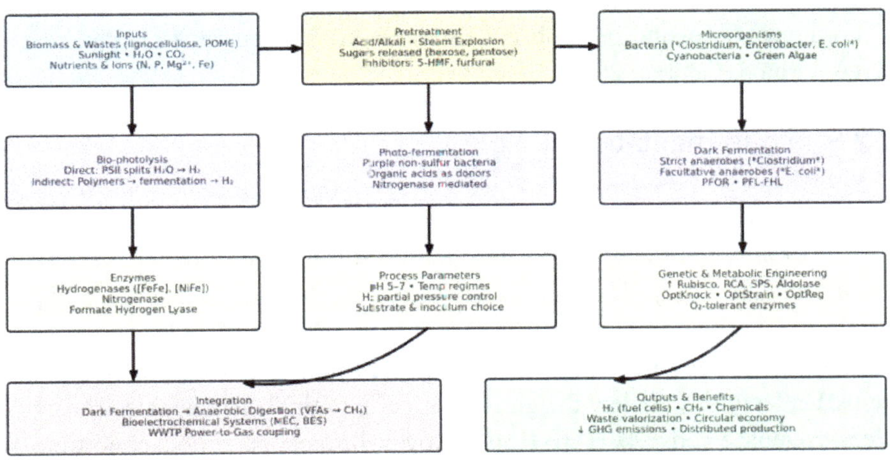

This chapter reviews recent advances in sustainable biohydrogen production methods that decrease carbon emission focusing on (i) microbial and enzymatic pathways, (ii) substrate types and pretreatment strategies, (iii) key process parameters governing yield and stability, and (iv) emerging technologies for scale-up and integration. Bio-H_2 occupies a distinct niche within the hydrogen economy by converting organic and wastewater feedstocks into fuel via biological routes, providing low-cost, decentralised hydrogen generation while simultaneously remediating waste (V) coupling biohydrogen production with carbon capturing.

2. GREEN SUSTAINABLE TECHNOLOGIES AND PATHWAYS OF BIOHYDROGEN PRODUCTION WITH CARBON CAPTURE

2.1 Biological Pathways for Hydrogen Production

2.1.1 Bio-photolysis

Cyanobacteria and green algae perform photosynthesis and, under certain conditions, bio-photolysis—using solar energy to split water into protons and oxygen via photosystem II (PSII) and transfer electrons into the photosynthetic chain (Shevela et al., 2023; Oliver et al., 2022). Bio-photolysis occurs as:

- **Direct bio-photolysis:** green algae derive electrons and protons directly from water during photosynthesis (Azizan & Jalil, 2024).
- **Indirect bio-photolysis:** CO_2 is fixed into intracellular polymers with concomitant O_2 evolution; subsequent (an)aerobic fermentation generates H_2 (Ashitha & George, 2022).

2.1.2 Photo-fermentation

Only a subset of photosynthetic microbes evolve or consume molecular hydrogen. *Allochromatium vinosum* was the first red sulfur bacterium shown to fix CO_2 under illumination using H_2 as an electron donor (Alarcon, 2023). In many purple non-sulfur bacteria, nitrogenase reduces protons to H_2 when fixed-nitrogen sources are absent. H_2 productivity depends on light regime, substrate type, and enzyme systems. Under long-wavelength light, bacteriochlorophyll mediates electron transfer, while nitrogenase converts H^+ to H_2 in nitrogen-limited, anaerobic conditions with adequate carbon/electron donors. Glutamate is often an optimal organic nitrogen source; organic acids are effective carbon/electron donors (Wu et al., 2023).

2.1.3 Dark (Anaerobic) Fermentation

Non-photosynthetic microbes generate H_2 in the absence of light by fermenting organic/inorganic substrates (Cha et al., 2025). Strains such as *Escherichia coli* and others convert complex carbohydrates (e.g., starch, xylan, pectin, mannitol, sorbitol, glycerol, cellobiose, sucrose) into H_2 and organic acids. Obligately anaerobic

Clostridium spp. produce H_2 via NAD(P)H oxidation mediated by ferredoxin and hydrogenase with pyruvate as an intermediate; facultative anaerobes (e.g., *Enterobacter, E. coli, Bacillus*) typically use the formate pathway via formate hydrogen lyase (FHL) and cytochrome *c* (Tripathi & Tripathi, 2024; Buckel, 2021).

2.1.4 Key Enzymes

Hydrogenases catalyze reversible $H_2 \leftrightarrow 2H^+ + 2e^-$, whereas nitrogenases catalyze $N_2 \rightarrow NH_3$ and release H_2 as a by-product under anaerobic, N-limited conditions (Kumar et al., 2024).

3. MICROORGANISMS FOR BIOHYDROGEN

Hydrogen's high energy content makes it a compelling clean alternative. Biohydrogen leverages biological conversion of wastes, aligning with circular-economy goals (Akram et al., 2024). Despite progress, industrial deployment is constrained by low biomass productivities and downstream processing costs (Bolatkhan et al., 2019). Biological hydrogen production (BHP) employs photosynthetic bacteria, cyanobacteria, microalgae, and certain heterotrophs under defined conditions (Table 1) (Bolatkhann et al., 2019). Biomass can be efficiently transformed into energy via multiple bioprocesses, supporting H_2's role in future transport and power applications; solar-driven biological routes offer CO_2 mitigation and waste valorization (Habib et al., 2024).

Table 1.

Group	Organism	Mechanism of Production	Advantages	Disadvantages	Reference
Bacteria	*Clostridium butyricum*	Dark fermentation via pyruvate: ferredoxin oxidoreductase & (Fe)-hydrogenase	High H_2 yield; simple substrates (sugars, starch).	Strict anaerobe; organic acid accumulation inhibits growth.	Das & Veziroglu, 2008 (Int J Hydrogen Energy 33:6046–6057)
	Enterobacter aerogenes	Dark fermentation using the formate hydrogen lyase (FHL) pathway	Facultative anaerobe; tolerates O_2 better than *Clostridium*.	Lower H_2 yield; mixed by-products (ethanol, acids).	Nath & Das, 2004 (Biotechnol Lett 26:13–19)
	Escherichia coli (engineered strains)	Engineered dark fermentation via FHL/ hydrogenase systems	Easy genetic manipulation; model organism.	Native H_2 yield low; requires genetic engineering.	Maeda et al., 2007 (Appl Microbio Biotechnol 73:130–135)

continued on following page

Table 1. Continued

Group	Organism	Mechanism of Production	Advantages	Disadvantages	Reference
Cyanobacteria	*Anabaena variabilis*	Nitrogenase-driven H_2 evolution under N_2-fixing conditions (heterocysts)	Produces H_2 photobiologically; heterocysts protect nitrogenase.	O_2-sensitive nitrogenase; slow growth.	Tamagnini et al., 2002 (FEMS Microbiol Rev 26:67–81)
	Nostoc punctiforme	Nitrogenase activity in heterocysts	H_2 linked to nitrogen fixation; microaerobic zones protect the enzyme.	Low efficiency in O_2-rich environments; slow growth.	Ghirardi et al., 2000 (Trends Biotechnol 18:506–511)
	Synechocystis sp. PCC 6803	Bidirectional (NiFe)-hydrogenase linked to photosynthetic electron transport	Genetically tractable model; direct link to photosynthesis.	Hydrogenase is highly O_2-sensitive; it needs anaerobic induction.	Miyake et al., 2013 (Photosynth Res 116:409–420)
Algae	*Chlamydomonas reinhardtii*	Direct biophotolysis (PSII → PSI → ferredoxin → (Fe)-hydrogenase); sulfur deprivation enhances H_2	Well-studied; strong model for algal H_2 research.	O_2 evolution inhibits hydrogenase; stress is needed.	Melis et al., 2000 (Plant Physiol 122:127–135)
	Scenedesmus obliquus	Biophotolysis + indirect fermentation under stress	High biomass productivity; usable in wastewater systems.	Sensitive to O_2; unstable H_2 production.	Kapdan & Kargi, 2006 (Int J Hydrogen Energy 31:1664–1673)
	Chlorella vulgaris	Biophotolysis (PSII-driven, stress-induced H_2)	Robust growth; can use wastewater; widely distributed.	Low sustained H_2 yield; needs stress induction.	Ueno et al., 1998 (Biotechnol Lett 20:195–199)

3.1 Bacteria: *Clostridium*, *Enterobacter*, *Escherichia coli*

A wide range of genera—*Clostridium*, *Enterobacter*, *E. coli*, *Klebsiella*, *Halanaerobium*, *Pseudomonas aeruginosa*, *Desulfuromondales* sp., *Dysgomonas* sp., among others—can produce H_2, with *Clostridium* most extensively studied for dark fermentation (Beschkov et al., 2023; Wang et al., 2021). *Enterobacter aerogenes* exhibits fast growth, broad substrate use, and genetic tractability, but observed yields are typically below the theoretical 4 mol H_2/mol glucose, necessitating metabolic optimization (Jiang et al., 2023).

3.2 Non-oxygenic Photosynthetic Bacteria

Purple non-sulfur photosynthetic (PNSP) bacteria produce H_2 primarily via nitrogenase, using ATP from photosynthesis; organic acids (formate, acetate, lactate, butyrate) serve as electron donors, with butyrate often the most effective inducer.

Key bottlenecks are H_2 uptake and the need to impose nitrogen deficiency (Fakhimi et al., 2020).

3.3 Heterotrophic Bacteria

Two major fermentative routes are prominent: the PFOR pathway in strict anaerobes (*Clostridium*) linking pyruvate oxidation to H_2 evolution, and the PFL pathway in facultative anaerobes (*E. coli*) coupling formate cleavage to H_2 via FHL. Competing fermentation pathways and product accumulation (ethanol, formate, acetate, lactate, succinate, glycerol, butyrate) restrict H_2 yields. Co-cultivation with algae/bacteria has potential but remains underexplored (Fakhimi et al., 2020).

3.4 Algae and Cyanobacteria

Microalgae/cyanobacteria dominate global photosynthesis and are promising bio-H_2 feedstocks as shown in Fig. 1 Light-harvesting complex is the antenna pigments (chlorophylls, carotenoids) that capture photons and funnel to PSII/PSI, then PSII (Photosystem II) splits $H_2O \rightarrow O_2 + 4e^- + 4H^+$; typically produces O_2 which inactivates algal hydrogenase, and PQ pool / Cyt b6f / PC plastoquinone, cytochrome b6f and plastocyanin — electron carriers in chain, then PSI (Photosystem I) reduces ferredoxin (Fd) which is considered as soluble electron carrier that donates electrons to hydrogenase when conditions allow. (Fe)-hydrogenase is an enzyme that catalyzes $2H^+ + 2e^- \rightarrow H_2$. Active only under low O_2 (and often after PSII downregulation) (Fakhimi et al., 2020). Candidate microalgae include *Scenedesmus obliquus*, *Chlorella vulgaris*, *Oscillatoria*, *Keratococcus*, and *Spirogyra*; cyanobacteria include *Arthrospira platensis*, *Microcystis aeruginosa*, *Microcystis* spp.; seaweeds include *Sargassum* sp., *Laminaria japonica*, *Saccharina japonica* (Sharma et al., 2023). H_2 is linked to photosynthetic activity and mediated by nitrogenase and hydrogenases; enzyme O_2-sensitivity is a key limitation. Heterocystous cyanobacteria spatially segregate O_2 evolution (vegetative cells) and N_2ase-driven H_2 production (heterocysts), mitigating O_2 inhibition. Well-studied genera include *Synechocystis*, *Synechococcus*, *Cyanothece*, *Oscillatoria*, *Anabaena*, and *Nostoc* (Bozieva et al., 2023).

Figure 2. Schematic representation of microalgal H2 production

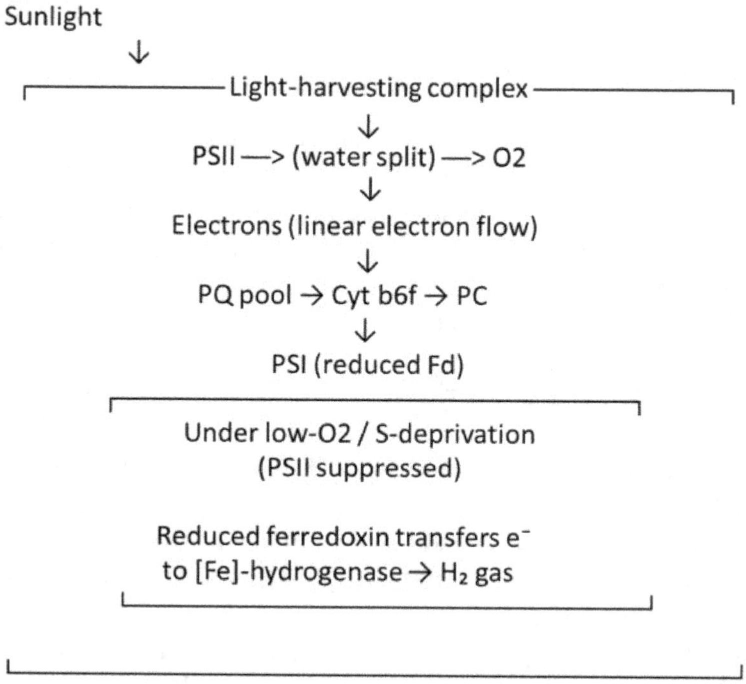

4. GENETIC AND METABOLIC ENGINEERING

Natural photosynthesis/respiration/fermentation interactions are insufficient for industrial H_2 rates. Directed genetic transformation can enhance productivity (Zhang et al., 2023). Increasing intracellular carbohydrate pools via overexpression (e.g., SPS in *Anabaena* sp. PCC 7120), boosting Rubisco activity and activation (RCA), or elevating enzymes like fructose-1,6-bisphosphate aldolase can raise photosynthetic capacity and biomass (Zhang et al., 2023; Barati et al., 2021; Priyadharsini et al., 2022). Because H_2 evolution diverts cellular energy, regulatory networks must be understood to reroute reducing equivalents toward nitrogenase/hydrogenase. Metabolic engineering (gene knockouts, heterologous expression, computational design) can relieve bottlenecks and improve flux to H_2 (An et al., 2022; Kar et al., 2021; Addissouky, 2024; Alamiery, 2024; Aminian-Dehkordi et al., 2023; Intasian et al., 2021; Ananthi et al., 2024).

4.1 Metabolic Pathway Engineering as a Strategy

Metabolic engineering applies quantitative pathway analysis, molecular biology, and genomics to redirect metabolism for higher H_2 yields, broader substrate utilization, and O_2-tolerant enzyme performance. Tools such as OptKnock, OptStrain, and OptReg support optimal deletion/modulation strategies (Aminian-Dehkordi et al., 2023). Achieving competitive bio-H_2 rates will likely require integrated strategies combining gene knockouts, nutrient limitation studies, heterologous pathways, and controlled environments (An et al., 2022; Kar et al., 2021; Addissouky, 2024; Alamiery, 2024; Intasian et al., 2021).

5. SUBSTRATES AND FEEDSTOCKS

Biomass is a sustainable, non-fossil feedstock that should be abundant, renewable, and non-competitive with food (Kim et al., 2021). Potential sources include agricultural residues, energy crops, food/municipal/industrial wastes, and algal biomass.

5.1 Food Waste

Food waste—rich in carbohydrates, proteins, lipids, nitrogen, and salts—varies by region and culture, typically containing 15–30% organics, 75–85% water, and up to 1.5% salts (Wani et al., 2024). Valorization yields energy, reduces waste and emissions, and mitigates groundwater contamination. Substrate composition (carbohydrate/fat/protein/cellulose/hemicellulose) and hydrolysis kinetics can limit H_2 yields (Xu et al., 2022; Rawindran et al., 2025).

5.2 Industrial Waste

Industrial effluents offer carbon-rich substrates. Vinasse, a high-sugar distillery by-product, is widely available and suitable for fermentative H_2 (Tsouko et al., 2024; Zavala-Mendez et al., 2022). Crude glycerol from biodiesel production—containing methanol/biodiesel traces—serves as an energy-dense carbon source for diverse microbes (Lima et al., 2022).

5.3 Other Renewable Sources

Algal biomass productivity can reach 150–600 $t \cdot ha^{-1} \cdot yr^{-1}$, enabling H_2 (and co-products like biodiesel) via photo- or dark-fermentation (Khan et al., 2023; Hitam & Jalil, 2023). Lignocellulosic residues (straws, stalks, wood chips, yard waste) supply hexose/pentose sugars but require pretreatment to overcome recalcitrance (Hassan et al., 2023).

6. PROCESS PARAMETERS AND OPTIMIZATION

Bio-H_2 production is governed by medium composition, pH, temperature, H_2 partial pressure (HPP), substrate characteristics, enzyme systems, and inoculum—key to scale-up and economic viability (figure reference) (Shanmugam et al., 2020; Asrul et al., 2022).

6.1 pH

pH strongly influences cellular metabolism, enzyme activity, and redox balance. Reported optima lie between pH 5–7 (76), with pH central to enzymatic efficiency and cellular redox potential (Tang et al., 2022). Immobilized *Synechocystis* sp. PCC 6803 reportedly achieves maximal H_2 production at an initial pH of 7.4 (Homburg et al., 2019).

6.2 Temperature

Temperature alters metabolic routing and hydrogenic activity. While higher temperatures can enhance production, excessive heat suppresses H_2 (Ananthi et al., 2024). Operational regimes include ambient (15–30 °C), mesophilic (30–39 °C), thermophilic (50–64 °C), and hyperthermophilic (>64 °C) (Patel et al., 2022). Mesophilic conditions are typically used for cost reasons but may favour non-hydrolytic microbes; thermophiles/hyperthermophiles can directly hydrolyse cellulose without exogenous cellulase and exhibit higher hydrolysis rates. Frequently used strains include *Thermoanobacterium* (thermophiles), *Clostridium* and *Enterobacter* (mesophiles) (Ivanenko et al., 2024). Heat pretreatment (e.g., 95 °C, 15 min) can suppress methanogens (Ananthi et al., 2024).

6.3 Hydrogen Partial Pressure (HPP)

Excess H_2 inhibits fermentation. HPP >60 Pa disfavors H_2 formation and shifts metabolism toward alcohols. Elevated HPP lowers the H^+/H_2 ratio and impedes electron transfer from reduced ferredoxin to H_2 via hydrogenase (Ananthi et al., 2024). Reported favorable HPP/temperature combinations include ~2 kPa at 98 °C, 20 kPa at 70 °C, and 50 kPa at 60 °C; inert-gas sparging (N_2) reduces HPP (Ananthi et al., 2024).

6.4 Enzymes

FHL and (FeFe)-hydrogenases are pivotal. Under anaerobic, mildly acidic conditions, FHL converts formate to H_2 and CO_2 via its hydrogenase, formate dehydrogenase, and electron-transfer subunits; (FeFe)-hydrogenases can exhibit 10–100× higher activities than other classes (Baeyens et al., 2020; Ananthi et al., 2024).

6.5 Inoculum

Obligate anaerobes (e.g., *Clostridium*) can reach ~4 mol H_2/mol glucose; facultative anaerobes typically yield ~2 mol H_2/mol glucose (Akhlaghi & Najafpour-Darzi, 2020). Facultative anaerobes rapidly scavenge O_2, facilitating anaerobiosis and lowering costs. Mixed cultures (soil, compost, sludge, and manure) are attractive industrially due to their robustness and low operating costs, but they risk H_2 consumption by methanogens/acetogens. Pretreatments (acid/base, chloroform, iodopropane, 2-bromoethanesulfonate, heat shock, aeration, freeze–thaw) can suppress H_2-consuming microbes; combinations may be most effective (Ahmad et al., 2024; Amjith & Bavanish, 2022). Both batch and continuous modes are used, with ongoing screening for efficient phototrophs and engineered strains (Li et al., 2024).

6.6 Substrate Composition

- **Carbohydrates:** Preferred substrates; simple sugars arise from hydrolysis of carbohydrate-rich wastes (e.g., rice slurry, potato waste, sugar and distillery residues) (Ananthi et al., 2024). Lignocellulosics require pretreatment (acid/alkali, steam explosion). Municipal, paper-mill, molasses/glycerol wastewater, dairy/distillery wastewater, and other wastes are viable. Reported yields include 4708 mL $\cdot L^{-1}$ of H_2 from POME and 880 $mL \cdot L^{-1} \cdot d^{-1}$ from tapioca wastewater (Ayodele et al., 2023).
- **Nitrogen:** Essential for growth and enzyme synthesis. Optimal levels enhance H_2, whereas excess nitrogen perturbs intracellular pH and enzyme activity. ~0.1 g $N \cdot L^{-1}$ is often reported as beneficial; ammonia is common, with dose tailored to inoculum. Organic wastes can supplement nitrogen (Ji & Wang, 2021; Ananthi et al., 2024).
- **Lipids:** Hydrolyze more slowly than carbohydrates. Lipase-mediated hydrolysis yields fatty acids and glycerol; subsequent β-oxidation and NADH oxidation generate H_2. VFA accumulation can be inhibited by lowering pH. Industry wastewaters (dairy, brewery, POME) and molasses effluents may contain 13–32% lipids (Abdelfattah et al., 2023; Ananthi et al., 2024).

- **Proteins:** Protease-driven hydrolysis produces amino acids, which ferment to VFAs, CO_2, and H_2. Proteinaceous wastes (casein, whey, eggs, fish) are applicable (Ayodele et al., 2023).
- **Phosphate:** Serves nutritional and buffering roles; adequate levels enhance H_2, but excessive phosphate can inhibit production (Ananthi et al., 2024).
- **Ions:** Mg^{2+} is a cofactor for >10 glycolytic/fermentative enzymes; Fe is critical for ferredoxin/hydrogenase function (Ananthi et al., 2024).

6.7 Substrate Pretreatment

Pretreatment of lignocellulosics disrupts lignin–cellulose–hemicellulose interactions, increasing the production of fermentable monomers (cellobiose, glucose, xylose, arabinose). Inhibitors (furfural, 5-HMF) may form; their impact depends on concentration and microbial tolerance (Ananthi et al., 2024).

7. INTEGRATION WITH OTHER TECHNOLOGIES

7.1 Two-Stage Dark Fermentation–Anaerobic Digestion

Coupling dark fermentation with anaerobic digestion converts residual metabolites into methane, thereby increasing overall energy recovery and improving process viability relative to thermochemical routes (Al-Dailami et al., 2022; Sarkar et al., 2021).

7.2 Bioelectrochemical Systems (BES) and Microbial Electrolysis Cells (MECs)

BES enable the production of fuels/chemicals by supplying external electrons to electroactive microbes (Gautam et al., 2023; Kumar et al., 2025). MECs require modest external energy and can be integrated with wastewater treatment, providing simultaneous remediation and energy recovery (Chandrasekhar et al., 2021; Zhu et al., 2024). Cathodic potentials ≤ -0.414 V are needed to drive hydrogen evolution; MECs approach near-stoichiometric conversion, whereas dark fermentation typically achieves ~33% of theoretical potential. Performance depends on electrodes, potential, microbiome, reactor design, and feedstock; readily biodegradable substrates yield the highest rates. Electrohydrogenesis overcomes endothermic barriers by combining small external inputs with anode-respiring bacteria (Hemashenpagam et al., 2024; Jalil et al., 2025).

7.3 Wastewater Treatment Synergy

Hydrogen-integrated energy systems (power-to-gas, methanation, fuel cells) at WWTPs can reduce costs and emissions under by optimizing H_2 injection ratios and carbon-trading schemes thereby improving low-carbon performance and offering practicable implementation pathways (Raut et al., 2025).

Figure 3. Integrated roadmap for biohydrogen production.

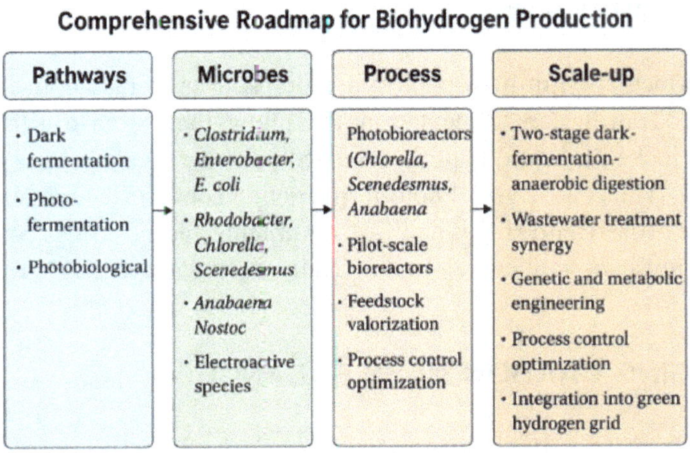

8. CARBON MASS BALANCE IN BIOHYDROGEN PATHWAYS

Carbon mass balance in biohydrogen pathways accounts for the input of carbon in biomass and tracks its conversion across gaseous, liquid, and solid fractions. The core challenge of biological hydrogen production is managing the inherent carbon trade-offs, where maximum hydrogen yields correspond directly to specific metabolic byproducts rather than complete substrate oxidation (Song, et al., 2024). Understanding the carbon mass balance requires analyzing the primary biological production methods and how carbon distributes across their respective outputs.

8.1 Dark Fermentation

during dark fermentation carbohydrates (e.g., glucose) are converted into hydrogen, carbon dioxide, and various soluble microbial products (Kaliyaperumal, et al.,

2025), because hydrogen gas does not contain carbon, the carbon from the hexose sugars must be accounted for in other products. The governing reactions and their carbon distribution follow several metabolic pathways such as Acetate Pathway where 67% of carbon converted to volatile fatty acids (VFAs), 33% released as $Co_{2,}$ under these ideal conditions, while Butyrate Pathway 67% carbon converted to butyrate, 33% released as Co_2. The maximum biological yield is limited to 4 mol H2 per mol of hexose (e.g., glucose) via the acetate pathway, or (e.g., glucose) via the acetate pathway or 2 mol H2 per mol via the butyrate pathway (Song, et al., 2024)

8.2 Photo-Fermentation

Photo-fermentation utilizes photosynthetic bacteria to metabolize the organic acids produced during dark fermentation. This allows the system to utilize the "missing" carbon locked in liquid byproducts (such as acetate and butyrate) to generate additional H_2 (Hu, et al., 2020). Photo-fermentation converts the remaining organic acids almost entirely into H_2 gas and Co_2, optimizing carbon-to-gas conversion but requiring significant light energy inputs (Kaliyaperumal, e al., 2025)

9. MANAGING CARBON MASS BALANCE

To ensure a successful carbon mass balance in biohydrogen, researchers typically measure carbon across three phases (Liu, et al,. 2018):

Gas Phase: Quantification of H_2 and Co_2

Liquid Phase: Quantification of metabolites like ethanol, acetic acid, propionic acid, and butyric acid.

Solid Phase: Carbon utilized for microbial biomass synthesis (roughly approximated by the formula $C_{60}H_{87}N_{12}O_{23}P$ for microbes)

To increase the hydrogen output and shift the carbon mass balance away from liquid acids and biomass growth, researchers optimize operational parameters like Organic Loading Rate (OLR), hydraulic retention time, and pH control (Song, et al., 2024).

10. MICROALGAE BASED CARBON DIOXIDE PRODUCTION AND THE CAPTURING BASED STRATEGIES

Microalgal carbon capture uses photosynthetic microorganisms to naturally assimilate atmospheric Co_2 into biomass. It is 10 to 50 times more efficient than terrestrial trees and converts carbon into valuable bioproducts like biofuels, proteins, and biochemicals, contributing to a sustainable bio-economy (Dayanath, et al., 2026).

10.1 Primary Capture Strategies

Post-Combustion Flue Gas Sequestration: Microalgae are directly cultivated utilizing concentrated Co_2 streams emitted from industrial chimneys and boilers (e.g., coal-fired power plants). The strains convert Co_2-rich flue gas into biomass. Strains are typically chosen for high tolerance to elevated temperatures and gas contaminants (Thomas, et al., 2016).

10.2 Direct Air Capture (DAC)

Microalgae-based DAC systems bypass the need for expensive chemical absorbent regeneration. By bubbling atmospheric air through photobioreactors, the algae sequester Co_2 and typically capture up to 1.83 kg of Co_2 per kilogram of biomass produced (Dayanath, et al., 2026).

10.3 Wastewater Integration

By combining microalgal cultivation with wastewater treatment, the system absorbs Co_2 while concurrently removing excess nitrogen and phosphorus from municipal or industrial runoff (Sadvakasova, et el., 2025).

11. CHALLENGES IN HYDROGEN PRODUCTION AND THEIR SOLUTIONS

Hydrogen can be produced from fossil sources (gasification, steam reforming, partial/auto-thermal oxidation) and renewables (solar/wind electrolysis, biomass/biofuel gasification) (Mansoori et al., 2021). Global dependence on coal-based H_2 (~21.5 billion tons per year) must be displaced by renewable routes. Even as transport and portable power demand rise, conventional H_2 expansion risks increased greenhouse emissions without CCS. Renewable, waste-derived H_2 improves the economics of distributed/semi-centralized production with minimal net emissions (Dash et al., 2023). However, each route faces distinct limitations in feedstock handling, conversion efficiency, and systems integration (Hooshmand Zaferani et al., 2022).

The production of hydrogen is essential for reducing carbon emissions and promoting the switch to sustainable energy. It is a flexible and eco-friendly energy source that helps lower greenhouse gas emissions and makes it easier to integrate renewable energy sources (Serik, et al., 2026). The importance of hydrogen production in relation to decarbonization and the shift to clean energy is demonstrated by the following points:

1. Reducing Greenhouse Gas Emissions: Due to its low greenhouse gas emissions during production, green hydrogen, which is made from renewable sources, provides an environmentally benign substitute for conventional fuels. We can help combat climate change by drastically lowering carbon emissions in a variety of industries, such as transportation, manufacturing, and heating, by using sustainable electricity to power electrolysis machines that split water into hydrogen and oxygen (Serik, et al., 2026).
2. Energy Storage and Grid Balancing: Intermittent renewable energy sources, such as solar and wind power, can be integrated into the power grid by using hydrogen as an energy storage option. Hydrogen can be produced via electrolysis using excess electricity produced during times of peak renewable energy production. The energy supply and demand on the power grid can be balanced by using fuel cells to transform this stored hydrogen back into electricity when needed (Khan, et al., 2026).
3. Sector Coupling and Electrification: By bridging historically disparate energy sectors including transportation, industry, and power generation, hydrogen has the potential to advance sector integration. Hydrogen can facilitate the electrification of industries that are difficult to decarbonize using only direct power by using hydrogen fuel cells in electric vehicles, for heating, and in high-temperature industrial operations (Phogat, et al., 2026).
4. Industrial Decarbonization: Using clean hydrogen in place of coal or natural gas can have a significant positive impact on the industrial sector, which is mostly dependent on fossil fuels for the production of heat and electricity. Industries can reduce their carbon footprint and move toward more sustainable and clean production methods by replacing fossil fuels with hydrogen in processes like steelmaking, ammonia generation, and chemical manufacturing (Johnson, et al., 2025).
5. International Energy Trade: The interchange of clean energy between areas with an abundance of renewable resources and those in need of energy imports could be facilitated by hydrogen's potential to become an internationally traded commodity. Countries can diversify their energy sources, improve energy security, and foster global collaboration in accomplishing climate goals by setting up hydrogen supply chains and infrastructure for international trade (Singh, e al., 2025).
6. Innovation and Technological Advancements: The growing interest in producing hydrogen has spurred innovation in fuel cell applications, electrolysis

technology, and storage techniques. The shift to a cleaner and more sustainable energy infrastructure is being accelerated by research and development projects that lower costs, increase efficiency, and broaden the potential applications of hydrogen in many industries (Albalawi, 2025).

12. RECENT ADVANCES IN HYDROGEN PRODUCTION AND EMERGING TRENDS

12.1 Synthetic Biology

(FeFe), (NiFe), and (Fe) hydrogenases mediate biological H_2 evolution. The (FeFe) HydA1 from *Chlamydomonas reinhardtii* (A1 subclass) exhibits very high turnover, e.g., ~21,000 ± 12,000 H_2 s^{-1} for *Clostridium acetobutylicum* HydA. Yet O_2 sensitivity and competing pathways limit yields in *C. reinhardtii* (Xuan et al., 2023).

12.2 Immobilization Techniques

Photofermentative bio-H_2 benefits from PNSB operating across a broad range of light wavelengths and diverse substrates. Cell immobilization, choice of supports, and photobioreactor design significantly affect productivity. Techno-economic analyses and AHP-based design selection highlight hydrodynamics and mass/heat transfer as critical scale-up challenges warranting pilot and large-scale studies (Khosravitabar & Spetea, 2024).

12.3. Integration of CCS into Hydrogen Production

A viable strategy for reducing carbon emissions and improving the environmental sustainability of hydrogen production is the integration of CCS into hydrogen production processes (Khalili, et al., 2025). The following are some effective ways that CCS can be included into various hydrogen production techniques:

1. Steam methane reforming (SMR) with CCS:

CO2 is a byproduct of the synthesis of hydrogen in SMR. CO2 emissions can be captured, compressed, and stored underground when CCS is implemented in SMR facilities. Before being sent to appropriate storage locations for sequestration, captured CO2 from the SMR process is purified and compressed. Environmental sustainability is promoted by combining CCS with SMR, which dramatically lowers

CO_2 emissions associated with hydrogen production (Bayramoğlu & Bayramoğlu, 2025).

2. Electrolysis with CCS:

Carbon emissions may occur when electrolysis is driven by fossil fuels. CO_2 released during the process can be captured and stored by integrating CCS with electrolysis facilities. The carbon footprint of producing hydrogen can be significantly decreased by adopting CCS technologies to capture and store CO_2 produced as a byproduct of electrolysis underground. Climate-friendly hydrogen production is encouraged and emissions are further reduced when electrolysis and CCS are combined (Ahmed, et al., 2025).

3. Biomass Gasification with CCS:

CCS can help with hydrogen production through biomass gasification. Carbon-neutral or even carbon-negative hydrogen can be created by capturing and storing the CO_2 produced during biomass gasification. After being captured during the gasification process, CO_2 is compressed and purified before being injected into appropriate geological storage formations. By combining biomass gasification with CCS, hydrogen production becomes more environmentally sustainable and provides a low-carbon, renewable method of producing hydrogen (Guler & Yumurtaci, 2025).

4. Synergy with Renewables:

CCS can be used in conjunction with renewable energy-powered hydrogen production techniques like electrolysis. A circular and sustainable energy system can be promoted by using excess renewable energy to capture and store CO_2 emissions from other energy-intensive processes. Industries can drastically cut their carbon emissions by incorporating CCS into several hydrogen production processes, supporting sustainable energy practices and attempts to mitigate climate change. In addition to reducing greenhouse gas emissions, the combination of hydrogen generation and CCS helps enterprises move toward cleaner energy production techniques and meet decarbonization targets (Madadi Avargani, et al., 2025). While there are several advantages to integrating CCS with hydrogen generation, there are drawbacks as well. By lowering CO_2 emissions, CCS integration into hydrogen production aids industry in meeting emissions reduction goals, adhering to regulations, and supporting international efforts to fight climate change. Additionally, using CCS guarantees that the hydrogen generated is cleaner and more ecologically friendly, increasing its marketability and sustainability as a low-carbon energy source. By reducing emissions and encouraging sustainable energy habits, the combination of CCS and hydrogen production speeds forward the decarbonization of transportation, power generation, and industrial operations (Khalili, et al., 2025).

13. CHALLENGES COUPLING OF CCS WITH HYDROGEN PRODUCTION

Adoption of CCS in hydrogen production facilities is financially hampered by the high upfront capital expenditures of adopting CCS technologies. The energy requirements of hydrogen production processes may rise as a result of CCS integration, which could have an impact on overall efficiency and operating costs (Alizadeh, e al., 2024). It can be difficult to set up the infrastructure needed for CO2 collecting, transportation, and storage; it involves stakeholder cooperation, regulatory compliance, and storage site investment. Ongoing research and innovation are required since the development and application of CCS technologies for hydrogen production may face challenges related to technolcgy readiness, scalability, and economic viability (Khalili, et al., 2025).

14. CONCLUSION

Biohydrogen production encompasses diverse biological routes, such as dark fermentation, photofermentation, and photobiological water-splitting, all guided by the common principle of directing electron flow toward hydrogenases and nitrogenases under controlled redox conditions. Its progress relies on enhancing enzyme systems for oxygen tolerance, utilizing waste-derived substrates to ensure economic viability, and integrating biological pathways with engineered platforms like anaerobic digestion and microbial electrolysis cells to improve overall energy recovery. Process intensification through immobilized cells, efficient gas stripping, and improved reactor design can address kinetic and mass-transfer limitations. Coupling hydrogen production with carbon capture and storage made this process mor sustainable and ecofriendly. In the coming decade, the synergy of metabolic engineering, bioprocess optimization, and supportive policy frameworks is anticipated to propel biohydrogen from laboratory research to decentralized, industrial-scale production systems capable of contributing meaningfully to the global hydrogen economy.

REFERENCES

Abdelfattah, A., Ali, S. S., Ramadan, H., El-Aswar, E. I., Eltawab, R., Ho, S. H., Elsamahy, T., Li, S., El-Sheekh, M. M., Schagerl, M., Kornaros, M., & Sun, J. (2023). Microalgae-based wastewater treatment: Mechanisms, challenges, recent advances, and future prospects. *Environmental Science and Ecotechnology*, *13*, 100205. DOI: 10.1016/j.ese.2022.100205

Addissouky, T. A. (2024). Precision medicine for personalized cholecystitis care: Integrating molecular diagnostics and biotherapeutics. *Bulletin of the National Research Center*, *48*(1), 89. DOI: 10.1186/s42269-024-01244-9

Ahmad, A., Rambabu, K., Hasan, S. W., Show, P. L., & Banat, F. (2024). Biohydrogen production through dark fermentation: Recent trends and advances in transition to a circular bioeconomy. *International Journal of Hydrogen Energy*, *52*, 335–357. DOI: 10.1016/j.ijhydene.2023.05.161

Ahmed, A. T. A., Seol, J. H., Seok, J. H., Jana, A., Meena, A., Cho, S., Sree, V. G., Park, Y., Lee, S. U., & Im, H. (2025). Boosting the electrocatalytic performance of CuCo2S4 via surface-state engineering for ampere current water electrolysis applications. *Applied Surface Science*, *680*, 161353. DOI: 10.1016/j.apsusc.2024.161353

Akhlaghi, N., & Najafpour-Darzi, G. (2020). A comprehensive review on biological hydrogen production. *International Journal of Hydrogen Energy*, *45*(43), 22492–22512. DOI: 10.1016/j.ijhydene.2020.06.182

Akram, F., Fatima, T., Ibrar, R., & ul Haq, I. (2024). Biohydrogen: Production, promising progressions and challenges of a green carbon-free energy. *Sustainable Energy Technologies and Assessments*, *69*, 103893. DOI: 10.1016/j.seta.2024.103893

Al-Dailami, A., Ahmad, I., Abdullah, N., Koji, I., & Yuzir, A. (2022, April). Feasibility and viability of procuring biohydrogen from microalgae: An emerging and sustainable energy resource technology. [). IOP Publishing.]. *Journal of Physics: Conference Series*, *2259*(1), 012014. DOI: 10.1088/1742-6596/2259/1/012014

Alamiery, A. (2024). Advancements in materials for hydrogen production: A review of cutting-edge technologies. *ChemPhysMater*, *3*(1), 64–73. DOI: 10.1016/j.chphma.2023.09.002

Alarcon, H. (2023). *Growth of Purple Sulfur Bacteria Allochromatium Vinosum on Solid Phase Metal Sulfides as Sulfur and Electron Sources* (Doctoral dissertation, The University of Texas at El Paso).

Albalawi, F. (2025). Hydrogen Energy Review: Challenges, Innovations, and Pathways to a Sustainable Future.

Alizadeh, S. M., Khalili, Y., & Ahmadi, M. (2024). Comprehensive review of carbon capture and storage integration in hydrogen production: Opportunities, challenges, and future perspectives. *Energies*, *17*(21), 5330. DOI: 10.3390/en17215330

Aminian-Dehkordi, J., Rahimi, S., Golzar-Ahmadi, M., Singh, A., Lopez, J., Ledesma-Amaro, R., & Mijakovic, I. (2023). Synthetic biology tools for environmental protection. *Biotechnology Advances*, *68*, 108239. DOI: 10.1016/j.biotechadv.2023.108239

Amjith, L. R., & Bavanish, B. (2022). A review on biomass and wind as renewable energy for sustainable environment. *Chemosphere*, *293*, 133579. DOI: 10.1016/j.chemosphere.2022.133579

An, B., Wang, Y., Huang, Y., Wang, X., Liu, Y., Xun, D., Church, G. M., Dai, Z., Yi, X., Tang, T.-C., & Zhong, C. (2022). Engineered living materials for sustainability. *Chemical Reviews*, *123*(5), 2349–2419. DOI: 10.1021/acs.chemrev.2c00512

Ananthi, V., Ramesh, U., Balaji, P., Kumar, P., Govarthanan, M., & Arun, A. (2024). A review on the impact of various factors on biohydrogen production. *International Journal of Hydrogen Energy*, *52*, 33–45. DOI: 10.1016/j.ijhydene.2022.08.046

Ananthi, V., Ramesh, U., Balaji, P., Kumar, P., Govarthanan, M., & Arun, A. (2024). A review on the impact of various factors on biohydrogen production. *International Journal of Hydrogen Energy*, *52*, 33–45. DOI: 10.1016/j.ijhydene.2022.08.046

Ashitha, S., & George, S. C. (2022). Splitting of water: biological and non-biological approaches. In *Handbook of Biofuels* (pp. 453–469). Academic Press. DOI: 10.1016/B978-0-12-822810-4.00023-3

Asrul, M. A. M., Atan, M. F., Yun, H. A. H., & Lai, J. C. H. (2022). A review of advanced optimization strategies for fermentative biohydrogen production processes. *International Journal of Hydrogen Energy*, *47*(38), 16785–16804. DOI: 10.1016/j.ijhydene.2022.03.197

Ayodele, D. T., Ogunbiyi, O. D., Akamo, D. O., Otun, K. O., Akinpelu, D. A., Adegoke, J. A., Fapojuwo, D. P., & Oladoye, P. O. (2023). Factors affecting biohydrogen production: Overview and perspectives. *International Journal of Hydrogen Energy*, *48*(71), 27513–27539. DOI: 10.1016/j.ijhydene.2023.04.001

Azizan, A., & Jalil, R. (2024). 9 Biophotolysis Process. *Hydrogen Production from Renewable Resources and Wastes*, 201.

Baeyens, J., Zhang, H., Nie, J., Appels, L., Dewil, R., Ansart, R., & Deng, Y. (2020). Reviewing the potential of bio-hydrogen production by fermentation. *Renewable & Sustainable Energy Reviews, 131*, 110023. DOI: 10.1016/j.rser.2020.110023

Barati, B., Zeng, K., Baeyens, J., Wang, S., Addy, M., Gan, S. Y., & Abomohra, A. E. F. (2021). Recent progress in genetically modified microalgae for enhanced carbon dioxide sequestration. *Biomass and Bioenergy, 145*, 105927. DOI: 10.1016/j.biombioe.2020.105927

Bayramoğlu, K., & Bayramoğlu, T. (2025). Integrated Modeling of Steam Methane Reforming and Carbon Capture for Blue Hydrogen Production. *Hydrogen, 6*(4), 94. DOI: 10.3390/hydrogen6040094

Beschkov, V., Parvanova-Mancheva, T., & Vasileva, E. (2023). Experimental Study of Bio-Hydrogen Production by Clostridium beijerinckii from Different Substrates. *Energies, 16*(6), 2747. DOI: 10.3390/en16062747

Bhagat, S. P., Wagh, B. D., Satpute, S. M., Gunjal, M. B., Kasar, S. K., & Roylawar, P. B. (2025). Fourth-Generation Biofuels: Genetic Engineering for Sustainable Energy. In *Recent Trends in Lignocellulosic Biofuels and Bioenergy: Advancements and Sustainability Assessment* (pp. 419–438). Springer Nature Singapore. DOI: 10.1007/978-981-96-4636-4_15

. Bolatkhan, K., Kossalbayev, B. D., Zayadan, B. K., Tomo, T., Veziroglu, T. N., & Allakhverdiev, S. I. (2019). Hydrogen production from phototrophic microorganisms: reality and perspectives. *international journal of hydrogen energy, 44*(12), 5799-5811.

Bozieva, A. M., Khasimov, M. K., Voloshin, R. A., Sinetova, M. A., Kupriyanova, E. V., Zharmukhamedov, S. K., Dunikov, D. O., Tsygankov, A. A., Tomo, T., & Allakhverdiev, S. I. (2023). New cyanobacterial strains for biohydrogen production. *International Journal of Hydrogen Energy, 48*(21), 7569–7581. DOI: 10.1016/j.ijhydene.2022.11.198

Buckel, W. (2021). Energy conservation in fermentations of anaerobic bacteria. *Frontiers in Microbiology, 12*, 703525. DOI: 10.3389/fmicb.2021.703525

Cha, M., Park, M. S., & Kim, S. J. (2025). Biological Routes for Biohydrogen Production: A Clean and Carbon-Free Fuel. *Biotechnology Journal, 20*(7), e70074. DOI: 10.1002/biot.70074

Cha, M., Park, M. S., & Kim, S. J. (2025). Biological Routes for Biohydrogen Production: A Clean and Carbon-Free Fuel. *Biotechnology Journal, 20*(7), e70074.

Chandrasekhar, K., Kumar, A. N., Raj, T., Kumar, G., & Kim, S. H. (2021). Bio-electrochemical system-mediated waste valorization. *Systems Microbiology and Biomanufacturing*, *1*(4), 432–443. DOI: 10.1007/s43393-021-00039-7

Chandrasekhar, T., Reddy, P. C. O., Swapna, B., Veeranjaneya Reddy, L., Anuprasanna, V., Dakshayani, L., Ramachandra Reddy, P., & Reddy, M. C. (2025). Algae: The game-changers in biohydrogen sector. *Critical Reviews in Biotechnology*, *45*(3), 509–529. DOI: 10.1080/07388551.2024.2387176

Das, D., & Veziroglu, T. N. (2008). Advances in biological hydrogen production processes. *International Journal of Hydrogen Energy*, *33*(21), 6046–6057. DOI: 10.1016/j.ijhydene.2008.07.098

Dash, S. K., Chakraborty, S., & Elangovan, D. (2023). A brief review of hydrogen production methods and their challenges. *Energies*, *16*(3), 1141. DOI: 10.3390/en16031141

Dayanath, W. M. S., Ilankoon, I. M. S. K., Ali, S. A. S., Chong, M. N., & Foo, S. C. (2026). Microalgae-Based Direct Air Capture Systems for Carbon Dioxide Emission Reductions in Developing Megacities: A Techno-Economic and Policy Analysis. *Journal of Environmental Chemical Engineering*, *14*(3), 122763. DOI: 10.1016/j.jece.2026.122763

Demirkaya, C., & De la Hoz Siegler, H. (2024). Dark Fermentation of Microalgae and Cyanobacteria for Hydrogen Production. In *Biotechnological Processes for Green Energy, and High Value Bioproducts by Microalgae, and Cyanobacteria Cultures* (pp. 99-111). Cham: Springer International Publishing.

Fakhimi, N., Gonzalez-Ballester, D., Fernández, E., Galván, A., & Dubini, A. (2020) Algae-bacteria consortia as a strategy to enhance H2 production. *Cells*, *9*(6), 1353 DOI: 10.3390/cells9061353

Ganguly, A., Sun, P., Liu, X., Delgado, H. E., Sun, L., & Elgowainy, A. (2025). Techno-economic and life cycle analysis of bio-hydrogen production using bio-based waste streams through the integration of dark fermentation and microbial electrolysis. *Green Chemistry*, *27*(21), 6213–6231. DOI: 10.1039/D4GC05020G

Gautam, R., Nayak, J. K., Talapatra, K. N., & Ghosh, U. K. (2023). Assessment of different organic substrates for bio-electricity and bio-hydrogen generation in an integrated bio-electrochemical system. *Materials Today: Proceedings*, *80*, 2255–2259. DOI: 10.1016/j.matpr.2021.06.223

Ghirardi, M. L., Zhang, L., Lee, J. W., Flynn, T., Seibert, M., Greenbaum, E., & Melis, A. (2000). Microalgae: A green source of renewable H_2. *Trends in Biotechnology, 18*(12), 506–511. DOI: 10.1016/S0167-7799(00)01511-0

Guler, N. U., & Yumurtaci, Z. (2025). Techno-economic evaluation of hydrogen production via biomass gasification: The role of carbon capture and storage in steam and oxygen gasification routes. *International Journal of Hydrogen Energy, 188*, 152134. DOI: 10.1016/j.ijhydene.2025.152134

Habib, M. A., Abdulrahman, G. A., Alquaity, A. B., & Qasem, N. A. (2024). Hydrogen combustion, production, and applications: A review. *Alexandria Engineering Journal, 100*, 182–207. DOI: 10.1016/j.aej.2024.05.030

Hassan, N. S., Jalil, A. A., Vo, D. V. N., & Nabgan, W. (2023). An overview on the efficiency of biohydrogen production from cellulose. *Biomass Conversion and Biorefinery, 13*(10), 8485-8507.

Hemashenpagam, N., Selvajeyanthi, S., & Gunasheela, N. (2024). Effect of Substrate on the Efficiency of Microbial Electrolysis Cells. In *Microbial Electrolysis Cell Technology: Fundamental to Sustainable Applications* (pp. 75–91). Springer Nature Singapore. DOI: 10.1007/978-981-97-3356-9_5

Hitam, C. N. C., & Jalil, A. A. (2023). A review on biohydrogen production through photo-fermentation of lignocellulosic biomass. *Biomass Conversion and Biorefinery, 13*(10), 8465–8483. DOI: 10.1007/s13399-020-01140-y

Homburg, S. V., Kruse, O., & Patel, A. V. (2019). Growth and photosynthetic activity of Chlamydomonas reinhardtii entrapped in lens-shaped silica hydrogels. *Journal of Biotechnology, 302*, 58–66. DOI: 10.1016/j.jbiotec.2019.06.009

Hooshmand Zaferani, S., Sams, M. W., Shi, X. L., Mehrabian, N., Ghomashchi, R., & Chen, Z. G. (2022). Applications of thermoelectric generators to improve catalytic-assisted hydrogen production efficiency: Future directions. *Energy & Fuels, 36*(15), 8096–8106. DOI: 10.1021/acs.energyfuels.2c01754

Hu, B., Li, Y., Zhu, S., Zhang, H., Jing, Y., Jiang, D., He, C., & Zhang, Z. (2020). Evaluation of biohydrogen yield potential and electron balance in the photo-fermentation process with different initial pH from starch agricultural leftover. *Bioresource Technology, 305*, 122900. DOI: 10.1016/j.biortech.2020.122900

Intasian, P., Prakinee, K., Phintha, A., Trisrivirat, D., Weeranoppanant, N., Wongnate, T., & Chaiyen, P. (2021). Enzymes, in vivo biocatalysis, and metabolic engineering for enabling a circular economy and sustainability. *Chemical Reviews, 121*(17), 10367–10451. DOI: 10.1021/acs.chemrev.1c00121

Ivanenko, A. A., Laikova, A. A., Zhuravleva, E. A., Shekhurdina, S. V., Vishnyakova, A. V., Kovalev, A. A., Kovalev, D. A., Trchounian, K. A., & Litti, Y. V. (2024). Biological production of hydrogen: From basic principles to the latest advances in process improvement. *International Journal of Hydrogen Energy*, *55*, 740–755. DOI: 10.1016/j.ijhydene.2023.11.179

Jalil, A., Ahmadi, H., Ndayisenga, F., Khan, S., Ahmad, A., Wang, X., & Yu, Z. (2025). Integrating dark fermentation and electrohydrogenesis for enhanced biohydrogen production from food waste. *Sustainable Energy & Fuels*, *9*(20), 5432–5457. DOI: 10.1039/D5SE00571J

Ji, M., & Wang, J. (2021). Review and comparison of various hydrogen production methods based on costs and life cycle impact assessment indicators. *International Journal of Hydrogen Energy*, *46*(78), 38612–38635. DOI: 10.1016/j.ijhydene.2021.09.142

Jiang, K., Bai, R., Gao, T., Lu, P., Zhang, J., Zhang, S., Xu, F., Wang, S., & Zhao, H. (2023). Optimization of hydrogen production in Enterobacter aerogenes by Complex I peripheral fragments destruction and maeA overexpression. *Microbial Cell Factories*, *22*(1), 137. DOI: 10.1186/s12934-023-02155-6

. Johnson, N., Liebreich, M., Kammen, D. M., Ekins, P., McKenna, R., & Staffell, I. (2025). Realistic roles for hydrogen in the future energy transition. *Nature Reviews Clean Technology*, 1-21.

Kaliyaperumal, G., Nagabhooshanam, N., Rathore, S., Kulshreshta, A., Sheela, C. S., Beulah, D., Maranan, R., Srinivasan, R., & Sathiyamurthy, S. (2025). Converging pathways in biomass-to-biohydrogen conversion: A review on hybrid systems, nanotechnological interventions, and sustainable integration frameworks. *Environmental Progress & Sustainable Energy*, •••, 70295. DOI: 10.1002/ep.70295

Kapdan, I. K., & Kargi, F. (2006). Bio-hydrogen production from waste materials. *International Journal of Hydrogen Energy*, *31*(12), 1664–1673.

Kar, S., Sanderson, H., Roy, K., Benfenati, E., & Leszczynski, J. (2021). Green chemistry in the synthesis of pharmaceuticals. *Chemical Reviews*, *122*(3), 3637–3710. DOI: 10.1021/acs.chemrev.1c00631

Khalili, Y., Yasemi, S., Abdi, M., Ghasemi Ertian, M., Mohammadi, M., & Bagheri, M. (2025). A Review of Integrated Carbon Capture and Hydrogen Storage: AI-Driven Optimization for Efficiency and Scalability. *Sustainability (Basel)*, *17*(13), 5754. DOI: 10.3390/su17135754

Khan, A. H., Bahar, A. N., Islam, A., & Wahid, K. (2026). A Review of Battery Storage Technology in Renewable Energy Utilization. *Arabian Journal for Science and Engineering*, *51*(8), 1–23. DOI: 10.1007/s13369-025-11038-4

Khan, S., Das, P., Abdul Quadir, M., Thaher, M. I., Mahata, C., Sayadi, S., & Al-Jabri, H. (2023). Microalgal feedstock for biofuel production: recent advances, challenges, and future perspective. *Fermentation, 9*(3), 281.

Khosravitabar, F., & Spetea, C. (2024). Evaluating technical parameters for microalgae immobilization to optimize green hydrogen photoproduction: A comparative review. *International Journal of Hydrogen Energy*, *67*, 925–932. DOI: 10.1016/j.ijhydene.2024.01.299

Kim, S. H., Kumar, G., Chen, W. H., & Khanal, S. K. (2021). Renewable hydrogen production from biomass and wastes (ReBioH2-2020). *Bioresource Technology*, *331*, 125024. DOI: 10.1016/j.biortech.2021.125024

Kumar, K., Anand, A., & Moholkar, V. S. (2024). Molecular hydrogen (H2) metabolism in microbes: a special focus on biohydrogen production. In *Biohydrogen-advances and processes* (pp. 25–58). Springer Nature Switzerland. DOI: 10.1007/978-3-031-49818-3_2

Kumar, P. V., Gobikannan, K., Nandakumar, S., & Manavaalan, G. (2025). Enhancing hydrogen production and wastewater treatment through an integrated bioelectrochemical system combining dark fermentation and galvanic cells. *Biomass and Bioenergy*, *201*, 108127. DOI: 10.1016/j.biombioe.2025.108127

Lak, S. Z., Rezaei, J., & Rahimpour, M. R. (2024). Health and pollution challenges of fossil fuels utilization. *Encycl. Renew. Energy Sustain. Environ*, *8*, 155.

Li, S., Tabatabaei, M., Li, F., & Ho, S. H. (2024). A review of green biohydrogen production using anoxygenic photosynthetic bacteria for hydrogen economy: Challenges and opportunities. *International Journal of Hydrogen Energy*, *54*, 218–238. DOI: 10.1016/j.ijhydene.2022.11.014

Lima, P. J. M., da Silva, R. M., Neto, C. A. C. G., Gomes e Silva, N. C., Souza, J. E. D. S., Nunes, Y. L., & Sousa dos Santos, J. C. (2022). An overview on the conversion of glycerol to value-added industrial products via chemical and biochemical routes. *Biotechnology and Applied Biochemistry*, *69*(6), 2794–2818. DOI: 10.1002/bab.2098

Liu, Y., Wang, Y., Wen, X., Shimizu, K., Lei, Z., Kobayashi, M., Zhang, Z., Sumi, I., Yao, Y., & Mogi, Y. (2018). Enhanced bioconversion of hydrogen and carbon dioxide to methane using a micro-nano sparger system: Mass balance and energy consumption. *RSC Advances*, *8*(47), 26488–26496. DOI: 10.1039/C8RA02924E

Ma, N., Zhao, W., Wang, W., Li, X., & Zhou, H. (2024). Large scale of green hydrogen storage: Opportunities and challenges. *International Journal of Hydrogen Energy, 50*, 379–396. DOI: 10.1016/j.ijhydene.2023.09.021

Madadi Avargani, V., Habibzadeh, M., Abdlla Maarof, H., Zendehboudi, S., & Duan, X. (2025). Harnessing renewable energy for hydrogen production: Advances, challenges, and opportunities. *Industrial & Engineering Chemistry Research, 64*(25), 12368–12418. DOI: 10.1021/acs.iecr.5c00061

Maeda, T., Sanchez-Torres, V., & Wood, T. K. (2007). Metabolic engineering to enhance bacterial hydrogen production. *Applied Microbiology and Biotechnology, 73*(1), 130–136.

Malode, S. J., Pandiaraj, S., Alodhayb, A., & Shetti, N. P. (2025). Review on Sustainable Biohydrogen Production from Biomass: Recent Developments and Prospects toward a Promising Source. *Energy & Fuels, 39*(17), 7967–7989. DOI: 10.1021/acs.energyfuels.5c00315

Mansoori, G. A., Agyarko, L. B., Estevez, L. A., Fallahi, B., Gladyshev, G., Santos, R. G. D., . . . Yen, J. (2021). Fuels of the future for renewable energy sources (ammonia, biofuels, hydrogen). *arXiv preprint arXiv:2102.00439*.

Melis, A., Zhang, L., Forestier, M., Ghirardi, M. L., & Seibert, M. (2000). Sustained photobiological hydrogen gas production upon reversible inactivation of oxygen evolution in the green alga Chlamydomonas reinhardtii. *Plant Physiology, 122*(1), 127–135. DOI: 10.1104/pp.122.1.127

Miyake, M., Miyake, J., & Asada, Y. (2013). Photosynthetic hydrogen production and its potential for practical application. *Photosynthesis Research, 116*(2–3), 409–420.

Nath, K., & Das, D. (2004). Improvement of fermentative hydrogen production: Various approaches. *Biotechnology Letters, 26*(1), 13–19.

Niescioruk, M. J., Bandrow, P., Szufa, S., Wozniak, M., & Siczek, K. (2025). Biomass-Based Hydrogen Extraction and Accompanying Hazards—Review. *Molecules, 30*(3).

Oliver, N., Avramov, A. P., Nürnberg, D. J., Dau, H., & Burnap, R. L. (2022). From manganese oxidation to water oxidation: Assembly and evolution of the water-splitting complex in photosystem II. *Photosynthesis Research, 152*(2), 107–133. DOI: 10.1007/s11120-022-00912-z

Patel, A. K., Singhania, R. R., Albarico, F. P. J. B., Pandey, A., Chen, C. W., & Dong, C. D. (2022). Organic wastes bioremediation and its changing prospects. *The Science of the Total Environment, 824*, 153889. DOI: 10.1016/j.scitotenv.2022.153889

Phogat, P., Sharma, S., & Rawat, S. (2026). *Next-Generation Hydrogen Economy: Production, Storage, and Fuel Cell Technologies*. John Wiley & Sons.

Pivac, I., Šimunović, J., Barbir, F., & Nižetić, S. (2024). Reduction of greenhouse gases emissions by use of hydrogen produced in a refinery by water electrolysis. *Energy*, *296*, 131157. DOI: 10.1016/j.energy.2024.131157

Priyadharsini, P., Nirmala, N., Dawn, S. S., Baskaran, A., SundarRajan, P., Gopinath, K. P., & Arun, J. (2022). Genetic improvement of microalgae for enhanced carbon dioxide sequestration and enriched biomass productivity: Review on CO2 bio-fixation pathways modifications. *Algal Research*, *66*, 102810. DOI: 10.1016/j.algal.2022.102810

Qian, Z., Yanqiu, S., Lin, G., Hongmei, D., Lihan, Z., Shu, Y., & Qi, L. (2024). Sulfur source promotes the biosorption and bioprecipitation of Cd in purple non-sulfur bacteria. *International Biodeterioration & Biodegradation*, *188*, 105742. DOI: 10.1016/j.ibiod.2024.105742

Raut, N., Al Saadi, A., Ghasemi, M., & Vaidya, R. (2025). Enhancing Biohydrogen Yield and Slaughterhouse Wastewater Treatment Efficiency through Microalgae and Bacterial Synergy. *Periodica Polytechnica. Chemical Engineering*, *69*(2), 313–328. DOI: 10.3311/PPch.38818

Rawindran, H., Khoo, K. S., Satpati, G. G., Maity, S., Chandran, K., Lim, J. W., Tong, W.-Y., Setiabudi, H. D., & Yunus, N. M. (2025). Composition of carbohydrate, protein and lipid derived from microalgae using thermally pretreated solid waste. *Journal of the Science of Food and Agriculture*, *105*(9), 4672–4679. DOI: 10.1002/jsfa.14038

Sadvakasova, A. K., Balouch, H., Bauenova, M. O., Kossalbayev, B. D., Allakhverdiev, S. I., & Shabala, S. (2025). Strategies to intensify CO2 capture by microalgae for the circular bioeconomy. *Trends in Plant Science*.

Sarkar, O., Santhosh, J., Dhar, A., & Mohan, S. V. (2021). Green hythane production from food waste: Integration of dark-fermentation and methanogenic process towards biogas up-gradation. *International Journal of Hydrogen Energy*, *46*(36), 18832–18843. DOI: 10.1016/j.ijhydene.2021.03.053

Serik, A., Kuspanov, Z., & Daulbayev, C. (2026). Cost-effective strategies and technologies for green hydrogen production. *Renewable & Sustainable Energy Reviews*, *226*, 116242. DOI: 10.1016/j.rser.2025.116242

Shaban, M., Sadeq, A. M., Rajhi, W., & Hajlaoui, K. (2026). Multi-objective optimization of a biomass-fired supercritical CO2 Brayton–electrolysis system for high-efficiency hydrogen, power, and cogenerative thermal services. Journal of CO2 Utilization, 103(10328), 1.

Shanmugam, S., Hari, A., Pandey, A., Mathimani, T., Felix, L., & Pugazhendhi, A. (2020). Comprehensive review on the application of inorganic and organic nanoparticles for enhancing biohydrogen production. *Fuel*, *270*, 117453. DOI: 10.1016/j.fuel.2020.117453

Sharma, P., Jain, A., Bora, B. J., Balakrishnan, D., Show, P. L., Ramaraj, R., Ağbulut, Ü., & Khoo, K. S. (2023). Application of modern approaches to the synthesis of biohydrogen from organic waste. *International Journal of Hydrogen Energy*, *48*(55), 21189–21213. DOI: 10.1016/j.ijhydene.2023.03.029

Shevela, D., Kern, J. F., Govindjee, G., & Messinger, J. (2023). Solar energy conversion by photosystem II: Principles and structures. *Photosynthesis Research*, *156*(3), 279–307. DOI: 10.1007/s11120-022-00991-y

Singh, V., Aishwarya, V. M., Sriprasath, V. J., Pranavi, M., & Singh, T. (2025). Green hydrogen value chain challenges and global readiness for a sustainable energy future. *iScience*, *28*(7). Advance online publication. DOI: 10.1016/j.isci.2025.112900

Song, S., Ginige, M. P., Cheng, K. Y., Peacock, C. S., & Kaksonen, A. H. (2024). Ultrasonication-induced metabolic pathway shifts and reduced electron carrier washout with biomass enhanced hydrogen yield in a continuous stirred tank reactor. *Chemical Engineering Journal*, *493*, 152594. DOI: 10.1016/j.cej.2024.152594

Tamagnini, P., Axelsson, R., Lindberg, P., Oxelfelt, F., Wunschiers, R., & Lindblad, P. (2002). Hydrogenases and hydrogen metabolism of cyanobacteria. *FEMS Microbiology Reviews*, *26*(1), 67–81.

Tang, T., Chen, Y., Liu, M., Du, Y., & Tan, Y. (2022). Effect of pH on the performance of hydrogen production by dark fermentation coupled denitrification. *Environmental Research*, *208*, 112663. DOI: 10.1016/j.envres.2021.112663

Thomas, D. M., Mechery, J., & Paulose, S. V. (2016). Carbon dioxide capture strategies from flue gas using microalgae: A review. *Environmental Science and Pollution Research International*, *23*(17), 16926–16940. DOI: 10.1007/s11356-016-7158-3

Tripathi, D. M., & Tripathi, S. (2024). Microbial advancements in dark fermentative biohydrogen production: applications and innovations. In *Emerging trends and techniques in biofuel production from agricultural waste* (pp. 57–80). Springer Nature Singapore. DOI: 10.1007/978-981-99-8244-8_4

Tsouko, E., Pilafidis, S., Kourmentza, K., Gomes, H. I., Sarris, G., Koralli, P., ... & Sarris, D. (2024). A sustainable bioprocess to produce bacterial cellulose (BC) using waste streams from wine distilleries and the biodiesel industry: evaluation of BC for adsorption of phenolic compounds, dyes and metals. *Biotechnology for Biofuels and Bioproducts, 17*(1), 40.

Ueno, Y., Otsuka, S., & Morimoto, M. (1998). Hydrogen production from industrial wastewater by anaerobic microflora in chemostat culture. *Biotechnology Letters, 20*(3), 195–199.

Wang, J., & Yin, Y. (2021). Clostridium species for fermentative hydrogen production: An overview. *International Journal of Hydrogen Energy, 46*(70), 34599–34625. DOI: 10.1016/j.ijhydene.2021.08.052

Wani, N. R., Rather, R. A., Farooq, A., Padder, S. A., Baba, T. R., Sharma, S., Mubarak, N. M., Khan, A. H., Singh, P., & Ara, S. (2024). New insights in food security and environmental sustainability through waste food management. *Environmental Science and Pollution Research International, 31*(12), 17835–17857. DOI: 10.1007/s11356-023-26462-y

Wu, P., Ding, P., Cao, Q. H., Zhang, C., Fu, B., Liu, H. B., Chen, C., & Liu, H. (2023). Amino acids as in-situ electron donors drive medium chain fatty acids production from sludge acidogenic fermentation liquid by electro-fermentation enhancement. *Chemical Engineering Journal, 476*, 146537. DOI: 10.1016/j.cej.2023.146537

Xu, Z., Qi, H., Yao, D., Zhang, J., Zhu, Z., Wang, Y., & Cui, P. (2022). Modeling and comprehensive analysis of food waste gasification process for hydrogen production. *Energy Conversion and Management, 258*, 115509. DOI: 10.1016/j.enconman.2022.115509

Xuan, J., He, L., Wen, W., & Feng, Y. (2023). Hydrogenase and nitrogenase: Key catalysts in biohydrogen production. *Molecules (Basel, Switzerland), 28*(3), 1392. DOI: 10.3390/molecules28031392

Zavala-Mendez, M., Vargas, A., & Carrillo-Reyes, J. (2022). Maximization of bio-hydrogen production from winery vinasses using on-line feedback control. *International Journal of Hydrogen Energy, 47*(78), 33259–33271. DOI: 10.1016/j.ijhydene.2022.07.196

Zhang, J., Xue, D., Wang, C., Fang, D., Cao, L., & Gong, C. (2023). Genetic engineering for biohydrogen production from microalgae. *iScience, 26*(8), 107255. DOI: 10.1016/j.isci.2023.107255

Zhang, J., Xue, D., Wang, C., Fang, D., Cao, L., & Gong, C. (2023). Genetic engineering for biohydrogen production from microalgae. *iScience*, *26*(8), 107255. DOI: 10.1016/j.isci.2023.107255

Zhu, J., Li, M., Yu, H., Zheng, Y., Yuan, L., Cao, Y., Liu, X., Sun, F., & Chen, C. (2024). Magnetic biochar enhanced microbial electrolysis cell with anaerobic digestion for complex organic matter degradation in landfill leachate. *The Science of the Total Environment*, *949*, 175013. DOI: 10.1016/j.scitotenv.2024.175013

Chapter 10
Diatoms as Ecological Engineers, Carbon Sequestration, and Emerging Biotechnologies

Muhammad Nauman Khan
http://orcid.org/0000-0003-3790-0902
University of Chakwal, Punjab, Pakistan

Hansa Gul
http://orcid.org/0009-0001-5554-617X
University of Mianwali, Punjab, Pakistan

Alevcan Kaplan
http://orcid.org/0000-0001-6738-7527
Dicle University, Diyarbakır, Turkey

ABSTRACT

Diatoms are dominant microalgae distinguished by intricately patterned silica frustules and vital roles in aquatic ecosystems. As major contributors to global primary productivity, they drive carbon sequestration through the biological carbon pump due to efficient carbon fixation and rapid sinking. This chapter examines their evolutionary diversity, physiological traits, silica biomineralization, photosynthetic efficiency, and nutrient-driven bloom dynamics. It also explores diatom-based biotechnologies, including biofuels, biosilica engineering, carbon-neutral materials, biosensors, and nanotechnology. Advances in genomics and metabolic engineering to enhance carbon capture are discussed, highlighting their ecological and industrial significance.

DOI: 10.4018/979-8-3373-7630-1.ch010

1. INTRODUCTION

Diatoms comprise a group of unicellular microalgae that occur in marine, freshwater and brackish environments (Salleh et al., 2024). They are one of the most productive photosynthetic organisms on Earth, accounting for about 20% of global net primary production and nearly 40% of marine primary production (Field et al., 1998). Fossil records suggest that the diatoms appeared 250 to 190 million years ago and since then had become greatly diversified, giving rise to the current domination of ecological and biogeochemical features by this phytoplankton group (Medlin, 2016). Diatoms, as crucial element of marine phytoplankton communities, dominate the oceanic carbon cycles and are considered a major natural reservoir for CO_2 sequestration.

Their extraordinary capacity to fix carbon aids in the sequestration of more than 25% of atmospheric CO_2, making them one of the most vital natural intermediates for climate change mitigation strategies (Tiwari et al., 2022). Besides, their ubiquity and responses to environmental changes make diatoms ideal bioindicator organisms of the health status of aquatic ecosystems. Carbon sequestration is one of the key ecological services performed by diatoms (Salleh et al., 2024). Diatoms are also natural reservoirs for atmospheric CO_2 through photosynthesis, and it has been suggested that large-scale culture of diatoms could serve as a sustainable means to increase carbon sequestration. In addition, their metabolic plasticity also allows for the redirection of assimilated carbon into high value-added molecules such as lipids, pigments and omega-3 fatty acids providing an environmentally friendly route for bioresource production (Sethi et al., 2020).

Diatoms bloom dynamics is another essential ecological process which is driven by abiotic drivers such as light, nutrients and water content. These blooms largely impact nutrient dynamics, primary productivity and carbon fluxes within ecosystem. Additionally, bloom conditions may promote the production of commercially valuable metabolites or substances so linking ecological processes to biotechnological opportunity (Avrahami et al., 2025; Zhuang et al., 2025). In addition to their ecological significance, diatoms show remarkable biosilica-forming skills. The fine-patterned silica cell walls called frustules are found to be highly porous in the nanometer range with high mechanical strength and biocompatible, which can be utilized for the development of new drug delivery systems, biosensor and in the nanotechnology. Recent developments have shown that the purified bio silica from diatom is a potential precursor to prepare metal-doped silica particles, showing their versatility for advanced materials (Terracciano et al., 2018) (Salleh et al., 2024).

In biotechnology, diatoms are also being tapped as source of implementing sustainable production technology for biofuels, pharmaceuticals and nutraceuticals along with other value added products, the biosilica–PVA microcomposites obtained from SIMV-5 algae showed obvious unexpected crack lines at different load or stress modes under controlled parameters as can be observed in these AFM micrographs which were clearly noticed on both tension and compression images. The fast growth rates, genetic manipulative ease and ability to bio-synthesize a variety of high value compounds make them highly suitable as attractive production platforms for green biotechnology, industrial etc (Salleh et al., 2024).

Notwithstanding their great potential, a number of hurdles still need to be overcome for the extensive utilization of diatoms in ecologic and biotechnological applications. The optimization of culture systems, the control of environmental stresses and sustainable culture media development are key for large-scale applicability. Furthermore, in order to make their application more specific, the detailed research studies are necessary on the diversity and species-specific morphological characters of diatoms for its applications in various industrial sectors (Salleh et al., 2024).

This chapter will first delve into the fundamental and unique biological features of diatoms, establishing the scientific basis for their diverse roles. Following this, we will explore their profound impact as ecological engineers, before transitioning to a detailed examination of how these intrinsic biological and ecological attributes are being harnessed for a wide array of biotechnological and industrial applications. Finally, we will address the future perspectives and key challenges that must be overcome to fully realize the potential of these remarkable microalgae.

2. THE UNIQUE BIOLOGY OF A DIATOM

2.1 The Silica Frustule and Cytoskeleton

The most defining biological feature of diatoms is their ornate, silica-based cell wall, known as the frustule. This rigid, nanopatterned structure is central to diatom taxonomy and their ecological success. However, this robust barrier has historically posed a significant challenge to cell biologists because it renders diatoms impermeable to many dyes and antibodies, limiting detailed studies of their internal ultrastructure (Flori et al., 2025).

The internal structure of diatoms reflects the presence of a well-organized cytoskeleton which is essential for maintaining cell shape, organizing organelles, and for producing the new half-valve during cell division. One important aspect is the a centriolar, multilayered microtubule-organizing center (MTOC) which mediates cell division in a way mediating for many other Eukaryotes. In spite of the huge range in morphology between centric (radially symmetric) and pennate (bilaterally symmetric) diatoms, our cryo-ExM investigations have demonstrated that key features of interphase microtubule organization are highly conserved. Microtubules seem to be arranged in all species with their minus-end anchored at or near the nuclei and their plus-ends reaching towards the cell periphery, indicating a conserved foundation of architecture through over 80 million years of evolution (Flori et al., 2025). Moreover, these high-resolution studies demonstrate that microtubules do penetrate along the setae into the long spines of diatom species such as Chaetoceros and carry out a morphogenetic role in these complex silica structures.

2.2. A Plastic Photosynthetic Apparatus

(INL), contrary to the conserved cytoskeleton, the diatom photosynthetic apparatus demonstrates extraordinary plasticity. Diatoms have complex plastids of secondary endosymbiotic origin, which contain stacked thylakoids where light reactions of photosynthesis take place. Central to carbon fixation is the pyrenoid, a dense phase in the chloroplast that sequesters ribulose-1,5-bisphosphate carboxylase/oxygenase (RuBisCO) with CO_2. The architecture of these organelles is species-specific and indicates adaptation to distinct ecological niches. For instance, *C. granii* possesses several oval chloroplasts and tetrahedral pyrenoids; the pennate diatom *Cylindrotheca sp.* has a pair of long, large chloroplasts (Flori et al., 2025). This phenotypic plasticity enables each species to construct different photosynthetic machinery that is specialized for its particular environment in terms of light absorption and CO_2 assimilation.

The intricate biological machinery of diatoms, from their robust silica frustules to their highly efficient carbon concentrating mechanisms, underpins their success in aquatic environments. These unique traits not only define their physiological resilience but also position them as pivotal players in global biogeochemical cycles, as will be explored in the subsequent section on their roles as master ecological engineers.

3. DIATOMS AS MASTER ECOLOGICAL ENGINEERS

3.1. The Biological Carbon Pump: The Engine of Global Carbon Sequestration

Carbon is found in several inorganic species in natural aquatic systems, and its uptake by diatoms is influenced by a combination of chemical equilibria and biology. In marine systems, when atmospheric CO_2 dissolves in water it forms carbonic acid as a result; carbonic acid dissociates into bicarbonate (HCO_3^-) and carbonate (CO_3^{2-}). In contrast to the terrestrial plants which use predominantly atmospheric CO_2, aquatic photosynthetic organisms depend for a large fraction of their carbon requirements on dissolved inorganic carbon (DIC), ~90% of which in oceanic waters is bicarbonate (Poschenrieder et al., 2018). Before the industrial revolution, atmospheric CO_2 levels were around 280 ppm (Battin et al., 2009), but have increased drastically to about 420 ppm in 2020, which led to enhanced sequestration of CO_2 from the atmosphere in terrestrial and marine environments. Seawater bicarbonate concentrations, that were already estimated at 1757 μmol kg^{-1} under pre-industrial conditions have risen with the consequence of ocean acidification and changes in carbonate chemistry (Poschenrieder et al., 2018).

Diatoms are excellently adapted to the uptake of dissolved inorganic carbon (DIC) from their natural aquatic habitat. They can utilize both CO_2 and bicarbonate (HCO_3^-), the latter being the main form of DIC in seawaterAnother useful source of carbon is sodium bicarbonate, which is more convenient than gaseous CO_2 as a carbon source and its uptake can be achieved through Na^+-dependent transporter systems (Flori et al., 2025; Rees, 1984). These sophisticated CCMs mean that RuBisCO, the enzyme found inside the pyrenoid, is provided with high levels of CO_2 – something not observed in plants – and thus diatoms serve as a potent engine for global carbon capture.

A large number of diatoms are known to positively respond to higher CO_2 with enhanced lipids yield. For example, *Thalassiosira weissflogii* and *Cyclotella cryptica* had higher lipid contents when 10% (v/v) CO_2 was added than control treatments, respectively (Botte et al., 2018b). In response to elevated CO_2 levels, lipid production increases. For instance, lipid accumulation of *Thalassiosira weissflogii* and *Cyclotella cryptica* was increased when supplemented with 10% (v/v) CO_2; triacylglycerols were the most common glycerolipid (>60%) in *C. cryptica* (Botte et al., 2018a). A significant uptake of carbon dioxide as bicarbonate has with the model diatom *Phaeodactylum tricornutum* been demonstrated, suggesting that it is an important inorganic carbon source (Hopkinson et al., 2016). CO_2 fixation rates of 35 g m^{-2} d^{-1} have been recorded for this species (Buono et al., 2016). Air-sparged culture experiments showed CO_2 consumption rates of around 1 g g^{-1} dry weight (DW) at pH 7.2 and 0.8 g g^{-1} DW at pH 9, which translates to a daily CO_2 uptake of about 0.06–0.08 g, and removal efficiencies between 50% and 65% for the supplied air (Piiparinen et al., 2018). Approaches such as the generation of microbubbles have even enhanced CO_2 solubilisation and uptake efficiency (Zimmerman et al., 2008). Also, predictive control algorithms in photobioreactors have demonstrated beneficial results in term of biomass productivity with low CO_2 costs consumed (García Sánchez et al., 2003). (Negoro et al., 1991).

Cross-species comparison of CCM performance suggests that species-specific characteristics have a greater influence on CCM efficiency than environmental conditions. Although higher CO_2 has generally had stimulatory effects, inhibitory responses have also been observed. For example, growth of *Attheya longicornis* was suppressed with high CO_2 availability (Artamonova et al., 2017). Temperature also interacts with CO_2 availability; higher temperature and CO_2 caused a decrease in cell size in *Navicula distans*, impacting metabolic rates (Baragi et al., 2015). In contrast, some diatom species, such as *Chaetoceros brevis*, show little physiological acclimation in response to elevated CO_2 (Boelen et al., 2011) (Figure 1).

Figure 1. Carbon Sequestration Phenomena in Diatoms (At the Cellular Level).

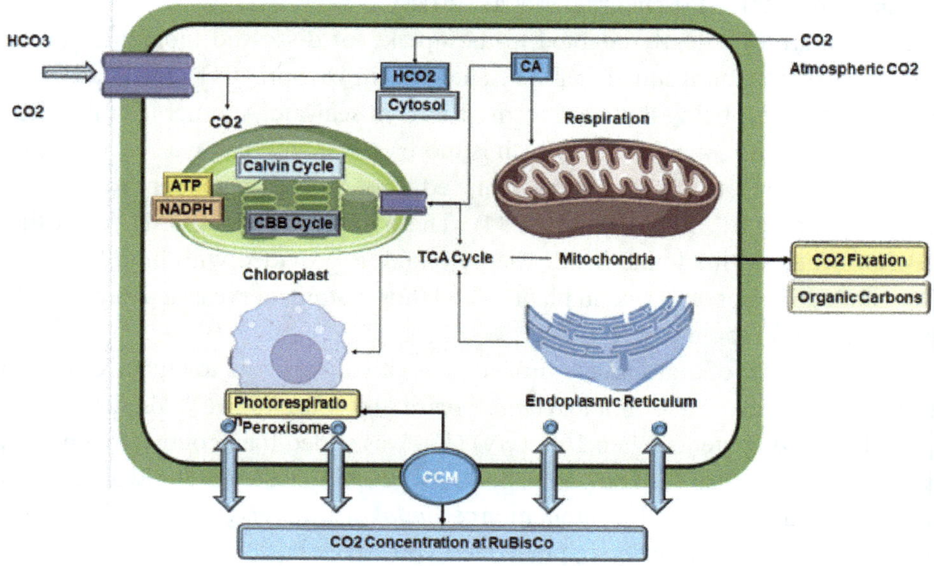

At the cellular level, carbon acquisition is controlled by complex mechanisms that enable CO_2 permeation and/or compartmentalization in multiple cellular compartments, such as cytosol, chloroplasts, mitochondria, peroxisomes and endoplasmic reticulum to finally accumulate CO_2 around RuBisCO for efficient fixation (Fettke & Fernie, 2015; Smith & Stitt, 2007). The molecular actors of the diatom carbon metabolism include chloroplasts, carbonic anhydrases, RuBisCO, enzymes from the Calvin–Benson–Bassham (CBB) cycle, and transporters proteins (Launay et al., 2020). Interestingly, members of the bZIP14 family have been suggested to be involved in CO_2 sensing and blue-light responses (Matthijs et al., 2017). Despite recent advances, advanced knowledge of enzyme localization, regulatory networks, and metabolic fluxes under different environmental conditions is required for improved CO_2 sequestration strategies using diatoms and will contribute to increasing their biotechnological potential.

3.2 Architects of Aquatic Food Webs

The diatoms are vital primary producers, forming the base of marine and aquatic food webs. Attachment of their biofilms to novel or surrogate surfaces is a pivotal process that governs the recruitment and settlement of subsequent assemblages such as larvae of fish and invertebrates (Ferguson et al., 2021; Tuck et al., 2022). The critical function of this factor is underscored that "primary productivity determines the diversity, succession, and abundance of attendant functional communities (Salamone et al., 2016).

In this research, nature-based imizi substrate was found to be very good in supporting this initial phase on primary production. Diatom colonisation was evident within 12 hours of deployment and resulted in an approximately 4.6-fold increase in species diversity after one week. The study revealed a distinct successional trend inherent in a food web base in the form of early colonizers like *Fragilaria pulchella*, and *Grammatophora undulata* at the initial stage leading to well-established and stable community precedence by genera like Cocconeis at 1 month after exposure (Ndaba et al., 2025). This observed evolution is a validation of the fact that the substrate developed was able to sponsor the installation of an efficient primary producer community, indispensable as a first compartment for more complicated aquatic food web.

3.3 Sentinels of Ecosystem Health (Biomonitoring)

Diatom communities were found to be a strong and sensitive bioindicator of ecosystem health, as the species composition reflects habitat quality, and accumulation through chain food is directly proportional to ambient pollution level Applied diatom community indices for the assessment of environmental fitness of substrate. There were significantly higher species richness, abundance and Shannon diversity index detected after one month deployment. The formation of this highly species-rich community which was similar to those found on other natural substrates such as wood and bamboo (Khatoon et al., 2007; Nenadović et al., 2015) was a clear indicator that the Indigencus material offered a suitable healthy habitat for marine life settlement (Ndaba et al., 2025).

Status of diatom biofilms as indicators of metal contamination. Biofilm, formed on imizi structures was able to bioaccumulate a number of metals in the urban coastal environment (Al, Fe and Zn). Most notably, the biofilm concentrated the toxic heavy metal arsenic (As) at levels that were "orders of magnitude higher" than in ambient water (Ndaba et al., 2025). Such a high potential for bioremediation was expressed as bioconcentration factors (BCF-1) useful to assess the efficiency of the diatom-biofilm enriched depositional area as sink of environmental pollutants. Such combined effects of biofilm growth and metal accumulation highlight the interest to using diatoms as bioindicators to evaluate habitat quality, along with the accuracy of certain pollutants (Ndaba et al., 2025; Tien & Chen, 2013).

3.4 Role in Coastal Restoration and "Greening Grey Infrastructure"

Ecological engineering is an approach combining ecological theory and engineering principles to create self-sustaining ecosystems, a concept of increasing relevance in coastal areas where natural systems have been transformed by human activity (Firth et al., 2024). Approaches such as "greening the grey infrastructure" are increasingly used to increase biodiversity and functionality for human use, of artificial structures such as seawalls and jetties (Ndaba et al., 2025). The success of these interventions frequently relies on the establishment provided to initial colonisers, including diatom biofilms that are pivotal to a healthy ecosystem.

The study by Ndaba et al. (2025) illustrates this principle through the investigation of an Indigenous Knowledge-based material, *Cyperus textilis* (imizi), as a nature-based prototype for coastal ecological engineering. The potential of this plant for inducing diatom succession along the natural colonisation cascade was tested by using trials built with it. The results suggested that the imizi structures had promoted a diverse diatom community and Indigenous materials can be succesfully incorporated into eco-engineering approaches to enhance primary production on urbanized coastlines. This perspective emphasises diatoms as more than passive squatters, but important targets and a barometer for success in engineering toward ecological restoration (Ndaba et al., 2025).

4. HARNESSING DIATOM ENGINEERING FOR BIOTECHNOLOGY AND INDUSTRY

4.1 Biofuels and Bioenergy

The carbon sequestration ability of diatoms arises from the functionality of their carbon concentrating mechanisms (CCMs) and CCM components such as carbonic anhydrases and RuBisCO. Some organisms are able to use both CO_2 and HCO_3^-, which are especially appropriate for industrial purposes, as in *Nitzschia palea* and *Chaetoceros muelleri* (Hu & Gao, 2008; Ihnken et al., 2011). Light availability also is critical in CCM regulation; one such example exists where external carbonic anhydrase activity and CO_2 affinity increase proportionally to high irradiance levels in *Skeletonema costatum*, enabling support for photosynthesis even when carbon-limited (Chen & Gao, 2003).

Diatoms have highly-ordered porous biosilica microshells and great potential as biotechnology resources. They are potential candidates for biofuel production as they can accumulate high levels of lipids under nutrient stress. Quite distinct from the traditional soybean or palm oil source, diatoms grow faster, have a larger yield of biomass and can be cultivated in non-arable land fields with little competition for food crops.

Diatoms are also investigated for biophotovoltaics, which can produce electric power by their photosynthetic performance and see potential application such as providing sustainable low-energy consumption power sources, wearable electronic or remote-sensing equipment.

4.2 Nanotechnology and Nanomedicine

In medicine, diatom frustules are used as biocompatible and bioresorbable carriers for the drug delivery. Modified frustules can be used as a vehicle for drugs, offering stimuli=response release, improved bioavailability, and limited side effects, thus opening up potential avenues in the eld of nanomedicine. Diatoms possess unique micro- to nanoscale properties that have enabled their integration into advanced nanotechnology and biotechnology applications (Rabiee et al., 2021)

4.2.1 Nanomedicine and Drug/Gene Delivery

The nanoporous structure of diatom frustules makes them highly valuable for drug and gene delivery systems. Genetically engineered diatom-derived biosilica has been successfully employed for targeted anticancer drug delivery, including studies on neuroblastoma and B-lymphoma cells, as well as in mouse xenograft models of neuroblastoma.

4.2.2 Bioimaging and Biosensing

The optical and structural properties of diatoms facilitate their use in bioimaging and biosensing applications, allowing sensitive detection and imaging at the cellular level.

4.2.3 Feed and Aquaculture

Diatom-based compounds can also be incorporated into animal feed and aquaculture, potentially improving animal health and growth (Phong et al., 2023).

4.2.4 Other Technological Applications

Beyond medicine, diatoms have shown potential in diverse technological fields, including photodynamic therapy, microfluidics, biophotonics, solar cells and molecular filtration systems (Rabiee et al., 2021).

Table 1. Types of bioactive compounds derived from diatoms, their examples, biological activities, and potential applications across food, health, and aquaculture sectors

Type of Bioactive Compound	Examples	Biological Activities	Applications
Polysaccharides	Chrysolaminarin, Chitin	Antioxidant, Immunomodulatory, Antimicrobial	Functional foods, Nutraceuticals, Pharmaceuticals
Proteins & Bioactive Peptides	Specific peptides from diatom proteins	Anti-inflammatory, Antioxidant, Antimicrobial	Drug development, Nutraceuticals
Pigments	Chlorophylls, Carotenoids, Marennine	Antioxidant, Antimicrobial, Coloring agent	Food additives, Cosmetics, Pharmaceuticals
Lipids / Fatty Acids	Omega-3 PUFAs, Eicosapentaenoic acid (EPA)	Cardioprotective, Anti-inflammatory	Nutraceuticals, Functional foods, Aquaculture
Phenolic Compounds	Polyphenols	Antioxidant, Anti-aging, Antimicrobial	Cosmetics, Food preservatives, Nutraceuticals
Vitamins & Minerals	Vitamin E, Selenium	Antioxidant, Immune support	Nutritional supplements, Functional foods

4.3 Environmental Remediation

Diatoms are important in the application of bioremediation and sustainability. Their frustules have the ability to adsorb heavy metals, such as lead, mercury and cadmium which make them suitable for wastewater treatment. Photosynthesis carried out by diatoms helps absorb large amounts of CO_2 from the atmosphere. Some species also have the ability to degrade hydrocarbons, increasing the natural biodegradation of oil spills and, as part of other bioengineered bacteria, hasten bioremediation in freshwater and marine habitats (Barua, 2025).

Beyond heavy metals and hydrocarbons, diatoms are being utilized to address the growing environmental challenge of mariculture wastewater, which often leads to coastal eutrophication. Recent research by Shen et al. (2024) demonstrated that the targeted cultivation of diatoms like *Chaetoceros muelleri* and *Conticribra weissflogii* can effectively treat nutrient-rich wastewater while simultaneously producing valuable algal biomass for aquaculture feed. This study established a precise nutrient regulation framework, identifying optimal concentrations—specifically 500 μmol/L silicate, 0.6 mg/L phosphate, and 4 mg/L ammonium—to maximize growth and nutrient uptake efficiency. This approach represents a transition from simple pollution mitigation to a circular economy model of resource recovery (Shen et al., 2024).

4.4 Sustainable Agriculture and Industrial Materials

The large-scale agricultural use of diatoms boosts soil properties, promotes plant growth and protects plants from pests. Using diatomaceous earth also has positive effects on soil aeration, water retention, and nutrient availability with its silicon content leading to strengthening of the plant structure as well as reducing pest stress. Diatom biofertilizer increases, microbial diversity and yields in crop production with the help of DE as non-poisonous pest control agent for controlling aphids, mites and weevils to support organic, integrative farming approaches (Barua, 2025).

Diatomaceous earth (DE), which was derived from fossilized diatom frustules, has been employed in industrial applications as a result its porosity, light weight, and abrasive characteristics. It is widely used in filtration, particularly for water treatment, beverages and pharmaceuticals, as it can effectively remove impurities, bacteria and suspended solids (Hamano et al., 2021).

DE is also applied as a pest control treatment, because the fine powder penetrates the exoskeletons of insects by entering their joints. It also acts as a refractory, high-temperature insulation material which is lightweight and fire resistant for furnace and kiln lining.

Diverse natural sources, including diatoms, are increasingly recognized for their rich profiles of bioactive compounds with multifaceted activities. These compounds offer significant potential across various sectors (Phong et al., 2023), (Phong et al., 2023). Table 1 summarizes the key types of bioactive compounds derived from diatoms, outlining their examples, biological activities, and applications in food, health, and aquaculture.

4.6 Genetic Engineering and the Path to Synthetic Diatoms

The transition of diatoms from ecological subjects to industrial production platforms is underpinned by their high degree of genetic tractability. Modern genetic toolkits now allow for the precise reprogramming of diatom metabolism to enhance the yield of lipids, pigments, and mineral structures (Y. Li et al., 2025). The implementation of CRISPR/Cas9 and Cas9-nickase systems has revolutionized genome editing, enabling stable, biallelic gene knockouts that were previously difficult to achieve in diploid microalgae (Yixuan Li et al., 2025).

Beyond simple gene edits, the delivery of large genetic episomes via bacteria-mediated conjugation allows for the introduction of entire metabolic pathways into the diatom cell. These extrachromosomal episomes can be "cured" or removed once the desired modification is stable, minimizing the risk of ongoing Cas9 expression and off-target mutations. Current frontiers in this field involve the assembly of synthetic chromosomes in yeast (*Saccharomyces cerevisiae*) for subsequent transfer into diatoms. This "synthetic biology" approach aims to create designer diatoms with minimal genomes, optimized exclusively for high-efficiency carbon capture or the secretion of novel pharmaceuticals (Yixuan Li et al., 2025). By utilizing inducible promoters—sensitive to environmental triggers like CO_2 levels or nutrient depletion—biotechnologists can exert dynamic control over these biological factories, ensuring that high-value metabolites are produced only at optimal growth stages.

5. FUTURE PERSPECTIVES AND CHALLENGES

5.1 Cultivation and Scale-up Challenges

The performance of culture density, growth stage and reactor type affect the efficiency of carbon sequestration. CO_2 fixation is not always enhanced with an increase in CO_2 supply, and considerable gas losses have been reported (Buono et al., 2016). Close PBR systems often have greater CO_2, uptake efficiencies than open ponds with ca 44.2% reduction in the efficiency (Buono et al., 2016; Ho et al., 2011). Although open ponds have economic benefits, they are also characterised by low productivity, poor CO_2 retention and contamination issues. In spite of these benefits tubular, flat-plate or air-lift photosynthetic photobioreactors offer superior light capture and mass transfer but are challenged by oxygen build-up and complexity of the operation (Ho et al., 2011; Kaewpintong et al., 2007; Pulz, 2001). Potential for diatom carbon sequestration and biorefining Large-scale growth of *P. tricornutum* in various reactor designs has proven the potential for diatom CO_2 capture and biomass production (Branco-Vieira et al., 2020; Gao et al., 2017; Hamilton et al., 2015).

Amongst others, diatoms are typically grown submerged-force fed with CO_2 via bubbling. Nevertheless, because CO_2 has a lower solubility in water, a significant percentage of the supplied gas is lost to the atmosphere, especially at higher flow rates (Meiser et al., 2004). CO_2 utilization is strongly impacted by bubble size, aeration rate, mixing efficiency and retention time. Maximizing the gas liquid interface is crucial for enhanced mass transfer efficiency (Ho et al., 2011). CO_2 can also be added directly into culture medium, where in case of high CO_2 levels that may lead to lowered culture pH (Piiparinen et al., 2018), growth will decrease.

5.2 Sourcing, Economic Feasibility, and Ecological Risks

CO_2 can be provided to diatoms as a solid (carbonate or bicarbonate salts) or concurrent with gas. Gaseous CO_2 is typically sourced from industrial flue gases, purify CO_2 cylinders or direct air capture (DAC). Flue gases from combustion, e.g., in industry, typically contain 6–15% CO_2 and product streams such as from biogas or ethanol production can have contents of 20–40%. Use of flue gas from cement production as a CO_2 source to cultivate *Skeletonema marinoi* has been demonstrated, leading to enhanced biomass productivity and superior biochemical composition in relation to atmospheric CO_2 aeration (Olofsson et al., 2015).

Direct air capture has been identified as a promising but currently costly approach for sustainable CO_2 sourcing. Commercial efforts like Climeworks, Carbon Engineering, and the SOLETAIR project have shown that capturing atmospheric CO_2 is possible and could be combined with diatom-based cultivation systems for sustainable carbon storage.

5.3 The Future Outlook: Genetic and Metabolic Engineering

Carbon dioxide is an important CO_2 source for autotrophic organisms such as diatoms, and efficient use of CO_2 is crucial for the construction of a carbon sequestration/diatom-based biomanufacturing system. Appropriate characterization of CO_2 sources for implementation, and efficient delivery and an understanding of the species-specific carbon-uptake mechanisms are also important to ensure successful application

Although diatoms have immense potential for biotechnology and industrial applications, several challenges remain in their large-scale implementation. For biofuel production, it is essential to optimize the growth conditions, improve harvesting efficiency, and develop cost-effective lipid extraction methods. Despite their high lipid yields, economic and technological constraints continue to limit their commercial feasibility.

In environmental remediation, careful management is required to avoid ecological imbalance. Although diatom blooms can enhance carbon sequestration, excessive nutrient fertilization may trigger harmful algal blooms that disrupt aquatic ecosystems (Lavoie et al., 2016). Maintaining a balance between maximizing the benefits of diatoms and ensuring ecological stability is a critical research priority.

Advances in genetic engineering and synthetic biology offer promising opportunities. Engineered diatoms with higher lipid content, improved heavy metal adsorption, and customized biosilica structures for nanotechnology could expand applications in sustainable energy, environmental cleanup, and biomedical research. Continued innovation in these areas may unlock new practical solutions that leverage the unique properties of diatoms.

6. CONCLUSION

Diatoms with their complex silica frustules and their complex physiology are much more than mere microalgae; they are the best ecological engineers, determining the basic biogeochemical cycle of the world. Their important contribution to the biological carbon pump has been revealed in this chapter, since they are the most efficient carbon concentrating systems and a high rate of sinking, and thus, they sequester a large share of atmospheric CO_2, which places them as natural partners in mitigating climate change. In addition to their global carbon effect, diatoms are the backbone of aquatic food webs, the indicators in the health of the ecosystem, and the pioneers in regulatory ecological revival initiatives. Their fast colonisation of surfaces, and their ability to create effective biofilms, are supporting the health and succession of whole aquatic communities, which reflect a multi-faceted engineering capability, which acts on the micro-scale of a biofilm, to the macro-scale of oceanic basins. It is the very biological properties that render diatoms such an effective ecological engineer that render them an exceptionally promising platform of a new generation of biotechnologies. The process of the discovery of the way it works in nature up to the possibility to use it in human innovations is not a new one. Their nanopatterned biosilica frustules are undergoing reuse in new directions in nanomedicine such as targeted drug delivery and biosensing. Their metabolic plasticity is under exploitation in the manufacture of high-value products, including omega-3 fatty acids and biofuels, pigments and other bioactive compounds, in the name of sustainable manufacture. Moreover, their bioremediation ability and the application of diatomaceous earth in farming and industry makes them multifaceted. Nevertheless, this huge potential is very difficult to translate into scalable economically viable solutions, such as optimization of cultivation systems, sustainable resource inputs, and ecological risks mitigation. The future is in the intersection between the ecological knowledge and the state of the art science. Innovations in genetic and metabolic engineering are the key to the improvement of their natural performance, developing more engineered strains with better carbon capture or bioproduct synthesis. We can also fully harness the potential of these microbial engineers to create a more carbon-neutral future by keeping on exploring and innovating.

REFERENCES

Artamonova, E., Vasskog, T., & Eilertsen, H. C. (2017). Lipid content and fatty acid composition of Porosira glacialis and Attheya longicornis in response to carbon dioxide (CO2) aeration. *PLoS One*, *12*(5), e0177703. DOI: 10.1371/journal.pone.0177703 PMID: 28542308

Avrahami, Y., Koplovitz, G., & Frada, M. J. (2025). Diatom community succession and bloom variability as a function of winter-mixing depth in the subtropical Gulf of Aqaba, Red Sea. *Marine Ecology Progress Series*, *760*, 39–54. Advance online publication. DOI: 10.3354/meps14848

Baragi, L. V., Khandeparker, L., & Anil, A. C. (2015). Influence of elevated temperature and pCO2 on the marine periphytic diatom Navicula distans and its associated organisms in culture. *Hydrobiologia*, *762*(1), 127–142. DOI: 10.1007/s10750-015-2343-9

Barua, P. J. M. B. (2025). Diatoms as Microbial Architects of the Silica Cycle. *Ecological Significance, Biogenic Bioactives, and Emerging Biotechnological Applications.*, *8*(1), 1–12.

Battin, T. J., Luyssaert, S., Kaplan, L. A., Aufdenkampe, A. K., Richter, A., & Tranvik, L. J. (2009). The boundless carbon cycle. *Nature Geoscience*, *2*(9), 598–600. DOI: 10.1038/ngeo618

Boelen, P., van de Poll, W. H., van der Strate, H. J., Neven, I. A., Beardall, J., & Buma, A. G. (2011). Neither elevated nor reduced CO2 affects the photophysiological performance of the marine Antarctic diatom Chaetoceros brevis. *Journal of Experimental Marine Biology and Ecology*, *406*(1-2), 38–45. DOI: 10.1016/j.jembe.2011.06.012

Botte, P., d'Ippolito, G., Gallo, C., Sardo, A., & Fontana, A. (2018a). Combined exploitation of CO2 and nutrient replenishment for increasing biomass and lipid productivity of the marine diatoms Thalassiosira weissflogii and Cyclotella cryptica. *Journal of Applied Phycology*, *30*(1), 243–251. DOI: 10.1007/s10811-017-1221-4

Botte, P., d'Ippolito, G., Gallo, C., Sardo, A., & Fontana, A. J. J. o. A. P. (2018b). Combined exploitation of CO2 and nutrient replenishment for increasing biomass and lipid productivity of the marine diatoms Thalassiosira weissflogii and Cyclotella cryptica. *30*(1), 243-251.

Branco-Vieira, M., San Martin, S., Agurto, C., Freitas, M. A. V., Martins, A. A., Mata, T. M., & Caetano, N. S. (2020). Biotechnological potential of Phaeodactylum tricornutum for biorefinery processes. *Fuel*, *268*, 117357. https://doi.org/https://doi.org/10.1016/j.fuel.2020.117357. DOI: 10.1016/j.fuel.2020.117357

Buono, S., Colucci, A., Angelini, A., Langellotti, A. L., Massa, M., Martello, A., Fogliano, V., & Dibenedetto, A. (2016). Productivity and biochemical composition of Tetradesmus obliquus and Phaeodactylum tricornutum: Effects of different cultivation approaches. *Journal of Applied Phycology*, *28*(6), 3179–3192. DOI: 10.1007/s10811-016-0876-6

Chen, X., & Gao, K. (2003). Effect of CO2 concentrations on the activity of photosynthetic CO2 fixation and extracelluar carbonic anhydrase in the marine diatom Skeletonema costatum. *Chinese Science Bulletin*, *48*(23), 2616–2620. DOI: 10.1360/03wc0084

Ferguson, R. M., O'Gorman, E. J., McElroy, D. J., McKew, B. A., Coleman, R. A., Emmerson, M. C., & Dumbrell, A. J. (2021). The ecological impacts of multiple environmental stressors on coastal biofilm bacteria. *Global Change Biology*, *27*(13), 3166–3178. DOI: 10.1111/gcb.15626 PMID: 33797829

Fettke, J., & Fernie, A. R. (2015). Intracellular and cell-to-apoplast compartmentation of carbohydrate metabolism. *Trends in Plant Science*, *20*(8), 490–497. DOI: 10.1016/j.tplants.2015.04.012 PMID: 26008154

Field, C. B., Behrenfeld, M. J., Randerson, J. T., & Falkowski, P. J. s. (1998). Primary production of the biosphere: integrating terrestrial and oceanic components. *281*(5374), 237-240.

Firth, L. B., Bone, J., Bartholomew, A., Bishop, M. J., Bugnot, A., Bulleri, F., Chee, S.-Y., Claassens, L., Dafforn, K. A., & Fairchild, T. P. (2024). Coastal greening of grey infrastructure: An update on the state of the art. Proceedings of the Institution of Civil Engineers-Maritime Engineering, Flori, S., Mikus, F., Flaum, E., Moog, K., Guessoum, S., Beavis, T., Zwahlen, S. M., Romero-Brey, I., Oorshot, V., Olivetta, M., Steele-Ogus, M., Yeh, E., Dudin, O., Schwab, Y., Dey, G., & Vincent, F. (2025). Diatom ultrastructural diversity across controlled and natural environments. *Current Biology : CB*, *35*(23), 5709–5720.e5704. https://doi.org/https://doi.org/10.1016/j.cub.2025.10.024

Gao, B., Chen, A., Zhang, W., Li, A., & Zhang, C. (2017). Co-production of lipids, eicosapentaenoic acid, fucoxanthin, and chrysolaminarin by Phaeodactylum tricornutum cultured in a flat-plate photobioreactor under varying nitrogen conditions. *Journal of Ocean University of China*, *16*(5), 916–924. DOI: 10.1007/s11802-017-3174-2

García Sánchez, J., Berenguel, M., Rodríguez, F., Fernández Sevilla, J., Brindley Alias, C., & Acién Fernández, F. (2003). Minimization of carbon losses in pilot-scale outdoor photobioreactors by model-based predictive control. *Biotechnology and Bioengineering*, *84*(5), 533–543. DOI: 10.1002/bit.10819 PMID: 14574687

Hamano, R., Shoumura, S., Takeda, Y., Yamazaki, T., Hirayama, K., Hanada, Y., Mayama, S., Takemura, M., Lin, H.-J., Umemura, K. J. M., & Microanalysis. (2021). Sinking of four species of living diatom cells directly observed by a "tumbled" optical microscope. *27*(5), 1154-1160.

Hamilton, M. L., Warwick, J., Terry, A., Allen, M. J., Napier, J. A., & Sayanova, O. (2015). Towards the Industrial Production of Omega-3 Long Chain Polyunsaturated Fatty Acids from a Genetically Modified Diatom Phaeodactylum tricornutum. *PLoS One*, *10*(12), e0144054. DOI: 10.1371/journal.pone.0144054 PMID: 26658738

Ho, S.-H., Chen, C.-Y., Lee, D.-J., & Chang, J.-S. (2011). Perspectives on microalgal CO2-emission mitigation systems — A review. *Biotechnology Advances*, *29*(2), 189–198. https://doi.org/https://doi.org/10.1016/j.biotechadv.2010.11.001. DOI: 10.1016/j.biotechadv.2010.11.001 PMID: 21094248

Hopkinson, B. M., Dupont, C. L., & Matsuda, Y. (2016). The physiology and genetics of CO2 concentrating mechanisms in model diatoms. *Current Opinion in Plant Biology*, *31*, 51–57. DOI: 10.1016/j.pbi.2016.03.013 PMID: 27055267

Hu, H., & Gao, K. (2008). Impacts of CO2 enrichment on growth and photosynthesis in freshwater and marine diatoms. *Chinese Journal of Oceanology and Limnology*, *26*(4), 407–414. DOI: 10.1007/s00343-008-0407-7

Ihnken, S., Roberts, S., & Beardall, J. (2011). Differential responses of growth and photosynthesis in the marine diatom Chaetoceros muelleri to CO2 and light availability. *Phycologia*, *50*(2), 182–193. DOI: 10.2216/10-11.1

Kaewpintong, K., Shotipruk, A., Powtongsook, S., & Pavasant, P. (2007). Photoautotrophic high-density cultivation of vegetative cells of Haematococcus pluvialis in airlift bioreactor. *Bioresource Technology*, *98*(2), 288–295. https://doi.org/https://doi.org/10.1016/j.biortech.2006.01.011. DOI: 10.1016/j.biortech.2006.01.011 PMID: 16516464

Khatoon, H., Yusoff, F., Banerjee, S., Shariff, M., & Bujang, J. S. (2007). Formation of periphyton biofilm and subsequent biofouling on different substrates in nutrient enriched brackishwater shrimp ponds. *Aquaculture (Amsterdam, Netherlands)*, *273*(4), 470–477. DOI: 10.1016/j.aquaculture.2007.10.040

Launay, H., Huang, W., Maberly, S. C., & Gontero, B. (2020). Regulation of carbon metabolism by environmental conditions: A perspective from diatoms and other chromalveolates. *Frontiers in Plant Science*, *11*, 1033. DOI: 10.3389/fpls.2020.01033 PMID: 32765548

Li, Y., Deng, L., Walker, E. J. L., Karas, B. J., & Mock, T. (2025). Genetic engineering in diatoms: Advances and prospects. *The Plant Journal : For Cell and Molecular Biology*, *121*(6), e70102. DOI: 10.1111/tpj.70102 PMID: 40089910

Li, Y., Deng, L., Walker, E. J. L., Karas, B. J., & Mock, T. (2025). Genetic engineering in diatoms: Advances and prospects. *The Plant Journal : For Cell and Molecular Biology*, *121*(6), e70102. DOI: 10.1111/tpj.70102 PMID: 40089910

Matthijs, M., Fabris, M., Obata, T., Foubert, I., Franco-Zorrilla, J. M., Solano, R., Fernie, A. R., Vyverman, W., & Goossens, A. (2017). The transcription factor bZIP14 regulates the TCA cycle in the diatom Phaeodactylum tricornutum. *The EMBO Journal*, *36*(11), 1559–1576. DOI: 10.15252/embj.201696392 PMID: 28420744

Medlin, L. K. J. P. (2016). Evolution of the diatoms: major steps in their evolution and a review of the supporting molecular and morphological evidence. *55*(1), 79-103.

Meiser, A., Schmid-Staiger, U., & Trösch, W. (2004). Optimization of eicosapentaenoic acid production byPhaeodactylum tricornutumin the flat panel airlift (FPA) reactor. *Journal of Applied Phycology*, *16*(3), 215–225. DOI: 10.1023/B:-JAPH.0000048507.95878.b5

Ndaba, J., Cotiyane-Pondo, P., Human, L., Puccinelli, E., Pieterse, P., Pattrick, P., & Porri, F. (2025). Diatom colonisation and biofilm metal bioaccumulation: Can Indigenous Knowledge Systems aid the ecological engineering of urban coastlines? *Ecological Engineering*, *219*, 107696. DOI: 10.1016/j.ecoleng.2025.107696

Negoro, M., Shioji, N., Miyamoto, K., & Micira, Y. (1991). Growth of microalgae in high CO2 gas and effects of SOx and NOx. *Applied Biochemistry and Biotechnology*, *28*(1), 877–886. DOI: 10.1007/BF02922657 PMID: 1929389

Nenadović, T., Šarčević, T., Čižmek, H., Godrijan, J., Marić Pfannkuchen, D., Pfannkuchen, M., & Ljubešić, Z. (2015). Development of periphytic diatoms on different artificial substrates in the Eastern Adriatic Sea. *Acta Botanica Croatica*, *74*(2), 377–392. DOI: 10.1515/botcro-2015-0026

Olofsson, M., Lindehoff, E., Frick, B., Svensson, F., & Legrand, C. (2015). Baltic Sea microalgae transform cement flue gas into valuable biomass. *Algal Research*, *11*, 227–233. https://doi.org/https://doi.org/10.1016/j.algal.2015.07.001. DOI: 10.1016/j.algal.2015.07.001

Phong, W. N., Regmi, R., Adhikari, K. C., Bhattarai, R. R., & Show, P. L. (2023). Bioactive Compounds from Diatoms. In *Diatoms Biotechnology* (pp. 20–37). CRC Press. DOI: 10.1201/9781003436553-2

Piiparinen, J., Barth, D., Eriksen, N. T., Teir, S., Spilling, K., & Wiebe, M. G. (2018). Microalgal CO2 capture at extreme pH values. *Algal Research, 32*, 321–328. DOI: 10.1016/j.algal.2018.04.021

Poschenrieder, C., Fernández, J. A., Rubio, L., Pérez, L., Terés, J., & Barceló, J. (2018). Transport and use of bicarbonate in plants: Current knowledge and challenges ahead. *International Journal of Molecular Sciences, 19*(5), 1352. DOI: 10.3390/ijms19051352 PMID: 29751549

Pulz, O. (2001). Photobioreactors: Production systems for phototrophic microorganisms. *Applied Microbiology and Biotechnology, 57*(3), 287–293. DOI: 10.1007/s002530100702 PMID: 11759675

Rabiee, N., Khatami, M., Jamalipour Soufi, G., Fatahi, Y., Iravani, S., Varma, R. S. J. A. b. s., & engineering. (2021). Diatoms with invaluable applications in nanotechnology, biotechnology, and biomedicine: recent advances. *7*(7), 3053-3068.

Rees, T. J. J. o. E. B. (1984). Sodium dependent photosynthetic oxygen evolution in a marine diatom. *35*(3), 332-337.

Salamone, A. L., Robicheau, B. M., & Walker, A. K. (2016). Fungal diversity of marine biofilms on artificial reefs in the north-central Gulf of Mexico. *Botanica Marina, 59*(5), 291–305. DOI: 10.1515/bot-2016-0032

Salleh, S., Mohammad, M., Cheah, W., Palaniveloo, K., Fadhullah, W., Khare, V., Zhongwei, H., & Ishak, S. D. (2024). The Biology of Diatom and Potential Usage for Value-Added Products. In Riti Thapar, K., Mohd, R., & Norli, I. (Eds.), *Marine Biomass* (pp. 341–366). De Gruyter., https://doi.org/doi:10.1515/9783111353951-015 DOI: 10.1515/9783111353951-015

Sethi, D., Butler, T. O., Shuhaili, F., & Vaidyanathan, S. J. B. (2020). Diatoms for carbon sequestration and bio-based manufacturing. *9*(8), 217.

Shen, J., Zheng, X., Liu, M., Xu, K., He, L., & Lin, Z. (2024). Targeted cultivation of diatoms in mariculture wastewater by nutrient regulation and UV-C irradiation [Original Research]. *Frontiers in Microbiology, Volume 15 - 2024*. DOI: 10.3389/fmicb.2024.1371855

Smith, A. M., & Stitt, M. (2007). Coordination of carbon supply and plant growth. *Plant, Cell & Environment, 30*(9), 1126–1149. DOI: 10.1111/j.1365-3040.2007.01708.x PMID: 17661751

Terracciano, M., De Stefano, L., & Rea, I. (2018). Diatoms Green Nanotechnology for Biosilica-Based Drug Delivery Systems. *Pharmaceutics, 10*(4), 242. Advance online publication. DOI: 10.3390/pharmaceutics10040242 PMID: 30463290

Tien, C.-J., & Chen, C. S. (2013). Patterns of metal accumulation by natural river biofilms during their growth and seasonal succession. *Archives of Environmental Contamination and Toxicology, 64*(4), 605–616. DOI: 10.1007/s00244-012-9856-2 PMID: 23247559

Tiwari, A., Marella, T. K., & Saxena, A. (2022). Diatom biorefinery: from carbon mitigation to high-value products. In *Biomass, biofuels, biochemicals* (pp. 401–420). Elsevier. DOI: 10.1016/B978-0-12-823500-3.00018-2

Tuck, B., Watkin, E., Somers, A., & Machuca, L. (2022). A critical review of marine biofilms on metallic materials. *npj Materials degradation, 6*(1), 25.

Zhuang, Y., Zhang, Y., Jin, H., Qi, D., & Chen, J. (2025). Unexpected ice-edge diatom blooms occur as thin phytoplankton layers in Barrow Canyon, Chukchi Sea. *Marine Environmental Research, 206*, 107035. https://doi.org/https://doi.org/10.1016/j.marenvres.2025.107035. DOI: 10.1016/j.marenvres.2025.107035 PMID: 40024173

Zimmerman, W. B., Tesar, V., Butler, S., & Bandulasena, H. C. (2008). Microbubble generation. *Recent Patents on Engineering, 2*(1), 1–8. DOI: 10.2174/187221208783478598

Chapter 11
Algal Nanotechnology:
Diatom Biosilica and Functional Nanomaterials

Wafaa F. Zohir
http://orcid.org/0000-0002-7635-6965
Botany and Microbiology Department, Faculty of Science, Damanhour University, Egypt

Rehab M. Mohamed
http://orcid.org/0000-0002-7635-6965
Biological and Geological Sciences Department, Faculty of Education, Ain Shams University, Egypt

Ahmed Hamdy Zabady
Botany and Microbiology Department, Faculty of Science, Damanhour University, Egypt

Mofida E. M. Makhlof
http://orcid.org/0000-0002-8636-7430
Botany and Microbiology Department, Faculty of Science, Damanhour University, Egypt

ABSTRACT

Diatoms produce intricate silica-based cell walls (frustules) with nanoscale precision, representing millions of years of evolutionary optimization in biomineralization. These structures offer unprecedented opportunities for developing functional nanomaterials with applications spanning drug delivery, biosensors, photonics, catalysis, and environmental remediation. This chapter will explore the emerging field of algal nanotechnology, with particular focus on diatoms as natural nanofabrication platforms.

DOI: 10.4018/979-8-3373-7630-1.ch011

1. INTRODUCTION

1.1 What is Algal Nanotechnology?

Nanotechnology focuses on synthesis, arrangement, and technical applications of nanosized materials (1-100nm). Nanoscale materials behave differently, and their properties and behaviors are modified with respect to shape, size, and dimensional aspects when compared to all other macro- and micro-scaled materials (Yaqoob et al., 2020). They thereby exhibit unique chemical, physical, and mechanical properties. So, it has applications in various fields of technology and science, including agriculture, wastewater treatment, medicine, biotechnology, energy, cosmetics, textiles, optical sensors, and electrical engineering (Maran et al., 2023). To date, understanding the fundamentals of nanostructure has been based on the synthesis of nanomaterials using top-down and bottom-up approaches. Up to now, nanomaterial synthesis using top-down and bottom-up methods has underpinned understanding of nanostructure principles (Mekuye et al., 2023; Kirubakaran et al., 2026). These approaches for nanoscale materials synthesize chemical, physical, and biological methods. In the case of physical methods of nanoparticle synthesis, the process focuses on evaporating, abrading, melting, or condensing material sources using mechanical forces, high-energy emissions, thermal, and electrical energy (Jagdeo 2023). Chemical methods and techniques, such as microemulsion, sol-gel, chemical vapor deposition, and hydrothermal-enhanced chemical vapor deposition, were employed for the synthesis of nanoscale materials (Rafique et al., 2020). However, these physical and chemical methods have disadvantages, including high costs, health hazards, and severe environmental risks. So, researchers have turned their focus towards less toxic and cost-effective biogenic methods for the fabrication of nanomaterials using biological sources such as plants, animals, viruses, and fungi (Srivastav and fungi, 2021). These essential technical findings show that capping, reducing, and stabilizing agents for the green synthesis of nanosized materials were achieved using various bioactive compounds present in most biological systems (Rai et al., 2025; El-Hosari et al., 2024). Sharma et al. (2016) and Chaudhary et al. (2020) reported on nanoparticle synthesis by algae belonging to the Rhodophyceae, Phaeophyceae, and Chlorophyceae and the mechanism behind the phycosynthesis. The first report on the biosynthesis of nanoparticles was the synthesis of Au nanoparticles by the marine algae *Sargassum wightii* (Singaravelu et al., 2007; Shera & Banik, 2022).

1.2 Diatoms as Nature's Nanotechnologists

Diatoms are classified by frustule shape to help select a diatom for a specific device application. Frustules come in three shapes: flake-shaped, rod-shaped, and a generic 3D shape (or valve shapes), **as in Fig. 1** (Babenko et al., 2022). They can be found as single cells, colonies, ribbons, zigzags, filaments, fans, or stellate structures. The pennate diatoms are motile, with two raphe slits in each valve (Kulichová & Urbánková, 2020). Diatoms are commonly known as producers in the food chain (Bouazzara et al., 2025). They have complex siliceous shells, which give them many potential uses. Their naturally occurring nanoscale structures combine sustainability, biocompatibility, and complex porosity, making them ideal for a variety of applications, including targeted drug delivery, environmental monitoring, and biosensing (Bouazzara et al., 2025).

In nanotechnology, the application of diatom frustules was first reported by Gordon and Drum in 1994 (Gholami et al., 2020). The diatomite silica exoskeleton can be obtained by various methods, such as cell culture or mining fossilized diatoms (Razzak et al., 2024). The fascinating structural architecture of diatoms, which are composed of silica, is one of their many unique characteristics that can be affected by genetics. The diatom's frustule has a distinct nanostructure and is characterized by numerous features, including channels, pores, ridges, spikes, and spines (Cvjetinovic et al., 2023).

Diatoms naturally produce porous silica with a range of shapes, particle sizes, and distributions, including mesoporous and macroporous silica. Diatoms have repeating holes and structures that can be as large as 1 micron across and as small as mesoporous scale (> 50 nm) (Willis et al., 2010). Furthermore, due to the size of its pores, diatom porous silica has been proposed as a potential material for nanotechnology (Hamed, 2023). Because different varieties of mesoporous, macroporous, and microporous silica are very helpful in nanotechnology applications, they provide materials with a very high surface area and fascinating physical and chemical properties (Hamed, 2023).

In relation to the great majority of unidentified diatom species, these microorganisms can provide a vast catalog of shapes, pore sizes, and other structures and parts that enable them to modify their morphogenesis to produce the required design. These discoveries made diatom frustules more useful and helped the field of "Diatom Nanotechnology" emerge (Hamed, 2023). Diatom nanotechnology is a new area of multidisciplinary study that involves partnerships between engineers and scientists across biochemistry, biology, physics, biotechnology, and materials science. It focuses on using diatoms or how they produce silica to make nanodevices (Hamed, 2023). Diatom frustules are useful in nanotechnology because they can create fascinating three-dimensional silica structures with finely inherited Nano details (Hamed, 2023). There are two methods for diatom nanotechnology: bionanotechnology, which directly uses living organisms or their constituent parts, and biomimetic, in which we attempt to improve or duplicate what we believe organisms are doing (Hamed, 2023).

1.3 Why Diatom Frustules Matter

More than 200,000 species of diatoms, various in size and shape, are estimated to exist, and the single-celled diatoms are widely distributed. The diatom's cell wall is composed of amorphous silica and is transparent, as shown **in Fig. 1** (Rogato et al., 2020). The mechanical strength of the frustule is characterized by a variety of three-dimensional (3D) shapes, multilevel microstructures, nanopores, unique optical properties, and a large surface area, making it a potential new functional material for bio-manufacturing (Li et al., 2021).

Figure 1. The various diatom diversity, frustule shapes and applications.

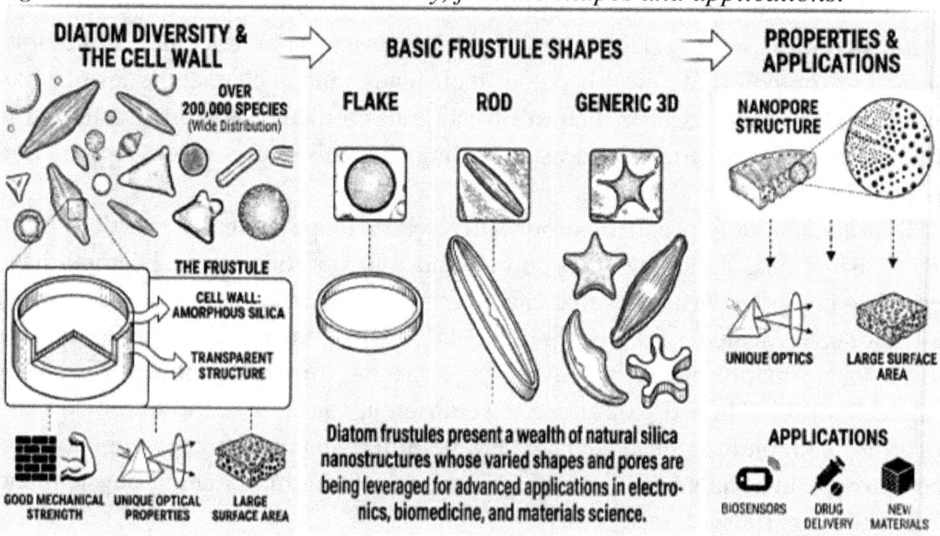

Over the past 200 years, diatom frustules have become increasingly important in the fields of micro- and nanotechnology. The frustules have been extensively explored for commercial applications and by academics. Recently reported processing methods for modifying the structure, bonding, and assembly of diatom frustules may enhance their functionality to meet the requirements of microdevice design (Sun et al., 2024).

The larger top half of the diatom cell wall covers the smaller bottom half, giving the cell wall a Petri dish-like form. Each theca is made up of an elaborate valve with intricate pore patterns and a succession of connecting siliceous bands with more uniform pore patterns. An array known as an areola, which can be circular, polygonal, or elongated and spread into the frustule depth, is produced by the valve pore pattern, which is genome-encoded and species-specific. The pores, which range in size from 10 to 100 nm, can be partially blocked by flat or domed cribra, which are biosilica flaps that occlude the areolae.

2. DIATOM BIOLOGY AND BIOSILICIFICATION

Diatoms are a major contributor to global carbon fixation and the largest contributor to biosilicification worldwide (Reid et al., 2021).

2.1 Diatom Structure and Classification

Diatoms are protected by a mineral shell, on which silica forms on the cell surface (Agarwal, 2025). Diatoms are enclosed in a siliceous cell wall (frustule) composed of two overlapping valves. The valve portions of diatoms are called the epitheca and hypotheca, respectively. The hypotheca and epitheca fit together like pill boxes, with the epitheca slightly larger (Bueno et al., 2017). The upper epigirdle band can traverse the inferior one because the lower valve, along with its associated hypogirdle, is slightly smaller than the upper assembly. The two connecting bands, though often mentioned independently, are collectively termed the girdle, which is unified by a growing connective zone. The valves have elaborate ornamentation (Hamed, 2023). The valve pore pattern is species-specific and genome-encoded, resulting in an array called an areola. It can be polygonal, round, or elongate, and it extends into the depth of the frustule (Parikh, 2021). Depending on the diatom species and growth conditions, frustules can have a variety of shapes. The formation and generation of certain frustule shapes have potential applications in nanotechnology. Because the silica cage of atoms is flexible, it can accommodate deformation, enabling material loading. The silica cage of diatoms is not rigid, allowing it to deform under material loading (Hussein et al., 2022; Sung et al., 2025).

Based on the symmetry of valves in diatoms, the centric group and pennate group are the two important diatom kinds (Laxa, 2021; Hamed, 2023). The pennate diatoms have an elongated structure and bilateral symmetry, live by attaching to other organisms or sediment surfaces, and are the primary constituents of benthic biofilms. There are many different valve shapes found in pennate diatoms, including dorsoventral (like *Amphora* and *Seminavis*), heteropolar (like *Licomophora*), and naviculoid (like *Navicula*). In addition, bilaterally symmetrical pennate diatoms can be divided, according to the presence of a raphe, into araphid and raphid. Depending on whether the raphe is present, raphid pennates have a raphe visible on at least one valve (Hamed, 2023). On the other hand, centric diatoms are the primary component of phytoplankton communities and are radially symmetrical and rounded, due to their tremendous porosity and detailed pore structure (Hamed, 2023).

Electron microscopy revealed amazing nanoscale ornamentation on diatom girdle bands and valves (Round et al., 1990; Ghobara et al., 2023). These structures are often porous, with pore diameters ranging from a few nm to a few μm, depending on the species (Ghobara et al., 2021a). These pores can have periodic or quasi-periodic configurations with varying degrees of flaws and symmetries, and they can be simple (punctate) or complex (loculate) (Ghobara et al. 2021b).In addition to pores, the literature reports a wide range of ultrastructural details for many species, including central nodules, spines, ocelli, fibulae, portulae, fasciae, and more (Spaulding et al. 2021).

The frustule's form and ornamentation are the primary taxonomic characteristics employed in diatom identification. Raphe, frustule ornamentation, stigmata, and other characteristics are crucial for distinguishing these species (Guiry, 2024). Most species can be observed with an optical microscope because diatom size ranges from 2 to 2000 μm (Cristóbal et al., 2020). Only half of the estimated 200,000 diatom species have been described, according to Min et al. (2024). The results of diatom biological indices are highly sensitive to the accuracy of taxonomic classification, as shown by numerous intercalibration studies (Li et al., 2025). The task of identification is very difficult because of the huge number of species estimated to exist (Kahlert et al., 2021). These micrographs show the multilayer, multiscalar porosity of girdle bands and diatom valves. Additionally, recent research using atomic force microscopy revealed novel nanoscale features in the valve topography (Ghobara et al., 2023). Also, there are two primary views of diatom frustules via an optical light microscope: valve and girdle views, and it is unable to show the frustules' ultrastructure, so optical microscopy is insufficient to show the 3D structure of diatom frustules, so scanning and transmission electron micrographs are used to describe the details of the frustule and appear its nanofeatures (Hamed, 2023). Diatom frustules have been identified as multiscalar pores with different sizes, arrangements, and periodicity patterns in porous silica via ultrastructure studies (Luís et al., 2017). The strong, porous silica frustules of diatoms may contribute to their success across various aquatic habitats. The frustule design, frustule size, pore distribution patterns, pore size, and material properties were determined by the optical properties, mechanical strength, and filtration abilities. As a result, recent research using atomic force microscopy revealed novel nanoscale features in the valve topography (Hamed, 2023).

2.2 The Biosilicification Process

The process by which organisms assimilate monomeric silicic acid (soluble Si(OH)4) into polymerized, insoluble silica (SiO2) is known as biosilicification (Ikeda, 2021). Biosilicification is the process by which diatoms form their intricate, ornate siliceous frustules. For diatoms to form frustules, they require a supply of silica from their surroundings. Soluble silicon is in the form of orthosilicic acid (Si(OH)4). It makes up 97% of all dissolved silicon (Kolbe & Brunner, 2022). Diatoms primarily use silicic acid, although it's unclear whether they also use other forms of silicon. Reid et al. (2021) were the first to report that diatoms absorb silicates from their surroundings and incorporate them into cells. The material enters first through the frustule pores before reaching the plasmalemma (Babenko et al., 2022). Recent research indicates that there may be several channels for silica uptake; therefore, the exact process remains unknown. Although they can vary from 19 to 340 mM within the diatom cell body, silicic acid concentrations in oceanic saltwater typically range from 1 to 70 mM (Kolbe & Brunner, 2022). This extraordinarily high gradient implies that the process is not dependent on simple passive diffusion across the cell membrane (Zhu et al., 2026). Instead, it is a result of active transport. The absorption of protein mediated by specialized silicon transfer proteins (SITs) is one possibility (Maurya et al., 2025). Three distinct uptake models were proposed by chemostat investigations, with an environment-dependent dominating mode. The three recommended modes are surge uptake, externally controlled uptake, and internally controlled uptake (Kolbe & Brunner, 2022).

2.3 Frustule Structure at Multiple Scales

Diatom frustules arise from the process of biomineralization (silicic acid polymerization in the deposition vesicles (low pH), with the deposition of nanospheres in dimensional and geometric patterns (Woodard, 2023). The role of frustules is to help diatoms in mechanical stability and filtration of nutrients, protection from attachment to substrates, the interaction and exchange between the surrounding environment and diatom cells, regulation of light penetrating the cell (and sinking rate, and DNA protection from UV radiation (Rogato & De Tommasi, 2020 ; V et al., 2023).

Frustules resist mechanical loads (predators) and internal (turgor and reproductive stages) (Rai et al., 2025). The liquid-solid phases found around the structure and the frustules' pores are of special interest. The diatoms' two external porous layers strengthen their resistance to liquid pressure and minimize friction with the surrounding water. According to their morphology, frustules act as self-lubricating and self-repairing media (Zuluaga-Astudillo et al., 2025). The size, stiffness, and pores are characteristics of frustules that distinguish them from other inorganic particles or soil, with typical diameters ranging from 30 to 100 μm (Zuluaga-Astudillo et al., 2022). The evolutionary adaptations of frustules have made them dominant in cellular morphogenesis and physiology (Babenko et al., 2022). The extent of frustule silicification depends on the silica fixation rate, which is influenced by the efficacy of silica transporters and genetic factors, as well as by the availability of environmental silica (Götz et al., 2019). As a result, the distinctive multiscale design of diatom frustules has attracted interest as a natural model for sophisticated nanostructured materials and bioinspired engineering applications (Musenich, 2024).

3. PROPERTIES OF DIATOM BIOSILICA

Another kind of silica is diatom biosilica, which has the highest specific strength and is a reactive silica extracted from diatom biomass using a unique method (Saoud et al., 2022). Biosilica is composed primarily of hydrated amorphous SiO2 (silica), with a small proportion of organic macromolecules long speculated to control silica deposition and nanopatterning (Saoud et al., 2022). Shells of biosilica last long after the diatom's death and after their internal organic matter has decayed (Saoud et al., 2022). The frustule's design, pore size, and pore distribution patterns could dramatically affect properties, leading to the special properties of these frustules. They have been extensively used across various fields (Rabiee et al., 2021).

3.1 Physical Properties

Diatom-derived silica, originating from the microshells of diatoms, is considered to have an impressive array of chemical and physical properties that make it a formidable candidate for various applications, particularly in nanotechnology and drug delivery (Zobi, 2022). In contrast to traditional silica particle production, which frequently requires harsh synthesis conditions and surfactant templates to achieve identical textural characteristics, diatom biosilica offers an environmentally friendly, low-cost alternative option (Saoud et al., 2022).

Diatom-derived biosilica exhibits several features typical of ceramic materials yet is composed of amorphous silica (Panwar & Dutta, 2019). The general definition of ceramics and bioceramics is inorganic, non-metallic solids that exhibit high mechanical strength, low thermal conductivity, and thermal resistance, properties that diatom biosilica inherently possesses (Choi, 2022). Because of its dense silicified structure, it is mechanically solid, resistant to chemical degradation across a wide pH range, and thermally stable up to about 1000 °C (Saoud et al., 2022). As with sintered ceramic components, these qualities enable biosilica to maintain structural integrity in challenging environmental conditions (Kapoor et al., 2023). Furthermore, its low thermal conductivity reinforces its electrical insulation capacity and its classification within the broader family of ceramic materials, particularly in the context of functional ceramics used in filtration, catalysis, and various biomedical applications (Saoud et al., 2022).

3.2 Chemical Properties

Biosilica consists primarily of hydrated amorphous SiO2 (silica), containing a small proportion of organic macromolecules that have long been speculated to control the silica deposition and nanopatterns (Saoud et al., 2022). It has biocompatibility, high chemical stability, hierarchical nano/microporosity, a homogeneous porous structure with a large specific surface, and chemical durability (Choi, 2022).

4. HARVESTING, PROCESSING, AND MODIFICATION

4.1 Diatom Cultivation

Diatoms are unicellular photosynthetic microalgae that belong to the phylum Bacillariophyta. They are distinguished by their silica cell walls, or frustules. These complex structures offer several advantages, such as defense against grazers, control over sinking rates, and protection against photoinhibition. Understanding cultivation methods, such as xenic and axenic cultures, provides vital resources for studying the physiology and behavior of diatoms, enabling the development of novel applications across a variety of industrial domains (Medlin, 2009).

4.2 Xenic and axenic culture development methods

They are frequently employed in biotechnology and microbiology to grow and examine a variety of microorganisms, including diatoms. Diatoms are cultivated alongside other species in xenic cultures, which offer a more environmentally appropriate setting. Axenic cultures, on the other hand, provide a controlled environment for examining the pure physiology and behavior of diatoms by separating them from other related metabolizing cells. It is essential to understand the subtleties, benefits, and challenges of these culture methods to advance diatom research and effectively harness their biotechnological potential (Parikh et al., 2025).

Diatoms exist within intricate ecosystems where their growth, distribution, and ecological functions are influenced by environmental factors such as nutrient availability, light levels, temperature, and interactions with other organisms. Often, these interactions yield context-dependent responses, underscoring the need for in-depth investigation. Nonetheless, even though lab research is thorough, it might not fully replicate real-world conditions. This can lead to restrictions and possible biases. Diatom physiology can be affected by fluctuations in cultural parameters such as nutrients or light, leading to difficult-to-interpret results. The ability to replicate natural conditions in laboratory environments has been greatly enhanced by technological and methodological advances, including precise environmental control and advanced analytical tools. Researchers can improve the alignment of experimental conditions with ecological realities by optimizing culture parameters, thus bridging the gap between lab and field studies and enhancing our understanding of diatom ecology and applications (Behrenfeld et al., 2021).

4.3. Cultural media and associated parameters

Some of the various culture media available are regarded as standards for isolating and cultivating diatoms, each meticulously designed to meet the specific needs of these microorganisms for their intended use. The f/2-Si culture medium is very effective in promoting diatom growth and is a favored option. PM medium is ideal for diatoms in acidic conditions, whereas WC medium supports diatoms that flourish in alkaline environments. Each medium is formulated with a precise balance of nutrients tailored to diatoms' requirements. Marine diatoms require f/2Si, which should be made with seawater. In contrast, PM and WC are appropriate for freshwater diatoms and should be prepared with distilled water. To maintain healthy diatom cultures and facilitate their study, it is essential to select the appropriate medium and prepare it correctly (Eissler, et al., 2025). Table 1 shows the exact chemical compositions of these three frequently used media. The optimal growth and productivity of diatom cultures rely on carefully regulated parameters such as light intensity, salinity, nutrient concentration, pH, temperature, and photoperiod. These elements are linked to one another and need to mimic natural conditions for diatom proliferation to succeed. The intensity of light controls photosynthesis, and salinity influences osmotic balance and nutrient absorption. The concentration of nutrients, especially nitrogen, phosphorus, and silicon, affects metabolism and biomass production. Temperature influences cellular metabolism, while pH affects both enzymatic activity and nutrient availability. Growth and reproduction are regulated by the photoperiod. To achieve optimal culture conditions, it is crucial to customize these parameters according to the specific requirements of each diatom species (Peng, et al., 2025).

4.4 Purification procedure of diatom biosilica

Diatoms possess inflexible cell walls known as frustules, which feature intricate structures composed of amorphous silica. They feature well-organized areolae on their outer or inner surfaces, along with a multitude of sieve pores about 40 nm in diameter for the exchange of nutrients and gases. Additionally, the frustules' porous structure enhances their surface area, which can potentially attain 200 m^2/g. While biosilica frustules lack the superior electrical conductivity and mechanical properties of carbon nanomaterials, their unique porous structures and various optical characteristics, along with their good biocompatibility, make them suitable core components for devices like photoluminescence-based gas sensors, filters, drug delivery systems, photovoltaic devices, and electrochemical/optical biomolecule diagnostic tools (Agarwal, 2025).

Table 1. Detailed chemical formulation of three standard culture media used for diatom cultivation

Chemical component	Culture media (mg/L)		
	f/2-Si	PM	WC
Macro elements	-	0.37	36.67
$CaCl_2$	-	2.90	08.71
K_2HPO_4	-	3.70	36.97(mm1)
$MgSO_4.7H_2O$	30.00	14.21	28.42
$Na_2SiO_3.9H_2O$	5.00	-	-
$NaH_2PO_4.H_2O$	-	3.15	12.60
$NaHCO_3$	75.00	56.70	85.10
$NaNO_3$			
Micro elements	4.360	2.180	4.360
Na_2EDTA	3.150	1.580	3.150
$FeCl_3.6H_2O$	0.010	0.005	0.010
$CuSO_4.5H_2O$	0.180	0.090	0.180
$MnCl_2.4H_2O$	0.22	0.011	0.022
$ZnSO_4.7H_2O$	0.010	0.005	0.010
$CoCl_2.6H_2O$	0.006	0.003	0.006
$NaMoO_4.2H_2O$	-	0.500	1.00
H_3BO_3			
Vitamins	1.0×10^{-1}	1.0×10^{-1}	1.0×10^{-4}
B1 (Thiamine)	5.0×10^{-4}	5.0×10^{-4}	5.0×10^{-4}
B7 or H (biotin)	5.0×10^{-4}	5.0×10^{-4}	5.0×10^{-4}
B12 (Cobalamin)			

To extract inorganic diatom biosilica from diatom cells for use as a biomedical material, it is essential to remove most organic substances, carbohydrates, and similar components. Currently, there are three widely used methods for processing purified diatom biosilica. (i) The first method involves filtering and washing diatom cells with deionized water, followed by treatment with a mixture of HCl and H2O2 to remove organic matter. After several cycles of this process and thorough rinsing, the purified diatom biosilica is vacuum-dried and stored. (ii) In the second method, algal mud is collected and mixed with a NaOH-urea lye solution, then washed and frozen. The treated slurry is then mixed with a piranha solution (sulfuric acid-H2O2) at 75°C, stirred, and the pH is adjusted to yield diatom biosilica. (iii) For the third method, collected diatom cells are washed and treated with nitric acid and sulfuric acid at 60°C. The resulting diatom biosilica is washed with water and ethanol, then vacuum-dried for 12 hours at 60°C. The selection of the method depends on experimental requirements and the properties of the diatom samples. Regardless of the method chosen, it is essential to implement appropriate safety measures when handling diatom samples (da Trindade et al., 2025).

5. FUNCTIONAL NANOMATERIALS FROM DIATOM FRUSTULES

Diatom silica frustules serve as natural templates for creating nanomaterials with unique properties for drug delivery and sensing in nanotechnology. Furthermore, diatoms play a vital role in wastewater remediation. Their efficient nutrient uptake and biomass production aid in pollutant removal. These actions also improve water quality and ecosystem vitality. Diatoms, with varied biochemical characteristics and therapeutic potential, provide an abundant source of bioactive substances in pharmaceuticals and cosmeceuticals. These bioactives can be utilized in skincare, drug development, and nutraceuticals. Diatoms have shown potential as a "cell factory" for generating products such as fucoxanthin, Omega-3 fatty acids, and eicosapentaenoic acid at the laboratory scale. Moreover, various sectors use diatom biomass to produce bio-oils and bio-crudes. Use in animal feed for aquaculture remains a leading application. Additionally, interdependencies with other microorganisms may lead to innovative biotechnologies for agriculture, wastewater treatment, and ecosystem restoration (Rinku et al., 2026). In summary, these diverse applications highlight the importance of diatoms as adaptable tools for addressing modern challenges. They help promote sustainable practices across multiple industries, **as shown in Fig. 2.**

Figure 2. Diatom diverse applications with great importance.

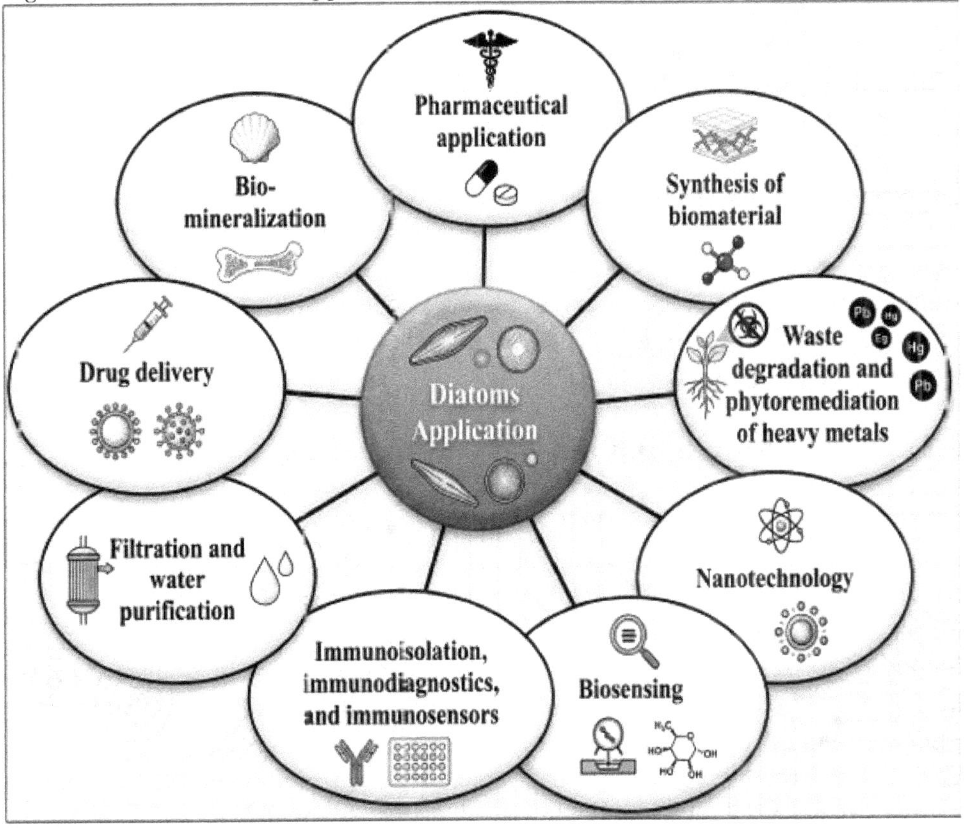

5.1 Drug Delivery Systems

Diatoms have opened new avenues for inventive systems in drug delivery, transforming therapeutic approaches. Their porous microshells made of silica, which are derived from biosilica, offer an excellent platform for controlled drug release. Diatom-based drug delivery systems facilitate the targeted delivery of therapeutic agents to specific cells or tissues through surface functionalization and chemical modification. As an illustration, diatom frustules that have undergone functionalization can act as templates for the arrangement of DNA-functionalized gold nanoparticles, demonstrating interactions based on specific sequences to improve drug delivery. These systems represent major progress in personalized medicine, with applications ranging from cancer treatment to chronic disease management (Angmo, et al., 2026).

5.2 Biosensors and Detection Systems

Thanks to the functionalization of diatom biosilica, it is now possible to create cutting-edge biosensors with unmatched sensitivity and specificity. Biosilica-based biosensors can achieve remarkable precision in detecting biomolecules by incorporating recognition elements like antibodies or DNA aptamers. For example, consider biosensors derived from Chaetoceros sp. diatoms that have been chemically modified with iron oxide nanoparticles. These biosensors enable the selective capture and separation of cancerous cells, demonstrating their promise for early cancer detection. Additionally, diatom biosilica serves as a substrate for ultrasensitive immunoassay biosensors, aiding precise diagnosis and disease surveillance across multiple biomedical contexts (Rinku et al., 2026).

5.3 Diatoms for Lithium Ion Battery Materials

Silicon is a widely used anode material for lithium-ion batteries and has a capacity of 4200 mAh g1. A range of silicon materials—including silicon films, nanoparticles, and electrochemically generated porous silicon—have been studied for potential applications in lithium batteries. Its highly accessible surface area for liquid electrolytes enables rapid lithium-ion transport, resulting in excellent rate performance and good electronic conductivity during charging and discharging cycles. This was found to be one of its features. This porous silicon possessed a distinctive hierarchical structure and a high surface area (Nzereogu et al., 2026). To enhance the performance of porous silicon electrodes, it is essential to develop electrodes with innovative porous nanoarchitectures that markedly improve performance by minimizing crystal strain and increasing the surface area available for ion transport. The effectiveness of silica as an anode material is greatly hindered by significant volume expansion and rapid capacity loss during lithium-ion insertion and extraction processes. This leads to the destruction of the electrode structure and poor cycling performance. To enhance the surface area and porosity of the silicon anode, it is crucial to incorporate new nanoscale morphologies, such as nanowires, nanotubes, nanosheets, and nanospheres (Kiai et al., 2026).

5.4 Catalysis and Energy Applications

Diatoms have surfaced as promising candidates for biofuel production in the search for sustainable energy sources, owing to their rapid growth rates and lipid-rich cell contents. Species such as Cyclotella cryptica and Thalassiosira weissflogii have been utilized for their oil-generating potential, especially in nitrogen-limited environments that promote lipid accumulation. These diatoms produce triacylglycerols, an essential element of biodiesel, positioning them as promising candidates for large-scale biofuel production (Rahul & Arumugam, 2026)

5.5 Environmental Remediation

The unique cellular and metabolic characteristics of diatoms render them valuable for remediating environments contaminated with heavy metals. Diatom species effectively remove toxic metals from wastewater via biosorption, providing sustainable solutions to environmental pollution. Diatoms such as Navicula pelliculosa and Phaeodactylum tricornutum have been used for the bioremediation of dichromate, demonstrating their potential to reduce heavy-metal toxicity. Moreover, modifying the surface of diatom-like biosilica increases its capacity to adsorb harmful heavy metal pollutants, highlighting the importance of diatoms in protecting the environment and managing water quality (Akram et al., 2026).

5.6 Antimicrobial Materials

Diatoms produce a range of bioactive substances that strongly inhibit pathogens, indicating their antimicrobial potential. Compounds such as monoacylglycerides, oxylipins, chrysolaminaran polysaccharides, and fucoxanthin have been extracted from diatoms and demonstrated antimicrobial properties. These substances can suppress the growth of bacteria, fungi, and other microorganisms, rendering them useful in the fight against infections. For example, it has been found that monoacylglycerides can disrupt microbial cell membranes, leading to cell death. In contrast, the chrysolaminaran polysaccharide demonstrates immunomodulatory effects that bolster the body's defense against infections. In addition, fucoxanthin—a carotenoid pigment found in high quantities in diatoms—exhibits antioxidant and anti-inflammatory effects, which add to its antimicrobial activity (Tiwari et al., 2026).

6. GENETIC ENGINEERING APPROACHES

6.1 Molecular Control of Frustule Formation

The frustules of diatoms are among the most complex examples of nanostructured and microstructured materials in nature, as a a result of biomineralization refined over tens of millions of years of evolution (Rogato & De Tommasi, 2020). Biosilica of diatoms is a hybrid material that consists of several organic compounds and amorphous silica. Diatom frustule formation is dependent on and coordinated by both self-assembly processes and silica polymerization in the SDV, which involve various molecules and proteins (Heintze et al., 20).

Frustule formation is under the control of a species-specific, tuned mechanism that, at each division, is transferred from the mother to the two daughter cells. In fact, many aspects of the process still need to be explained, such as how this higher-order structure is created and the precise temporal and spatial factors that influence it. The process of silicification manipulation and change in frustule architecture through gene modification is challenging. Furthermore, in diatoms, recognizing and modifying the biomineralization process may provide a powerful tool for a range of nanotechnological applications (Heintze et al., 2020). Diatom genetic engineering is still in its early stages compared with other organisms (Rogato & De Tommasi, 2020).

Genetic reporter systems is the study expression of gene, modulation of expression of genes methods and regulation and spatial localization of proteins, over-expression, gene knockdown, knockout, the CRISPR/Cas9 system (Rogato& De Tommasi (2020)) and tailored TALEN endonucleases are now available, allowing the modifications of genetic and mutated strains generation (Nymark et al., 2016 ; Rogato& De Tommasi, 2020). These molecular tools enable the study of the characteristics and functions of the proteins involved in cell wall synthesis (Fabris et al., 2020). *Phaeodactylum tricornutum and Thalassiosira pseudonana* are the molecular model species, since *P. tricornutum* is characterized by a cell wall poor in silica and fusiform cells (Dhaouadi et al., 2020), while *Thalassiosira pseudonana* and *Cylindrotheca fusiformis* have been used as model species to study silica cell wall formation (Poulsen& Kröger, 2023 ;Petrucciani et al., 2023).

6.2 Diatom genomic resources

Thalassiosira pseudonana and Phaeodactylum tricornutum were the first diatoms to have their genomes sequenced, serving as models for studying diatom evolution and ecological roles (Mock et al., 2022). The limited number of sequenced diatom genomes prompted further sequencing efforts, leading to discoveries about adaptations in harsh environments, such as how Fragilariopsis cylindrus adapts to polar oceans, Nitzschia putrida evolves from a phototropic to a heterotrophic lifestyle (Li et al., 2025), and how Fistulifera solaris and C.cryptica provide a foundation for improving lipid synthesis through genetic modifications (Traller et al., 2016), which is relevant for diverse biotechnological applications. The '100 Diatom Genomes Project' https://jgi.doe.gov/csp-2021-100-diatom-genomes) aims to enhance understanding of diatom functional and evolutionary genomics, which is funded by the Joint Genome Institute (JGI) and the Department of Energy (USA), this project represents a significant advancement in the field. These rapidly growing diatom genomic resources will yield extensive biological data available through public databases like NCBI, JGI, ENA, Ensembl, DiatOmicBase, and PLAZA. To exploit this information, developing synthetic biology tools for genome engineering is essential, alongside the use of multi-omics strategies to gain comprehensive insights into gene functions, genetic networks, and their roles in various biological processes (Li et al. 2025).

6.3 Genetic engineering

There are two primary approaches in diatom genetic engineering: forward and reverse genetics. Forward genetics focuses on phenotypes resulting from random mutagenesis, while reverse genetics deliberately modifies specific genetic loci to create and characterize new phenotypes. Reverse genetics plays a crucial role in synthetic biology that aims to develop synthetic diatoms with engineered chromosomes containing novel gene combinations. Although synthetic diatoms are not yet developed, progress in large-scale chromosome assemblies and transformation protocols suggests future possibilities.

Forward genetics: Forward genetics utilizes random mutations induced by chemical or radiation methods to introduce genetic changes, including DNA integrations from exogenous sources (Goold et al., 2024; Moosburner et al., 2022). High-throughput screening of many treated cell lines is used to identify phenotypic variations relative to wild-type controls. Notable examples include the identification of a new uridine-5-monophosphate synthase in *Phaeodactylum tricornutum* through ENU mutagenesis and the increase in eicosapentaenoic acid (EPA) levels to 44% via UV mutagenesis. UV mutagenesis also helps discover selectable markers linked to herbicide resistance, highlighting the method's utility for generating dominant mutations (Li et al. 2025). However, the diploid nature of diatom genomes complicates the achievement of homozygous (biallelic) gene inactivation through untargeted mutagenesis, which relies on single-strand breaks (Moosburner et al., 2022), a challenge similarly faced by reverse genetics approaches.

Reverse genetics: To modify genetic loci in diatoms, effective methods for delivering genetic material are essential and need to be established. Transformation protocols typically use marker genes such as green and yellow fluorescent proteins (GFP and YFP) to assess genetic modification stability, which is vital for reverse genetics applications (Faktorova et al., 2020). Common genetic modifications include gene overexpression, knockdown, and knockout, with species like T. pseudonana also enabling gene replacement via homology-directed repair (HDR). In 20 diatom species, gene overexpression can create new phenotypic traits and help elucidate gene functions, first successfully implemented in 1995 by Dunahay (Li et al. 2025).

Gene knockdown, the reversible process of reducing or abolishing gene expression by targeting mRNA rather than DNA, was first reported in P. tricornutum using antisense RNA technology (Li et al. 2025). Its application has become common in diatoms (Wang et al., 2023). The exact molecular mechanism remains unclear, as it does not strictly adhere to the classic RNA interference (RNAi) pathways, which typically involve small non-coding RNAs and the Dicer endoribonuclease; for example, T. pseudonana still retains mRNA of silenced genes (Li et al., 2025). Recent analyses indicate that diatoms have diversified RNAi effectors, suggesting distinct RNAi pathways, with P. tricornutum DCR processing double-stranded sRNAs mostly from transposons with associated epigenetic marks (Grypioti et al., 2024).

In the past decade, advancements in genome modification of diatoms have been driven by techniques such as meganucleases (MNs), transcriptional activator-like effector nucleases (TALEN), and CRISPR/Cas-assisted genome editing (Li et al., 2025). These methods create double-strand DNA breaks (DSB) at targeted locations, which are repaired through either homology-directed repair (HDR) or non-homologous end joining (NHEJ). Notably, while MNs are the oldest technology for cleaving long DNA sequences, TALENs offer high specificity and low off-target activity, albeit at the cost of more complex design. However, CRISPR/Cas technology has become the preferred method due to its efficiency and simpler design. The native CRISPR-Cas9 system utilizes guide RNAs to direct the Cas9 protein to cleave DNA (Fortunato et al., 2016). Landmark studies established CRISPR/Cas-based editing in diatom species like *P. tricornutum and T. pseudonana*, followed by developments in homologous recombination and DNA-free gene knockouts (Li et al., 2025). Recent innovations include the Cas9 nickase (nCas9) to reduce off-target effects while editing diatoms (Li et al., 2025; Nigishi et al., 2024).

Recent research has introduced dead (d)Cas9 in diatom studies for the first time, marking a significant development. Unlike Cas9, dCas9 cannot cut DNA but can still bind to it, enabling CRISPR interference (CRISPRi) for gene knockdown in P. tricornutum (Guo et al., 2024). While dCas9 can be utilized for CRISPR activation (CRISPRa) by fusing it with transcriptional activators to upregulate target genes, this application remains unexplored in diatom research. Additionally, base editors, which enable precise single-base modifications to introduce amino acid changes conferring substance resistance, have yet to be applied in this field (Lee et al., 2023; Li et al., 2025).

7. COMMERCIALIZATION AND SCALE-UP CHALLENGES

7.1 Production Economics

Large-scale diatom cultivation is capital-intensive and cost-sensitive. Open ponds and photobioreactors require significant CAPEX (land, reactors) and OPEX (nutrients, energy, CO_2). In practice, nutrient inputs (notably silica), CO_2 and electricity dominate operating costs (Wang & Seibert, 2017). For example, cheap plastic-lined ponds (~\$20/m^2 materials) produced ~120 ton ha^{-1} yr^{-1} of dry diatoms (Wang & Seibert, 2017). Even so, converting biomass to fuel (e.g., biocrude) remains expensive – early estimates suggest diatom biocrude could cost ~\$240–330/barrel (5–6× current oil) without co-products (Wang & Seibert, 2017). By contrast, high-value co-products (nutraceuticals, cosmetics, biosensors) can improve economics. Techno-economic studies of microalgae show small-scale costs of €300–600 kg^{-1} biomass, but state-of-the-art, large-scale systems (≥1500 m^2) can cut costs by 60–80%, to €40–80 kg^{-1}(Oostlander et al. 2020). In summary, diatom biomass costs currently range from tens to hundreds of €/US dollars per kg (dry), depending on scale and technology. Key cost drivers include nutrients (especially silicates), CO_2 supply, mixing and harvesting, and reactor materials.

7.2 Technical Challenges

Scaling diatoms poses the usual algal challenges plus diatom-specific issues. Light and mixing: Diatoms are photoautotrophs limited by light penetration and self-shading, so reactors must ensure uniform illumination and mixing (Lebeau & Robert, 2003). Open ponds are cheap but reach low biomass (<0.5–1 g/L) and suffer heat loss; closed photobioreactors (PBRs) achieve higher densities but at much higher CAPEX (Novoveská et al. 2023). Contamination: Open systems are vulnerable to grazers, competing algae and pathogens. Closed PBRs reduce invasions but require air/CO_2 sparging and clean conditions (Novoveská et al. 2023). Using extreme conditions (e.g., high salinity or pH) can deter contaminants (Novoveská et al. 2023) Nutrient supply: Diatoms demand silica (SiO_2) for frustule growth. Continuous culture must supply soluble silicate, which adds cost. Recycling or sourcing silicate (e.g., from industrial waste streams) is an area of research. Harvesting and processing: After growth, dewatering micron-sized diatoms is energy- and equipment-intensive (sedimentation, flocculation, filtration, centrifugation). Harvesting can account for ~20–25% of total cultivation costs (Novoveská et al. 2023). Downstream cleaning (burning/milling frustules, etching organic material) adds further cost. Infrastructure and maintenance: Large PBRs require corrosion-resistant, transparent materials and heating and cooling systems; these increase capital costs. For example, rigid transparent PBRs often cost too much for fuel-targeted production (Wang & Seibert, 2017).

7.3 Regulatory and Safety Considerations

Commercial diatom operations must navigate industrial and biosafety regulations. Worker safety: Diatom biosilica is amorphous silica; inhalation of fine silica dust can damage lungs. OSHA and EU regulators set exposure limits (OSHA TLV ~3–6 mg/m^3 respirable dust (Novoveská et al., 2023). Facilities must implement dust controls, ventilation and PPE to comply with workplace safety limits. Product safety: If diatom-derived products enter food, feed or cosmetics, they fall under those safety frameworks. In the EU, any novel food ingredient (e.g., a new algal biomass) requires Novel Food approval (Reg. 2015/2283), unless the species has QPS (Qualified Presumption of Safety) status. A few microalgae (including some diatoms such as Phaeodactylum) are on EFSA's list, facilitating entry (Novoveská et al. 2023). Environmental and GMO regulations: Cultivation sites (ponds or raceways) may require permits for water use and effluent discharge (nutrient-rich media). Directive 2001/18/EC (EC—European Commission, 2001) sets out a step-by-step approach to the introduction of a GMO into the environment, including evaluation of impacts on human health and the environment. Information is required on the parental/donor/GM organism and the receiving environment. Risk assessment follows a case-by-case and step-by-step approach (Fig 3) (Mitchell & Bartsch, 2020). It is important to identify characteristics that may cause adverse effects, e.g., on the dynamics of species populations in the receiving environment and on the genetic diversity of each of these populations. Nanomaterials oversight: If diatom frustules are marketed as engineered nanomaterials (e.g., for drug delivery), additional nano-specific regulations (e.g., EU Cosmetics Reg. 1223/2009 labeling, or FDA's nanotechnology guidance) could apply. Overall, companies must engage with regulators (food safety authorities, workplace safety agencies, and environmental agencies) early to ensure compliance.

Figure 3. The six steps in the analysis of Environmental Risk Assessment (ERA) in the European Union according to Directive 2001/18/EC (EC—European Commission, 2001) (Mitchell & Bartsch, 2020).

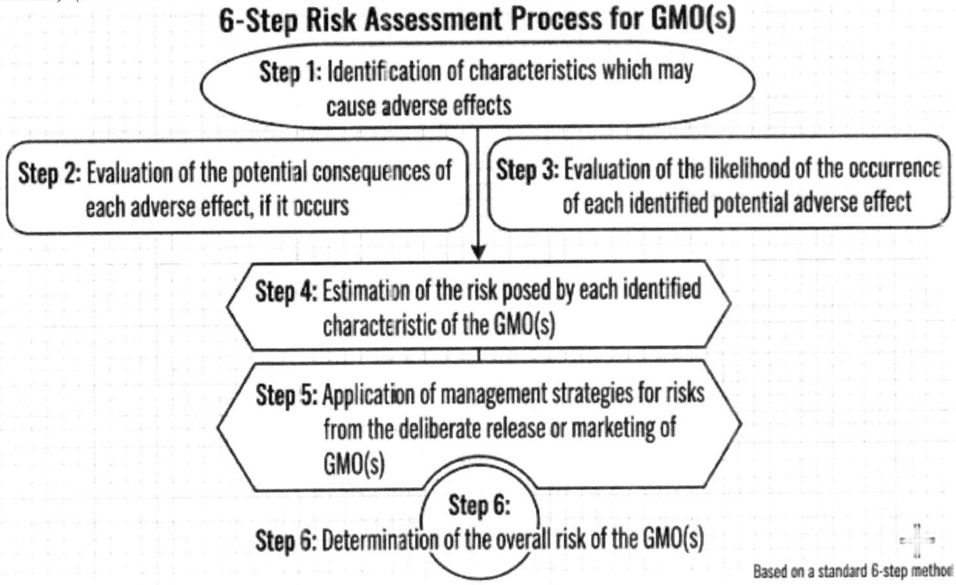

7.4 Current Commercial Applications

Today, most commercial uses of diatom silica derive from diatomaceous earth (DE) – fossilized diatoms mined from sedimentary deposits. DE is a mature industry (~$1–1.5 billion global market) used as filter aids (beer, water, pharmaceuticals), absorbents (spill cleanup, pet litter), agricultural soil amendment and pest control, and cement fillers (Novoveská et al. 2023). By contrast, applications of cultured diatom biosilica are nascent. In aquaculture, live diatoms (e.g., *Chaetoceros*, *Skeletonema*) are used as feed for oyster, shrimp and mollusk larvae (Novoveská et al. 2023). In high-tech materials, companies are beginning to exploit diatom frustules as *functional additives*: for example, Swedish Algae Factory markets Algica®, a diatom-derived silica ingredient for skincare (moisturizing, UV protection) and is testing it in solar cells and batteries. Research prototypes include diatom-based biosensors and catalysts, and diatom frustules with deposited metals (e.g., Ag, Au) for plasmonic optics (Barber, 2025), but these are largely laboratory demonstrations. In summary, current commercial applications include:

- Filtration and absorbents: Diatomite (calcined or raw) in beverage, water and air filters, and as insecticidal dust or absorbent cakes (pet litter, oil cleanup) (Novoveská et al. 2023)
- Aquaculture feed: Cultivated diatoms (e.g., *Chaetoceros*, *Navicula*) for bivalve and shrimp hatchery feed (Novoveská et al. 2023).
- Skincare and cosmetics: Algica® and similar products – diatom silica sold as natural SPF boosters, exfoliants or actives in beauty products (Barber, 2025).

7.5 Success Stories and Ongoing Ventures

The most notable success to date is Swedish Algae Factory (Gothenburg, Sweden). Founded in 2016, it has become the first to commercialize *large-scale* diatom cultivation for functional materials. Its product "Algica" (dried diatom biosilica) is now incorporated into over 60 skincare formulations worldwide. In early 2025, the company raised €3 million to expand its 3,500 ft^2 facility to 10,000 ft^2, reflecting strong investor confidence (Spherical Insights, 2023). The founders emphasize a circular model: diatoms are grown on recycled nutrients, and CO_2, and organic biomass byproducts are used for biogas or fertilizers (with potential for animal feed or food use) (Barber, 2025). This venture demonstrates that partnerships with established industries (beauty and personal care) can drive near-term revenue, easing the path to profitability. Beyond Swedish Algae Factory, a few start-ups are exploring diatom platforms (e.g., for biosensing or bioremediation), but none have yet reached notable scale or revenue. In a broader algal context, other companies (e.g., Provectus Algae, AlgoDyne) are tackling scale-up for algae-derived products (feeds, fuels), and their experiences in photobioreactor design and cost-reduction offer lessons for diatoms as well.

8. FUTURE DIRECTIONS AND EMERGING APPLICATIONS

The future of algal nanotechnology, particularly diatom-based biosilica, lies in the convergence of synthetic biology, advanced materials science, and sustainable manufacturing. Emerging tools such as CRISPR-Cas9 will enable precise engineering of frustule architecture and functionality, allowing the design of tailored nanostructures for applications in targeted drug delivery, biosensing, and catalysis. At the same time, integration with Machine Learning is expected to accelerate the discovery and optimization of structure–function relationships, improving both efficiency and scalability. Future innovations will likely focus on smart, stimuli-responsive materials, hybrid biomimetic–synthetic composites, and eco-friendly large-scale production systems using renewable resources. Additionally, expanding applications in biomedicine, environmental remediation, and energy technologies will position diatom biosilica as a key platform in next-generation nanomaterials, with successful translation depending on interdisciplinary collaboration and industrial investment.

9. CONCLUSION AND PERSPECTIVES

Overall, diatoms offer many applications that can address pressing global challenges. Therefore, in the current era, where sustainable and innovative solutions are increasingly sought, a comprehensive understanding of diatom biology and culture techniques is essential to unlock their biotechnological potential and advance sustainable practices across various industries.

REFERENCES

Agarwal, N. (2025). *Diatoms: basic and applied research.* Educohack Press.

Akram, S., Guan, G., Hu, B., Saddique, M. A. B., Xi, Y., Luo, X., & Ren, M. (2026). Microalgae-based strategies for cadmium remediation: Insights, challenges, and future directions. *Critical Reviews in Biotechnology, 46*(2), 318–337. DOI: 10.1080/07388551.2025.2585681 PMID: 41260883

Angmo, S., Srivastava, A., Saxena, A. K. R., & Tiwari, A. (2026). Diatoms nano-factories: Synthesis, properties, and biomedical applications. *Systems Microbiology and Biomanufacturing, 6*(2), 30. DOI: 10.1007/s43393-025-00427-3

Babenko, I., Friedrich, B. M., & Kröger, N. (2022). Structure and morphogenesis of the frustule. In *The molecular life of diatoms* (pp. 287–312). Springer International Publishing. DOI: 10.1007/978-3-030-92499-7_11

Barber, L. (2025). *Swedish Algae Factory raises €3 million to expand algae production and boost beauty partnerships.* Cosmetics Business. https://cosmeticsbusiness .com/swedish-algae-factory-raises-%E2%82%AC3million-expand-algica-production-boost-beauty-partnerships

Behrenfeld, M. J., Halsey, K. H., Boss, E., Karp-Boss, L., Milligan, A. J., & Peers, G. (2021). Thoughts on the evolution and ecological niche of diatoms. *Ecological Monographs, 91*(3), e01457. DOI: 10.1002/ecm.1457

Bouazzara, H., & Chaibi, R. (2025). Diatoms and nanotechnology, current trends in nanomaterials applications and environmental monitoring. *Environmental Monitoring and Assessment, 197*(7), 793. DOI: 10.1007/s10661-025-14227-8 PMID: 40550980

Bueno, G., Deniz, O., Pedraza, A., Ruiz-Santaquiteria, J., Salido, J., Cristóbal, G., Borrego-Ramos, M., & Blanco, S. (2017). Automated diatom classification (Part A): Handcrafted feature approaches. *Applied Sciences (Basel, Switzerland), 7*(8), 753. DOI: 10.3390/app7080753

Chaudhary, R., Nawaz, K., Khan, A. K., Hano, C., Abbasi, B. H., & Anjum, S. (2020). An overview of the algae-mediated biosynthesis of nanoparticles and their biomedical applications. *Biomolecules, 10*(11), 1498. DOI: 10.3390/biom10111498 PMID: 33143289

Choi, A. H. (2022). Biomaterials and bioceramics—part 1: traditional, natural, and nano. In *Innovative Bioceramics in Translational Medicine I: Fundamental Research* (pp. 1–45). Springer Singapore. DOI: 10.1007/978-981-16-7435-8_1

Cristóbal, G., Blanco, S., & Bueno, G. (Eds.). (2020). *Modern trends in diatom identification: fundamentals and applications*. Springer Nature. DOI: 10.1007/978-3-030-39212-3

Cvjetinovic, J., Luchkin, S. Y., Statnik, E. S., Davidovich, N. A., Somov, P. A., Salimon, A. I., Korsunsky, A. M., & Gorin, D. A. (2023). Revealing the static and dynamic nanomechanical properties of diatom frustules—Nature's glass lace. *Scientific Reports*, *13*(1), 5518. DOI: 10.1038/s41598-023-31487-x PMID: 37015973

da Trindade, L. G., Bürck, M., de Souza, E. L., Zanchet, L., Assis, M., & Braga, A. R. C. (2025). Diatomaceous biosilica: A multifunctional resource for biomedicine and sustainable applications. *Ceramics*, *8*(2), 62. DOI: 10.3390/ceramics8020062

Dhaouadi, F., Awwad, F., Diamond, A., & Desgagné-Penix, I. (2020). Diatoms' breakthroughs in biotechnology: Phaeodactylum tricornutum as a model for producing high-added value molecules. *American Journal of Plant Sciences*, *11*(10), 1632–1670. DOI: 10.4236/ajps.2020.1110118

Essler, Y., Anabalon, J., Celis-Plá, P. S., Kieft, B., Junier, P., Rivera, P., Cruces, F., Ascencio, E., Dorador, C., & Molina, V. (2025). High nutrient requirement on the photosynthetic performance of a diatom species of the genus Nitzschia (Bacillariophyta) isolated from the high-altitude wetland, Salar de Huasco, Chile. *Phycological Research*, *73*(2), 134–147. DOI: 10.1111/pre.12581

El-Hosari, D. G., Mokhtar, F. A., Khalaf, H. A., Ibrahim, A. R., Mohamed, R. M. & Makhlof, M. E. (2024). Minimum-Run resolution IV design for optimized bio removal of Fe2+ Using enteromorpha intestinalis aqueous extract and its extract-Coated silver nanoparticles. *Plants*, *14*(1), 40. DOI: 10.3390/plants14010040 PMID: 39795301

Fabris, M., Abbriano, R. M., Pernice, M., Sutherland, D. L., Commault, A. S., Hall, C. C., Labeeuw, L., McCauley, J. I., Kuzhiuparambil, U., Ray, P., Kahlke, T., & Ralph, P. J. (2020). Emerging technologies in algal biotechnology: Toward the establishment of a sustainable, algae-based bioeconomy. *Frontiers in Plant Science*, *11*, 279. DOI: 10.3389/fpls.2020.00279 PMID: 32256509

Faktorova, D., Nisbet, R. E. R., Fernandez Robledo, J. A., Casacuberta, E., Sudek, L., Allen, A. E., Ares, M.Jr, Aresté, C., Balestreri, C., Barbrook, A. C., Beardslee, P., Bender, S., Booth, D. S., Bouget, F.-Y., Bowler, C., Breglia, S. A., Brownlee, C., Burger, G., Cerutti, H., & Lukeš, J. (2020). Genetic tool development in marine protists: Emerging model organisms for experimental cell biology. *Nature Methods*, *17*(5), 481–494. https://doi.org/10. DOI: 10.1038/s41592-020-0796-x PMID: 32251396

Fortunato, A. E., Jaubert, M., Enomoto, G., Bouly, J. P., Raniello, R., Thaler, M., Malviya, S., Bernardes, J. S., Rappaport, F., Gentili, B., Huysman, M. J. J., Carbone, A., Bowler, C., d'Alcalà, M. R., Ikeuchi, M., & Falciatore, A. (2016). Diatom phytochromes reveal the existence of far-red- light-based sensing in the ocean. *The Plant Cell, 28*(3), 616–628. DOI: 10.1105/tpc.15.00928 PMID: 26941092

Ghobara, M., El-Sheekh, M., Hamed, A. F., Abdelhamid, M. A., & Pack, S. P. (2023). Diatom nanostructured biosilica. In *Value-Added Products from Algae: Phycochemical Production and Applications* (pp. 461–492). Springer International Publishing.

Ghobara, M. M., Gordon, R., & Reissig, L. (2021a). The mesopores of Raphid pennate diatoms: toward natural controllable anisotropic mesoporous silica microparticles. In *Diatom morphogenesis* (pp. 383–409). Wiley., DOI: 10.1002/9781119488170.ch16

Ghobara, M. M., Tiffany, M. A., Gordon, R., & Reissig, L. (2021b). Diatom pore arrays' periodicities and symmetries in the euclidean plane: nature between perfection and imperfection. In *Diatom morphogenesis* (pp. 117–158). Wiley., DOI: 10.1002/9781119488170.ch6

Gholami, P., Khataee, A., & Bhatnagar, A. (2020). Environmentally superior cleaning of diatom frustules using sono-Fenton process: Facile fabrication of nanoporous silica with homogeneous morphology and controlled size. *Ultrasonics Sonochemistry, 64*, 105044. DOI: 10.1016/j.ultsonch.2020.105044 PMID: 32146334

Goold, H. D., Moseley, J. L., & Lauersen, K. J. (2024). The synthetic future of algal genomes. *Cell Genomics, 4*(3), 100505. DOI: 10.1016/j.xgen.2024.100505 PMID: 38395701

Götz, W., Tobiasch, E., Witzleben, S., & Schulze, M. (2019). Effects of silicon compounds on biomineralization, osteogenesis, and hard tissue formation. *Pharmaceutics, 11*(3), 117. DOI: 10.3390/pharmaceutics11030117 PMID: 30871062

Grypioti, E., Richard, H., Kryovrysanaki, N., Jaubert, M., Falciatore, A., Verret, F., & Kalantidis, K. (2024). Dicer-dependent heterochromatic small RNAs in the model diatom species Phaeodactylum tricornutum. *The New Phytologist, 241*(2), 811–826. DOI: 10.1111/nph.19429 PMID: 38044751

Guiry, M. D. (2024). How many species of algae are there? A reprise. Four kingdoms, 14 phyla, 63 classes and still growing. *Journal of Phycology, 60*(2), 214–228. DOI: 10.1111/jpy.13431 PMID: 38245909

Guo, W., Weng, Y., Ma, W., Chang, C., Gao, Y., Huang, X., & Zhang, F. (2024). Improving lipid content in the diatom Phaeodactylum tricornutum by the knockdown of the enoyl-CoA hydratase using CRISPR interference. *Current Issues in Molecular Biology*, *46*(10), 10923–10933. DOI: 10.3390/cimb46100649 PMID: 39451529

Hamed, A. (2023). A brief and general overview on diatoms and their applications. *Egyptian journal of phycology, 24*(1), 1-53.

Heintze, C., Formanek, P., Pohl, D., Hauptstein, J., Rellinghaus, B., & Kröger, N. (2020). An intimate view into the silica deposition vesicles of diatoms. *BMC Materials*, *2*(1), 11. DOI: 10.1186/s42833-020-00017-8

Hussein, H. A., Nazir, M. S., Azra, N., Qamar, Z., Seeni, A., Tengku Din, T. A. D. A. A., & Abdullah, M. A. (2022). Novel drug and gene delivery system and imaging agent based on marine diatom biosilica nanoparticles. *Marine Drugs*, *20*(8), 480. DOI: 10.3390/md20080480 PMID: 36005484

Ikeda, T. (2021). Bacterial biosilicification: A new insight into the global silicon cycle. *Bioscience, Biotechnology, and Biochemistry*, *85*(6), 1324–1331. DOI: 10.1093/bbb/zbab069 PMID: 33877302

Insights, S. (2023). *Global diatomite market size, share, and COVID-19 impact analysis, by type (calcined, flux-calcined, and natural), by application (filter aids, cementitious materials, fillers, absorbents, and others), and by region (North America, Europe, Asia-Pacific, Latin America, Middle East, and Africa), analysis and forecast 2023 – 2033*. https://www.sphericalinsights.com/reports/diatomite-market

Jagdeo, K. R. (2023). Physical methods for synthesis of nanoparticles. In *Nanochemistry* (pp. 66–76). CRC Press. DOI: 10.1201/9781003081944-4

Kahlert, M., Bailet, B., Chonova, T., Karjalainen, S. M., Schneider, S. C., & Tapolczai, K. (2021). Same same, but different: The response of diatoms to environmental gradients in Fennoscandian streams and lakes–barcodes, traits and microscope data compared. *Ecological Indicators*, *130*, 108088. DOI: 10.1016/j.ecolind.2021.108088

Kapoor, S., Singh, M., Paul, S., Kar, S., Bagchi, T., Chavali, M., & Chandrasekhar, K. (2023). Sustainable bio-applications of diatom silica as nanoarchitectonic material. *Algae refinery*, 161-186.

Kiai, M. S., Aslfattahi, N., Karakoc, H., Baydogan, N., Samylingam, L., Kadirgama, K., Kok, C. K., & Schmirlerova, H. (2026). A Comprehensive Review, Future Possibilities and the Impact of Sustainable Battery Technologies for Practical Applications. *Korean Journal of Chemical Engineering*, *43*(5), 1–31. DOI: 10.1007/s11814-025-00642-7

Kirubakaran, D., Wahid, J. B. A., Karmegam, N., Jeevika, R., Sellapillai, L., Rajkumar, M., & SenthilKumar, K. J. (2026). A comprehensive review on the green synthesis of nanoparticles: Advancements in biomedical and environmental applications. *Biomedical Materials & Devices*, *4*(1), 388–413. DOI: 10.1007/s44174-025-00295-4

Kolbe, F., & Brunner, E. (2022). Silicic acid uptake and storage by diatoms. In *The molecular life of diatoms* (pp. 345–365). Springer International Publishing. DOI: 10.1007/978-3-030-92499-7_13

Kulichová, J., & Urbánková, P. (2020). Symmetric and Asymmetric Components of Shape Variation in the Diatom Genus Frustulia (Bacillariophyta). *Symmetry*, *12*(10), 1626. DOI: 10.3390/sym12101626

Laxa, R. A. (2021). Construction of Dichotomous Taxonomic Keys for San Francisco Bay Planktonic Diatoms.

Lebeau, T., & Robert, J. M. (2003). Diatom cultivation and biotechnologically relevant products. Part I: Cultivation at various scales. *Applied Microbiology and Biotechnology*, *60*(6), 612–623. DOI: 10.1007/s00253-002-1176-4 PMID: 12664139

Lee, J., Lim, K., Kim, A., Mok, Y. G., Chung, E., Cho, S. I., Lee, J. M., & Kim, J.-S. (2023). Prime editing with genuine Cas9 nickases minimizes unwanted indels. *Nature Communications*, *14*(1), 1786. DOI: 10.1038/s41467-023-37507-8 PMID: 36997524

Li, J., Gao, M., Liu, B., Fan, Y., Wei, J., Zhang, Y., Li, F., Zhang, N., & Hu, Z. (2025). Seasonal Dynamics of Phytoplankton Community Structure and Environmental Drivers in the Coastal Waters of the Leizhou Peninsula, China. *Diversity*, *17*(12), 867. DOI: 10.3390/d17120867

Li, Q., Gluch, J., Liao, Z., Posseckardt, J., Clausner, A., Łępicka, M., Grądzka-Dahlke, M., & Zschech, E. (2021). Morphology and mechanical properties of fossil diatom frustules from genera of Ellerbeckia and Melosira. *Nanomaterials (Basel, Switzerland)*, *11*(6), 1615. DOI: 10.3390/nano11061615 PMID: 34202999

Li, Y., Deng, L., Walker, E. J. L., Karas, B. J., & Mock, T. (2025). Genetic engineering in diatoms: Advances and prospects. *The Plant Journal : For Cell and Molecular Biology*, *121*(6), e70102. DOI: 10.1111/tpj.70102 PMID: 40089910

Luís, A. T., Hlúbiková, D., Vaché, V., Choquet, P., Hoffmann, L., & Ector, L. (2017). Atomic force microscopy (AFM) application to diatom study: Review and perspectives. *Journal of Applied Phycology*, *29*(6), 2989–3001. DOI: 10.1007/s10811-017-1177-4

Maran, F. N., Türk, İ., Akkoyun, B., Meligy, M. T. F., Lubiçeva, E., İkballı, D., & Şen, F. (2023). Nanotechnology and its versatile applications in medicine, environment, energy, textiles, food industry. *International Journal of Boron Science and Nanotechnology*, (1), 18–49.

Maurya, A. K., Rathi, S., Saksena, Y. P., Awasthi, A., Kaushik, B., & Sheetal. (2025). Uptake, Translocation and Consequences of Application of Si/SiO 2 Nanoparticles (SiNPs) in Plants. *Silicon Nanoparticles: Synthesis, Uptake, and Applications*, 213-261.

Medlin, L. K. (2009). Diatoms (Bacillariophyta). *Timetree Life*, *1*, 127–130. DOI: 10.1093/oso/9780199535033.003.0010

Mekuye, B., & Abera, B. (2023). Nanomaterials: An overview of synthesis, classification, characterization, and applications. *Nano Select*, *4*(8), 486–501. DOI: 10.1002/nano.202300038

Min, K. H., Kim, D. H., Youn, S., & Pack, S. P. (2024). Biomimetic diatom biosilica and its potential for biomedical applications and prospects: A review. *International Journal of Molecular Sciences*, *25*(4), 2023. DOI: 10.3390/ijms25042023 PMID: 38396701

Mitchell, H. J., & Bartsch, D. (2020). Regulation of GM organisms for invasive species control. *Frontiers in Bioengineering and Biotechnology*, *7*, 454. DOI: 10.3389/fbioe.2019.00454 PMID: 32039172

Mock, T., Hodgkinson, K., Wu, T., Moulton, V., Duncan, A., & van Oosterhout, C. (2022). Structure and evolution of diatom nuclear genes and genomes. In Falciatore, A., & Mock, T. (Eds.), *The molecular life of diatoms* (pp. 111–145). Springer. DOI: 10.1007/978-3-030-92499-7_5

Moosburner, M., Allen, A. E., & Daboussi, F. (2022) Genetic engineering in marine diatoms: current practices and emerging technologies. In: Falcia- tore, A. & Mock, T. (Eds.) The molecular life of diatoms. London: Springer Press, pp. 743–773. DOI: 10.1007/978-3-030-92499-7_25

Musenich, L. (2024). Design of Multifunctional Bioinspired Structures.

Nigishi, M., Shimakawa, G., Yamagishi, K., Amano, R., Ito, S., Tsuji, Y., Nagasato, C., & Matsuda, Y. (2024). Low-CO2-inducible bestrophins outside the pyrenoid sustain high photosynthetic efficacy in diatoms. *Plant Physiology*, *195*(2), 1432–1445. DOI: 10.1093/plphys/kiae137 PMID: 38478576

Novoveská, L., Nielsen, S. L., Eroldoğan, O. T., Haznedaroglu, B. Z., Rinkevich, B., Fazi, S., Robbens, J., Vasquez, M., & Einarsson, H. (2023). Overview and challenges of large-scale cultivation of photosynthetic microalgae and cyanobacteria. *Marine Drugs*, *21*(8), 445. DOI: 10.3390/md21080445 PMID: 37623726

Nymark, M., Sharma, A. K., Sparstad, T., Bones, A. M., & Winge, P. (2016). A CRISPR/Cas9 system adapted for gene editing in marine algae. *Scientific Reports*, *6*(1), 24951. DOI: 10.1038/srep24951 PMID: 27108533

Nzereogu, P. U., Omah, A. D., Ezema, F. I., Iwuoha, E. I., & Nwanya, A. C. (2026). Silica derived from rice husk waste as anode material for lithium-ion battery: A comprehensive study. *Next Materials*, *11*, 101615. DOI: 10.1016/j.nxmate.2026.101615

Oostlander, P. C., van Houcke, J., Wijffels, R. H., & Barbosa, M. J. (2020). Microalgae production cost in aquaculture hatcheries. *Aquaculture (Amsterdam, Netherlands)*, *525*, 735310. DOI: 10.1016/j.aquaculture.2020.735310

Panwar, V., & Dutta, T. (2019). Diatom biogenic silica as a felicitous platform for biochemical engineering: Expanding frontiers. *ACS Applied Bio Materials*, *2*(6), 2295–2316. DOI: 10.1021/acsabm.9b00050 PMID: 35030724

Parikh, H. (2021). Diatom Biosilica as a source of Nanomaterials. [IJARESM]. *International Journal of All Research Education and Scientific Methods*, *9*(11).

Parikh, H. S., Dave, G., Singh, P. K., & Tiwari, A. (2025). Exploring the significance of xenic and axenic cultures for diatoms: Methodological insights, scope, and biotechnological implications. *Systems Microbiology and Biomanufacturing*, *5*(3), 969–989. DOI: 10.1007/s43393-025-00363-2

Peng, M., Liu, P., Peng, R., Jiang, X., Li, S., & Jiang, M. (2025). Effects of nitrogen, phosphorus, iron, and silicon on the growth, total lipid content, and fatty acid composition of the diatom Corethron sp. *Journal of Applied Phycology*, •••, 1–14.

Petrucciani, A., Moretti, P., Ortore, M. G., & Norici, A. (2023). Integrative effects of morphology, silicification, and light on diatom vertical movements. *Frontiers in Plant Science*, *14*, 1143998. DOI: 10.3389/fpls.2023.1143998 PMID: 37056507

Poulsen, N., & Kröger, N. (2023). Thalassiosira pseudonana (Cyclotella nana) (Hustedt) Hasle et Heimdal (Bacillariophyceae): A genetically tractable model organism for studying diatom biology, including biological silica formation. *Journal of Phycology*, *59*(5), 809–817. DOI: 10.1111/jpy.13362 PMID: 37424141

Rabiee, N., Khatami, M., Jamalipour Soufi, G., Fatahi, Y., Iravani, S., & Varma, R. S. (2021). Diatoms with invaluable applications in nanotechnology, biotechnology, and biomedicine: Recent advances. *ACS Biomaterials Science & Engineering*, *7*(7), 3053–3068. DOI: 10.1021/acsbiomaterials.1c00475 PMID: 34152742

Rafique, M. S., Rafique, M., Tahir, M. B., Hajra, S., Nawaz, T., & Shafiq, F. (2020). Synthesis methods of nanostructures. In *Nanotechnology and photocatalysis for environmental applications* (pp. 45–56). Elsevier. DOI: 10.1016/B978-0-12-821192-2.00003-6

Rahul, S., & Arumugam, A. (2026). Current Advancements in Algal Feedstock Cultivation and Processing Aiming 3G Biorefinery. In *Biorefineries in Circular Bio-economy: Reactor Configurations Analysis* (pp. 57–80). Springer Nature Singapore.

Rai, I., Achankunju, J., Gordon, R., & Vinayak, V. (2025). Terraforming Mars with Microalgae, Especially Diatoms. *Diatom Cultivation for Biofuel, Food and High-Value Products*, 203-230.

Razzak, S. A., Bahar, K., Islam, K. O., Haniffa, A. K., Faruque, M. O., Hossain, S. Z., & Hossain, M. M. (2024). Microalgae cultivation in photobioreactors: Sustainable solutions for a greener future. *Green Chemical Engineering*, *5*(4), 418–439. DOI: 10.1016/j.gce.2023.10.004

Reid, A., Buchanan, F., Julius, M., & Walsh, P. J. (2021). A review on diatom biosilicification and their adaptive ability to uptake other metals into their frustules for potential application in bone repair. *Journal of Materials Chemistry. B*, *9*(34), 6728–6737. DOI: 10.1039/D1TB00322D PMID: 34346480

Rinku, Angmo, S., Srivastava, A., Merina, K., Sindhwal, S., & Tiwari, A. (2026). Diatoms Nanotechnology: Prospects and Challenges for Sustainable Environment. *Advanced Nanotechnology Applications for Environmental Sustainability*, 41-70.

Rogato, A., & De Tommasi, E. (2020). Physical, chemical, and genetic techniques for diatom frustule modification: Applications in nanotechnology. *Applied Sciences (Basel, Switzerland)*, *10*(23), 8738. DOI: 10.3390/app10238738

Round, F. E., Crawford, R. M., & Mann, D. G. (1990). *Diatoms: biology and morphology of the genera*. Cambridge University Press.

Saoud, H. A. A., Sprynskyy, M., Pashaei, R., Kawalec, M., Pomastowski, P., & Buszewski, B. (2022). Diatom biosilica: Source, physical-chemical characterization, modification, and application. *Journal of Separation Science*, *45*(17), 3362–3376. DOI: 10.1002/jssc.202100981 PMID: 35652201

Shera, S. S., & Banik, R. M. (2022). Algal nanoparticles: synthesis and characterization. In *Bioprospecting Algae for Nanosized Materials* (pp. 25–69). Springer International Publishing.

Singaravelu, G., Arockiamary, J. S., Kumar, V. G., & Govindaraju, K. (2007). A novel extracellular synthesis of monodisperse gold nanoparticles using marine alga, Sargassum wightii Greville. *Colloids and Surfaces. B, Biointerfaces*, *57*(1), 97–101. DOI: 10.1016/j.colsurfb.2007.01.010 PMID: 17350236

Spaulding, S. A., Potapova, M. G., Bishop, I. W., Lee, S. S., Gasperak, T. S., Jovanoska, E., Furey, P. C., & Edlund, M. B. (2021). Diatoms.org: Supporting taxonomists, connecting communities. *Diatom Research*, *36*(4), 291–304. DOI: 10.1080/0269249X.2021.2006790 PMID: 35958044

Sun, X., Zhang, M., Liu, J., Hui, G., Chen, X., & Feng, C. (2024). The art of exploring diatom biosilica biomaterials: From biofabrication perspective. *Advanced Science (Weinheim, Baden-Württemberg, Germany)*, *11*(6), 2304695. DOI: 10.1002/advs.202304695 PMID: 38044309

Sung, C. H., Hao, T., Fang, H., Nguyen, A. T., Perricone, V., Yu, H., Huang, W., Sarmiento, E., Ornelas, A. F. D., Lublin, D., Wehling, R., Farajollahi, S., Arakaki, A., Nepal, D., Lord, N. P., & Kisailus, D. (2025). Biological and Biologically Inspired Functional Nanostructures: Insights into Structural, Optical, Thermal, and Sensing Applications. *Advanced Materials*, *37*(51), e09281. DOI: 10.1002/adma.202509281 PMID: 40981639

Tiwari, A., Saini, G. K., Parikh, H. K. R., & Saxena, A. (2026). Diatom-Derived Biochemicals: An In-Depth Analysis of Polysaccharides, Extraction Methodologies, and Diverse Applications. *Applied Biochemistry and Biotechnology*, *198*(1), 1–28. DOI: 10.1007/s12010-025-05414-x PMID: 41182542

Traller, J. C., Cokus, S. J., Lopez, D. A., Gaidarenko, O., Smith, S. R., McCrow, J. P., Gallaher, S. D., Podell, S., Thompson, M., Cook, O., Morselli, M., Jaroszewicz, A., Allen, E. E., Allen, A. E., Merchant, S. S., Pellegrini, M., & Hildebrand, M. (2016). Genome and methylome of the oleaginous diatom Cyclotella cryptica reveal genetic flexibility toward a high lipid phenotype. *Biotechnology for Biofuels*, *9*(1), 258. DOI: 10.1186/s13068-016-0670-3 PMID: 27933100

Wang, J. K., & Seibert, M. (2017). Prospects for commercial production of diatoms. *Biotechnology for Biofuels*, *10*(1), 16. DOI: 10.1186/s13068-017-0699-y PMID: 28115988

Wang, L., Sun, Y., Zhang, R., Pan, K., Li, Y., Wang, R., Zhang, L., Zhou, C., Li, J., Li, Y., Zhu, B., & Han, J. (2023). Enhancement of hemostatic properties of Cyclotella cryptica frustule through genetic manipulation. *Biotechnology for Biofuels and Bioproducts*, *16*(1), 136. DOI: 10.1186/s13068-023-02389-x PMID: 37710352

Willis, L., Page, K. M., Broomhead, D. S., & Cox, E. J. (2010). Discrete free-boundary reaction-diffusion model of diatom pore occlusions. *Plant Ecology and Evolution*, *143*(3), 297–306. DOI: 10.5091/plecevo.2010.415

Yaqoob, A. A., Ahmad, H., Parveen, T., Ahmad, A., Oves, M., Ismail, I. M., Qari, H. A., Umar, K., & Mohamad Ibrahim, M. N. (2020). Recent advances in metal decorated nanomaterials and their various biological applications: A review. *Frontiers in Chemistry*, *8*, 341. DOI: 10.3389/fchem.2020.00341 PMID: 32509720

Zhu, J., Cheng, J., Zeng, J., Zhang, W., Liu, C., Effiong, K. S., & Hao, Q. (2026). A Review of Phytoplankton Sinking Rates: Mechanisms, Methodologies, and Biogeochemical Implications. *Biology (Basel)*, *15*(2), 130. DOI: 10.3390/biology15020130 PMID: 41594864

Zobi, F. (2022, June). Diatom biosilica in targeted drug delivery and biosensing applications: Recent studies. In *Micro* (Vol. 2, No. 2, pp. 342-360). MDPI.

Zuluaga-Astudillo, D., Ruge, J. C., & Caicedo-Hormaza, B. (2025). Micromechanical Characterization of Diatom Frustules of Multiple Origin. *Applied Sciences (Basel, Switzerland)*, *15*(2), 749. DOI: 10.3390/app15020749

Zuluaga-Astudillo, D., Ruge, J. C., Camacho-Tauta, J., Reyes-Ortiz, O., & Caicedo-Hormaza, B. (2022). Diatomaceous soils and advances in geotechnical engineering—Part I. *Applied Sciences (Basel, Switzerland)*, *13*(1), 549. DOI: 10.3390/app13010549

Chapter 12

Synthesis and Application of Microalgal Nanoparticles for Wastewater Treatment

K. Gayathri
http://orcid.org/0009-0007-3786-2009
Department of Microbiology, Sacred Heart College, Tirupattur, India

M. Manigandan
http://orcid.org/0000-0002-8115-2713
Department of Microbiology, Sacred Heart College, Tirupattur, India

A. Mubarak Ali
http://orcid.org/0009-0009-6367-7864
Department of Microbiology, Sacred Heart College, Tirupattur, India

B. Lokeshwari
http://orcid.org/0009-0005-6612-5134
Department of Microbiology, Sacred Heart College, Tirupattur, India

P. Saranraj
http://orcid.org/0000-0002-5052-6532
Department of Microbiology, Sacred Heart College, Tirupattur, India

K. Kesavardhini
http://orcid.org/0009-0005-8235-042X
Department of Microbiology, Sacred Heart College, Tirupattur, India

Ashish Kumar Nayak
http://orcid.org/0000-0002-3211-7065
Prasad V. Potluri Siddhartha Institute of Technology, Vijayawada, India

ABSTRACT

The reason is that sustainable remediation methods are necessary since the water pollution caused by the heavy metals, organic pollutants, and industrial effluent entails severe risks to both the environment and human health. Microalgae is a

DOI: 10.4018/979-8-3373-7630-1.ch012

green platform to produce Nanoparticles (NPs) because it is fast growing, biocompatible, and metabolically versatile. Algal enzymes, proteins and polysaccharides by intracellular and extracellular pathways are reducing and capping agents in the production of microalgal nanoparticles, which are predominantly metallic (including silver, gold and iron oxide) and metal oxide NPs. These aspects are also aligned to the principles of the circular economy. The problems such as scalability, NP leaching, and long-term stability are overcome by immobilization methods and genetic modification of algae to hyperproduce it. The future perspective is an integration of pilot-scale reactors to practical use with AI-optimized processes. Therefore, microalgal NPs is an indication of a paradigm shift in wastewater treatment all over the world to green nanotechnology.

1. INTRODUCTION

The problem of water scarcity and pollution retains existential threats to the global ecosystems and human health, and the wastewater of the industries, agriculture, and cities contributes more than 80 percent of untreated effluents worldwide, as described by the United Nations Environment Programme (UNEP, 2023). The traditional treatment processes such as activated sludge processes, chemical coagulation and membrane filtration are ineffective because of high energy requirements, second sludge, and failure to remove recalcitrant pollutants such as heavy metals, dyes, pharmaceuticals, and nutrients (Abbasi et al., 2025). Here, nanotechnology is proposed as an innovative paradigm, with high surface-area-to-volume proportions and adjustable surface chemistries to allow the capture of pollutants efficiently. Microalgal nanoparticles are one of the rare sources of nanomaterials, whose synthesis is eco-friendly, and can serve as multifunctional nanoparticles, which connects biological remediation with advanced materials science. Microalgae are unicellular non-heterotrophic photosynthetic microorganisms, which grow in a variety of aquatic habitats, and represent a very promising renewable biomass power source. Metals and organics can be bios orbed and bioaccumulated in species like Chlorella vulgaris, Spirulina platensis and Scenedesmus obliquus using cell wall constituents, i.e., polysaccharides, proteins and lipids (Abdelfattah et al., 2022). It is these innate bioremediation abilities that lead to the creation of Microalgal Nanoparticles (MNPs) in which algal biomass is used as a reducing and stabilizing agent in creating green nanoparticles. Use of microalgal-mediated synthesis versus chemical methods to reduce metal salts to zero-valent or metal oxide nanoparticles. Microalgal enzymes (e.g., nitrate reductase), polyphenols, and flavonoids are extracellular enzymes used in microalgal-mediated synthesis instead of the toxic reductants used in chemical methods. This biological process reduces ecological footprints, and stable nanoparticles are obtained with a size between 5-100 nm as was established by various methods such as TEM, XRD, and FTIR. The development of the synthesis process occurs in stages. First, biomass of algae (cultures or wastewater itself) is pretreated (e.g., through ultrasonication or autoclaving) to free biomolecules. Metal precursors (e.g., silver nitrate ($AgNO_3$) to BadgNPs or chloroauric acid ($HAuCl_4$) to AuNPs are then incubated with the extract at controlled pH (usually 6-8) and temperature (Acién et al., 2023).

The reduction process is high in rate as witnessed by change of color (e.g. yellow to brown in the case of AgNPs) and the process is then stabilized through algal capping agents, which prevent aggregation. Recent technological advances use waste algal biomass in production of biofuels, which forms a circular economy model. As an example, Chlorella-derived Fe_3O_4 nanoparticles of 95 percent magnetite purity were demonstrated with magnetic separability on reuse. These MNPs can be utilised in the treatment of wastewater in various ways: adsorption, photocatalysis, antimicrobial activity and redox reactions. Functionalized surfaces are bound by heavy metals such as Cd^{2+}, Pb^{2+} and Cr^{6+} using electrostatic attraction and chelation and can be removed at high percentages (>90) with low doses (0.1-1 g/L) (Ahmed et al., 2023). Organic dyes (e.g., methylene blue) decay under the visible light through the actions of the reactive oxygen species produced by the plasmonic Ag/Au MNPs. Selective uptake of nutrients (e.g. phosphates and nitrates) suppresses eutrophication. One of such remarkable examples is Spirogyra-synthesized ZnO nanoparticles that eliminated 98.00 percent of Congo red dye in 120 minutes in the presence of solar irradiation. In addition, MNPs have antibacterial synergy on wastewater pathogens such as E. coli and Vibrio cholerae decreasing the biofouling of treatment systems. Microalgal nanoparticles have superior benefits as compared to synthetic ones. Bioengineered MNPs are better biocompatible, less toxic, and more stable in multifaceted matrices thanks to natural capping (Anvar et al., 2023). They are consistent with UN Sustainable Development Goal 6 (Clean Water and Sanitation) and circular bioeconomy and use low-cost algal feedstocks (less than USD 1/kg dry weight). Investigations like the incorporation of MNPs in algal ponds as tertiary treatment have been reported to result in 70-85% COD reduction and biomass valorisation to bioenergy. Issues continue to exist, such as scalability, stability in the long term at different pH/ionic strength and the leaching of algal organics. Continued research is focusing on these through genetic engineering of hyper-accumulating algal strains and hybrid nanocomposites (e.g. MNP-graphene oxides). Laws such as those of the EPA and EU REACH have called on ecotoxicological profiling to guarantee safe usage. Nanoparticles of microalgae promise to represent a sustainable nano revolution in wastewater treatment as it integrates the biorefinery of nature with nanotechnology. Their production using renewable algal materials and the wide range of applications are a great assurance of cost-effective, green technologies to treat contaminated waters that can lead to strong agri-food systems and urban sanitation (Aragón et al., 2024).

2. MICROALGAL BIOLOGY AND NANOPARTICLE SYNTHESIS POTENTIAL

The extraordinary capability of nanoparticles synthesis in wastewater treatment is based on microalgal biology, which takes advantage of the exceptional physiology of such eukaryotic microalgae as Chlorella, Spirulina and Scenedesmus as natural bio factories. These photosynthetic giants have firm cell walls endowed with alginates, chitosan-like polysaccharides, metal binding sulphated groups that help in the binding of metal ion and intracellular enzymes such as nitrate reductase, superoxide dismutase, and dehydrogenases that are involved in the redox nature of creating nanoparticles (Armah et al., 2020). Reducing and capping agents can be produced using flavonoids, terpenoid and proteins extracellular secretion, and thus, one-pot synthesis of metal (Ag, Au), metal oxides (ZnO, TiO2, Fe3O4) and bimetallic nanoparticles can be synthesized without harsh chemicals. It begins by harvesting the algal biomass, be it by grinding, sonication or thermal lysis, to release biomolecules that destabilize precursor salts at moderate conditions (pH 6-9, 25-80 DegC) to generate monodisperse nanoparticles (5-50 nm) that are biologically stable because of biogenic coatings that hinder aggregation and enhance dispersion in aqueous media (Arteaga-Castrejón et al., 2024).

Due to the superior qualities of wastewater remediation these biological machineries possess: a high surface area to adsorb heavy metals (Pb, Cd, Cr) by chelation and ion exchange; photochemical to break down dyes and organics by generating electron-hole pairs under light; antimicrobial to inhibit bacterial membrane of effluent pathogens. Magnets Magnetically recoverable iron oxide MNPs, e.g., specifically sequester phosphates and nitrates, inhibit eutrophication and plasmonic Ag/Au variants enhance radical-mediated dismantling of recalcitrant pollutants such as pharmaceuticals (Ayyadurai and Ragavendran, 2025). The high accumulation characteristics of microalgae that have been developed to absorb nutrients in nutrient-deficient conditions enhance production efficiency, and the biomass of the cultures cultured in wastewater transforms biomass into nutrient cycles within the circular economy. Tailored nanoparticle designs can be achieved by strain-specific differences, including lipid-rich Butyraceous which is to be used as a hydrophobic capping agent or Arthrospore which is heavy in polysaccharide as its steric stabilizer. Such complications as the optimization of biomass pretreatment and scale-up are counterbalanced by the fact that microalgae have a short doubling time (less than 24 hours) and captures CO 2, which makes MNPs a more sustainable alternative to synthetic nanomaterials. Finally, via microalgal biology, wastewater treatment will be converted into energy-saving, biologically inspired and multifunctional nanotechnology to ensure efficient, eco-benign pollutant abatement (Bisheh et al., 2022).

2.1 Overview of microalgal diversity and biomass production

The synthesis of nanoparticles to treat wastewater is based on the microalgal diversity and biomass production which harbors more than 50,000 recognized species grouped under major categories such as Chlorophyta (green algae, e.g., Chlorella vulgaris, Scenedesmus obliquus), Cyanophyta (blue-green algae, e.g., Spirulina platensis, Anabaena), Bacillariophyta and Euglenophyta with a diverse biochemical profile Green algae have the most protein- and chlorophyll-rich cell walls, which are best in stabilizing Ag and Au nanoparticles through amino acid-based reduction; cyanobacteria are good providers of exopolysaccharides to make the best magnetic Fe3O4 particles with better chelation of heavy metals; diatoms can provide silica frustules to make hierarchical nanostructures of ZnO and extremophiles such as Dunaliella salina can provide carotenoid-capped particles which are stable in saline effluents. It allows the selection of strain depending on the target pollutants, such as lipid-accumulators such as Botryococcus braunii to hydrophobic organic remediation or metal-tolerant Stigeoclonium to Cr (VI) reduction to optimise nanoparticle yield (up to 90% conversion efficiency) and functionality (Bisheh et al., 2023).

This potential is scaled by biomass production using either photobioreactors (PBRs), open raceway ponds, or hybrid systems with productivities of 0.5-5 g/L/day under controlled light (100-300 umol photons m -2 s -1), CO2 supplementation and wastewater source itself, doubling as cost-free medium recycling N/P and pre-treating effluents. The mixotrophic or heterotrophic life form enhances yields by 2-3 times, and the resulting biomass is obtained through flocculation, centrifugation, or bio flocculants with 10-50 g/L of the dry biomass (Braga et al., 2026). The synergy in the example of wastewater-grown cultures is the precursors of nanoparticle nanoparticles (trace metals) are provided in industrial effluents, and the BOD 0 -90 initial is removed by the uptake of the algal. The release of biomolecules (polysaccharides to cap and enzymes to reduce) by post-harvest drying and milling allows a biomass-friendly synthesis of nanoparticles with a cost of less than USD/kg biomass. Other measures to enhance productivity and make metabolites more diverse to defeat season fluctuations are genetic engineering and consortia (e.g. Chlorella-Spirulina blends). It is not just a low-input, scalable platform that can produce nanoparticles feedstocks but also adds biomass valorisation-residuals to biofuels or biofertilizers with the objective of achieving the creation of circular wastewater treatment economies with a small ecological footprint (Chaudhary et al., 2020).

2.2 Key strains for NP biosynthesis (*Chlorella*, *Spirulina*, *Chlamydomonas*, etc.)

Microalgae like Chlorella, Spirulina, and Chlamydomonas are of significant interest as key strains in the biosynthesis of Nanoparticles (NPs) to treat wastewater as they use their inherent metabolism to synthesize eco-friendly nanomaterials that that also have the capability to treat pollutants. Chlorella vulgaris is a robust green microalga that is effective in synthesizing silver (AgNPs) and gold (AuNPs) nanoparticles by extracellular and intracellular processes wherein enzymes such as nitrate reductase and phytochemicals reduce metal ions in the process of its growth in nutrient-rich effluents. This strain prospers in municipal and industrial wastewater, takes up heavy metals such as cadmium, lead, and chromium, and produces NPs, which improve flocculation and disinfection, up to 90% of the Chemical Oxygen Demand (COD) and phosphates in high-rate algal ponds (El-Sheekh et al., 2025). Spirulina platensis, one of the cyanobacteria, a high-protein hydrogel, contributes to the production of iron oxide (Fe3O4) and zinc oxide (ZnO) as NPs by phycocyanin-mediated bio reduction, which is effective against textile dyes and antibiotics in anaerobic digests, the helical structure of this cyanobacterium helps it to form biofilms, which enhances nutrient uptake and stabilization of NPs to eliminate antibiotic-resistant Chlamydomonas reinhardtii is also rare, as selenium (SeNPs) and titanium dioxide (TiO2) NP biosynthesis uses its flagella to chelate and biosorb metal ions in swine and dairy wastewaters, with genetically engineered strains overexpressing its metallothionein gene to survive high copper and arsenic concentrations, and as a result of which it expresses NPs capable of photocatalytic activity to degrade organics in the presence of UV light. The other possible strains would be Scenedesmus obliquus to copper oxide (CuO) NPs in pharmaceutical effluents (they have a better capacity to hyperaccumulate phosphorus), and Daniella salina to silica (SiNPs) in saline industrial discharges (they use carotenoid antioxidants to cap the NPs, and to form antiviral surfaces on NPs) (Ennaceri et al., 2025). These micro algae proliferate in the photobioreactors or open raceways under the optimum conditions of controlling the pH 7-9, temperature of 25-30 deg C as well as the redox reactions and photosynthesis transform the nutrient components present in the wastewater to the biomass and the NPs. The biosynthesized NPs of 10-50 nm size and high monodispersing have multifunctional properties: AgNPs by Chlorella have bactericidal activity against E. coli and Vibrio cholerae, AuNPs by Spirulina catalyze the decolorization of dyes by the Fenton-like processes, and TiO2 by Chlamydomonas entails Sono photo catalysis of micropollutants to mineralize them. The introduction of these strains in hybrid systems- such as algal-bacterial consortia - NP is given 2-3-folds increase in yield and biomass recycled to generate biofertilizer or biofuels at 40-60 less than chemical processes. Challenges like

NP aggregation are addressed using exopolysaccharide coats and scalability using continuous-flow reactors make microalgal NP biosynthesis important to achieve sustainable wastewater treatment to, allow circular economies in urban and agro-industrial locations with high pollutant concentrations (Fetisoa et al., 2024).

Besides the fundamental strains, other microalgae like Scenedesmus sp. and Nano chlorosis oculate are exploited to enhance the NP biosynthesis in wastewater treatment by binding the metal ions to their cell wall polysaccharides. Scenedesmus obliquus forms palladium (PdNPs) and nickel (NiNPs) through extracellular release of flavins and quinones and is a good target of pesticide-contaminated agricultural runoffs; NPs are separated by its self-flocculation effect, which promotes the settlement of suspended solids by 85 percent in membrane bioreactors. Having lipid profile, nano chlorosis can synthesize cerium oxide (CeO2NPs) during nutrient stress that induces oxidative degradation of endocrine disruptors like bisphenol an in the pulp mill effluents, where omega-3 fatty acids adsorb the NP surfaces, leading to long-term catalysis (Garlapati et al., 2025). The benefit of butyraceous braunii is that hydrocarbon-mediated platinum (PtNPs) formation is most convenient in case of using the refinery-generated oily wastewater since the extracellular matrix does not leach NPs during the hydrodynamic shear. These strains co-exist with synergy between Chlorella-Spirulina consortia and NP in the mixed cultures wherein quorum sensing works with NP to form hybrid Ag-ZnO core-shells with synergistic antimicrobial effects against biofilms in cooling tower waters. The optimization methods include the nutrition of the nutrients using urea and bio trace elements to achieve stress reaction, the 150 percent rise in the reductase activity, and the 450 nm LED light to achieve bio reduction through photosystem II-controlled transfer of electrons. Down processing involves the use of ultra-filtration to extract NPs and it is then amassed through the process of lyophilization where the recovery is guaranteed at 95%. Such systems have been shown to provide 98 per cent. nitrogen reduction and 92%. pathogen elimination in test pilots including the industrial belts of Chennai and even more worth sludge biochemical NPs composites that are utilized to treat soil. Genetic engineering when CRISPR-Cas9 targets the genes of the nitrate reduction process increases yields further and spectroscopic analysis (XRD, TEM) determines NP crystallinity. The paradigm shift minimizes the use of chemical precipitants as well as sequestration of CO2 at 1.8 kg/kg biomass which corresponds to net-zero targets of decentralized treatment in developing regions (Goyal et al., 2023).

2.3 Biochemical mechanisms: Enzymes, proteins, and metabolites as reductants agents

The biochemical processes underpinning the biosynthesis of microalgal Nanoparticle (NP) to treat wastewater depends on an ensemble of enzymes, proteins and metabolites as reductants, capping agents and stabilizers to convert the toxic metal ions into functional nanomaterials using the green pathways. Nitrate reductase can become a cornerstone in strains such as Chlorella vulgaris and Spirulina platensis and catalyses the delivery of electrons to metal precursors (Ag +, Au3 +, or Zn 2 +) by electrons donated by NADH or NADPH co factors in six-electron transfers to reduce them to zerovalent NPs, in a pH-dependent manner (best at 7-9). It is a cytoplasmic enzyme, which is upregulated by nitrogen assimilation in nutrient-stressed wastewater, which accelerates intracellular accumulation with NPs being nucleated around vacuoles as in the case of Chlamydomonas reinhardtii where Superoxide Dismutase (SOD) and ascorbate peroxidase counteract Reactive Oxygen Species (ROS), generated during bio reduction and preventing oxidative cellular damage and stabilizing NP surfaces with proteins to form monodispersed (Gupta et al., 2024). In Spirulina, extracellular synthesis is the dominant mode operating via cell wall secreted NADH-dependent nitrate reductase and dissimilatory sulphate reductase which reduce ions at the surface by Photosynthetic Electron Transport Chains (PETC) channelling electrons of photosystem II (PSII) water-splitting complexes through plastoquinone to flavins such as FMNH2 to produce 10-50 nm AgNPs stabilized by Exopolysaccharides (EPS). Metabolites are also important: phycocyanin and phycoerythrin in Spirulina release electrons via tetrapyrroles rings, which serve as mild reductants to change $Fe3+$ to Fe3O4, and chlorophyll derivatives in Chlorella release ligands of porphyrin and chelate metals, increasing quantum yield in TiO2 NP formation using blue light (450nm) (Hamida et al., 2022).

The polysaccharides (alginate and fucoidan) of Scenedesmus obliquus have two roles: reduction through the aldehyde groups and capping through the hydroxyls, to create a biolayer which confers negative zeta potential (-20 to -40 mV) to the colloidal certainty in turbid effluents. Heavy metals are bound by proteins such as metallothionein (MTs) in Nano chlorosomes through cysteine thiolates, consequently trapping Cu2+ or Cd2+, which are reduced by enzymes using glutathione-S-transferase (GST), and Gluathione (GSH) attaching to the peptides capping NPs, giving them ROS-reduction potential (Indhur et al., 2023). Dunaliella salina provides alkaloids and terpenoids that provide the phenolic hydroxyls to be used in the ZnO NP synthesis with lipoxygenases peroxidizing lipids to hydroperoxides to drive non-enzymatic cascades of auto-reduction. Yields are increased 200 percent in mixed consortia because the expression of reductase is synchronized by the quorum-sensing autoinducers such as N-acyl Homoserine Lactones (AHLs), by electrostatic communication mediated between cells (Jabri et al., 2020). These processes are regulated by the constituents of wastewater: phosphates stimulate alkaline phosphatase, increasing the reduction processes that are ATP-dependent, and organics such as humic acid bind the ion with aculate bio reduction. The quantum mechanical observations demonstrate the overlaps of the orbital between the metal d-bands and algal biomolecule p-systems, which reduce the activation energies (Ea 20-40 kJ/mol) in comparison to chemical reactions. After synthesis, Fourier-Transform Infrared (FTIR) signatures identify amide I/II bands of proteins and C-O vibrations of carbohydrates on NP coronas, allowing, e.g., AgNP-mediated ROS burst to lyse bacteria (log reduction of *E. coli* >5 folds) or Gold NP-facilitated peroxyl mono sulphate activation to mineralize dyes (95 percent decolorization). These processes do not only cleanse effluents, which has been shown to eliminate 90 per cent of Cr (VI) through chromate reductase, but also value biomass wherein NPs are shown to increase algal flocs to recover phosphorus. Genetic modifications such as ferredoxin-NADP+ reductase overexpression are to fine-tune electron flux positioning microalgae as viable bio factories to make zero-waste production of NPs on a circular wastewater economy (Karabulut et al., 2024).

2.3 Intracellular vs extracellular synthesis pathways

Intracellular and extracellular synthesis pathways are the two paradigms of microalgal Nanoparticle (NP) synthesis to treat wastewater because each of them uses different cellular compartments to transform metal ions into environmentally effective nanomaterials with functional aspects specific to pollutant remediation. Intracellular synthesis is done in the algal cell, where the live cultures of such strains as Chlorella vulgaris or Chlamydomonas reinhardtii actively incorporate metal precursors (Ag+, Au3+, Zn2+) through the negatively charged cell wall sites containing carboxyl and phosphoryl groups and then takes them across the plasma membrane via ion channels or endocytosis (Karnwal et al., 2023).

Photosynthetic or respiratory metabolism reductants of the photochemical nature, such as NADH/NADPH-dependent nitrate reductase, ferredoxin-NADP+ oxidoreductase and superoxide dismutase, donate electrons to reduce ions to zerovalent NPs (5-20 nm) which nucleate around granules of polyphosphates or chloroplasts, and enzymatic reductants are stabilized through the degradation of intracellular This Pathway has been By comparison, it is extracellular in Spirulina platensis and Scenedesmus obliquus, where the electrostatic interaction of the metal ions with Exopolysaccharides (EPS), lipopolysaccharides and glycoproteins leads to their adsorption onto the cell surfaces and is bio reduced by released enzymes such as sulphate reductase or alkaline phosphatase, with electrostatic photosynthetic energy obtained by photosystem II through photosystem II to flavins (FMNH2) and phen It to form larger (Spirulina has a helical structure that enhances the creation of biofilm in raceways pond systems where extra-cellular AgNPs burst the membranes of E. coli by bursts of ROS, leading to eliminating 99 percent coliform, unlike ZnO NPs of Scenedesmus photocatalyzed tetracycline under sunlight. Continuous flow photobioreactors are scalable to extracellular methods, which allow high metal concentrations (50-200 mg/L) to be used without toxicity, as the NPs form extracellularly, light (450-650 nm) and pH 8-10 help in electron shuttling and quorum-sensing signals to organize the EPS secretion in consortia using hybrid Au-Ag NPs (Khanashyam et al., 2024).

The Hybrid approaches are a combination of the two: at the beginning of the process, extracellular nucleation, and subsequently intra-cellular maturation in Dunaliella salina enhances PtNPs uniformity in the process of demulsifying oily wastewater. The intracellular pathways generate biocompatible enzyme-filled NP-based bioaugmentation to enhance interaction between algal-bacterial in activated sludge to remove 95 per cent phosphorus and the extracellular dominate industrial pilot to omit chemical coagulant specifications (70 per cent reduction). The cause of mechanistic disparities is transport kinetics between intracellular (limited by diffusion 10 6 cm/s) and extracellular (convective mass transfer) in the absence of spectroscopic markers (TEM-EDX) of cytoplasmic gold aggregates and surface-bound silver fractals. Selective strain: Chlorella on intracellular AuNPs in dairy effluents, Spirulina on extracellular Fe3O4 in textile dyes optimizes remediation: intracellular NPs on selective biosorption (Cd2+ >90), extracellular on photocatalytic mineralization (COD <50 mg/L). Staged bioreactors: maximisation is optimised by maximising the conversion of wastewater to NP-laden biomass which can be applied as biofertilizers and 2-4 g/L biofertilizer productivity is realised in ambient conditions (Khandelwal et al., 2023).

3. SYNTHESIS PROTOCOLS FOR MICROALGAL NANOPARTICLES

Synthesis protocols of microalgal nanoparticles are a pillar of green nanotechnology whereby it is proposed to combine algal growth with in-situ synthesis of Nanoparticle (NP) to remediate pollutants and simultaneously produce value-added nanomaterials through scalable bioprocesses, which is eco-effective. These protocols take advantage of the metabolic capabilities of such strains as Chlorella vulgaris, Spirulina platensis, Chlamydomonas reinhardtii, Scenedesmus obliquus, and Nano chlorosis oculate which biosynthesize metallic (Ag, Au, Cu), metal oxide (Fe3O4, TiO2, ZnO) and hybrid NP syntheses through intracellular pathways or extracellular pathways utilizing enzymes like nitrate reductase, superoxide (Kundu et al., 2024). This is carried out through the preparation of algal biomass: live cells in logarithmic growth, cell-free extracts in ultrasonication or autoclave, or dried powders reconstituted in minimal media (BG-11, f/2) in dilution with 10-50% real wastewater to use inherent nutrients to grow at cost-neutral conditions. AgNPs, CuNPs, Fe3O4, TiO2, ZnO Precursors Silver nitrate, chloroauric acid, copper sulfate, iron salts, titanium isopropoxide, zinc acetate, AgNPs, CuNPs, Fe_3O_4, TiO_2, ZnO Precursors AgNPs, CuNPs, Fe_3O_4, TiO_2, ZnO Introduced AgNPs, CuNPs, Fe_3O_4, TiO_2, ZnO in millimolar Incubation in either Erlenmeyer flasks, bubble columns, or closed photobioreactors takes 12-96 hours, and is characterized by calorimetrical changes (yellow-to-brown AgNPs, ruby-red AuNPs) and confirmed by UV-Vis spectroscopy following surface plasmon resonance peaks (Lakkakula et al., 2025).

The enzyme secreted lowers the amount of surface adsorbed ions to form harvest-ready NPs: simple filtration (common in Spirulina) or lysis (sonication, Triton X-100) and centrifugation and dialysis in Chlorella (participates in intracellular action). Hybrid techniques enhance functionality, such as core-shell Ag-ZnO sequentially. It implies that wastewater is contained in it: textile effluents offer dyes as complex enhances yield of boosting, dairy wastes offer dyes calcium to use in doping, swine manure induces reductase expression twofold in times of stress. Recovery 80-95 percent Fe_3O_4 magnetic separation, ultrafiltration with other and lyophilization is used as downstream processing with NP sizes of 10-100 nm, polydispersity indices below 0.2, and zeta potentials ranging between -20 and -40 mV. Characterization based on TEM-EDX, XRD, FTIR, and DSS demonstrates the existence of crystallinity, biomolecular coronas (amide I/II bands) as well as bandgap tuning through photocatalysis. Microwave irradiation, ultrasound cavitation or immobilized algal beads are used to intensify processes in packed-bed reactors 3 times and treat COD, nitrogen and phosphates with 85-98% (Lens et al., 2020). To scale-up to 2-10 g/L NPs and biomass valorization to biofuels or biofertilizers 1000 L PBRs or pilot raceways fed-out to pilot raceways (real time monitoring using pH probes, DO sensors, etc.) are used. Research glows: AgNPs kill pathogens by disturbing membranes using ROS, AuNPs kill nitroarenes, Fe_3O_4 flocculate TSS, TiO_2 /ZnO mineralize recalcitrant organics under sun/UV and CuNPs chelate heavy metals. Through glycerol or alginate stabilizers, toxin efflux overexpression and variability, these barriers such as aggregation are removed by quorum sensing strain consortia. The protocols costs are estimated to save 50-70 percent of their chemical's expenditure, and the carbon footprint is less than 0.5 kg CO_2/kg NP, which is transformative to circular urban-industrial belt economies, and commercial biorefineries will be able to turn waste into wealth (Li et al., 2023).

3.1 Precursors and reaction conditions (pH, temperature, light, agitation)

Microalgal nanoparticle synthesis protocols start with the selection of suitable metal precursor (usually water-soluble salts such as silver nitrate ($AgNO_3$, 1-5 mM) in the case of AgNPs, chloroauric acid ($HAuCl_4$, 0.5-2 mM) in the case of AuNPs, copper sulphate ($CuSO_4$, 2-4 mM) in the case of CuNPs, iron chloride ($FeCl_3$/

FeCl2, 1:2 ratio, Reaction conditions are monitored to stimulate algal reductases and metabolites: pH 7-9 (at alkaline: where Chlorella and Spirulina are active: monitored using HEPES buffer), temperature 25-40degC (mesophilic optima with Chlorella and Spirulina, and 37degC generating all possible optima: 24-72 h), light intensity 100-500 0mol photons m -1 s -1 (blue LED 4Wastewater combination involves the application of actual effluents as diluents (10-30 percent v/v) wherein nutrients (N/P 16:1) cause stress reactions that enhance EPS release to capping. Enzymes are sterilized by pre-treatment of extracts by autoclaving (121degC, 15 min) or UV exposure (254 nm, 30 min), which does not inactivate enzymes. ICP-MS ion depletion-monitored size evolution (10-100 nm) is tracked by monitoring through dynamic light scattering (DLS). The parameters provide 1-5 g/L NP productivity, which can be used in fed-batch modes, avoiding the use of energy (0.5-1 kWh/m3) yet treating COD with 70-90 percent minimizing protocols, and positioning protocols as plug-and-play with decentralized plants (Lu and Q, 2025).

3.2 Synthesis of metallic NPs (AgNPs, AuNPs, CuNPs)

Metallic nanoparticles (AgNPs, AuNPs, CuNPs) synthesis through microalgae takes advantage of a fast process of bio reduction in the disinfection of wastewater with antimicrobials. In the case of AgNPs, Chlorella vulgaris biomass (2 g/L wet) reacts with 1-3 mM AgNO3 at pH 8-10, 30-37degC, 150 rpm shaking in continuous white light (200 mol m -2 -1 s -1), 24-48h; extracellular nitrate reductase and EPS reduce Ag+ to 10-50nm spheres, which are harvested by centrifugation (8000 AuNPs of Scenedesmus obliquus live cells utilize 2 mM CuSO4 at pH 6-8, 28degC, blue light 400 nm, 180 rpm 36-72 h; phycocyanin and chlorophyll chelate Au3 + to generate 15-40 nm red-ruby particles (520 nm peak), purified by dialysis (MWCO 10 kDa), and produce 300 mg/L. CuNPs of Sc The wastewater: 20 percent textile effluent medium yields increase because through dye-complexation, it uses consortia (Chlorella-Spirulina) as the bimetallic Ag-Au (Mahlangu et al., 2024). Characterization is used to confirm FCC lattices (TEM, XRD), amide bands at the FTIR edges. The NPs find their applications in the following ways: AgNPs (MIC 5-10 7g/mL) eliminate 99.9 percent of pathogens, AuNPs eliminate 95 percent of nitroaromatic, CuNPs eliminate 85 percent of TSS. Different variants have been tested including microwave-assisted (300W, 5 min) acceleration or immobilization on alginate beads to be incorporated in continuous reactors to enhance the stability (shelf-life 6 months). Optimal UV-pretreated Chlamydomonas strain-specific modifications with smaller AgNPs maximize PDI = 0.2, and renders batches reproducible (CV = 5-10) (Malik et al., 2022).

3.3 Synthesis of metal oxide NPs (Fe_3O_4, TiO_2, ZnO)

Microalgae metal oxide nanoparticles (Fe3O4, TiO2, ZnO) can offer strong photocatalytic and adsorptive platforms of sophisticated wastewater treatment and use biomolecules of algal organisms to regulate crystallization and phase-purity in sustainable procedures. Preparation The Fe3O4 synthesis is usually made by combining Chlorella sorokiniana or Spirulina platensis dried biomass (5-8 g/L, pre-treatment by boiling 30-45 min at 100degC to release phenolics and reductases) with an equivalent of molar concentration of ferric (FeCl3.6H2O, 6 mM) and ferrous (FeCl2.4H2O, 3 mM) salts in an inert (Mantovani et al., 2022). Titanium(IV) isoprenoid (TTIP, 2-4 mM) or TiCl C (1-3 mM) is incubated at pH 7-9 (ammonia stabilization), 80-100degC in hydrothermal autoclaves (150 rpm), and with the presence of UV-A (24-48 h) to make the TiO2 by a reaction of Spirulina platensis (supernatant 20-30%, centrifuged 6000 rpm Sequential Fe3O4 at TiO2 core shell via electrostatic assembly or ZnO-Fe3O4 composites are hybrid protocols that do not only improve functions: magnetic-photocatalytic demulsification of oily sludge (95 percent oil removal). The process may be intensified by either microwave (acceleration yields 3 times) or ultrasound (40 Khz) cavitation (uniform nucleation) or flow chemistry (capillary reacters; residence 1-2 h). Downstream utilizes tangential ultrafiltration (10 kDa) or magnetic decant and is 92-97% recovers (Mérida et al., 2023).

4. PHYSICOCHEMICAL PROPERTIES OF MICROALGAL NPS

Physicochemical properties of microalgal nanoparticles dictate their efficacy, stability, and environmental compatibility in wastewater treatment, arising from the synergy between algal biomolecules and inorganic metal or metal oxide cores. These nanoparticles typically exhibit a wide size range, from 10 to 100 nano meters, with metallic variants such as silver, gold, and copper nanoparticles often forming spherical or quasi-spherical morphologies, while metal oxides like iron oxide, titanium dioxide, and zinc oxide may crystallize as rods, cubes, or hierarchical structures depending on the synthesis pathway and biomolecular templating. The specific crystalline phase is critical: face-cantered cubic lattices dominate in metallic nanoparticles, imparting high electrical conductivity and localized surface plasmon resonance that enables antimicrobial activity and visible light–driven catalysis, whereas spinel magnetite and wurtzite zinc oxide crystallites confer magnetic responsiveness and semiconductor bandgaps ideal for photocatalytic degradation of organic pollutants (Moghri et al., 2023).

Surface charge, expressed as zeta potential, usually falls in the negative regime due to abundant carboxyl, phosphate, and sulfate groups from algal exopolysaccharides and proteins, promoting colloidal stability in aqueous media and minimizing aggregation in complex wastewater matrices. High surface area to volume ratios, often exceeding several hundred square meters per gram, amplify adsorption capacity for heavy metals, dyes, and emerging contaminants through electrostatic interactions, ligand exchange, and coordination with sulphur and nitrogen donor sites. The biomolecular corona, comprising proteins, polysaccharides, and lipids, not only caps the particle surface but also modulates surface hydrophilicity, biocompatibility, and functional group availability, enabling easy functionalization with additional ligands or polymers for targeted pollutant binding. Optical properties are equally pivotal: silver and gold nanoparticles display strong absorption in the visible region attributable to collective electron oscillations, with peak position and bandwidth governed by size, shape, and local dielectric environment, allowing real-time monitoring of growth and aggregation behaviour during wastewater exposure. Photocatalytic metal oxide nanoparticles, on the other hand, harness bandgap energies that facilitate electron–hole pair generation under solar or ultraviolet irradiation, driving redox reactions that decompose organic micropollutants and inactivate microbes. Thermal stability is generally robust, especially for metal oxides, where the inorganic lattice retains integrity under moderate heating and fluctuating pH, while the biocompatible envelope often imparts resistance to harsh ionic strength and pH gradients encountered in industrial effluents (Mohsenpour et al., 2020).

Table 1. Physicochemical Properties of Microalgal NPs

Physicochemical Property	Description	Method of Characterization	Significance / Application
Particle Size	Average diameter of nanoparticles typically ranging from 1–100 nm	Dynamic Light Scattering (DLS), Transmission Electron Microscopy (TEM)	Influences reactivity, stability, and biological interactions
Shape and Morphology	Structural form such as spherical, triangular, rod-shaped, or irregular	Scanning Electron Microscopy (SEM), TEM	Affects catalytic activity and cellular uptake
Surface Charge (Zeta Potential)	Electrical charge present on the nanoparticle surface	Zeta Potential Analyzer	Determines stability and dispersion in aqueous solutions
Surface Area	Total exposed area of nanoparticles	Brunauer–Emmett–Teller (BET) Analysis	Higher surface area enhances adsorption and catalytic efficiency
Crystalline Structure	Atomic arrangement and crystal phase of nanoparticles	X-Ray Diffraction (XRD)	Determines optical, catalytic, and mechanical properties
Chemical Composition	Elements and functional groups present on nanoparticle surfaces	Energy Dispersive X-ray Spectroscopy (EDX), FTIR	Identifies biomolecules involved in nanoparticle synthesis

continued on following page

Table 1. Continued

Physicochemical Property	Description	Method of Characterization	Significance / Application
Optical Properties	Interaction of nanoparticles with light	UV–Visible Spectroscopy	Useful for monitoring nanoparticle formation and stability
Stability	Ability of nanoparticles to remain dispersed without aggregation	DLS, Zeta Potential Measurement	Important for biomedical and environmental applications
Surface Functionalization	Presence of biomolecules such as proteins, lipids, or polysaccharides from microalgae	Fourier Transform Infrared Spectroscopy (FTIR)	Enhances biocompatibility and targeted applications
Thermal Stability	Resistance of nanoparticles to temperature changes	Thermogravimetric Analysis (TGA)	Important for industrial and catalytic applications

Iron oxide nanoparticles are especially attractive in magnetic properties, and act as superparamagnetic nanoparticles at room temperature, which allows the convenient recovery of the nanoparticles in treated water using external magnetic fields and reduces secondary contamination. These physicochemical characteristics interact to control transport, adsorption, and reactivity in wastewater environments: small size and high surface area facilitate diffusion and contact with contaminants, surface charge facilitates electrostatic attraction between pollutants of opposite charge, and crystalline defects serve as catalytic hotspots in advanced oxidation reactions. Increased ecological benignity due to the algal source of the capping layer further decreases phytotoxicity of aquatic biota and allows ultimate biodegradation or safe sedimentation. Dynamic light scattering, scanning and transmission electron microscopy, X ray diffraction, Fourier transform infrared spectroscopy, UV visible-near infrared spectroscopy and other characterization tools are all useful in providing detailed information on size distribution, morphology, crystallinity, functional group composition, and optical behaviour, which is why rational design of microalgal nanoparticles can be designed on a case-by-case basis to achieve particular wastewater treatment goals, including heavy metal sequestration, organic pollutant degradation, or microbial disinfection (Moraes et al., 2021).

4.1 Stability, dispersity, and functionalization

The long-term stability, dispersibility and surface functionalization of microalgal nanoparticles are the main determinants that define their long-term performance and reuse in wastewater treatment systems. Microalgal NPs usually have high colloidal stability than their chemically synthesized counterparts, which is mainly related to a biomolecular corona of exopolysaccharides, proteins, glycoproteins and lipids that coat the inorganic core to produce steric and/or electrostatic repulsion. This gives the biopolymers a negative surface charge (seen as a zeta potential of -20 to -40 mV in most instances), which allows aggregation of the particles to be avoided in even complicated, high ion strength effluents. The resulting small values of polydispersity index (PDI) usually less than 0.2 under ideal conditions reflect a small size distribution and homogeneous populations which are needed to give predictable adsorption dynamics and reproducible catalytic performance. There are a lot of chances to functionalize biomolecular layer because it has many reactive sites, including amine, hydroxyl, carboxyl and thiol groups, and allows to graft more ligands, polymers or targeting moieties without any harsh chemical manipulations (Morgan et al., 2023). In the case of wastewater, it can be easily conjugated with the selective chelators of heavy metals, affinity peptides of a particular dye or anti-microbial agents that leads to an increase in disinfection. Mechanical stability and NP encapsulation may be further enhanced with alginate, chitosan or silica-based matrices that prevent the leaching or corrosion of NPs when placed in variable pH conditions. These functionalized micro algae NPs have a high capacity of working through many adsorption, photocatalysis or magnetic recovery cycles and some have been shown to show little degradation or activity change despite repeated reuse cycles, which is essential in the context of realistic cost-effective treatment plants. The natural self-functionalization imparted by microalgae in this case is a great benefit compared to traditional nanoparticles which usually must undergo post synthetic modifications to attain similar stability and dispersity and thereby lower the process complexity and improve the overall sustainability (Mukherjee et al., 2021).

4.2 Comparative advantages over chemically synthesized NPs (eco-toxicity, yield)

Microalgal nanoparticles have several persuasive benefits when compared to chemically synthesized NPs, especially in eco toxicity, process sustainability as well as biomass integrated yield that are vital in wastewater treatment processes. In comparison to standard NPs synthesized through harsh reductants (e.g. sodium borohydride), potent stabilizing agents or high energy methods, microalgal NPs are synthesized in mild and aqueous conditions using algal metabolites, enzymes and polysaccharides as natural reductants and capping agent. The biogenic pathway reduces the utilization of toxic chemicals and organic solvents that encourages the decrease of the ecological footprint and prevents the formation of dangerous by products. The resulting NPs are also less acutely toxic to aquatic life and aquatic plants, as the surfaces are covered with biocompatible biomolecules rather than synthetic surfactants or stabilizers that do not interact with cell membranes or create oxidative stress. Conversely, most chemically synthesized NPs are revealed to be more ecotoxic in microalgae, bacteria, and higher trophic organisms, which implies their release to receiving waters is of concern (Narwal et al., 2025). Microalgal NPs can be more productive: since wastewater treatment is a low-cost growth medium that can grow accumulations of algal biomass, at the same time acting as a source of metal ions that can be used in the process of NP synthesis, making pollutant remediation a value addition process. The yields may be equal or even higher than that of purely chemical pathways, especially when the strains and conditions are set to maximize the reductase activity and the EPS production. Also, the capping layer can be of algal origin, possibly increasing biocompatibility and biodegradability, allowing a safer end of life, such as sedimentation or controlled biodegradation, but chemically produced NPs may last longer in the environment. Continuous or semi continuous processes can also be undertaken through the synthesis of NP with cultivation of algae, where the biomass is periodically recovered and reintroduced into several cycles of NP synthesis, which enhances the utilization of resources and lowers the operational expenses. Altogether, the green, closed loop microalgal NP production, its positive ecotoxicological profile and high process efficiency put it in a more sustainable position in comparison to conventional production of NP by chemicals that should be used in large scale wastewater treatment (Ndubuisi et al., 2025).

5. MECHANISMS OF WASTEWATER TREATMENT USING MICROALGAL NPS

The wastewater treatment mechanisms through microalgal nanoparticles involve a complex combination of adsorption, catalysis, oxidation and biotic interactions which together eliminate or convert pollutants at chemical, physical and biological levels. The Microalgal NPs are essentially utilized as nanoscale adsorbents with a high surface area and functionalized biomolecular corona exploiting their electrostatic interactions with cationic and anionic contaminants, including heavy metals, dyes, and organic micropollutants, via a ligand exchange, coordination with carboxyl, phosphate, and amine groups, or hydrogen bonding with hydrocarbon rich regions. Simultaneously, numerous microalgal NPs are photocatalysts or Fenton type catalysts under the addition of light or oxidants to generate reactive oxygen species, including hydroxyl radical and superoxide ions, which are used to mineralize recalcitrant organics, like pharmaceuticals, pesticides, and endocrine disruptors, into smaller, less toxic molecules or fully mineralized products. Silver, copper and some metal oxide NPs also have antimicrobial activity capability to disrupt microbial membranes, disrupt cellular respiration, cause oxidative stress, and hence inactivate pathogenic bacteria and other microorganisms found in wastewater. Microalgal NPs may also be used to increase the flocculation and coagulation processes by binding the particles together or by serving as the nucleus of aggregation to cause the sedimentation and also eliminate the suspended solids, turbidity, and other related contaminants. As a component of algal bacterial consortia, or in immobilized matrix, microalgal NPs concomitantly promote the growth and activity of useful microorganisms, which collectively synergistically eliminates nutrients, including nitrogen and phosphorus, by biological uptake and denitrification. Biocompatibility of algal derived capping layer normally lowers the toxicity of the nearby biota, and therefore, allows long-term operation in dynamic wastewater systems with minimal ecological hazards. All in all, the synergistic effect between adsorption, photocatalysis, biocidal activity and microbial support is what makes microalgal nanoparticles a viable green and sustainable remediation technology as opposed to traditional remediation methods (Nguy n et al., 2020).

5.1 Adsorption kinetics and isotherms (Langmuir, Freundlich models)

Microalgal nanoparticles have short adsorption kinetics with regards to wastewater pollutants, and the rates tend to adhere to pseudo-second-order models due to the predominance of chemisorption over physisorption, and the rate constant of the reactions achieve equilibrium in 30-120 minutes under ambient conditions. External mass transfer and intraparticle diffusion are the regulating factors of initial adsorption as indicated by multi-linearity of Weber-Morris plots, with surface diffusion being dominant at low concentration and pore diffusion at saturation. The Langmuir and Freundlich models are the most common equilibrium isotherms, but in terms of the monolayer assumption of Langmuir, nanoparticles based on biomolecular corona generate higher maximum capacities when compared to chemically functionalized nanoparticles because the biomolecular corona contains a high concentration of binding sites, such as carboxyl and phosphate groups (Noreen et al., 2024). The Langmuir parameters indicate preferential adsorption and the heterogeneity factor of Freundlich indicates the diversity of energy sites by algal polysaccharides. In the case of heavy metals such as Pb2+ and Cd2+, microalgal AgNPs or Fe3O4 NPs have capacities of 100-300 mg/g, that is, their selectivity is higher than that of activated carbon and this is due to the electrostatic attraction and complexation. Dye adsorption e.g. that of methylene blue follows Freundlich at low loads (multilayer) and then Langmuir at high equilibrium concentrations. The preceding is confirmed by thermodynamic favourability of spontaneous, exothermic processes, which were enhanced by $pH > 6$ where deprotonated sites prevailed. The models allow predictive design to be performed on fixed-bed or batch reactors where contact time and dosage are optimized to provide 90-99% removal efficiencies with real effluents (Pachouri et al., 2024).

Figure 1. Mechanisms of Wastewater Treatment Using Microalgal NPs

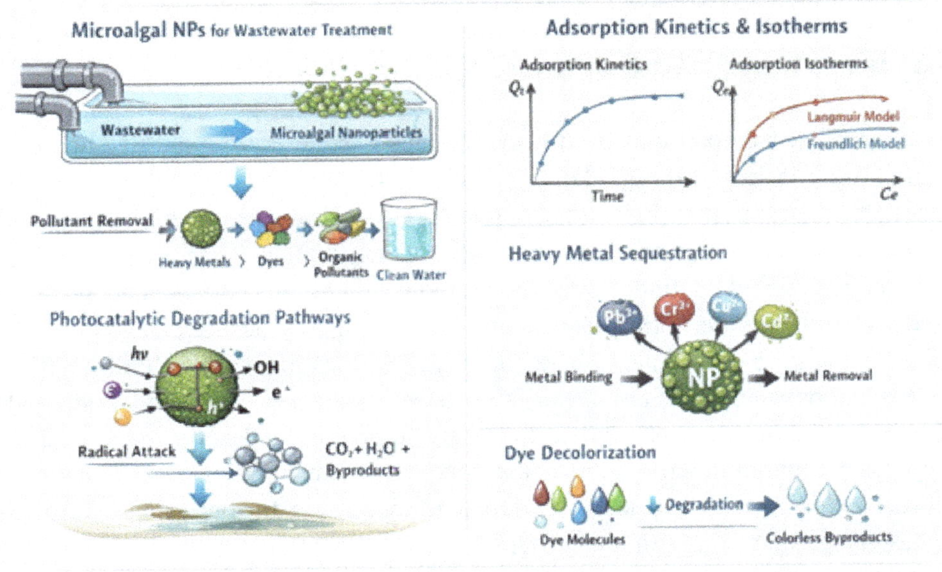

5.2 Photocatalytic degradation pathways

Photocatalytic degradation Photocatalytic degradation by metal oxide nanoparticles (TiO2, ZnO) of microalgal cells begins with the absorption of photons that surpass bandgap energy resulting in the formation of electron-hole pairs that can propagate redox cascades in wastewater. Valence band holes turn H2O or OH - into hydroxyl radicals (- OH), the main mineralizers of organics such as dyes, pharmaceuticals and pesticides, and the electrons of the conduction bands reduce O2 to superoxide radicals (O2 -). The metabolic routes include successive hydroxylation, ring cleavage, and decarboxylation to give CO 2, H2O and inorganic ions; an example is the tetracycline breakdown which follows a metabolic pattern of demethylation, dihydroxylation and breaking up of the central ring. Algal capping dopes NPs with C/N, narrows bandgaps to enable visible-light activity, and prevents recombination through electron traps. The kinetics of degradation are Langmuir-Hinshelwood models, pseudo-first-order rates, (k = 0.01-0.1 min -1), 80-95% mineralization in 60-180 min solar/UV degradation. Radical scavenging proves the hegemony of - OH/

$O2-^{-}$ on h+, and holes proximate the oxidation of electron-rich pollutants directly. Persulfates react with synergies to form SO 4 - on recalcitrant compounds. The solidity of 5-10 cycles is due to biocompatible coating that inhibits the corrosion of the photos. These routes allow effective treatment of complicated effluents and cut COD/TOC by 70-90% and valorise byproducts (Pande et al., 2023).

5.3 Heavy metal sequestration and dye decolorization

Microalgal nanoparticles used in heavy metal sequestration and dye decolorization are considered the golden applications in wastewater remediation where synergetic adsorption, redox, and catalytic processes lead to high efficiencies in a variety of effluents. Extraction of heavy metal mainly involves surface complexation and electrostatic interactions where the biomolecular corona, which is abundant in carboxyl, thiol, amine, and phosphate groups of algal polysaccharide and proteins, forms stable inner- and outer-sphere complexes with cations such as Pb2+, Cd2+, Cu2 + and oxyanions such as Cr(VI) and As(V), as an example, Fe3O 4 or AgNP surfaces coordinate CrO 42 - by electrostatic interaction at p The magnetic variants allow easy separation with the low-field magnets, allowing 95-99% recovery with reuse 10+ times with low leaching. Divalent metals are further trapped through ion exchange with algal sulphate groups and the colloidal metal oxides are trapped through size-exclusion pores. In practice, pH-dependent speciation maximizes selectivity, and EDTA-functionalized NPs chelate recalcitrant species better (Premarathna et al., 2024).

The dye decolorization occurs by dual adsorption-degradation mechanisms with cationic dyes (methylene blue, crystal violet) bound through p-p and electrostatic interactions with negatively charged coronas and anionic azo dyes (Congo red, reactive black) interacting with the hydrogen bonding and van der Waals forces resulting in the adsorption of 90-95% of the dye in 10-30 minutes. Fenton-like or photocatalytic activation splits chromophores azo (-N=N)- and aromatic rings: N-de-ethylation, desulfonation, and mineralization to CO2, NH 4 +, SO 4 2 - occur under UV/solar light by ZnO/TiO2 NPs, which produce hydroxyl radicals that activate the reaction with pseudo-first-order Kinetics (k=0.02-0.15 min -1). AuNPs facilitate dye reduction through electron relay where 200 mg/L rhodamine B was decolorized in 15 minutes (Song et al., 2025). The Ag-ZnO hybrid systems are improved via the sequential electrostatic preconcentration and ROS attack. Full effluent in textiles containing salts and surfactants undergo 85-98 per cent colour removal, TOC reduction of 70 and negligible toxicity of effluents post-treatment. Photo corrosion is alleviated by algal capping and performance is maintained by regenerating performance by mild elution (0.1 M NaOH) or magnetic rinsing. These reactions are incorporated in the membrane bioreactors or algal ponds, and these reactions integrate dye/metal separation and nutrient assimilation to the overall treatment process, lowering the volumes of sludge by 60% than in conventional coagulation, and locating microalgal NPs as multitasked steps to eliminating liquid discharge in polluted industrial belts (Su et al., 2024).

6. APPLICATIONS IN WASTEWATER TREATMENT

The use of microalgal nanoparticles in wastewater management to exploit their physicochemical characteristics is a continuum of pollutants, such as heavy metals and dyes to pathogens and nutrients, as part of composite bioremediation methods. These nanoparticles are effective in the chores of heavy metals removal when magnetic Fe3O 4 variants of Chlorella or Spirulina are used to sequester the ions such as Pb 2+, Cd 2+, and Cr (VI) on the surface by complexation of the particles and redox conversion with high efficiency of 95 percent in electroplating and mining effluents through a facile magnetic recovery that can be reused. Photocatalytic ZnO and TiO2 NPs synthesized by Scenedesmus have the advantage to de-color the azo bonds and mineralize chromophores in textile wastewaters with dyes to de-color the wastes, thereby reducing their color and COD as well as de-coloration under solar or UV light. The pharmaceutical and personal care products contaminants which are recalcitrant to biological treatment are treated by the Ag-Au hybrid NPs, hydroxyl radical-mediated hydroxylation and ring opening produce biodegradable intermediates. Bacterial biofilms, pathogens such as *E. coli* and *Vibrio* in municipal sewage are disrupted by antimicrobial action of AgNPs and CuNPs, disinfection ensured, and loop closure of the loop of fertilizer production occurs through nutrient recovery-through adsorption of phosphorus onto NP-algal flocs and nitrogen through enhanced denitrification in consortia (Sumathi et al., 2024). New applications are oily wastewater demulsification with hydrophobic AuNPs and micropollutants tracing in urban stormwater through plasmonic sensing. Improved performance through integration into hybrid systems, e.g. NP-enhanced algal ponds or membrane bioreactors, increases throughput: constant flow systems treat high-strength industrial effluents at throughputs of 10-50 L/m2/day, reducing operational costs by 50-70% compared to chemical coagulants. Scalability: Pilot tests in coastal areas such as Chennai highlight the importance of such biofuels, where bio-mass co-vaporization increases circularity. These applications make microalgal NPs the harness of versatility, eco-benign agents, and convert wastewater into resource as progressive goals of sustainable development become realized using the paradigms of zero-liquid discharge (Tabassum et al., 2023).

6.1 Removal of heavy metals (Cr, Pb, Cd, As)

Microalgal nanoparticles are superior in removing heavy metals such as Cr (VI), Pb (II), Cd (II), and As(V) in industrial effluent by adsorption and complexation (as well as by redox) processes. The superparamagnetic cores modified with algal polysaccharides to create inner-sphere complexes with Pb2 3+ and Cd2 3+ at pH 6-8 enables Fe3O 4 NPs of Chlorella sorokiniana to electrostatically capture and remove Pb2 3+ and Cd2 3+ at 95-99% in 60 minutes and with capacities up to 200 mg/g. The interaction of AgNPs or ZnO hybrids with Cr(VI) occurs through an initial stage of adsorption on anion and then reduction of Cr(VI) through enzymatic-like reduction by phenolic electron donors on the biomolecular corona to avoid re-release of Cr(VI) under varying redox conditions. As(V) specifically interacts with amine-coated surfaces of Spirulina-derived AuNPs, and phosphate competition is reduced by pH-dependent uptake at elevated pH (p. 7), resulting in 90 percent uptake in mining wastewaters. Magnetic separability allows 10+ reuse factors with less than 5% efficiency loss, with a higher level of selectivity than ion exchangers under co-ion situations such as SO 2 -. Fixed-bed columns, maintained at throughputs of 5-20 BV/h by such integration, reduce sludge (by 70 per cent) compared to precipitation, and algal capping decreases leaching and ecotoxicity thus NPs are now the ideal choice to treat tannery and battery effluents in a decentralized environment (Tonelli et al., 2023).

6.2 Textile dyes and organic pollutants (MB, RhB, phenols)

Rhodamine B (RhB) and Methylene Blue (MB) are textile dyes that are susceptible to microalgal NP through adsorption-photocatalysis synergy in high-color effluents. MB cations are adsorbed by TiO2/ZnO NPs of Nano chlorosis due to p-p interactions and deprotonated hydroxyls and hydroxyls radical of UV/solar attack decolorize 200 mg/L in 30-90 minutes with 85-95% TOC removed by p-p interaction. The process by which N-de-ethylation and lactone ring hydrolysis take place on Ag-ZnO hybrids catalysed by plasmonic hot electrons is described as RhB. Recalcitrant phenols in the wastes produced by paper mills degrade by hydroxylation and the formation of quinones on Fe3O 4 surfaces catalysed by persulfate, and algal lipids doped to respond to visible light. Kinetics follow pseudo-first-order models, which are enhanced 3-fold in saline matrices through EPS stabilization. Actual textile effluents have 92 per cent colour and 75 per cent COD reduction, and NP reusability of more than 8 cycles through mild elution. These reactions reduce AOP energy requirements by half and allow the decolorization of the solar-driven integration of a raceway to be sustainable (Uddandarao et al., 2023).

6.3 Pharmaceutical and pesticide residues

Some biodegradable substances in wastewater, such as pharmaceuticals (tetracycline, ibuprofen) and pesticide (atrazine, chlorpyrifos) are further degraded by microalgal NPs via ROS-mediated pathways. With solar irradiation, ZnO/AgNPs of Scenedesmus produce - OH and O2- ⁻ and hydroxylate tetracycline amide bonds, with its central ring cut into two byproducts, at 95% removal in 120 minutes with minimal byproducts. Ibuprofen decarboxylates and aromatizes on TiO2 surfaces and mineralizes (90) through direct oxidation by h+. Pd-doped AuNPs are used in atrazine dichlorination through hydrogenolysis, and cunp coronas are used on P-S bonds using chlorpyrifos hydrolysate. Fe3O 4Fenton systems increase production of SO 4 - for recalcitrant spikes to lessen EC50 to non-toxic concentrations. Abatement of hospital and agricultural runoffs is found to be 8598 percent with LC-MS confirming the disappearance of parent compounds. Biocompatible capping results in stability during 7 cycles and low ecotoxicity to Daphnia and allows membrane hybrid configurations to prune to the level of traces and meet regulation requirements (Usman et al., 2025).

7. CHALLENGES AND LIMITATIONS

Issues and constraints in the production and use of microalgal nanoparticles in wastewater treatment are scalability barriers, in which the use of flasks in laboratories is replaced by bioreactors in industry, which have uneven biomass production due to light inhibition and pollution in the open. Biomass harvesting requires much power through centrifugation or flocculation, to the extent of driving up treatment costs. The NP leaching of the deteriorating algal coronas poses a risk of secondary pollution by ion leaching and toxic sludge, and recovery by chemical washing is washed away by efficacy after a limited number of cycles. Shear aggregation decreases the active surface area, whereas the inconsistency of NP size and stability is due to the variation in algal strains. Regulatory challenges are high, with no nanomaterial-specific effluent guidelines and intensive ecotoxicological information requirements in the wake of apprehensions of aquatic bioaccumulation and microbial interference (Uzair et al., 2020).

7.1 Scalability and biomass harvesting issues

The scalability of microalgal nanoparticle production to industrial levels of photobioreactor or open raceways has major challenges in changing lab-scale flasks to industrial-scale and unregulated biomass productivity and inefficient Downstream harvesting. Microalgae such as Chlorella and Spirulina demand a very fine regulation of light penetration, CO_2 and nutrient dosage in large scale whereby light shading and oxygen accumulation restricts cell densities to less than 5 g/L reducing NP yields to less than 1 g/L/day. Open systems are contaminated and subject to weather variations whereas closed PBRs increase capital costs by 5-10-folds. The harvesting of biomass presents problems: centrifugation requires energy (1-2 kWh/m^3) comparable to treatment savings and auto flocculation through NP-algal interactions is not reliably applicable to turbid effluents. Foulage in the membrane of ultrafiltration reduces the life span, requiring chemical cleaning which is a weakness of green credentials. Other strategies such as genetic strain engineering of flocculation proteins or hybrid raceway-PBR cascades also have potential although they would require pilot validation to be economically feasible (Venkatesan and Sathishkumar, 2024).

8. ADVANCES AND FUTURE DIRECTIONS

Recent progress in the production and use of microalgal Nanoparticles (NPs) as wastewater treatment exploits green biosynthesis, in which microalgae such as Spirulina platensis, Chlorella and Euglena gracilis are used as green nano factories, which aids in reducing metal ions (e.g., Ag, Au, TiO_2) at low temperature with minimal toxicity and minimal energy consumption in comparison to chemical methods (Yadav et al., 2025). This Phyto nanotechnology provides extracellular and intracellular formation of NPs through algal enzymes and metabolites, where stable metallic, metal oxide and carbon-based NPs are produced that can increase microalgae natural biosorption, bioaccumulation and biodegradation of pollutants (e.g. heavy metals up to 65.8 per cent removal across cycles), dyes (e.g. Methylene Blue up to 65.8 per cent removal across cycles), antibiotics, nitrogen Optimized NP doses enhance photosynthetic efficiency, nutrient uptake, pathogen mitigation, and biomass vaporization to biofuels, pigments, biopolymers and biofertilizers, and magnetic NPs ensure >99% harvesting efficiency and recyclability, reducing expenditures on circular bioeconomy systems. Future trends are towards establishing safe dose limits, standardization of ecotoxicological tests, pilot to industrial translation, life cycle analysis and comparison with conventional methods to eliminate the toxicity uncertainties so that ecotoxicology can be easily applied to environmentally friendly sanitation. Combining genetic engineering in enzyme overexpression, and hybrid nanofiber immobilization will provide multifunctional, scaled ways of solving water scarcity problems around the world (Yuan et al., 2024).

8.1 Genetic engineering of microalgae for enhanced NP yield

Genetic engineering increases the performance of microalgal Nanoparticle (NP) by altering the major synthetic pathways. The use of CRISPR-Cas9 and RNA interference targeting nitrate reductase and cytochrome P450 in Chlorella vulgaris and Spirulina platensis, respectively, and enhances metal ion reduction to AgNPs and AuNPs by 3-5 fold. Photoheating and metallothionein overexpression enhance intracellular sequestration, and the resulting monodisperse NPs (5-20 nm) are more stable, whereas promoter optimization scales photobioreactor production. Artificial biology is used to express bacterial reductases through Agrobacterium-mediated transformation with 90 per cent conversion efficiency of precursors such as $AgNO_3$ at controlled salinity and pH. There are also off-target effects and genomic instability issues, which have been solved through base editing to make specific mutations. This increases the NP production of mg/L to g/L, which contributes to sustainable wastewater treatment (Zanoni et al., 2025).

8.2 Nanocomposites and multifunctional NPs

Microalgal nanocomposites combine NPs with biopolymers such as alginate-chitosan scaffolds, in which magnetic Fe_3O_4-core NPs capture heavy metals, whereas TiO_2 shells undergo visible light photocatalysis of dyes. Multifunctional designs inbuilt quantum dots to sense (real-time BOD monitoring) and antimicrobial AgNPs, which eliminates pathogens by 99.9% in antibiotic-charged effluents. The pH-responsive release of nutrients through core-shell structures improves microalgal regrowth after treatment. These function better than monofunctional NPs in the reduction of COD and in recyclability. In the future, the synergies with MOFs can be used to selectively sieve ions but this is only possible with toxicity profiling (Zhang et al., 2025).

9. CONCLUSIONS

The development and use of microalgal Nanoparticles (NPs) is a paradigm shift in the treatment of wastewater as it combines the green nanotechnology with the inherent bioremediation capabilities of the microalgae to tackle the current challenges of water on earth. Strides in the field of biosynthesis, the use of species such as Chlorella vulgaris and Spirulina platensis to produce Ag, Au, TiO_2 and magnetic NPs in an eco-friendly manner, has shown improved pollutant clearing, such as heavy metals dyes, nutrients antibiotics and COD in addition to biomass vaporization to biofuels and biofertilizers. 3-5-fold genetic engineering NP yields are boosted upon selective overexpression of enzymes with CRISPR-Cas9 and synthetic biology, which result in scalable photobioreactor systems. The hybrid scaffolds are also refined to nanocomposites and multifunctional NP with sequential adsorption photocatalysis, real-time sensing and recycling, being more cost-effective and sustainable than the conventional ones. Nevertheless, the issues have not been addressed: lack of knowledge about NP toxicity, failure to extrapolate between the laboratory and pilot/industrial (after 2025 data is scarce) and failure to normalize ecotoxicological tests and life-cycle evaluation. Future outlooks Hybrid integrations - gene engineering of MOFs/nanofibers - AI-controlled dosing - regulated deployment - cross-disciplinary input to make comparisons with activated sludge or membrane systems. Using the cyclic bio economies, microalgal NPs have the potential to provide resilient, zero-waste solutions to the augmented urbanization and pollution and restructure environmental sanitation by 2030, assuming that the translation gaps are overcome. This nexus of biological, nanotechnology, and engineering will deliver a blueprint of a sustainable water security that is clean.

REFERENCES

Abbasi, I. A., Đ c, A. N., Nam, S., & Yeom, I. T. (2025). Efficient removal of polystyrene nano plastics from aqueous solution by chitosan-coated black tea magnetic nanocomposite: A green synthesis, influencing factors and mechanisms. *Journal of Water Process Engineering*, *77*, 108650. DOI: 10.1016/j.jwpe.2025.108650

Abdelfattah, A., Ali, S. S., Ramadan, H., El-Aswar, E. I., Eltawab, R., Ho, S., Elsamahy, T., Li, S., El-Sheekh, M. M., Schagerl, M., Kornaros, M., & Sun, J. (2022). Microalgae-based wastewater treatment: Mechanisms, challenges, recent advances, and future prospects. *Environmental Science and Ecotechnology*, *13*, 100205. DOI: 10.1016/j.ese.2022.100205 PMID: 36247722

Acién, F. G., Gómez-Serrano, C., Morillas, A., Arbib, Z., Sanchez, A., Nordio, R., Rodríguez, E., Guzmán, J. L., & Fernández-Sevilla, J. M. (2023). Wastewater treatment by microalgae-based processes. In *IWA Publishing eBooks* (p. 77). UWA Publishing., DOI: 10.2166/9781789063547_0077

Ahmed, T., Luo, J., Noman, M., Ijaz, M., Wang, X., Masood, H. A., Manzoor, N., Wang, Y., & Li, B. (2023). Microbe-mediated nanoparticle intervention for the management of plant diseases. *Crop Health*, *1*(1), 3. Advance online publication. DOI: 10.1007/s44297-023-00006-9 PMID: 41652775

Anvar, S. A. A., Nowruzi, B., & Afshari, G. (2023). A Review of the Application of Nanoparticles Biosynthesized by Microalgae and Cyanobacteria in Medical and Veterinary Sciences [Review of A Review of the Application of Nanoparticles Biosynthesized by Microalgae and Cyanobacteria in Medical and Veterinary Sciences]. *Iranian Journal of Veterinary Medicine*, *17*(1), 1–18. DOI: 10.32598/IJVM.17.1.1005309

Aragón, D., García-Merino, B., Barquín, C., Bringas, E., Rivero, M. J., & Ortíz, I. (2024). Advanced green capture of microplastics from different water matrices by surface-modified magnetic nanoparticles. *Separation and Purification Technology*, *354*, 128813. DOI: 10.1016/j.seppur.2024.128813

Armah, E. K., Chetty, M., Adedeji, J., Kukwa, D. T., Mutsvene, B., Shabangu, K. P., & Bakare, B. F. (2020). Emerging Trends in Wastewater Treatment Technologies: The Current Perspective. In *IntechOpen eBooks*. IntechOpen. DOI: 10.5772/intechopen.93898

Arteaga-Castrejón, A. A., Agarwal, V., & Khandual, S. (2024). Microalgae as a potential natural source for the green synthesis of nanoparticles. *Chemical Communications (Cambridge)*, *60*(29), 3874–3890. DOI: 10.1039/D3CC05767D PMID: 38529840

Ayyadurai, P., & Ragavendran, C. (2025). Green nanotechnologies for removing cosmetic pollutants: Approach towards preserving and protecting aquatic health. *Nanotechnology for Environmental Engineering*, *10*(2), 21. Advance online publication. DOI: 10.1007/s41204-025-00416-w

Braga, M. Q., Lorentz, J. F., Pereira, A. S. A. de P., Magalhães, I. B., Silva, T. A., Avelar, N. V., Henriques, B. S., de Castro, A. P. M., Assemany, P. P., & Calijuri, M. L. (2026). Application of nanoparticles to microalgae-based wastewater treatment: Mechanisms, bioremediation potential and technological frontiers. *Journal of Environmental Management*, *401*, 128827. DOI: 10.1016/j.jenvman.2026.128827 PMID: 41650870

Chaudhary, R., Nawaz, K., Khan, A. K., Hano, C., Abbasi, B. H., & Anjum, S. (2020). An Overview of the Algae-Mediated Biosynthesis of Nanoparticles and Their Biomedical Applications. *Biomolecules*, *10*(11), 1498. DOI: 10.3390/biom10111498 PMID: 33143289

El-Sheekh, M. M., El-Kassas, H. Y., & Ali, S. S. (2025). Microalgae-based bioremediation of refractory pollutants: An approach towards environmental sustainability. *Microbial Cell Factories*, *24*(1), 19. DOI: 10.1186/s12934-024-02638-0 PMID: 39810167

Ennaceri, H., Mkpuma, V. O., Amo-Duodu, G., Menkiti, M. C., Ishika, T., Vadiveloo, A., & Moheimani, N. R. (2025). Microalgae as sustainable bio-factories for nanoparticle biosynthesis: Progress and challenges. *Bioresource Technology Reports*, *30*, 102089. DOI: 10.1016/j.biteb.2025.102089

Fetisoa, M. R., Acheampong, A., Bondzie-Quaye, P., Bashir, M. A., & Huang, Q. (2024). Synergistic strategies for optimizing microalgae-based wastewater treatment: A comprehensive review. *Journal of Water Process Engineering*, *68*, 106409. DOI: 10.1016/j.jwpe.2024.106409

. Garlapati, V. K., Sharma, S., Sharma, D. K., Kumar, S., Jacob, S., Kuila, A., Gupta, A. K., & Chaudhary, A. (2025). Sustainable production of microalgal nanoparticles through green synthesis towards cancer treatment [Review of *Sustainable production of microalgal nanoparticles through green synthesis towards cancer treatment*]. *Frontiers in Bioengineering and Biotechnology, 13*. Frontiers Media. DOI: 10.3389/fbioe.2025.1621876

. Goyal, V. K., Rani, D., Arora, R., Mehrotra, S., Deng, C., & Wang, Y. (2023). Unlocking the Potential of Nano-Enabled Precision Agriculture for Efficient and Sustainable Farming [Review of *Unlocking the Potential of Nano-Enabled Precision Agriculture for Efficient and Sustainable Farming*]. *Plants, 12*(21), 3744. Multidisciplinary Digital Publishing Institute. DOI: 10.3390/plants12213744

Gupta, R., Mishra, N., Singh, G., Mishra, S., & Lodhiyal, N. (2024). Microalgae cultivation and value-based products from wastewater: Insights and applications. *Blue Biotechnology, 1*(1), 20. Advance online publication. DOI: 10.1186/s44315-024-00019-1

Hamida, R. S., Ali, M. A., Alkhateeb, M. A., Alfassam, H. E., Momenah, M. A., & Bin-Meferij, M. M. (2022). Algal-Derived Synthesis of Silver Nanoparticles Using the Unicellular ulvophyte sp. MBIC10591: Optimisation, Characterisation, and Biological Activities. *Molecules (Basel, Switzerland), 28*(1), 279. DOI: 10.3390/molecules28010279 PMID: 36615473

Indhur, R., Amoah, I. D., Bux, F., & Kumari, S. (2023). Nanomaterials for Microplastic Removal from Wastewater: Current State of the Art Nanomaterials and Future Prospects. *ACS ES&T Water, 3*(12), 3741–3754. DOI: 10.1021/acsestwater.3c00203

Jabri, H. A., Das, P., Khan, S., & Thaher, M. (2020). Treatment of Wastewaters by Microalgae and the Potential Applications of the Produced Biomass—A Review. *Water (Basel), 13*(1), 27. DOI: 10.3390/w13010027

Karabulut, G., Purkiewicz, A., & Gökşen, G. (2024). Recent developments and challenges in algal protein and peptide extraction strategies, functional and technological properties, bioaccessibility, and commercial applications. *Comprehensive Reviews in Food Science and Food Safety, 23*(3), e13372. Advance online publication. DOI: 10.1111/1541-4337.13372 PMID: 38795380

. Karnwal, A., Dohroo, A., & Malik, T. (2023). Unveiling the Potential of Bioinoculants and Nanoparticles in Sustainable Agriculture for Enhanced Plant Growth and Food Security [Review of *Unveiling the Potential of Bioinoculants and Nanoparticles in Sustainable Agriculture for Enhanced Plant Growth and Food Security*]. *BioMed Research International, 2023*(1). Hindawi Publishing Corporation. DOI: 10.1155/2023/6911851

. Khanashyam, A. C., Shanker, M. A., Babu, K. S., Thorakkattu, P., Krishnan, R., Abdullah, S., Bekhit, A. E. A., McClements, D. J., Santivarangkna, C., & Nirmal, N. P. (2024). Emerging alternative food protein sources: production process, quality parameters, and safety point of view [Review of *Emerging alternative food protein sources: production process, quality parameters, and safety point of view*]. *Critical Reviews in Biotechnology, 45*(1), 1. Taylor & Francis. DOI: 10.1080/07388551.2024.2341902

Khandelwal, A., & Lens, P. N. L. (2023). Algal systems for resource recovery from waste and wastewater. In *IWA Publishing eBooks* (p. 3). UWA Publishing., DOI: 10.2166/9781789063547_0003

Kundu, P., Dutta, N., & Bhattacharya, S. (2024). Application of microalgae in wastewater treatment with special reference to emerging contaminants: A step towards sustainability. *Frontiers in Analytical Science, 4*, 1513153. Advance online publication. DOI: 10.3389/frans.2024.1513153

Lakkakula, J., Kalra, P., Mungaji, H., Suprasanna, P., & Kadam, U. S. (2025). Microalgae-Derived Metal Nanostructures: Biosynthesis, Characterization, and Applications. *Chemistry & Biodiversity, 23*(1), e03229. Advance online publication. DOI: 10.1002/cbdv.202503229 PMID: 41264401

Lens, P. N. L., Khandelwal, A., Khandelwal, A., Lens, P. N. L., Pessi, B. A., Rada-Ariza, A. M., López-Vázquez, C. M., Steen, P., Lens, P. N. L., Lawrence, J. F., Oliva, A., Murphy, J. D., Lens, P. N. L., Acién, F. G., Gómez-Serrano, C., Morillas, A., Arbib, Z., Sanchez, A., Nordio, R., & Hooshmand, M. (2023). Algal Systems for Resource Recovery from Waste and Wastewater. In *IWA Publishing eBooks*. UWA Publishing., DOI: 10.2166/9781789063547

Li, L., Gao, K., Yang, M., Zheng, Q., Zhang, M., & Deng, X. (2023). Challenges and potential solutions of microalgae-based systems for wastewater treatment and resource recovery. *Frontiers in Bioengineering and Biotechnology, 11*, 1210228. DOI: 10.3389/fbioe.2023.1210228 PMID: 37342510

Lu, Q. (2025). A State-of-the-Art Review of Microalgae-Based Food Processing Wastewater Treatment: Progress, Problems, and Prospects. *Water (Basel), 17*(4), 536. DOI: 10.3390/w17040536

Mahlangu, D., Mphahlele, K., Paola, F. D., & Mthombeni, N. H. (2024). Microalgae-Mediated Biosorption for Effective Heavy Metals Removal from Wastewater: A Review. *Water (Basel), 16*(5), 718. DOI: 10.3390/w16050718

. Malik, S., Dhasmana, A., Preetam, S., Mishra, Y. K., Chaudhary, V., Bera, S. P., Ranjan, A., Bora, J., Kaushik, A., Minkina, T., Jatav, H. S., Singh, R. K., & Rajput, V. D. (2022). Exploring Microbial-Based Green Nanobiotechnology for Wastewater Remediation: A Sustainable Strategy [Review of *Exploring Microbial-Based Green Nanobiotechnology for Wastewater Remediation: A Sustainable Strategy*]. *Nanomaterials, 12*(23), 4187. Multidisciplinary Digital Publishing Institute. DOI: 10.3390/nano12234187

Mantovani, M., Collina, E., Lasagni, M., Marazzi, F., & Mezzanotte, V. (2022). Production of microalgal-based carbon encapsulated iron nanoparticles (ME-nFe) to remove heavy metals in wastewater. *Environmental Science and Pollution Research International, 30*(3), 6730–6745. DOI: 10.1007/s11356-022-22506-x PMID: 35008581

Mérida, L. G. R., & Padrón, R. A. R. (2023). Application of microalgae in wastewater: Opportunity for sustainable development. *Frontiers in Environmental Science, 11*, 1238640. Advance online publication. DOI: 10.3389/fenvs.2023.1238640

Moghri, P., Sabahi, H., Rezayan, A. H., & Akbari, B. (2025). A green approach: Utilizing untreated pine pollen grains as a natural biosorbent for microplastic removal from water systems. *Journal of Contaminant Hydrology, 104675*, 104675. Advance online publication. DOI: 10.1016/j.jconhyd.2025.104675 PMID: 40680663

Mohsenpour, S. F., Hennige, S., Willoughby, N., Adeloye, A. J., & Gutiérrez, T. (2020). Integrating micro-algae into wastewater treatment: A review. *The Science of the Total Environment, 752*, 142168. DOI: 10.1016/j.scitotenv.2020.142168 PMID: 33207512

Moraes, L. C., Figueiredo, R. C., Ribeiro-Andrade, R., Pontes-Silva, A. V., Arantes, M. L., Giani, A., & Figueredo, C. C. (2021). High diversity of microalgae as a tool for the synthesis of different silver nanoparticles: A species-specific green synthesis. *Colloid and Interface Science Communications, 42*, 100420. DOI: 10.1016/j.colcom.2021.100420

Morgan, S. H., Mahmoud, M. A., Mottaleb, S. A., El-Bahbohy, R. M., & Mahmoud-Aly, M. (2023). A sustainable nanotechnology producing high-quality remediated sewage wastewater used for microalgal protein-rich biomass and biodiesel production. *Biomass Conversion and Biorefinery, 14*(24), 31021–31037. DOI: 10.1007/s13399-023-05016-9

Mukherjee, A., Sarkar, D., & Sasmal, S. (2021). A Review of Green Synthesis of Metal Nanoparticles Using Algae. *Frontiers in Microbiology, 12*, 693899. DOI: 10.3389/fmicb.2021.693899 PMID: 34512571

Narwal, N., Katyal, D., & Bathi, J. R. (2025). Green synthesis of Iron Oxide Nanoparticles using Momordica charantia: Kinetics of removal of heavy metal and microplastic pollutants. *Regional Studies in Marine Science, 104189*, 104189. Advance online publication. DOI: 10.1016/j.rsma.2025.104189

Ndubuisi, O. G. N. OBIORAH, C. A., Ugah, T. A., Agbakhamen, C. O., Ali, S. E., & Nesiama, O. (2025). *Integrated Water Resource Management in a Changing Climate: Assessing Adaptive Strategies, Technologies, and Policies for Resilient Water System*. DOI: 10.5281/zenodo.15769652

. Nguy n, T. T., Heimann, K., & Zhang, W. (2020). Protein Recovery from Underutilised Marine Bioresources for Product Development with Nutraceutical and Pharmaceutical Bioactivities [Review of *Protein Recovery from Underutilised Marine Bioresources for Product Development with Nutraceutical and Pharmaceutical Bioactivities*]. *Marine Drugs, 18*(8), 391. Multidisciplinary Digital Publishing Institute. DOI: 10.3390/md18080391

Noreen, S., Saleem, M. H., Ali, B., Khan, K. A., Hafeez, A., & Javed, M. A. (2024). Advancing Plant Resilience Against Microplastics and Metals Through Nanotechnology. *BioNanoScience*, *14*(2), 2065–2079. DOI: 10.1007/s12668-024-01301-7

Pachouri, V., Chandramauli, A., Singh, R., Gehlot, A., Priyadarshi, N., & Twala, B. (2024). Removal of contaminants by chlorella species: An effort towards sustainable remediation. *Discover Sustainability*, *5*(1), 19. Advance online publication. DOI: 10.1007/s43621-024-00199-1

Pande, R., & Patil, A. G. (2023). Assessing the feasibility of inorganic nanomaterials for nanohybrids formation. In *IWA Publishing eBooks* (p. 99). UWA Publishing., DOI: 10.2166/9781789063592_0099

Panbehkar Bisheh, M., & Amini Rad, H. (2023). Optimization of the culture of Chlorella sorokiniana PA. 91 by RSM: effect of temperature, light intensity, and MgAC-NPs. Environmental Science and Pollution Research, 30(17), 50896-50919.

Premarathna, K. S. D., Lau, S. Y., Chiong, T., Show, P. L., Vithanage, M., & Lam, M. K. (2024). Greening up the fight against emerging contaminants: Algae-based nanoparticles for water remediation. *Clean Technologies and Environmental Policy*. Advance online publication. DOI: 10.1007/s10098-024-02862-7

. Song, Y., Hu, Z., Liu, S., Luo, S., He, R., Yang, X., Li, S., Yang, X., An, Y., & Lu, Y. (2025). Utilization of Microalgae and Duckweed as Sustainable Protein Sources for Food and Feed: Nutritional Potential and Functional Applications [Review of *Utilization of Microalgae and Duckweed as Sustainable Protein Sources for Food and Feed: Nutritional Potential and Functional Applications*]. *Journal of Agricultural and Food Chemistry, 73*(8), 4466. American Chemical Society. DOI: 10.1021/acs.jafc.4c11610

. Su, B., Sun, S., Li, T., Yan, Y., & Wang, Z. (2024). Application research and progress of microalgae as a novel protein resource in the future [Review of *Application research and progress of microalgae as a novel protein resource in the future*]. *Critical Reviews in Food Science and Nutrition, 65*(28), 5790. Taylor & Francis. DOI: 10.1080/10408398.2024.2431208

Sumathi, Y., Dong, C., Singhania, R. R., Chen, C., Baskar, G., & Patel, A. K. (2024). Advancements in Nano-Enhanced microalgae bioprocessing. *Bioresource Technology, 401*, 130749. DOI: 10.1016/j.biortech.2024.130749 PMID: 38679239

Tabassum, H., Sailaja, A., Afreen, H., & Wani, M. (2023). Sewage Treatment Using Nanoparticles. In *IntechOpen eBooks*. IntechOpen., DOI: 10.5772/intechopen.109407

Tonelli, F. M. P., Silva, C. S., Delgado, V. M. S., Tonelli, F. C. P., Tonelli, F. C. P., & Tonelli, F. C. P. (2023). Algae-based green AgNPs, AuNPs, and FeNPs as potential nanoremediators. *Green Processing and Synthesis, 12*(1), 20230008. Advance online publication. DOI: 10.1515/gps-2023-0008

Uddandarao, P., & Lens, P. N. L. (2023). Introduction to wastewater treatment and energy recovery. In *IWA Publishing eBooks* (p. 3). UWA Publishing., DOI: 10.2166/9781789063592_0003

Usman, M., Saidu, M., Usman, G., Agboola, P. O., Onwuchekwa, C., Onwuchekwa, N., & Oniye, M. M. (2025). Eco-Friendly Nanomaterials for Remediation of Microplastics and Heavy Metals in Wetland Ecosystems. *Deleted Journal, 20*(1), 54–86. DOI: 10.62154/ajesre.2025.020.01014

Uzair, B., Liaqat, A., Iqbal, H., Menaa, B., Razzaq, A., Thiripuranathar, G., Rana, N. F., & Menaa, F. (2020). Green and Cost-Effective Synthesis of Metallic Nanoparticles by Algae: Safe Methods for Translational Medicine. *Bioengineering (Basel, Switzerland), 7*(4), 129. DOI: 10.3390/bioengineering7040129 PMID: 33081248

Venkatesan, L. S., & Sathishkumar, P. (2024). Recent advances in the removal of emerging pollutant tetracycline using green nanomaterials: An eco-friendly mitigation approach. *International Journal of Environmental Science and Technology, 22*(6), 5129–5142. DOI: 10.1007/s13762-024-06163-w

Yadav, C., Chu, K. H., Hassan, Z., Lee, J., & Jang, W. (2025). Nanocellulose sponges embedding metal oxide nanoparticles for adsorption and photodegradation of microplastics. *Chemosphere, 386*, 144614. DOI: 10.1016/j.chemosphere.2025.144614 PMID: 40779960

Yuan, W., Xu, E. G., Shabaka, S., Chen, P., & Yang, Y. (2024). The power of green: Harnessing phytoremediation to combat micro/nanoplastics. *Eco-Environment & Health, 3*(3), 260–265. DOI: 10.1016/j.eehl.2024.04.001 PMID: 39234422

Zanoni, I., Amadori, S., Brigliadori, A., Costa, A. L., Ortelli, S., Giacò, P., Baldisserotto, C., Pancaldi, S., & Blosi, M. (2025). Hybrid materials for wastewater treatment: Synergistic coupling of Neochloris oleoabundans and TiO2 nanoparticles. *Nanoscale Advances, 7*(12), 3803–3816. DOI: 10.1039/D5NA00236B PMID: 40395621

Zhang, X., Lv, D., Li, B., Liu, Z., Yang, F., Tang, Q., Gao, Y., Zhao, H., Tan, Z., Du, P., Li, J., & Wang, J. (2025). Polydopamine-Functionalized Magnetic Algae Composite for Efficient Removal of Polystyrene Microplastics: Mechanistic Insights and Performance. *Colloids and Surfaces. A, Physicochemical and Engineering Aspects, 137594*, 137594. Advance online publication. DOI: 10.1016/j.colsurfa.2025.137594

Chapter 13
Microalgae-Driven Carbon Sequestration and Microbial Biomineralization for Sustainable Cementitious Materials

Bharathi Priya Ravichandran
http://orcid.org/0009-0009-5667-4630
SASTRA (Deemed to be) University, India

Gowdhaman Dharmalingam
http://orcid.org/0000-0002-7316-9224
SASTRA (Deemed to be) University, India

ABSTRACT

Cement and concrete production represent one of the largest industrial sources of CO_2 emissions, contributing 8% of the global total. Addressing this challenge requires not only carbon reduction strategies but also active carbon capture and storage mechanisms within construction materials themselves. This chapter explores an innovative biotechnological approach: integrating microbial carbon sequestration with engineered cementitious systems to develop carbon-negative, self-healing construction materials. Wastewater grown microbial carbon sequestration harnesses autotrophic and heterotrophic microorganisms that fix CO_2 through photosynthesis or enzymatic pathways. Cyanobacteria and microalgae capture atmospheric CO_2, while urease- and carbonic anhydrase-producing bacteria catalyse biominerali-

DOI: 10.4018/979-8-3373-7630-1.ch013

sation reactions, converting CO_2 into stable calcium carbonate ($CaCO_3$). These two paradigms create a bio–material synergy wherein microbial metabolism and cement chemistry mutually enhance CO_2 fixation, material strength, and crack-healing capacity.

1. INTRODUCTION

Common greenhouse gases such as carbon dioxide (CO_2) are a major environmental concern. This contributes to 68% of total greenhouse gas emissions in the atmosphere (Bhola et al., 2014; Ighalo et al., 2022). Over the last three decades, carbon emissions have increased by more than 50% (Singh et al., 2013) due to the rise in industrialisation and urban growth worldwide. Global warming increases as greenhouse gas emissions increase. These emitted greenhouse gases are toxic to the environment, so one main way to reduce CO_2 emissions is to cut down on fossil fuel use and other limited resources, or to capture CO_2 through biological processes and also focus on biological products instead of chemical substances, as they can reduce the carbon footprint (Prasad et al., 2021). There are many reasons for carbon emissions, but one such source is the emission of CO_2 from building materials, deforestation, burning of fossil fuels, the release of toxic effluent from industries, and the emission of carbon from construction sites. As the construction industry grows rapidly, carbon emissions are also increasing. Developed countries have much lower carbon emissions from unsustainable building materials than developing countries and regions (Li et al., 2025). For instance, the rate is about 27% to 40% in the EU and 29% to 30% in the US. In contrast, it is much higher in India (50% to 61%) and China (43% to 73%). This may be because developed countries have stricter environmental laws, better access to new technologies, and more financial incentives to use sustainable building materials. Thus, the use of sustainable and eco-friendly construction materials is increasing and in demand worldwide. These factors help reduce the reliance on unsustainable options (Rajesh & Sumathi, 2024). As a result, there is a need for sustainable building materials to improve strength, durability, and CO_2 capture (Rajadesingu et al., 2024).

Capturing atmospheric CO_2 and using it for different environmental uses is one method of lowering CO_2 emissions. Carbon capture is conventionally done using various methods, which include biological, physical, chemical and geological routes (Nanda et al., 2016). One such way of reducing the emission of carbon from concrete structures is by using bio materials as a replacement for the traditional raw materials (Jaramillo et al., 2023). This might include replacement of cement with certain biological sources such as rice husk, bamboo, and so on, which contain almost a similar composition with respect to cement (Ramadan et al., 2025; Onuaguluchi &

Banthia, 2016). Certain plant extracts are also used as bio-admixture or self-binding agents for preparing eco-efficient cement-based composite materials (Hazarika et al., 2018). Plant-based natural fibres are cost-efficient, renewable materials that can be used for reinforcement in cement-based composites. Even though plant-based materials are used to increase their strength and durability, these natural materials can't be used if there is any damage to the concrete structures. Crack formation is a major deformation in the construction sector; it could even lead to the collapse of a major structure, just because a small crack was an inducing factor for erosion and also reduces the stability of the structure (Golewski, 2023). There are different techniques to stop and repair cracks in concrete structures to avoid that. One of those approaches is the use of self-healing concrete, which increases the concrete's strength and facilitates sequestration. The self-healing concrete applications make this possible by incorporating microbes like microalgae, cyanobacteria, and bacteria to reduce the formation of cracks and heal cracks if any are formed (Wong et al., 2024; Mokhtar et al., 2021; Elgendy et al., 2025). The organisms used in self-healing concrete applications are capable of producing carbonate precipitates by reacting with the calcium and magnesium content and the alkaline nature of the concrete (Li et al., 2024). Some organisms, like microalgae and cyanobacteria, not only heal the cracks but also capture atmospheric CO_2 and utilise it for their metabolism. Making a green life by employing cyanobacteria and microalgae as effective CO_2 collection agents is one such strategy (Sidhu et al., 2022).

2. CONVENTIONAL METHODS FOR CO_2 SEQUESTRATION AND REMOVAL

2.1 Natural and Anthropogenic Greenhouse Effect

Greenhouse gases (GHG) are gases present in the atmosphere that trap heat by absorbing and re-emitting infrared radiation. This effect caused by these gases is called the greenhouse effect (Hu et al., 2024). Some of the greenhouse gases include carbon dioxide, fluorinated gases, methane, nitrous oxide and water vapour. The greenhouse effect happens naturally, in which the amount of radiated heat escaped from the atmosphere is high when compared to the heat emitted back to the atmosphere (Nanda et al., 2016). Also, in the natural process, there is less GHG in the atmosphere. But in cases of anthropogenic greenhouse effect, this happens oppositely, that is, there will be more GHG in the atmosphere, and more heat will be emitted back into the atmosphere. Less heat will be lost, such that this phenomenon causes global warming, which influences major calamities by impacting the global climate and functioning of the ecosystem.

2.2 Carbon Dioxide Emission

One of the GHGs, carbon dioxide, contributes significantly to global warming, as CO_2 concentrations are increasing at 1-3% per year. Rapid urbanisation and industrialisation have also caused the emission of CO_2 into the atmosphere. Carbon dioxide emissions happen from different sources, ranging from respiration to the release of CO_2 from industries. The above-mentioned factors influence the total amount of CO_2 in the atmosphere. The increase in global warming is mainly because of the increasing rate of CO_2 in the atmosphere. India stands among the top countries emitting the highest amount of CO_2 worldwide.

2.3 Carbon Capture and Routes of Capture

Carbon capture, utilisation, and storage (CCUS) is a new field that is crucial to ecosystem maintenance because of these factors (Saxena et al., 2024; Shen et al., 2022). The primary goals of CCUS are to absorb CO_2 from major emitters, consume it, and store it to stop it from returning to the environment (Orr Jr, 2018). The Kyoto Protocol has a significant impact on those wealthy countries that release the most CO_2. This global agreement establishes global initiatives for reducing greenhouse gas emissions. Carbon can be captured using a variety of techniques, such as geological, biological, and physicochemical techniques (Nanda et al., 2016).

2.3.1 Physicochemical Capture

This method involves physical and chemical processes of carbon capture, such as adsorption, membrane separation/filtration, absorption, and cryogenic distillation (Nwali et al., 2024; Nanda et al., 2016). Both physical and chemical methods can be used to collect gas molecules. In the physical sorption method, a solvent mixture is absorbed together with the CO_2. The solvents employed in this process are mostly aqueous alkanolamines, polyethene glycol, and dimethyl ethers. However, these solvents are restricted to flue gases, also sulphur and nitric oxide form heat-stable complexes when combined with such solvents, which can be corrosive and pose serious risks. Additionally, the adsorption process helps in the adsorption of gaseous molecules onto adsorbent materials such as metal organic frameworks, chemically modified polymeric materials, and zeolites (Karimi & Ghaemi, 2025; Trickett et al., 2017). In fuel cells, certain membranes are employed to separate gaseous components. To obtain the required product through a condensation process with the aid of cryogenic distillation, flue gases are further separated according to the freezing points of the constituents (He et al., 2023).

2.3.2 Biological Capture

The biological method of CO_2 capture includes the use of living organisms for carbon sequestration and their utilisation to produce value-added products (Osman et al., 2022). Algal bacterial systems can be used for the capture of carbon and its utilisation (Gayathri et al., 2021; La et al., 2022; Kandasamy et al., 2025). Algae (Macroalgae, microalgae, and cyanobacteria) can thrive under CO_2-concentrating conditions, as it is needed for their metabolism, and as these organisms perform photosynthesis, in which the atmospheric CO_2 is utilised by these organisms for their metabolism to produce storage materials (carbohydrates, proteins), they release oxygen in the process. Such that these organisms act as carbon sinks in the ecosystem, conserving and capturing carbon. Algae take up the CO_2 and undergo biomass conversion to produce value-added products like biofuel, biomethane, bioplastics and so on. Some may give rise to medicinally important compounds, pigments, etc. (Gurau et al., 2025).

3. MICROBIAL CARBON SEQUESTRATION MECHANISMS

An important component of these CO_2 reduction strategies is the developing field of carbon capture, utilisation, and storage (CCUS). CO_2 capture, transportation/use, and storage are the three primary processes in this process (Singh & Ahluwalia 2013; Storrs et al., 2023). One way to sequester CO_2 is through the process called photosynthesis. The organisms that perform photosynthesis capture CO_2 and utilise it for their own metabolism and also convert this carbon into value-added products. Biological CO_2 capture can be performed by photosynthetic microorganisms like microalgae and cyanobacteria. Microbial carbon sequestration is far more efficient, as the CO_2 utilised can be converted into value-added products, which supports the goal of CCUS (Yaashikaa et al., 2024).

3.1 Photosynthetic Sequestration

3.1.1 Microalgae

They are photosynthetic eukaryotic microorganisms. The chloroplast is where photosynthesis occurs in microalgae. These green pigments, which include β-carotene, xanthophylls, and primarily chlorophyll a and chlorophyll b, are present in photosynthetic cells. Microalgae exist as individual cells or groups of individual cells and can vary in shape and size. Over thousands of microalgae species have been described. In the case of algae, less light is needed when compared to higher plants (Thoré et al., 2023). During photosynthesis, this organism utilises CO_2, sunlight and water to produce oxygen as its byproduct. By integrating the production of microbial biomass with the current infrastructure for wastewater treatment and power generation, microalgae-mediated carbon sequestration can be made more sustainable (Kumar et al., 2011).

3.1.2 Cyanobacteria

Cyanobacteria have been known to perform photosynthesis for over a million years. They are primitive prokaryotic organisms (also called as blue-green algae) that are the first to perform photosynthesis and carbon sequestration (Thoré et al., 2023). Gram-negative cyanobacteria are distinguished from other gram-negative bacteria by their thicker peptidoglycan coating. The cell of cyanobacteria is devoid of organelles. The thylakoid and plasma membranes contain the respiratory chain. Photosynthetic electron transport occurs in the membranes of thylakoids, which are cytoplasmic structures where photosynthesis occurs (Kamennaya et al., 2012). Green algae, plants' chloroplasts and cyanobacteria's photosynthetic machinery are comparable (Singh et al., 2013). However, the antenna system is a significant distinction.

Cyanobacteria rely mostly on chlorophyll a and a specific protein complex called phycobilisomes to collect light energy because they lack chlorophyll b. There are two types of light-harvesting complexes: hydrophilic and lipophilic. While hydrophilic antenna pigments are found in phycobilisomes attached to the outside of thylakoid membranes, lipophilic pigments like carotenoids and chlorophyll a are found inside these membranes (Stadnichuk et al., 2015; He et al., 2016). Similar to green algae and plants, cyanobacteria use oxygenic photosynthesis to produce oxygen. However, they can also do anoxygenic photosynthesis in anaerobic environments using Photosystem I (PSI). Cyclic photophosphorylation takes place near PSI, where anoxygenic photosynthesis produces ATP (Xiong et al., 2017).

3.1.3 Wastewater Valorisation of Microalgae and Cyanobacteria

Microalgal systems enable a dual-process synergy that accounts for both economic viability and environmental compliance when integrated into wastewater treatment infrastructure. Technically speaking, microalgae efficiently carry out secondary or tertiary treatment through biological uptake, taking advantage of the high concentrations of nitrogen and phosphorus in effluents as vital nutrients for metabolic growth (Abdel-Raouf et al., 2012). This metabolic process converts dissolved contaminants into a stable, concentrated cellular biomass, frequently in conjunction with the sequestration of carbon dioxide and the biosorption of heavy metals. The conversion of this biomass into a variety of products, including biofuels, bioplastics, and nitrogen-rich biofertilizers, is what gives this process its economic significance beyond mere cleanup. The overhead costs of microalgal production are greatly decreased by repurposing wastewater as an inexpensive growing medium, changing the perception of wastewater from an expensive burden to a strategic resource within a circular bioeconomy framework.

3.2 Carbon Concentration Mechanism (CCM)

The light-dependent reactions and the light-independent reactions are the two distinct phases of photosynthesis (Raven et al., 2012). CO_2 is transformed into triose phosphates, starch, sucrose, and other compounds by the light-independent process using ATP and NADPH. The reductive pentose pathway (RPP), often known as the C3 cycle, is the primary component of photosynthetic carbon metabolism (Kupriyanova et al., 2023; He et al., 2016). By actively pumping inorganic carbon into their cells and concentrating it around the photosynthetic enzyme RuBisCo (ribulose 1,5-bisphosphate carboxylase), cyanobacteria and microalgae sequester carbon utilising the Carbon Concentration Mechanism (CCM) (Aizawa & Miyachi, 1986). While low CO_2 concentrations induce RuBisCO to perform oxygenase activity, high CO_2 pressure activates its carboxylase activity. The CCM is mediated by specialised organelles such as pyrenoids in microalgae and carboxysomes in cyanobacteria. These organelles contain enzymes like carbonic anhydrase and RuBisCo, which facilitate the conversion of CO_2 into biomolecules, which are further used for their metabolism (An et al., 2023; Espie & Kimber, 2011).

3.2.1 Carboxysomes

They are bacterial microcompartments consisting of a polyhedral protein shell that encapsulates the enzyme RuBisCo and carbonic anhydrase (Fig 1) (Price et al., 2008; Jiang, 2022). These structures are a key part of CCM, as they increase the efficiency of carbon fixation by raising CO_2 levels around RuBisCo (Kumar et al., 2022). They are present in organisms that are generally found in environments with low CO_2 concentrations. To meet their need for carbon, they rely on CCM, which accumulates within the cytosol, mainly in the form of bicarbonates. However, RuBisCo requires CO_2 and cannot utilise bicarbonates; this problem is solved by the other enzyme found in the carboxysomes, which is carbonic anhydrase that converts cytosolic bicarbonate to CO_2 (Cannon et al., 2010). There are two types of carboxysomes: α carboxysomes and ß carboxysomes, which differ in their protein makeup and the specific form of RuBisCo they enclose. The protein shell is composed of various proteins, including shell-forming proteins, which help in shell assembly. RuBisCo, carbonic anhydrase, and scaffold proteins constitute the internal proteins, in which the scaffold proteins are essential for the assembly and structure of carboxysomes.

CO_2 and bicarbonate ions from the environment enter the cyanobacterial cytosol through the carbon pumps that are present in the membrane, which are then directed to the carboxysomes, where the conversion of carbonate and bicarbonate ions take place with the help of enzymes like RuBisCo and carbonic anhydrase, the CO_2 is converted into 3- phosphoglyceric acid which enters the calvin cycle and continues in the metabolic pathway (Kumar, Kar, & Shukla, 2022; Cannon et al., 2010; Salar et al., 2013).

Figure 1. CCM in cyanobacteria in carboxysomes

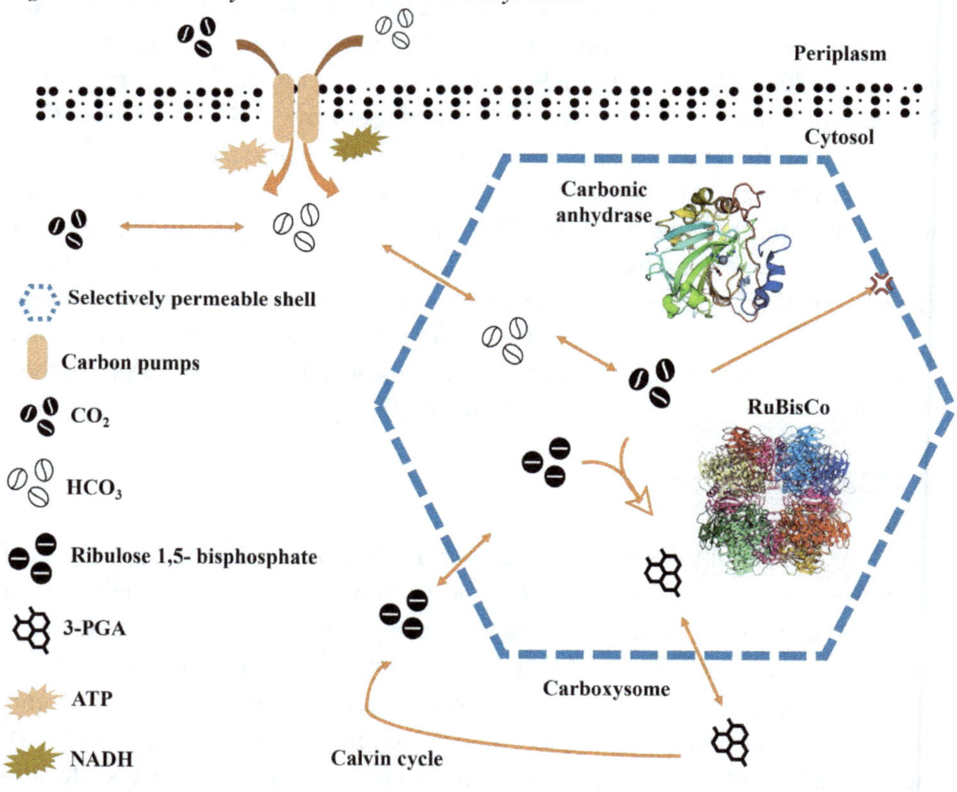

(modified from Espie & Kimber, 2011)

3.2.2 Pyrenoids

These are subcellular compartments in algae and hornworts that are central to their CCM. Pyrenoids are liquid-like structures containing a protein core that is rich in

RuBisCo, and surrounded by a starchy sheath that traps CO_2 and minimises leakage (Fig 2). They are larger than the cyanobacterial carboxysomes. They are considered the main sites in which inorganic carbon transfer takes place. In microalgae, the CCM actively pumps CO_2/bicarbonate into pyrenoids. These bicarbonates then travel to thylakoids and are carried to the pyrenoid via membrane tubules after entering the chloroplast via transporters. The carbonic anhydrase enzyme in these tubules helps convert bicarbonates to CO_2 (An et al., 2023; Catherall et al., 2025; Bradley, 2025). The converted CO_2 is metabolised with the help of the enzyme RuBisCo.

Figure 2. CCM mechanism in microalgae in pyrenoids

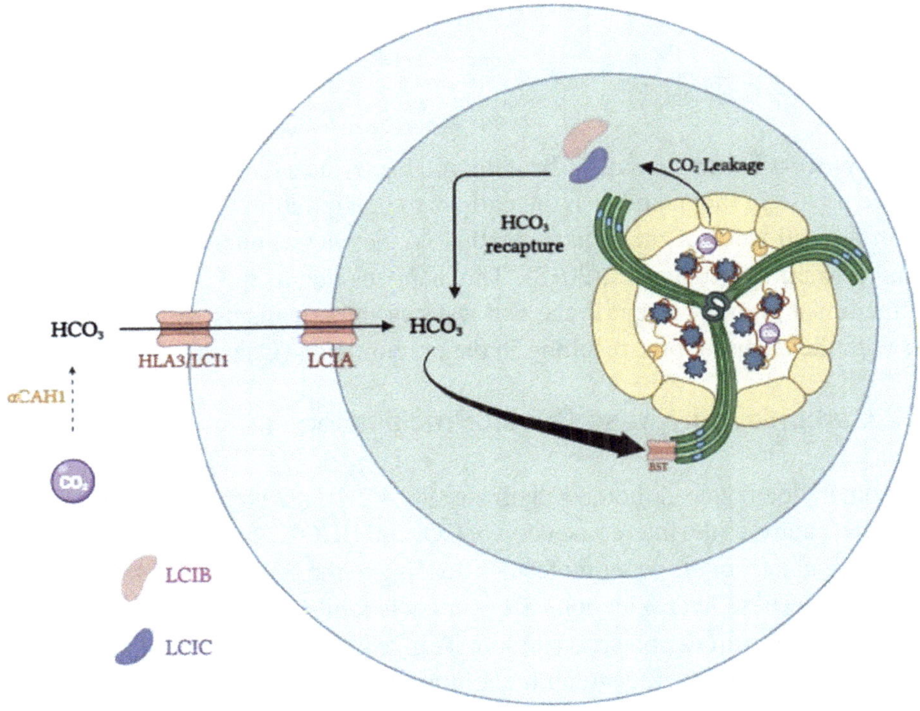

(created in https://BioRender.com)

3.3 Biochemical Pathways for Carbonate Precipitation

Urease and carbonic anhydrase are the key enzymes involved in the MICP process.

3.3.1 Urease-Driven Precipitation

Urease (EC 3.5.1.5) is a metalloenzyme that contains nickel and is classified as urea amidohydrolase (Krajewska, 2018). This enzyme is found to be present in many organisms and is also known to have a high molecular weight. Bacterial ureases are different from plant ureases. The hydrolysis of urea into ammonia and carbonate ions is catalysed by urease (Eq 1). This process releases ammonia, which raises the pH and produces an alkaline atmosphere. Calcium carbonate precipitates are created when carbonate ions released from the process mix with the calcium ions in the environment (Eq 2).

$$CO(NH_2)_2 \leftrightarrow NH_3 + + CO_3^{2-} \quad (1)$$

$$CO_3^{2-} + Ca^{2+} \leftrightarrow CaCO_3 \quad (2)$$

This process is carried out by the ureolytic bacteria as these types of organisms produce urease in them, and also the carbonate precipitation of crystallisation occurs in the cell walls of the organisms, that is, they act as nucleation sites for the crystal formation (Krajewska, 2018). This happens due to the negative charge of the functional groups present in the cell walls of the organisms, which attract and bind with the calcium ions, resulting in the precipitation (Chandra et al., 2023).

3.3.2 Carbonic Anhydrase-Driven Precipitation

The metalloenzyme carbonic anhydrase (EC.4.2.1.1), which contains zinc in its active site, catalyses the interconversion of CO_2 and HCO^{3-}. During this process, H+ ions are created from the water molecules, leading to the generation of the hydroxyl group (OH^-). These hydroxide ions act as a nucleophile, which attacks the carbon atom of CO_2 to form the bicarbonate ion. This reaction not only regulates the pH levels but also enhances the strength and durability of the cementitious materials by promoting the precipitation of carbonate as calcium carbonate (Heidarnezhad et al., 2024).

In this biochemical process, gaseous CO_2 is converted to hydrated aqueous CO_2 (aq) (Eq 3), which dissolves in water to make H_2CO_3 (Eq 4), whose ionisation in water produces H^+ and HCO^{3-} (Eq 5). The HCO3− further ionises to create CO_3^{2-} and H_2O in alkaline circumstances (Eq 6). The interaction of (Ca^{2+}) with CO_3^{2-} produces $CaCO_3$ precipitates in the presence of a calcium ion source (Eq 7). Although other divalent ions, such as Mg^{2+} or Fe^{2+}, can be utilised, Ca^{2+} remains the most commonly used; as a result, $CaCO_3$ is precipitated in the majority of biocementation investigations. Bacterial cells operate as nucleation sites when bacteria are utilised instead of the free enzyme (Mwandira et al., 2023).

$$CO_2(g) \leftrightarrow CO_2\ (aq) \quad (3)$$

$$CO2\ (aq) + H2O\ (aq) \leftrightarrow H2CO3(aq) \quad (4)$$

$$H_2CO_3(aq) \leftrightarrow H+ + HC^{-3}\ (aq) \quad (5)$$

$$2HCO^{-3} + 2OH^{-} \leftrightarrow 2CO_3^{-2} + 2H_2O \quad (6)$$

$$Ca^{2+}(aq) + CO_2^{-3}\ (aq) \leftrightarrow CaCO_3\downarrow \quad (7)$$

This enzyme not only promotes carbonate precipitation but also enhances the corrosion resistance and overall performance of the concrete structures (Heidarnezhad et al., 2024; Xue et al., 2025).

3.4 Role of Microbial Surface

Microorganisms are used in the biocementation process to create calcium carbonate for building (Ali & Almurshedi, 2024). The process by which bacteria react with chemical components to form minerals that function as a binding agent is also known as Microbially Induced Calcium Carbonate Precipitation (MICP/MICCP) (Iqbal et al., 2021). By altering the pH and carbonate ion availability of their surroundings, photosynthetic organisms cause calcium carbonate to precipitate. By using the calcium and carbonate ions found in the environment, certain microbes can remove these minerals from the environment and possibly produce useful products (Aceituno-Caicedo, 2025). Unlike other organisms that utilize/rely on urea hydrolysis or denitrification, microalgae and cyanobacteria utilise their photosynthetic process to induce MICP (Wong et al., 2024; Qian et al., 2021; Elgendy et al., 2025). The steps involving photosynthetically induced MICP are as follows;

Inorganic Carbon Uptake

Microalgae and cyanobacteria absorb inorganic carbon and calcium ions from the water and environment, leading to the formation of carbonate ions.

Alkaline Environment

Their photosynthetic activity increases the pH in the microenvironment, creating alkaline conditions that favour carbonate precipitation.

Nucleation Sites

The microorganisms themselves act as nucleation sites, providing surfaces where the mineral crystals can begin to form and grow.

4. INTEGRATION IN CEMENTITIOUS MATERIALS

4.1 Carbon Emission From Concrete

As the world is enlarging with rapid urbanisation, so is the use of concrete also increasing. Concrete is made up of cement, water, fine and coarse aggregates, and cement is the main source of carbon emissions, usually during concrete preparation and shipping (Pérez et al., 2024; Cong et al., 2024). To reduce this, modified concrete using sustainable agro-based products is used. Some reports show that modified concrete emits less amount of CO_2 when compared to traditional concrete.

4.2 Autogenous Crack Healing

Autogenous healing is the mechanism by which concrete's spontaneous process of partial crack repairing takes place. When the crack width is less than 100μm, autogenous crack healing takes place (Alengaram et al., 2021). But it's not possible to have such a crack width every time, thus this process of healing is fastened by adding external factors like chemical admixtures, biological composites.

4.2 Biomineralisation

Biomineralisation is the process of the production of minerals by living organisms (Ali & Almurshedi, 2024; Krajewska, 2018). Both biologically generated and biologically regulated mineralisation fall within this category. The most abundant mineral produced through biomineralisation is calcium carbonate. Various polymorphs of calcium carbonate are found in nature as minerals (Krajewska, 2018). Among these are three anhydrous polymorphs, such as vaterite, argonite, and calcite, the latter of which is more stable (Li et al., 2023). Microorganisms can also secrete calcium carbonate in a calcium-rich environment. There are different pathways for biomineralisation, which include

- Biomieralisation through photosynthesis (algae and cyanobacteria)
- Biomineralisation through sulphate-reducing bacteria
- Biomineralisation through the nitrogen cycle, utilising urease for the hydrolysis of urea
- Biomineralisation through the carbonic anhydrase enzyme

4.2.1 Biomineralisation in Bacteria

Certain enzymes, such as carbonic anhydrase, which is crucial for the conversion of CO_2 into bicarbonates, enable prokaryotic organisms, such as bacteria, to sequester CO_2. Carbonates are produced when these bicarbonates interact with environmental calcium or magnesium ions (Rusznyák et al., 2012). This type of mineralisation can be carried out with the help of urease, another enzyme that aids in the precipitation of carbonate (Achal & Mukherjee, 2015; Phillips et al., 2013). While many bacteria may precipitate carbonate, not all bacteria can live in the strong alkaline conditions found in concrete. For biocementation, spore-forming and alkali-tolerant bacteria are selected to increase the process's viability (Ivaškė et al., 2023). Several bacterial species have been reported to have the ability to perform biocementation, including organisms like *Sporosarcina pasteurii*, *Bacillus subtillis*, *Bacillus tropicus*, *Bacillus megaterium*, *Bacillus sphaericus*, and so on (Smitha et al., 2022; Mutitu et al., 2019; Rajesh et al., 2024; Mokhtar et al., 2021).

4.2.2 Biomieralisation Process by Microalgae and Cyanobacteria

Inorganic Carbon Uptake (Ci)

Microalgae and cyanobacteria have specialised membrane-bound transporters that actively pump inorganic carbon into the cell from the external environment (Yamaoka et al., 2025).

Inorganic Carbon Conversion

The inorganic carbon pumped into the cells is taken into specialised structures where it is converted into CO_2 by the enzyme carbonic anhydrase (Singh et al., 2022).

RuBisCo Compamentalization

The converted CO_2 is then fixed by RuBisCo into organic molecules during photosynthesis.

4.3 Carbonate Precipitation in Bacteria

Carbonate precipitation in bacteria can occur through three pathways, which include: ureolytic, non-ureolytic, and denitrifying bacteria (Luo et al., 2015). In self-healing concrete, the denitrifying bacteria break down the nitrates in the nitrogen cycle to create calcium carbonate crystals, which can function in oxygen-limited conditions. The ureolytic bacteria precipitate calcium carbonates with the help of urease-driven precipitation (Algaifi et al., 2020; Chayaamor et al., 2023). The main disadvantage of this pathway is that it results in the release of ammonia into the environment. Non-ureolytic bacteria, on the other hand, are able to convert environmental calcium into calcium carbonate without generating ammonium ions.

4.4 Strategies for Microbial Incorporation

The importance of these photosynthetic carbon-capturing organisms is now studied in various fields of science and engineering. One such is their integration into the cementitious materials. One of the main causes of CO_2 emissions in the environment is construction materials. Through the MICP process, the carbon-capturing organisms in these materials will help the environment and increase the strength and endurance of the concrete (Nayeem et al., 2025). The application of microorganisms in the manufacture of biocement and carbon capture offers a viable solution to environmental issues and the evolving demands of the building sector (Gebru et al., 2021). One of the biggest problems in building is the development of cracks in concrete structures. Microalgae and cyanobacteria address this problem in two main ways: i) by repairing fractures through microbial carbonate precipitation, and ii) by helping to collect CO_2, which lowers the carbon emissions linked to the manufacture of concrete. This dual functionality of algal species may improve the durability of concrete structures while promoting sustainability (Kamennaya et al., 2012).

Bacteria-based self-healing concrete follows the integration of bacterial cells while making the mortar mixture (Rajesh & Sumathi, 2024). The main trait is the selection of spore-forming bacterial species, as they tend to have a longer dormant period. Also, the strains that are to be used for this application should have tolerance to alkaline conditions, as the concrete environment is alkaline in nature. Thus, the isolated strain should first be evaluated for its ability to survive under high alkaline conditions. Along with the bacterial culture, nutrient solution is also added, which includes the nutrients for the spore-forming bacteria such as calcium lactate, urea and yeast extract (Shaaban et al., 2023). Sometimes, incorporation of bacterial culture can be performed in many ways, which include the addition of microorganisms directly into the mortar mixture, or encapsulating the organisms in any porous material and then incorporating it into the cement mixture (Soda et al., 2024; Cardoso et al., 2023). Encapsulation of these spore-forming bacteria enhances their survival rate in the concrete environment (Elgendy et al., 2025; Lee & Park, 2018). Once the crack forms, the water enters the cracked areas, reviving these dormant bacterial spores, and once revived, the enzymes urease and carbonic anhydrase present in these species capture water and CO_2 and convert them into carbonate or bicarbonate ions. The alkaline nature of the cement facilitates the carbonate precipitation. Thus, the formed precipitates heal the cracked areas.

Microalgal and cyanobacterial strains may be incorporated by submerging the cast material in culture media (Chandra et al., 2023) or by adding the culture directly during mortar production (Alengaram et al., 2024; Alengaram et al., 2021). In both methods, the production of carbonate precipitates within the porous cementitious matrix enhances concrete strength and reduces water permeability. Eventually, the precipitates help to heal the fractured area (Fig 3). The alkaline environment, which is favourable for carbonate production, or even the factors that affect the growth of the algal species, will also have an impact on the algal species capability for self-healing. These algal cultures can occasionally be injected into the fractured area or sprayed over time as crack development progresses; these incorporation techniques have also been reported to exhibit effective self-healing capabilities (Srinivas et al., 2024).

In certain cases, the combined effect of algal-bacterial symbiosis also results in higher healing efficiency, and this symbiotic approach supports the organism's own metabolism. That is, the CO_2 released by bacteria is utilised by the growth of microalgae and, in turn, it provides oxygen to the bacteria by photosynthesis (Zhang et al., 2025; Zong et al., 2025). Thus, both the organisms benefit in this process, and the integrated approach doubles the MICP process. Encapsulation of these microbes separately or as a consortium also helps in their viability in the alkaline environments (Espitia-Nery et al., 2019; Alazhari et al., 2018; Patil et al., 2025; Madhumithra & Sumathi, 2025). Sustainable bio-based cementitious materials require the replacement of cement with bio-based materials, which can be done with these microbes, and also certain plant species can be added to replace cement content; some of these materials act as a binding agent and enhance the strength and durability of the concrete (Fu et al., 2025; Yan et al., 2025; Kandasamy et al., 2025). This approach not only enhances the stability of the cementitious material but also serves as a cost-effective alternative for cement.

Figure 3. Algal-based application in self-healing concrete

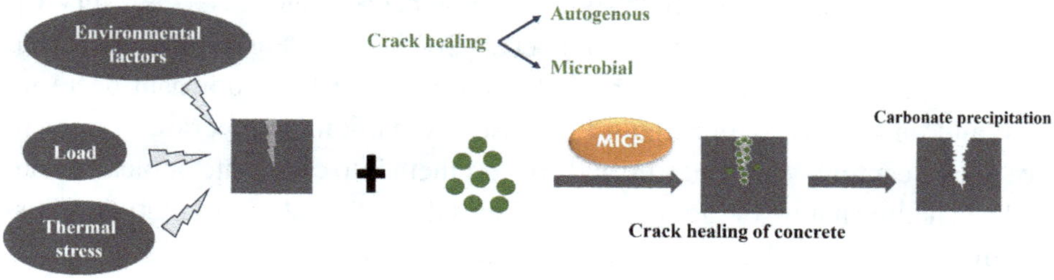

5. CHALLENGES AND FUTURE DIRECTIONS

Although biomineralisation is a sustainable way of replacing conventional concrete, there are certain limitations in this process, which include the viability of strain in corresponding to the harsh conditions of concrete, their effect in terms of lab scale and industrial scale, and other challenges are also seen in the biomineralisation or biocementation process

5.1 Scaling Up and Strain Selection

One of the main challenges in the production of biocement from microalgae is the challenge of scaling up microalgae cultivation to satisfy the construction industry's needs. Selecting and optimising the right microalgae strains is crucial for the manufacture of biocement. These strains have to exhibit quick development for the synthesis of $CaCO_3$, efficient CO_2 capture, and environmental condition adaptation (Nur & Dewi, 2024). Research on selecting specific strains suitable for this purpose is still in progress. Also, if the concentrations of bacterial and algal cells are high or low, it will lead to reduced calcite precipitation. Thus, the proper concentration of bacterial cells should be added (Elgendy et al., 2025).

5.2 Viability

It is difficult to keep microalgae viable over the long term in the concrete matrix. Microalgae and cyanobacteria, or any microorganism capable of carbonate precipitation, need to remain active and able to repair themselves throughout the life of the concrete structure, so that it can last for many years (Pawar & Parekar, 2018). Even though there are positive reports regarding the incorporation of microorganisms in biocement and for environmental benefits, the long-term viability of these algal species is still uncertain. It is also important to know how long the algae and other microbes can stay dormant before cracks form. The selected algal strain must withstand alkaline conditions and harsh environments. Such factors affect the ability of microorganisms like algae and bacteria in the mineralisation process and reduce their self-healing ability.

5.3 Practical Difficulties in Industrial Scale

Incorporation of microbial-based self-healing concrete with traditional materials can be complex, and it is important to make sure this does not harm the overall properties of the mixture, as they tend to affect the growth conditions of the organisms (Nur & Dewi, 2024). To take this idea from lab-scale to industrial scale, several key factors must be addressed. These include matching microalgae growth, validation and determination of integrating microorganisms and following regulations, managing costs, and providing education and training for people in the construction industry.

6. ENGINEERING DESIGN AND SCALABILITY

The cultivation system's design controls the shift from laboratory-scale experiments to wastewater valorisation at the industrial level. Due to its low initial cost and ease of upkeep, Open Raceway Ponds (ORPs) are the most popular for large-scale operations; nonetheless, they have significant evaporation rates and are prone to pollution. On the other hand, Photobioreactors (PBRs)—which come in tube, flat-plate, and column designs—offer greater biomass densities and better control over environmental factors, but they come at a much higher energy and capital cost. Scalability is still a challenge since expanding a system's volume frequently results in "scale-down" consequences, when mixing efficiency declines and nutrient gradients become erratic, potentially causing culture collapse (Srimongkol et al., 2022; Llamas et al., 2021).

6.1 Light Requirements and Photo-Limitation

In microalgal engineering, light is the main limiting substrate. The self-shading effect, in which cells near the surface prevent light from reaching those in the reactor's centre, happens when biomass concentration rises. This reduces total yield by creating a "dark zone" that stops photosynthesis and starts cell respiration. Optimising surface-to-volume ratios and applying high-frequency mixing are engineering techniques that guarantee cells move quickly between the "light" and "dark" zones (the flashing light effect). Water turbidity (cloudiness) in wastewater systems makes light penetration even more difficult, frequently necessitating a pre-treatment or dilution phase to guarantee sufficient photon flux reaches the culture (Saccardo et al., 2022).

6.2 Life-Cycle Assessment (LCA)

To verify the "green" promises of microalgal valorisation, a thorough Life-Cycle Assessment (LCA) is necessary. The energy intensity of harvesting (e.g., centrifugation) and drying has a significant impact on the environmental footprint, even though the process sequesters and purifies water. The energy generated (from biodiesel, for example) must be more than the energy used in the treatment and processing stages for there to be a net positive environmental impact. According to recent life cycle assessment (LCA) research, combining flue gas collection with wastewater treatment greatly reduces the system's global warming potential (GWP) when compared to traditional activated sludge procedures (Akinbuja et al., 2025; Arashiro et al., 2022).

6.3 Cost Implications and Economic Viability

High capital expenditures (CAPEX) and operating expenses (OPEX) now make wastewater-based microalgae systems economically unfeasible. Harvesting continues to be the "bottleneck," frequently contributing 20–30% of overall production expenses. A biorefinery strategy is required to reach a break-even point; low-value biofuel sales are insufficient to cover expenses. Rather, a diverse revenue stream is produced by extracting high-value co-products, including pigments or speciality proteins, before fuel conversion. To accurately represent the technology's true worth, the financial model must account for the "credits" obtained from avoiding conventional wastewater treatment expenses (such as decreased aeration and chemical use) (Nigam et al., 2022).

7. CONCLUSION

In today's world, global warming and greenhouse gas emissions are some of the major problems that need to be addressed. The construction sector is one source of CO_2 emissions worldwide; to reduce these, sustainable materials can be integrated into such commercial sectors. Microorganisms like microalgae, cyanobacteria and carbonic anhydrase and urease-producing bacteria offer a path toward carbon-negative cementitious materials due to their efficient CO_2 sequestration through photosynthesis, carbon-concentrating mechanisms, and their ability to promote the production of calcium carbonate. *In situ* mineralisation boosts durability and lowers permeability, and helps in crack repair. Carbon capture is also done by these microorganisms, thus achieving the goal to reduce carbon footprint in the industrial sector, as well as using bio-based sustainable cementitious materials. Photosynthetic microorganisms are a safer alternative to traditional self-healing methods that mostly depend on chemical additives or bacterial urease activity through their metabolic processes. Despite their wide range of potential, there are many factors that hinder the widespread use of these microbes at an industrial level. The availability of nutrients, and how they interact with the regular cement chemistry, their understudied viability and rate of dormancy once added into the concrete specimens are some factors to take into consideration. Modification of strain and selection of an appropriate strain can be performed to enhance the self-healing ability and their viability in such harsh conditions. To make their industrial application easier, scalability, cost-effectiveness, and strain selection should also be considered. Life cycle assessment, adaptation with existing construction practices, encapsulation and protection strategies, and strain selection can be focused on in the future. To expand microalgae biocementation from lab-scale research to more practical and industrial scales for sustainable construction and climate change mitigation, these challenges should be addressed.

REFERENCES

Abdel-Raouf, N., Al-Homaidan, A. A., & Ibraheem, I. (2012). Microalgae and wastewater treatment. *Saudi Journal of Biological Sciences*, *19*(3), 257–275. DOI: 10.1016/j.sjbs.2012.04.005 PMID: 24936135

Aceituno-Caicedo, D. F. (2025). *Stress Tested: Impact of Salinity on Cyanobacteria Surface Chemistry and Mineral Precipitation-Implications for Biotechnological Applications* (Doctoral dissertation, University of Toronto (Canada)).

Achal, V., & Mukherjee, A. (2015). A review of microbial precipitation for sustainable construction. *Construction & Building Materials*, *93*, 1224–1235. DOI: 10.1016/j.conbuildmat.2015.04.051

Aizawa, K., & Miyachi, S. (1986). Carbonic anhydrase and CO2 concentrating mechanisms in microalgae and cyanobacteria. *FEMS Microbiology Letters*, *2*(3), 215–233. DOI: 10.1111/j.1574-6968.1986.tb01860.x

Akinbuja, O., Orta, S. V., & Boodhoo, K. (2025). Life cycle assessment of microalgae-assisted microbial fuel cells. *The International Journal of Life Cycle Assessment*, *30*(10), 2133–2144. DOI: 10.1007/s11367-025-02518-8

Alazhari, M., Sharma, T., Heath, A., Cooper, R., & Paine, K. (2018). Application of expanded perlite encapsulated bacteria and growth media for self-healing concrete. *Construction & Building Materials*, *160*, 610–619. DOI: 10.1016/j.conbuildmat.2017.11.086

Alengaram, U. J., Ibrahim, S., Phang, S. M., Vello, V., Jun, H. K., & Alnahhal, A. M. (2021). Evaluation of crack healing potential of cement mortar incorporated with blue-green microalgae. *Journal of Building Engineering*, *44*, 102958. DOI: 10.1016/j.jobe.2021.102958

Alengaram, U. J., Ibrahim, S., Vello, V., & Phang, S. M. (2024). Investigation on the enhancement of crack restoration properties in cement incorporated with Arthrospira platensis cultured in modified medium. *Environmental Science and Pollution Research International*, *31*(17), 25538–25558. DOI: 10.1007/s11356-024-32784-2 PMID: 38478311

Algaifi, H. A., Bakar, S. A., Sam, A. R. M., Ismail, M., Abidin, A. R. Z., Shahir, S., & Altowayti, W. A. H. (2020). Insight into the role of microbial calcium carbonate and the factors involved in self-healing concrete. *Construction & Building Materials*, *254*, 119258. DOI: 10.1016/j.conbuildmat.2020.119258

Ali, M. S., & Almurshedi, A. D. (2024). Bio-cementations: A Review on Enzyme and Microbially Induced Calcite Precipitation Mechanisms and Applications in Geotechnical Engineering. *Journal of Engineering*, *30*(12), 167–188. DOI: 10.31026/j.eng.2024.12.11

An, Y., Wang, D., Du, J., Wang, X., & Xiao, J. (2023). Pyrenoid: Organelle with efficient CO2-Concentrating mechanism in algae. *Journal of Plant Physiology*, *287*, 154044. DOI: 10.1016/j.jplph.2023.154044 PMID: 37392525

Arashiro, L. T., Josa, I., Ferrer, I., Van Hulle, S. W., Rousseau, D. P., & Garfí, M. (2022). Life cycle assessment of microalgae systems for wastewater treatment and bioproducts recovery: Natural pigments, biofertilizer and biogas. *The Science of the Total Environment*, *847*, 157615. DOI: 10.1016/j.scitotenv.2022.157615 PMID: 35901897

Bradley, J. M. (2025). Transporting to the matrix: A pyrenoid-localized phosphate transporter required for optimal photoautotrophic growth in algae.

Cannon, G. C., Heinhorst, S., & Kerfeld, C. A. (2010). Carboxysomal carbonic anhydrases: Structure and role in microbial CO2 fixation. *Biochimica et Biophysica Acta (BBA)-. Biochimica et Biophysica Acta. Proteins and Proteomics*, *1804*(2), 382–392. DOI: 10.1016/j.bbapap.2009.09.026 PMID: 19818881

Cardoso, R., Borges, I., Vieira, J., Duarte, S. O., & Monteiro, G. A. (2023). Interactions between clay minerals, bacteria growth and urease activity on biocementation of soils. *Applied Clay Science*, *240*, 106972. DOI: 10.1016/j.clay.2023.106972

Catherall, E., Musial, S., Atkinson, N., Walker, C. E., Mackinder, L. C., & McCormick, A. J. (2025). From algae to plants: Understanding pyrenoid-based CO2-concentrating mechanisms. *Trends in Biochemical Sciences*, *50*(1), 33–45. DOI: 10.1016/j.tibs.2024.10.010 PMID: 39592300

Chandra, S., Bhattacharya, S., Pandit, S., Banerjee, S., Roy, A., & Rai, A. K. (2023). Cyanobacteria as an eco-friendly bioresource for EPS production with crack healing capacity in concrete. *Biocatalysis and Agricultural Biotechnology*, *54*, 102909. DOI: 10.1016/j.bcab.2023.102909

Chayaamor-Heil, N., Perricone, V., Gruber, P., & Guéna, F. (2023). Bioinspired, biobased and living material designs: A review of recent research in architecture and construction. *Bioinspiration & Biomimetics*, *18*(4), 041001. DOI: 10.1088/1748-3190/acd82e

Cong, P., Du, R., Gao, H., & Chen, Z. (2024). Comparison and assessment of carbon dioxide emissions between alkali-activated materials and OPC cement concrete. [English Edition]. *Journal of Traffic and Transportation Engineering*, *11*(5), 918–938. DOI: 10.1016/j.jtte.2023.07.011

Elgendy, I. M., Elkaliny, N. E., Saleh, H. M., Darwish, G. O., Almostafa, M. M., Metwally, K., Yahya, G., & Mahmoud, Y. A. G. (2025). Bacteria-powered self-healing concrete: Breakthroughs, challenges, and future prospects. *Journal of Industrial Microbiology & Biotechnology*, *52*, kuae051. DOI: 10.1093/jimb/kuae051 PMID: 39673695

Espie, G. S., & Kimber, M. S. (2011). Carboxysomes: Cyanobacterial RubisCO comes in small packages. *Photosynthesis Research*, *109*(1), 7–20. DOI: 10.1007/s11120-011-9656-y PMID: 21556873

Espitia-Nery, M. E., Corredor-Pulido, D. E., Castaño-Oliveros, P. A., Rodríguez-Medina, J. A., Ordoñez-Bello, Q. Y., & Pérez-Fuentes, M. S. (2019). Mechanisms of encapsulation of bacteria in self-healing concrete. *Dyna*, *86*(210), 17–22. DOI: 10.15446/dyna.v86n210.75343

Fu, X., Ye, W., Wang, C., Zhu, P., & Tian, J. (2025). A Review on the Improvement Technology of Plant-Based Bio-Cement (PBBC). *Geomicrobiology Journal*, *42*(8), 1–15. DOI: 10.1080/01490451.2025.2516468

Gayathri, R., Mahboob, S., Govindarajan, M., Al-Ghanim, K. A., Ahmed, Z., Al-Mulhm, N., Vodovnik, M., & Vijayalakshmi, S. (2021). A review on biological carbon sequestration: A sustainable solution for a cleaner air environment, less pollution and lower health risks. *Journal of King Saud University. Science*, *33*(2), 101282. DOI: 10.1016/j.jksus.2020.101282

Gebru, K. A., Kidanemariam, T. G., & Gebretinsae, H. K. (2021). Bio-cement production using microbially induced calcite precipitation (MICP) method: A review. *Chemical Engineering Science*, *238*, 116610. DOI: 10.1016/j.ces.2021.116610

Golewski, G. L. (2023). The phenomenon of cracking in cement concretes and reinforced concrete structures: The mechanism of cracks formation, causes of their initiation, types and places of occurrence, and methods of detection—a review. *Buildings (Basel, Switzerland)*, *13*(3), 765. DOI: 10.3390/buildings13030765

Gurau, S., Imran, M., & Ray, R. L. (2025). Algae: A cutting-edge solution for enhancing soil health and accelerating carbon sequestration–A review. *Environmental Technology & Innovation*, *37*, 103980. DOI: 10.1016/j.eti.2024.103980

Hazarika, A., Hazarika, I., Gogoi, M., Bora, S. S., Borah, R. R., Goutam, P. J., & Saikia, N. (2018). Use of a plant based polymeric material as a low cost chemical admixture in cement mortar and concrete preparations. *Journal of Building Engineering*, *15*, 194–202. DOI: 10.1016/j.jobe.2017.11.017

He, L. N., Rogers, R. D., Su, D., Tundo, P., & Zhang, Z. C. (2016). *Green chemistry and sustainable technology* (Doctoral dissertation, Dalian Institute of Chemical Physics, Chinese Academy of Sciences).

He, T., Liu, Z., Son, H., Gundersen, T., & Lin, W. (2023). Comparative analysis of cryogenic distillation and chemical absorption for carbon capture in integrated natural gas liquefaction processes. *Journal of Cleaner Production*, *383*, 135264. DOI: 10.1016/j.jclepro.2022.135264

Heidarnezhad, S., Wang, S., & Rahbar, N. (2024). Carbonic Anhydrase as a Sustainable Corrosion Inhibitor for Concrete. *ACS Applied Engineering Materials*, *3*(1), 128–141. DOI: 10.1021/acsaenm.4c00643

Hu, L., Li, Z., Kong, L., Wei, J., Chang, J., & Shi, W. (2024). Reassessing the greenhouse effect of biogenic carbon emissions in constructed wetlands. *Journal of Environmental Management*, *354*, 120263. DOI: 10.1016/j.jenvman.2024.120263 PMID: 38387360

Ighalo, J. O., Dulta, K., Kurniawan, S. B., Omoarukhe, F. O., Ewuzie, U., Eshiemogie S. O., Ojo, A. U., & Abdullah, S. R. S. (2022). Progress in microalgae application for CO2 sequestration. *Cleaner Chemical Engineering*, *3*, 100044. DOI: 10.1016/j.clce.2022.100044

Iqbal, D. M., Wong, L. S., & Kong, S. Y. (2021). Bio-cementation in construction materials: A review. *Materials (Basel)*, *14*(9), 2175. DOI: 10.3390/ma14092175 PMID: 33922871

Ivaškė, A., Gribniak, V., Jakubovskis, R., & Urbonavičius, J. (2023). Bacterial viability in self-healing concrete: A case study of non-ureolytic Bacillus species. *Microorganisms*, *11*(10), 2402. DOI: 10.3390/microorganisms11102402 PMID: 37894059

Jaramillo, H. Y., Vasco-Echeverri, O. H., Moreno-Pacheco, L. A., & García-León, R. A. (2023). Biomaterials in concrete for engineering applications: A bibliometric review. *Infrastructures*, *8*(11), 161. DOI: 10.3390/infrastructures8110161

Jiang, Q. (2022). *Synthetic engineering of biological carbon-fixing organelles shells*. The University of Liverpool.

Kamennaya, N. A., Ajo-Franklin, C. M., Northen, T., & Jansson, C. (2012). Cyanobacteria as biocatalysts for carbonate mineralization. *Minerals (Basel), 2*(4), 338–364. DOI: 10.3390/min2040338

Kandasamy, S., Shah, M. P., Subbiah, K., & Manickam, N. (Eds.). (2025). *Microbial Niche Nexus Sustaining Environmental Biological Wastewater and Water-Energy-Environment Nexus*. Springer Nature. DOI: 10.1007/978-3-031-62660-9

Karimi, K., & Ghaemi, A. (2025). A comprehensive review of the physicochemical properties and performance of novel carbon-based adsorbents for CO2 capture. *Environmental Science and Pollution Research International, 32*(34), 20248–20287. DOI: 10.1007/s11356-025-36803-8 PMID: 40836200

Krajewska, B. (2018). Urease-aided calcium carbonate mineralization for engineering applications: A review. *Journal of Advanced Research, 13*, 59–67. DOI: 10.1016/j.jare.2017.10.009 PMID: 30094083

Kumar, K., Dasgupta, C. N., Nayak, B., Lindblad, P., & Das, D. (2011). Development of suitable photobioreactors for CO2 sequestration addressing global warming using green algae and cyanobacteria. *Bioresource Technology, 102*(8), 4945–4953. DOI: 10.1016/j.biortech.2011.01.054 PMID: 21334885

Kumar, N., Kar, S., & Shukla, P. (2022). Role of regulatory pathways and multi-omics approaches for carbon capture and mitigation in cyanobacteria. *Bioresource Technology, 366*, 128104. DOI: 10.1016/j.biortech.2022.128104 PMID: 36257524

Kupriyanova, E. V., Pronina, N. A., & Los, D. A. (2023). Adapting from low to high: An update to CO2-concentrating mechanisms of cyanobacteria and microalgae. *Plants, 12*(7), 1569. DOI: 10.3390/plants12071569 PMID: 37050194

La, Y. H., Lee, K. S., Kim, T. W., & Song, J. Y. (2022). A bacteria-based carbon sequestration and waste recycling system. *Scientific Reports, 12*(1), 10236. DOI: 10.1038/s41598-022-14239-1 PMID: 35764782

Lee, Y. S., & Park, W. (2018). Current challenges and future directions for bacterial self-healing concrete. *Applied Microbiology and Biotechnology, 102*(7), 3059–3070. DOI: 10.1007/s00253-018-8830-y PMID: 29487987

Li, C., Pradhan, P., Chen, G., Kropp, J. P., & Schellnhuber, H. J. (2025). Carbon footprint of the construction sector is projected to double by 2050 globally. *Communications Earth & Environment, 6*(1), 831. DOI: 10.1038/s43247-025-02840-x PMID: 41163641

Li, X., He, X., Ren, K., Dong, H., & Lian, B. (2023). Mechanisms of carbonate precipitation induced by two model bacteria. *Chemical Geology*, *628*, 121461. DOI: 10.1016/j.chemgeo.2023.121461

Li, Z., Liu, A., Sun, C., Li, H., Kong, Z., & Zhai, H. (2024). Biomineralization process of CaCO3 precipitation induced by Bacillus mucilaginous and its potential application in microbial self-healing concrete. *Applied Biochemistry and Biotechnology*, *196*(4), 1896–1920. DOI: 10.1007/s12010-023-04634-3 PMID: 37440115

Llamas, B., Suarez-Rodriguez, M. C., Gonzalez-Lopez, C. V., Mora, P., & Acien, F. G. (2021). Techno-economic analysis of microalgae related processes for CO2 bio-fixation. *Algal Research*, *57*, 102339. DOI: 10.1016/j.algal.2021.102339

Luo, M., Qian, C. X., & Li, R. Y. (2015). Factors affecting crack repairing capacity of bacteria-based self-healing concrete. *Construction & Building Materials*, *87*, 1–7. DOI: 10.1016/j.conbuildmat.2015.03.117

Madhumithra, E., Elakkiya, K., & Sumathi, A. (2025). Enhancing Self-Healing Concrete Using Sodium Alginate Beads and Citrullus lanatus Seed Powder for Bacterial Encapsulation. *Arabian Journal for Science and Engineering*, •••, 1–21.

Mokhtar, N., Johari, M. A. M., Tajarudin, H. A., Al-Gheethi, A. A., & Algaifi, H. A. (2021). A sustainable enhancement of bio-cement using immobilised Bacillus sphaericus: Optimization, microstructural properties, and techno-economic analysis for a cleaner production of bio-cementitious mortars. *Journal of Cleaner Production*, *318*, 128470. DOI: 10.1016/j.jclepro.2021.128470

Morrow, R., Bridgens, B., & Mackenzie, L. (Eds.). (2023). *Bioprotopia: Designing the built environment with living organisms*. De Gruyter. DOI: 10.1515/9783035625806

Mutitu, K. D., Munyao, M. O., Wachira, M. J., Mwirichia, R., Thiong'o, K. J., & Marangu, M. J. (2019). Effects of biocementation on some properties of cement-based materials incorporating Bacillus species bacteria–a review. *Journal of Sustainable Cement-Based Materials*, *8*(5), 309–325. DOI: 10.1080/21650373.2019.1640141

Mwandira, W., Mavroulidou, M., Gunn, M. J., Purchase, D., Garelick, H., & Garelick, J. (2023). Concurrent carbon capture and biocementation through the carbonic anhydrase (CA) activity of microorganisms-a review and outlook. *Environmental Processes*, *10*(4), 56. DOI: 10.1007/s40710-023-00667-2

Nanda, S., Reddy, S. N., Mitra, S. K., & Kozinski, J. A. (2016). The progressive routes for carbon capture and sequestration. *Energy Science & Engineering*, *4*(2), 99–122. DOI: 10.1002/ese3.117

Nayeem, M. A., Saduwale, S., Vempada, S. R., & Rao, M. S. (2025, March). Bio-inspired Self-Healing Concrete using Algae-Based Healing Agents for Crack Repair. In *International Conference on Advanced Materials, Manufacturing and Sustainable Development (ICAMMSD 2024)* (pp. 1002-1016). Atlantis Press. DOI: 10.2991/978-94-6463-662-8_78

Nigam, H., Jain, R., Malik, A., & Singh, V. (2022). Comparative Life-Cycle assessment of microalgal biomass production in conventional growth media versus newly developed nanoemulsion media. *Bioresource Technology*, *352*, 127069. DOI: 10.1016/j.biortech.2022.127069 PMID: 35367603

Nur, M. M. A., & Dewi, R. N. (2024). Opportunities and challenges of microalgae in biocement production and self-repair mechanisms. *Biocatalysis and Agricultural Biotechnology*, *56*, 103048. DOI: 10.1016/j.bcab.2024.103048

Nwali, O. I., Oladunjoye, M. A., & Alao, O. A. (2024). A review of atmospheric carbon dioxide sequestration pathways; processes and current status in Nigeria. *Carbon Capture Science & Technology*, *12*, 100208. DOI: 10.1016/j.ccst.2024.100208

Onuaguluchi, O., & Banthia, N. (2016). Plant-based natural fibre reinforced cement composites: A review. *Cement and Concrete Composites*, *68*, 96–108. DOI: 10.1016/j.cemconcomp.2016.02.014

Orr, F. M.Jr. (2018). Carbon capture, utilization, and storage: An update. *SPE Journal*, *23*(06), 2444–2455. DOI: 10.2118/194190-PA

Osman, A. I., Fawzy, S., Farghali, M., El-Azazy, M., Elgarahy, A. M., Fahim, R. A., Maksoud, M. I. A. A., Ajlan, A. A., Yousry, M., Saleem, Y., & Rooney, D. W. (2022). Biochar for agronomy, animal farming, anaerobic digestion, composting, water treatment, soil remediation, construction, energy storage, and carbon sequestration: A review. *Environmental Chemistry Letters*, *20*(4), 2385–2485. DOI: 10.1007/s10311-022-01424-x PMID: 35571983

Patil, D., Mishra, A., Tyagi, G., Reddy, P. M., Lahoti, M., Pable, A., Jadhav, U. U., & Sachs, S. G. (2025). Comparative evaluation of bacterial strains and novel SCM-based biocarriers for self-healing and strength enhancement in mortar. *Clean Technologies and Environmental Policy*, *27*(12), 7611–7631. DOI: 10.1007/s10098-025-03312-8

Pawar, S. S., & Parekar, S. R. (2018). Bacteria based self-healing concrete. *International Research Journal of Engineering and Technology*, *5*(3), 1001–1004.

Pérez, O. F. A., Arrieta, V. S., Ospina, J. H. G., Herrera, S. H., Rojas, C. F. R., & Navarro, A. M. S. (2024). Carbon dioxide emissions from traditional and modified concrete. A review. *Environmental Development*, *52*, 101036. DOI: 10.1016/j.envdev.2024.101036

Phillips, A. J., Gerlach, R., Lauchnor, E., Mitchell, A. C., Cunningham, A. B., & Spangler, L. (2013). Engineered applications of ureolytic biomineralization: A review. *Biofouling*, *29*(6), 715–733. DOI: 10.1080/08927014.2013.796550 PMID: 23802871

Price, G. D., Badger, M. R., Woodger, F. J., & Long, B. M. (2008). Advances in understanding the cyanobacterial CO2-concentrating-mechanism (CCM): Functional components, Ci transporters, diversity, genetic regulation and prospects for engineering into plants. *Journal of Experimental Botany*, *59*(7), 1441–1461. DOI: 10.1093/jxb/erm112 PMID: 17578868

Qian, C., Zheng, T., Zhang, X., & Su, Y. (2021). Application of microbial self-healing concrete: Case study. *Construction & Building Materials*, *290*, 123226. DOI: 10.1016/j.conbuildmat.2021.123226

Rajadesingu, S., Palani, N., Mendonce, K. C., Vijayakumar, P., Monisha, P., & Ayyadurai, S. (2024). State-of-the-art review on advancements of eco-friendly bacterial-infused self-healing concrete for sustainable constructions. *Journal of Building Engineering*, *91*, 109669. DOI: 10.1016/j.jobe.2024.109669

Rajesh, A., Hariny, V. S., & Sumathi, A. (2024). Performance of Bacillus tropicus on Mechanical, Durable and Crack Remediation Properties in Sustainable Vermiculite Concrete. *Journal of Bionic Engineering*, *21*(4), 1987–1999. DOI: 10.1007/s42235-024-00546-y

Rajesh, A., & Sumathi, A. (2024). Improvement on strength, durability, and crack closure behavior of bacteria concrete under marine soil exposures. *Journal of Testing and Evaluation*, *52*(2), 1269–1279. DOI: 10.1520/JTE20230403

Rajesh, A., & Sumathi, A. (2024, January). Strength and self-healing behavior of bacteria biocomposite concrete in soil exposure condition. [). Elsevier.]. *Structures (Oxford)*, *59*, 105673. DOI: 10.1016/j.istruc.2023.105673

Ramadan, R., Ghanem, H., Khatib, J. M., & ElKordi, A. M. (2025). Effect of Plant-based natural fibers on the mechanical properties and volume change of cement paste. *International Journal of Building Pathology and Adaptation*, *43*(4), 693–711. DOI: 10.1108/IJBPA-11-2023-0166

Raven, J. A., Giordano, M., Beardall, J., & Maberly, S. C. (2012). Algal evolution in relation to atmospheric CO2: Carboxylases, carbon-concentrating mechanisms and carbon oxidation cycles. *Philosophical Transactions of the Royal Society of London. Series B, Biological Sciences*, *367*(1588), 493–507. DOI: 10.1098/rstb.2011.0212 PMID: 22232762

Rusznyák, A., Akob, D. M., Nietzsche, S., Eusterhues, K., Totsche, K. U., Neu, T. R., Frosch, T., Popp, J., Keiner, R., Geletneky, J., Katzschmann, L., Schulze, E.-D., & Küsel, K. (2012). Calcite biomineralization by bacterial isolates from the recently discovered pristine karstic Herrenberg cave. *Applied and Environmental Microbiology*, *78*(4), 1157–1167. DOI: 10.1128/AEM.06568-11 PMID: 22179248

Saccardo, A., Bezzo, F., & Sforza, E. (2022). Microalgae growth in ultra-thin steady-state continuous photobioreactors: Assessing self-shading effects. *Frontiers in Bioengineering and Biotechnology*, *10*, 977429. DOI: 10.3389/fbioe.2022.977429 PMID: 36032730

Salar, R. K., Gahlawat, S. K., Siwach, P., & Duhan, J. S. (Eds.). (2013). *Biotechnology: prospects and applications* (p. 315). Springer. DOI: 10.1007/978-81-322-1683-4

Saxena, A., Prakash Gupta, J., Tiwary, J. K., Kumar, A., Sharma, S., Pandey, G., Biswas, S., & Raghav Chaturvedi, K. (2024). Innovative pathways in carbon capture: Advancements and strategic approaches for effective carbon capture, utilization, and storage. *Sustainability (Basel)*, *16*(22), 10132. DOI: 10.3390/su162210132

Shaaban, S., Hammad, N., Elnemr, A., & Shaaban, I. G. (2023, June). Efficiency of bacteria-based self-healing mechanism in concrete. [). Trans Tech Publications Ltd.]. *Materials Science Forum*, *1089*, 135–143. DOI: 10.4028/p-tc6w54

Shen, M., Kong, F., Tong, L., Luo, Y., Yin, S., Liu, C., Zhang, P., Wang, L., Chu, P. K., & Ding, Y. (2022). Carbon capture and storage (CCS): Development path based on carbon neutrality and economic policy. *Carbon Neutrality*, *1*(1), 37. DOI: 10.1007/s43979-022-00039-z

Sidhu, N., Goyal, S., & Reddy, M. S. (2022). Biomineralization of cyanobacteria Synechocystis pevalekii improves the durability properties of cement mortar. *AMB Express*, *12*(1), 59. DOI: 10.1186/s13568-022-01403-z PMID: 35587839

Singh, A. K., Santos-Merino, M., Sakkos, J. K., Walker, B. J., & Ducat, D. C. (2022). Rubisco regulation in response to altered carbon status in the cyanobacterium Synechococcus elongatus PCC 7942. *Plant Physiology*, *189*(2), 874–888. DOI: 10.1093/plphys/kiac065 PMID: 35201348

Singh, S. K., Badgujar, G., Reddy, V. R., Fleisher, D. H., & Bunce, J. A. (2013). Carbon dioxide diffusion across stomata and mesophyll and photo-biochemical processes as affected by growth CO2 and phosphorus nutrition in cotton. *Journal of Plant Physiology*, *170*(9), 801–813. DOI: 10.1016/j.jplph.2013.01.001 PMID: 23384758

Singh, U. B., & Ahluwalia, A. S. (2013). Microalgae: A promising tool for carbon sequestration. *Mitigation and Adaptation Strategies for Global Change*, *18*(1), 73–95. DOI: 10.1007/s11027-012-9393-3

Smitha, M. P., Suji, D., Shanthi, M., & Adesina, A. (2022). Application of bacterial biomass in biocementation process to enhance the mechanical and durability properties of concrete. *Cleaner Materials*, *3*, 100050. DOI: 10.1016/j.clema.2022.100050

Soda, P. R. K., Mogal, A., Chakravarthy, K., Thota, N., Bandaru, N., Shukla, S. K., & Mini, K. M. (2024). Performance assessment of sustainable biocement mortar incorporated with bacteria-encapsulated cement-coated alginate beads. *Construction & Building Materials*, *411*, 134198. DOI: 10.1016/j.conbuildmat.2023.134198

Srimongkol, P., Sangtanoo, P., Songserm, P., Watsuntorn, W., & Karnchanatat, A. (2022). Microalgae-based wastewater treatment for developing economic and environmental sustainability: Current status and future prospects. *Frontiers in Bioengineering and Biotechnology*, *10*, 904046. DOI: 10.3389/fbioe.2022.904046 PMID: 36159694

Srinivas, M. K., Alengaram, U. J., Ibrahim, S., Vello, V., & Phang, S. M. (2024). Feasibility study on the use of microalgae as an external crack healing agent for cement mortar rehabilitation. *Journal of Sustainable Cement-Based Materials*, *13*(1), 17–32. DOI: 10.1080/21650373.2023.2249889

Stadnichuk, I. N., Krasilnikov, P. M., & Zlenko, D. V. (2015). Cyanobacterial phycobilisomes and phycobiliproteins. *Microbiology*, *84*(2), 101–111. DOI: 10.1134/S0026261715020150 PMID: 26263619

Storrs, K., Lyhne, I., & Drustrup, R. (2023). A comprehensive framework for feasibility of CCUS deployment: A meta-review of literature on factors impacting CCUS deployment. *International Journal of Greenhouse Gas Control*, *125*, 103878. DOI: 10.1016/j.ijggc.2023.103878

Thoré, E. S., Muylaert, K., Bertram, M. G., & Brodin, T. (2023). Microalgae. *Current Biology : CB*, *33*(3), R91–R95. DOI: 10.1016/j.cub.2022.12.032 PMID: 36750029

Trickett, C. A., Helal, A., Al-Maythalony, B. A., Yamani, Z. H., Cordova, K. E., & Yaghi, O. M. (2017). The chemistry of metal–organic frameworks for CO2 capture, regeneration and conversion. *Nature Reviews. Materials*, *2*(8), 1–16. DOI: 10.1038/natrevmats.2017.45

Wong, P. Y., Mal, J., Sandak, A., Luo, L., Jian, J., & Pradhan, N. (2024). Advances in microbial self-healing concrete: A critical review of mechanisms, developments, and future directions. *The Science of the Total Environment*, *947*, 174553. DOI: 10.1016/j.scitotenv.2024.174553 PMID: 38972424

Xiong, W., Cano, M., Wang, B., Douchi, D., & Yu, J. (2017). The plasticity of cyanobacterial carbon metabolism. *Current Opinion in Chemical Biology*, *41*, 12–19. DOI: 10.1016/j.cbpa.2017.09.004 PMID: 28968542

Xue, Q., Feng, X., Sun, Y., Si, Z., Zhang, L., Zhou, Z., Zhong, Y., Mei, K., Lin, G., & Zhao, H. (2025). A systematic review of concrete self-healing mechanism and analysis of new research directions. *Journal of Environmental Chemical Engineering*, *13*(6), 119009. DOI: 10.1016/j.jece.2025.119009

Yaashikaa, P. R., Saravanan, A., Kumar, P. S., Thamarai, P., & Rangasamy, G. (2024). Role of microbial carbon capture cells in carbon sequestration and energy generation during wastewater treatment: A sustainable solution for cleaner environment. *International Journal of Hydrogen Energy*, *52*, 799–820. DOI: 10.1016/j.ijhydene.2023.05.307

Yamaoka, Y., Petroutsos, D., Je, S., Yamano, T., & Li-Beisson, Y. (2025). Light, CO2, and carbon storage in microalgae. *Current Opinion in Plant Biology*, *84*, 102696. DOI: 10.1016/j.pbi.2025.102696 PMID: 39983365

Yan, Z., Nakashima, K., Takano, C., & Kawasaki, S. (2025). Strategies for cost-optimized biocement production: A comprehensive review. *World Journal of Microbiology & Biotechnology*, *41*(2), 67. DOI: 10.1007/s11274-025-04281-2 PMID: 39920409

Zhang, G. Z., Liu, Z. H., Zhang, J., Zhao, H., Xia, Y., Su, Y., Wang, H., & Wang, L. (2025). Algae-bacteria symbiosis for cracks active repair in cement mortar: A novel strategy to enhance microbial self-healing efficacy. *Construction & Building Materials*, *483*, 141815. DOI: 10.1016/j.conbuildmat.2025.141815

Zong, M., Wang, W., Ma, H., Cedrick, N., Sun, Y., Yan, X., Liu, H., Zhu, P., & Hua, M. (2025). Progress in Self-Repair Technology for Concrete Cracks via Biomineralization. *Materials (Basel)*, *18*(21), 5004. DOI: 10.3390/ma18215004 PMID: 41227962

KEY TERMS AND DEFINITIONS

Carbonic anhydrase-driven precipitation: This is one of the pathways for the precipitation of calcium carbonate using the carbonic anhydrase enzyme, which catalyses the interconversion of carbon dioxide and bicarbonate ions.

Carboxysomes: These are polyhedral organelles found in cyanobacteria that house major enzymes like carbonic anhydrase and RuBisCo, which play a major role in carbon capture and utilisation.

CCM: Carbon concentration mechanism – a mechanism in which the micro-organisms themselves capture the atmospheric carbon and concentrate around the specialised organelles like RuBisCo, which help in the breakdown of the obtained CO_2 into carbon sources and utilise them for their metabolism.

CCUS: Carbon capture, utilisation, and storage, which means that the excess carbon can be captured and can be utilised for various functional properties and stored as biomass, which can then be converted

MICP: Microbially induced Carbonate precipitation – Ability of organisms to precipitate carbonates / perform mineralisation with their own biological process

Pyrenoids: These are specialised structures present in microalgae and hornworts, which help in the conversion of CO_2 into value-added products.

Self-healing concrete: The ability of microorganisms to heal the cracks formed in the concrete structures by the mineralisation process (MICP)

Urease-driven precipitation: This pathway uses the urease enzyme for the bio-mineralisation process, in which urea is hydrolysed into ammonia and carbonate ions.

Chapter 14
Carbon Credits and Policy Perspectives for Algal CO2 Sequestration

Md. Mirazul Islam
http://orcid.org/0000-0002-4393-1396
Department of Chemical Engineering and Polymer Science, Shahjalal University of Science and Technology, Bangladesh

ABSTRACT

With a rapid accumulation of atmospheric CO_2, which has grown 47% throughout the industrial revolution and currently stands at 409.8 ppm in 2019, global warming is intensifying and needs innovative solutions. This chapter reviews algal-based sequestration of CO_2 as a sustainable method of carbon capture, utilization, and storage. The photosynthetic efficiency of microalgae ranges at 40-93.7% and fix up to 20% of the CO_2. Massive production on 13 million acres could capture 50 million tons of CO_2 and produce more than 3 million tons of biomass. Recent growth technologies comprise photobioreactors, hybrid systems, artificial carbon concentration systems, and genetic and metabolic engineering to enhance the rate of fixation. Carbon credit regimes, technological-economic analyses, and policy frameworks for enabling quantifiable and verifiable algal systems in carbon markets are also studied. Overall, the innovation of algal carbon sequestration blocks scientific, economic, and environmental policy progress to global net-zero carbon emissions and a circular bioeconomy

DOI: 10.4018/979-8-3373-7630-1.ch014

1. INTRODUCTION

The growing amount of atmospheric carbon dioxide (CO_2), which is mainly caused by humans' activities like the burning of fossil fuels, is a critical environmental issue with far-reaching consequences (Iqbal et al., 2022; Jaiswal et al., 2020).. Global CO_2 emissions have grown by about 47% since the industrial revolution to an average of 409.8 ppm in 2019, with a forecast that the global temperature could rise by an average of 1.50°C by 2030-2050 as the pattern of emission is projected to continue (Iqbal et al., 2022). High levels of greenhouse gas (GHG) are a significant cause of climatic change effects, such as global warming, acidification of oceans, and changes in weather patterns, thus causing significant risks to the stability of ecosystems, biodiversity, and human health (Jaiswal et al., 2020). These issues highlight the importance of having effective mitigation measures that are based on scientific, technological, and policy-based measures.

Carbon credit schemes, more generally market-based mechanisms, have been widely adopted to provide an incentive to reduce emissions by issuing tradable credits to the entities below specified emission limits (Cheong, 2025; Singh et al., 2014). The voluntary carbon offset market also enables the implementation of the cost-effective mitigation strategies (Fujii et al., 2024). Although the use of carbon credits has always been linked with renewable sources of energy as well as land-use projects, the use of carbon credits is undergoing a transformation with the introduction of newer technologies (Fujii et al., 2024).

One of these is the algal-based CO_2 sequestration that has been receiving interest as a biological alternative to the traditional method of carbon capture (Ighalo et al., 2022; X. Sun et al., 2023). Microalgae and macroalgae are able to effectively convert CO_2 into biomass and can be grown under a variety of conditions, such as wastewater and saline conditions (Hasnain et al., 2023; Tarafdar et al., 2023). Moreover, large-scale estimates indicate that cultivation in an area of about 13 million hectares would be able to sequester around 50 million tons of CO_2/year and produce more than 3 million tons of biomass, which underscores their potential to be scaled up (J. Ma et al., 2025). Algal systems have potential benefits compared to the conventional carbon capture and storage (CCS) in terms of sustainability and compatibility with resource recovery pathways (Dahai et al., 2024; X. Sun et al., 2023).

The algal systems are also effective as they can adapt to various environmental conditions such as wastewater streams and saline environments, thus not requiring fresh water resources and arable land (Hasnain et al., 2023; Tarafdar et al., 2023). High CO_2 uptake abilities have been shown in a number of microalgal species such as Chlorella sp., Dunaliella tertiolecta, Spirulina platensis, Desmodesmus sp., and Nannochloropsis sp. (Tarafdar et al., 2023). To a great extent, this is due to effective photosynthetic pathways, especially CO_2 reduction by the RuBisCO-mediated process that facilitates conversion of CO_2 to organic biomass (Jaiswal et al., 2020; Souza et al., 2024). In microalgal systems, CO_2 sequestration efficiencies are reported as 40-93.7% (Ighalo et al., 2022). Furthermore, the possibility to combine it with wastewater treatment systems can provide additional environmental advantages, as the treatment will eliminate nutrients and clean water at the same time (Hasnain et al., 2023; Iqbal et al., 2022).

In addition to carbon capture, algal CO_2 sequestration can be incorporated into larger carbon management systems, such as carbon capture, utilization, and storage (CCUS), and also negative emission systems, such as bioenergy with carbon capture and storage (BECCS) and direct air capture (DAC) systems (Z. Ma et al., 2022). Algal biomass also has the potential to be converted into various value-added products, such as biofuels, biochemicals, and biomaterials, therefore facilitating the implementation of circular bioeconomy models (Ashour et al., 2024; Dasan et al., 2025; J. Ma et al., 2025). This capability to be multipurpose makes algal systems more economically viable by converting CO_2 a waste product, into a useful resource. In example, Chlorella-based systems are under development in terms of capturing CO_2 and producing biohydrogen simultaneously, and this is evidence of being integrated with environmental and energy (Ashour et al., 2024).

Along with these encouraging trends, there are several barriers to large-scale implementation and commercialization of algal-based CO_2 sequestration technologies (K. Kumar et al., 2016). The available studies have been largely on the biological performance and cultivation optimization, although less emphasis has been on the economic feasibility, LCA, and integration of policies (Cheong, 2025; Fujii et al., 2024; Quiroz et al., 2023). The main issues are the limitation to scaling, high cost of operation, and lack of standardized methods of counting carbon (Hanson et al., 2025). Moreover, monitoring, reporting, and verification (MRV) systems that are unique to algal-based systems are yet to be developed, restricting their participation in the voluntary and compliance carbon markets (Chua & Palevsky, 2026). These loopholes point to a lack of tie-in between the potential of technology and its application in carbon credit systems.

In this regard, this chapter offers a holistic and combined evaluation of the carbon credit systems as compared to the algal-based CO_2 sequestration. It assesses the efficiency, scalability, and economic viability of the algal carbon capture technology, focusing on physiological processes, selection of species, and cultivation technologies. The chapter also discusses the approaches towards optimization of processes such as strain improvement, metabolic engineering, and coupling with industrial emission and wastewater treatment systems. Further, it discusses how the algal systems can be utilized within the carbon capture, utilization, and storage (CCUS) models and how they align with new negative emission technologies. Special focus is placed on the implementation of algae-based solutions into the carbon credit markets, such as policy framework, carbon accounting procedures, and verification issues. Overall, the goal of this chapter is to connect the advances in technology and market forces to facilitate scalable, cost-effective, and sustainable ways of mitigating CO_2.

2. FUNDAMENTALS OF ALGAL CO_2 SEQUESTRATION

2.1. Biological mechanisms of algal carbon fixation

Photosynthetic processes in algae and their efficiency: Photosynthesis in algae is the key process in the fixation of carbon, as it allows them to be known as 'carbon wealth generators' when converting CO_2 to generate valuable biomass (Y. Xie et al., 2022). This reaction is the key to biological carbon fixation, which includes the Calvin-Benson-Bassham (CBB) cycle, in which the first step of carboxylation of ribulose-1,5-bisphosphate (RuBP) is performed by the enzyme ribulose-1,5-bisphosphate carboxylase/oxygenase (RuBisCO). Photosynthetic efficiencies of algae are significant, with microalgae having photosynthetic efficiencies of between 6 to 8%, which is usually higher than terrestrial plants (David et al., 2025). This efficiency is partly explained by the fact that they are aquatic and are constantly hydrated and fed (Zeng et al., 2021). In ideal circumstances, microalgae are capable of fixing CO_2 at much higher rates than plants on land, and some strains can tolerate and effectively use high CO_2 levels up to 15-20% (David et al., 2025). Light strictly controls the activity of the CBB cycle via ionic signals and the ferredoxin-thioredoxin redox system so that carbon fixation happens at the moment of energy availability.

The biomass that is produced by these processes has direct economic value. Algal biomass has the potential to be converted into biofuels, bioplastics, nutraceuticals and other bioproducts, thus effectively turning a liability of a greenhouse gas into a resource (Farooqui et al., 2021; Mamilla et al., 2024). Combined with wastewater treatment, algal cultivation also has a social benefit, simultaneously removing nutrients and purifying water along with capturing CO_2, therefore having a triple-bottom-line impact encompassing environmental, economic and social outcomes (H. Li et al., 2025; Srimongkol et al., 2022). In contrast to the traditional carbon capture approaches, algal sequestration fits the models of Negative Emission Technologies (NETs) and is based on the principles of circular bioeconomy and scalable carbon removal and value generation (Guan et al., 2026; W. Xie et al., 2025).

Carbon concentration mechanisms in different algal species: Most species of algae have carbon-concentrating mechanisms (CCMs), which actively translocate and store inorganic carbon (CO_2 or bicarbonate, HCO_3^-) around RuBisCO, which improves the efficiency of photosynthesis and reduces photorespiration (An et al., 2023; R. Li et al., 2025). Photorespiration is an inefficient mechanism in which

RuBisCO binds O_2 rather than CO_2, which lowers photosynthetic production. CCMs in eukaryotic algae can also entail an organelle, the pyrenoid, a suborganelle located in the chloroplasts and accumulating RuBisCO (An et al., 2023; Fei et al., 2022). The pyrenoid-based CCM allows increasing the concentration of CO_2 locally and drives RuBisCO to carboxylase activity (Fei et al., 2022). As an example, it has been demonstrated that the assembling of metal-organic frameworks on *Chlorella pyrenoidosa* can greatly improve the photosynthetic carbon fixation through a higher CO_2 concentration and transport to RuBisCO (D. Li et al., 2023). Moreover, there are also microalgae, such as *Chlamydomonas reinhardtii*, which use a dynamic pyrenoid structure that is created by liquid-liquid phase separation of RuBisCO to fix CO_2 effectively (An et al., 2023). This enables the algae to be highly fixed on carbon even when the ambient CO_2 concentration is low (R. Li et al., 2025).

Factors affecting CO_2 uptake efficiency in algal systems: The nature of the environment and the operation conditions has a significant influence on the efficiency of CO_2 uptake in the algal systems. These relies on the concentration and flow rate of the feed gas, as well as the temperature, pH, light intensity, and humidity (A. Kumar et al., 2025). Light conditions are essential since light supply directly affects the rates of photosynthesis. A high intensity of light may stimulate CO_2 fixation, but it may also result in photoinhibition and reduce efficiency (Zeng et al., 2021) Enzymatic activity is influenced by temperature, and each species of algae possesses an optimal range of temperature in which it grows and absorbs CO_2 (Do et al., 2021) pH influences the presence of inorganic carbon (CO_2, HCO_3^-, CO_3^{2-}) in the culture to be taken up by algae (A. Kumar et al., 2025; Kusi et al., 2024). Chlorella vulgaris can proliferate and retain high carbon fixation capacity within closed systems that have the total range of CO_2 of 0.12-0.49 mol CO_2/L absorbent (Yan et al., 2024). The development of algae and its capacity to sequester CO_2 depends on the availability of nutrients such as nitrogen and phosphorus (Prasad et al., 2021). The mass transfer of CO_2 is a limiting critical concern, particularly in large-scale systems, because of poor solubility in water (Kanna Dasan et al., 2023). Enhancing the mass transfer via aeration as well as photobioreactor design is of great importance in increasing the uptake of CO_2 (Z. Sun et al., 2025).

2.2. Types of algae used for CO_2 sequestration

Microalgae and macroalgae are both involved in carbon sequestration, but they differ in nature and use. Microalgae are single-celled, microscopic organisms that have a high photosynthetic efficiency and a rapid growth rate, which makes them very efficient in the biological fixation of CO_2 (David et al., 2025; Udaypal et al., 2024). They may be grown in the controlled system and thus deliver CO_2 with the efficiency of industrial sources such as flue gases (D. Li et al., 2023). Microalgae are capable of capturing and storing carbon dioxide in efficiencies of 40 to 93.7% (Ighalo et al., 2022). Seaweeds or macroalgae are bigger, multicellular organisms, which constitute a major carbon sink in the ocean, effectively contributing to carbon cycling in the ocean (Hu et al., 2025; X. Sun et al., 2023). Although the role of macroalgae in carbon sequestration in the marine environment is significant, its use in industrial CO_2 sequestration encounters other challenges and opportunities than the microalgae (X. Sun et al., 2023). Nevertheless, newer studies also investigate the synergistic interactions of bacteria-algae consortia of both microalgae and macroalgae in order to increase the CO_2 fixation (Hu et al., 2025).

Key species and their characteristics for optimal CO_2 uptake: Several microalgal species were known to possess an excellent CO_2 uptake capacity. One of the most commonly studied types of species is *Chlorella vulgaris*, as it has a high photosynthetic potential and a large CO_2 concentration (Dasan et al., 2020; Tarafdar et al., 2023). It has been determined that it is capable of effectively absorbing CO_2 and retaining lipids during conditions of stress (Dasan et al., 2020). One of the promising species to be mentioned is the *Scenedesmus acuminatus TH04*, which has also been demonstrated as a competent bicarbonate fixer in both bicarbonate-based and CO_2-based systems (Do et al., 2021). Other notable bioindicators include *Dunaliella tertiolecta*, *Spirulina platensis*, *Desmodesmus sp.*, and *Nannochloropsis sp.*, which are also believed to have colossal CO_2 mitigation potential (Tarafdar et al., 2023). Resistant strains like the *Chlorella sp. AE1* that are more resistant to a high concentration of CO_2 and those with higher carbon-fixing capacity have been acquired through adaptive evolution in the laboratory (D. Li & Zhao, 2023). Specifically, the example of the genus Chlorella is mentioned as the one that can capture CO_2 and turn it into useful bioenergy (Ashour et al., 2024).

Selection criteria for algal strains in sequestration projects: Selection criteria of algal strains in sequestration projects: There are various criteria based on which the optimal strain of algae is chosen in a CO_2 sequestration project. CO_2 resistance and high CO_2 uptake have become important issues with the use of industrial flue gases (Ashour et al., 2024). The growth rates must be favorable enough to produce the maximum possible production of biomass and fixation ability of carbon (Udaypal et al., 2024). It is pertinent to tolerate and react to various environmental parameters, such as the temperature, pH, and light changes, and apply them in practice (Do et al., 2021). The economic sustainability and feasibility of the process can be enhanced by exploiting its capacity to produce useful co-products (e.g., biofuels, bioplastics, fertilizers) and also through carbon sequestration (D. Li et al., 2023). They are convenient due to their easy growth and harvesting, as well as their resistance to contamination (X. Sun et al., 2023). The focus is now on strain development, genetic engineering, and adaptive evolution in research to develop these qualities better (J. Ma et al., 2025).

2.3. Cultivation systems and technologies

Open pond systems: The most basic and the least costly microalgae cultivation system is the open pond system that encompasses both the raceway and circular pond (Atikah et al., 2025). The advantages include that they are less costly to construct and operate as well as could be utilized in large-scale industrial applications (Rios Pinto et al., 2021). However, they possess several disadvantages, including the inability to control parameters of the culture (temperature, pH, dissolved O_2, and CO_2), the present exposure to contamination by unwanted microorganisms, and dependence on the weather conditions of the ambient (Rios Pinto et al., 2021). This lack of regulation is likely to decrease the biomass productivity and the CO_2 fixation efficiency compared to closed systems (Rios Pinto et al., 2021).

Photobioreactors: Photobioreactors (PBRs) are closed cultivation systems, which provide a controlled environment to grow algae and result in increased biomass production and CO_2 uptake rate (Atikah et al., 2025; Islam et al., 2024). Typical ones are PBRs with tubes and PBRs with flat plates (Rios Pinto et al., 2021). PBRs offer an enhanced ability to regulate key parameters such as light intensity, temperature, pH, and CO_2 supply to reduce the risk of contamination and increase growth rates to the maximum (Rios Pinto et al., 2021). This is a controlled environment that allows the capture rate of carbon to be higher, usually 10 to 50 times more productive than an open pond system (David et al., 2025). Nevertheless, the construction and operation of PBRs are generally costlier, and the large scale can be an issue when it comes to very large-scale industrial applications (D. Li et al., 2023; Rios Pinto et al., 2021). The optimization of the PBR design to achieve an efficient CO_2 fixation is being optimized using Computational Fluid Dynamics (CFD) modeling (Ojaniemi et al., 2025).

Hybrid systems: Hybrid cultivation systems are meant to combine the strengths of open ponds and PBRs and reduce their shortcomings (Atikah et al., 2025). Several of these systems usually integrate the cost-effectiveness and scalability of open ponds with the increased control and efficiency of PBRs. An illustration is that there may be hybrid designs that involve the use of PBRs to produce inoculums or to keep dense cultures in the best possible conditions which are then transferred to open ponds to allow the large-scale growth of biomass. Improvements in hybrid systems also involve floating PBRs and integrated open pond-photovoltaic or PBR-photovoltaic systems, which will ensure minimal energy usage and efficiency in general (Atikah et al., 2025). Such systems are a move in the right direction towards more sustainable and economically viable large-scale algal growth for CO_2 sequestration.

Emerging technologies in algal cultivation for CO_2 sequestration: The algal CO_2 sequestration is being enhanced by emerging technologies. They are enhanced bioreactor designs, enhanced nutrient delivery, and process control using artificial intelligence (A. Kumar et al., 2025; J. Ma et al., 2025). Carbon sequestration through microalgae is also gaining momentum due to the creation of artificial CO_2 concentration systems, which increases the delivery of CO_2 and its use (Hou et al., 2025). Another novel method is the co-culture of microalgae with heterotrophic microorganisms because the synergy between bacteria and algae can enhance the level of CO_2 fixation and the resilience of the system (Hasnain et al., 2023; Hu et al., 2025). Moreover, the combination of algal farming and wastewater treatment offers a sustainable solution because of the ability to use wastewater nutrients and remove CO_2 at the same time (Hasnain et al., 2023; D. Li et al., 2023). Genetic engineering and metabolic engineering are also under consideration to create super-efficient algal strains with an increased CO_2 fixation capacity and biomass production (J. Ma et al., 2025).

2.4. Quantification and measurement of CO_2 sequestration

Methods for measuring carbon uptake in algal systems: The common ways to measure carbon uptake in algal systems are by observing CO_2 reduction in the gas phase or carbon accumulation in biomass (A. Kumar et al., 2025). The rate of CO_2 consumption is determined by gas analysis methods, such as infrared gas analysis (IRGA) (A. Kumar et al., 2025). Biomass is directly measured by dry weight or elemental analysis of carbon to quantify how much is sequestered in organic matter (D. Li et al., 2023). Carbon assimilation pathways can be tracked with isotopic labeling using $^{13}CO_2$, so partitioning of carbon in algal cells can be measured (D. Li & Zhao, 2023). Rates of CO_2 fixation are also estimated with stoichiometric calculations based on algal growth kinetics and biochemical composition (Souza et al., 2024).

Challenges in accurate quantification of sequestered carbon: Several difficulties are related to the precise measurement of sequestered carbon. These are the rates of CO_2 dissolution and desorption, which can lead to a significant part of the loss of CO_2 through open systems and also the inability to directly measure the uptake of algae (Kanna Dasan et al., 2023). Environmental factors such as light, temperature, and pH make it difficult to establish the exact amount of CO_2 that is used by the algae and the amount that is emitted into the atmosphere (A. Kumar et al., 2025). Moreover, CO_2 fixation and respiration may be the basis of the metabolic processes of algae, and the contrast between them is required to attain the values of the net sequestration (Zeng et al., 2021). Switching to industrial applications following the laboratory tests introduces further problems in maintaining the uniformity of measurements and the carbon balances in the system level (X. Sun et al., 2023).

Standardization efforts for CO_2 sequestration measurement: The use of the metric of CO_2 sequestration should be standardized to render the credit markets of carbon credible (Souza et al., 2024). This will entail the development of standard procedures that will oversee the CO_2 uptake, the growth of the biomass, and the carbon content of the various cultivation systems and algal species. It is necessary to prepare powerful protocols of life cycle assessment (LCA) as well to take the entire carbon footprint of algal CO_2 sequestration, including energy required to grow, harvest, and process algae (D. Li et al., 2023). The uniformity would help in the proper verification and reporting of carbon that has been sequestered that would induce better confidence among the policymakers and investors. These methodologies are still being optimized in the studies and are using a combination of modeling techniques to provide improved and universally accepted standards for quantifying algal carbon sequestration (Ighalo et al., 2022; A. Kumar et al., 2025). Figure 1 represents a schematic diagram of the algal CO_2 sequestration process and demonstrates the cultivation systems used and the mechanism through which CO_2 is sequestered in the algal biomass.

Figure 1. Schematic diagram of algal CO_2 sequestration process, including cultivation systems and carbon fixation mechanisms

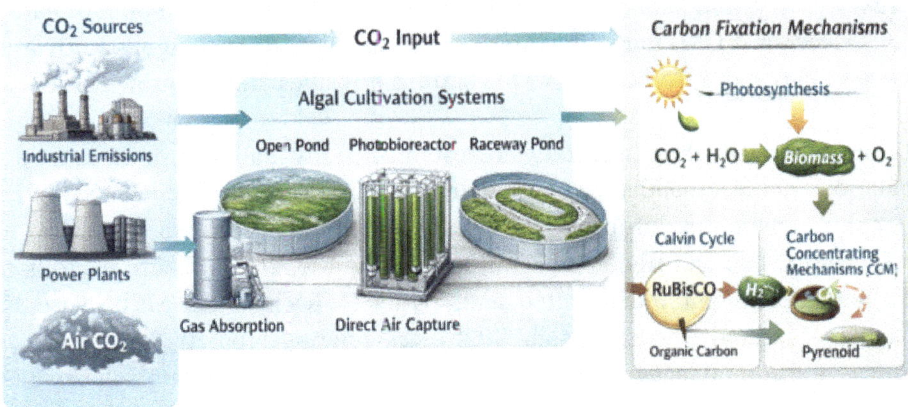

3. CARBON CREDIT SYSTEMS AND MARKETS

3.1. Overview of Carbon Credit Mechanisms

Carbon credit systems are created to handle the emission of greenhouse gases (GHGs) and are developed in various forms, including compliance markets, voluntary carbon credits, and cap-and-trade systems (L. Chen & Xie, 2023; Nandhini &

Venkataraman, 2025). The mechanisms make carbon dioxide (CO_2) cuts or removals commodifiable, which provides economic incentives to take action about climate (Nandhini & Venkataraman, 2025; Singh et al., 2014).

The cap-and-trade systems constitute a kind of control whereby a government sets a limit (cap) on GHG emissions within a particular industry or economy (L. Chen & Xie, 2023). The emitting parties purchase or offer the allowances, termed as the right to emit one tonne of CO_2 equivalent (Nandhini & Venkataraman, 2025). The one producing less than the quota is free to sell its unused quota, and the one producing over the quota is obligated to purchase additional quota and therefore creates a carbon market (Nandhini & Venkataraman, 2025). It is a system that commends cost-effective emission reduction among the regulated parties (L. Chen & Xie, 2023).

Voluntary carbon markets (VCMs) are systems through which people, companies, or institutions can buy and sell carbon credits voluntarily to offset their carbon emissions (L. Chen & Xie, 2023; Fujii et al., 2024). These markets operate based on corporate social responsibility, sustainability objectives, or brand image (Günther et al., 2020) and are not regulated. VCM project types vary, including renewable energy, forestry, and land management (Fujii et al., 2024). The amount of carbon credits issued in VCMs under the forestry and land use category significantly grew between 2006 and 2015 because of the shift in priorities and the growing demand (Fujii et al., 2024).

Compliance carbon markets are established by national or international regulations to meet mandatory emission reduction targets (L. Chen & Xie, 2023). Examples include the European Union Emissions Trading System (EU ETS) and the California cap-and-trade program (L. Chen & Xie, 2023). These markets are more regulated, monitored, reported, and verified (MRV) than VCMs because they are legally binding (Boettcher et al., 2025). The relationship between VCMs (soft regimes) and compliance markets (hard regimes) is not simple, compliance markets tend to eat up, overlap, or squeeze out voluntary efforts by institutional complementarities and market boundary changes (L. Chen & Xie, 2023).

3.2. Carbon Credit Verification and Certification Processes

The integrity of carbon credits, regardless of the market, depends greatly on the effective verification and certification (Arcusa & Sprenkle-Hyppolite, 2022). In the case of algal systems, these processes imply certain methodologies of calculating carbon credit, compliance with the third-party verification standards, and the special issues related to algal CO_2 sequestration.

Calculation methodologies of carbon credit through algal systems are generally based on the quantity of CO_2 taken up by algae in the process of photosynthesis and transferred to biomass (Alami et al., 2021; Udaypal et al., 2024). Microalgae need CO_2 as one of the main nutrients and transform it into other bioproducts and biomass, which means that they can be used as bio-CO_2 sequestration candidates (Ngo et al., 2022; Udaypal et al., 2024). Optimized culture conditions, such as hydrodynamics to promote the transfer of light and CO_2, can be used to maximize the efficiency of CO_2 fixation by microalgae (B. Zhang et al., 2025). Theoretical analysis indicates that massive microalgae culture may attain high CO_2 fixation rates, including about 50 million tons per year on 13 million acres of land (J. Ma et al., 2025).

The credibility and additionality of carbon credits cannot be achieved without third-party verification standards (Arcusa & Sprenkle-Hyppolite, 2022). These standards normally need to monitor, report, and verify (MRV) procedures to determine that the captured carbon is genuine, quantifiable, irreversible, and extra to that which otherwise would have been captured without the project. Project design documents are evaluated by certification bodies, data is monitored, and the emission reductions or removals are checked against the set baselines (Arcusa & Sprenkle-Hyppolite, 2022).

The issues peculiar to the algal CO_2 sequestration credit verification include the correct determination of the efficiency of CO_2 uptake, the permanence of the stored carbon in the algal biomass or its derived products, and the existence of the life-cycle emissions during cultivation, harvesting, and processing (Elouardi et al., 2023; X. Sun et al., 2023). Although algae have a high photosynthetic efficiency and a rapid growth rate, low biomass productivity and high harvesting costs may prevent scale-up and make generating carbon credits cost-effective (J. Ma et al., 2025; X. Sun et al., 2023). Moreover, the final use of algal biomass should also be taken into account in this case, it can be immediately burned, which again releases CO_2, and it cannot be considered long-term sequestration (Souza et al., 2024).

3.3. Current Status of Algal CO_2 Sequestration in Carbon Markets

The use of algal CO_2 sequestration is also becoming a popular biological method of carbon capture and utilization (CCU) (Daneshvar et al., 2022). Microalgae is also being examined as a potential CO_2 mitigation project and initiative in many parts of the world (Udaypal et al., 2024). They are combined systems of CO_2 capture of industrial flue gasses, wastewater treatment, and, consequently, the production of biofuels (Iqbal et al., 2022; D. Li et al., 2023). In 2025, the average CO_2 concentration in the atmosphere was 425.7 ppm, and the average growth rate was 2.3 ppm/year, which is why the effective use of sequestration technologies is urgent (Friedlingstein et al., 2025). Macroalgae such as *Kappaphycus alvarezii* also have the potential to be used to sequester CO_2, and the average carbon concentration is 1.7% of its live weight (Santos et al., 2024).

This has prompted the idea of national carbon market policies to measure GHG emissions and incorporate species such as algae (Santos et al., 2024). The EU is also considering the addition of blue carbon of seaweed to voluntary carbon removal certification schemes (van den Burg et al., 2023). Nevertheless, algal sequestration has major entry obstacles to the carbon markets. These are the facts that large-scale cultivation of algae, especially photobioreactors, is expensive in terms of capital and operation, and there are difficulties with efficient harvesting and processing of biomass (Xu et al., 2024). Market participation can also be discouraged by the absence of standardized and widely accepted methodologies of quantifying algal carbon sequestration and the perceived temporality of certain algal carbon storage (Arcusa & Sprenkle-Hyppolite, 2022). Addressing these economic challenges requires strain development, optimization of the culture system, control of the parameters, and metabolic engineering to increase the efficiency of microalgal carbon sequestration (J. Ma et al., 2025; Tripathi et al., 2023).

3.4. Comparison with Other Carbon Sequestration Technologies

Algal CO_2 sequestration has a future but needs to be compared with other established and emerging technologies of carbon sequestration.

Forestry and Land-Use Change (LUC): Forests are also important carbon sinks, and afforestation/reforestation is a popular practice in carbon markets (Fujii et al., 2024). Nevertheless, their performance may be constrained by the extended growth time to achieve significant carbon capture, the possibility of carbon re-emission through forest fires or diseases, and the plateau in the carbon-capturing rate of forests as they become older (Fujii et al., 2024). Although the forest and land management projects experienced a rise in the issuance of carbon credits within the voluntary markets, the efficiency of these projects can be affected by the factors of priority change and scale (Fujii et al., 2024).

Direct Air Capture (DAC): DAC technologies use chemical capture to remove CO_2 from ambient air (G. Li & Yao, 2024). While effective, these technologies face serious issues: high capital costs, energy demands, and large infrastructure needs (Baskaran et al., 2024; Cole et al., 2023). The thermodynamic limit for separating CO_2 and air sets a minimum energy cost. Current DACCS systems are promising, removing CO_2 up to the kiloton scale and below (Cole et al., 2023).

Ocean Fertilization: This will include the addition of nutrients (e.g., iron) to the oceanic areas to promote the growth of phytoplankton and hence the increased uptake of CO_2 in the atmosphere by the process of photosynthesis. Although it can be used on a large scale to remove carbon, ocean fertilization has some serious ecological issues, such as possible disturbances in marine life, changes in nutrient cycles, and unpredictable long-term ecological effects (El Zrelli et al., 2021). The sequestered carbon may be permanent in the long term and may have unintended side effects, which is a significant obstacle to the implementation of the sequestering process. The evaluation should be conducted thoroughly, including a life-cycle assessment that considers all emissions and impacts on the entire system. Comparative analysis of different carbon sequestration technologies has been done in Table 1 based on their efficiency in removing CO_2, costs involved, environmental effects, and permanence, which provide an insight into the potential of algal CO_2 sequestration.

Table 1. Comparison of different carbon sequestration technologies, including efficiency, cost, and environmental impact

Technology	Efficiency (CO_2 Removed per Unit Effort)	Cost (USD/ tonne CO_2)	Environmental Impact	Permanence (typical)	Reference
Algal CO2 Sequestration	High photosynthetic efficiency (6–8%), rapid CO2 fixation	Variable, high initial capital, operational challenges	Generally positive (biofuel/biomaterial production), but large-scale cultivation may have water/nutrient demands	Variable (days to years depending on biomass fate)	(Dasan et al., 2025; David et al., 2025; Souza et al., 2024)
Forestry and Land-Use Change	Moderate to high, dependent on species and growth rates	Generally low to moderate	Positive (biodiversity, ecosystem services), but susceptible to disturbances (fire, pests) and land use competition	Decades to centuries (risk of reversal)	(Fujii et al., 2024)
Direct Air Capture (DAC)	High, able to capture CO_2 from ambient air	High (capital, energy demands)	Energy intensive, potential for large physical footprint	Centuries to millennia (if geologically stored)	(Baskaran et al., 2024; Cole et al., 2023)
Ocean Fertilization	Potentially high, but uncertain and unverified	Research phase, cost highly speculative	Significant ecological risks, potential for ecosystem disruption, anoxia	Variable (risk of short-term reversal due to ocean dynamics)	(Boettcher et al., 2025)

4. POLICY PERSPECTIVES ON ALGAL CO_2 SEQUESTRATION

The increasing amount of carbon dioxide (CO_2) emissions in the world requires new and scalable forms of mitigation, and the use of algal CO_2 sequestration is one of the potential biological pathways (X. Sun et al., 2023). The effective deployment of the algal methods of carbon capture depends on the existence of the powerful policy structures, which encompass the international agreements, national legislation, and economic motivations (Otundo Richard, 2024).

4.1. International Policy Frameworks

The international agreements outline the general framework of the global climate action with a strong impact on the use of carbon sequestration technologies, and the algae-based ones are not exceptions. Among the pillars of the global fight against climate change, one can distinguish the Paris Agreement, which was presented within the framework of the United Nations Framework Convention on Climate Change (UNFCCC) and which was to reduce global warming to at least 2°C and, ideally, 1.5°C compared to the pre-industrial period (Green, 2024; Kuyper et al., 2018). This agreement compels all the signatories to ensure that by the second half of this century, balance will be achieved between anthropogenic greenhouse gas emissions and sink removals, with an implicit acknowledgment of the necessity of carbon removal technologies (Cook et al., 2024). Article 6 of the Paris Agreement also presents the tools of international collaboration, including the carbon markets, which could assist countries to execute their Nationally Determined Contributions (NDCs) on the basis of cooperative strategies (Cook et al., 2024; Workman et al., 2023). The UNFCCC supports the process of climate mitigation through multiple channels, one of which may include algal carbon sequestration, especially when countries are seeking new ways of decreasing emissions and increasing carbon sinks (Kuyper et al., 2018). The regional programs (e.g., the EU Emissions Trading System (EU ETS)) are also important policy tools that may have an effect on the algal CO_2 sequestration. Although the forest-based carbon credits are not currently recognized under the EU ETS, there is a discussion about the possibility of including different carbon removal credits, particularly with the rulebook of the Paris Agreement being finalized (Bleuel & Müller, 2024). The EU has also suggested a voluntary system of certifying carbon removals, which is yet to establish its position on blue carbon, including seaweed-based one (van den Burg et al., 2023). Seaweed is considered a new blue carbon alternative that has already been tested to have the potential for long-term carbon removal and the ability to store carbon for 100 years or longer and is certified in different EU communications (van den Burg et al., 2023). This implies that marine-based carbon sequestration solutions, such as algae-based ones, have become of interest to policymakers.

4.2. National Policies and Regulations

The national policies and regulations play a significant role in the translation of international climate objectives into domestic action plans. Nations are starting to investigate and, in certain instances, put into practice enabling policies regarding algal sequestration. As an example, the growing level of atmospheric CO_2 has caused panic on the planet, yet CO_2 is also considered a product that can be used to empower people and nations by means of carbon credit markets (Singh et al., 2014).

Regulatory issues are associated with the necessity of clear guidelines on measuring, reporting, and verifying (MRV) the carbon sequestered by algal systems to be credible and prevent the double count of carbon in carbon accounting schemes (Ighalo et al., 2022; D. Li et al., 2023). The prospects of the algal systems are linked to the potential to solve the CO_2 fixation and other environmental problems, including wastewater treatment, to facilitate a circular bioeconomy (Leong & Chang, 2023; Viswanaathan et al., 2022). As an example, the grid-connected solar capacity in India has grown greatly, yet the process also poses certain challenges associated with solar photovoltaic (PV) waste management, which might also be connected to the algae-based biomass utilization strategies (Jain et al., 2022).

4.3. Incentives and Support Mechanisms

The algal CO_2 sequestration projects should be scaled up using financial incentives. The introduction of new technologies can be greatly de-risked with the use of tax credits and subsidies and encouraged (Terjanika et al., 2023). These mechanisms can include direct payments for carbon removal, including Payment for Ecosystem Services (PES) programs, or tax exemptions for firms that construct algal cultivation plants (Aryanpur & Rogan, 2024; Király & Borovics, 2024).

Access to research and development (R&D) funds is needed to resolve the uncertainty of technologies and enhance the performance of algal systems in efficiency and cost-effectiveness (Hou et al., 2025). This financing can be introduced to develop strains, culture system optimization, and metabolic engineering to increase the CO_2 fixation levels and biomass valorization (J. Ma et al., 2025). The concept of public-private partnerships is also getting used as a necessity to fill the research-to-commercial divide (Green, 2024). Government agencies, the universities, and the private sector can also combine the resources and redistribute the risk and support the development and implementation of algal sequestration technologies in a faster manner (Workman et al., 2023).

4.4. Policy Challenges and Barriers

Despite the potential, there are several policy challenges and barriers that do not allow the large-scale application of algal CO_2 sequestration.

Technological uncertainties and their impact on policy development: Despite the high photosynthetic power and versatility of microalgae, issues of optimization of culture systems, strain development, and metabolic engineering continue to be challenging in achieving an improved rate of carbon sequestration and biomass production on the industrial scale (Hou et al., 2025; J. Ma et al., 2025). The policy must be adaptable to accommodate the fast changes in technology and give clear guidelines on the current usage.

Economic viability concerns and policy solutions: The established chemical and physical CO_2 sequestration methods are also expensive due to the large energy consumption (Ramanna et al., 2025). Although microalgae are also a good alternative, the economic barrier may continue to be a huge initial investment and the cost of operation (D. Li et al., 2023). The policy tools that will be employed to improve the economic viability include carbon pricing, carbon taxes, and subsidies that would result in the internalization of the social cost of carbon and the creation of market demand of algal products (Terjanika et al., 2023).

Regulatory requirements and environmental impact assessment: The large-scale introduction of the systems of algal cultivation must be supplemented by the thorough examination of the environmental impact to understand the potential effect on the local ecosystems, water resources, and biodiversity (Wang et al., 2023). The guidelines should ensure that the projects undertaken are sustainable and do not create new environmental issues. The necessity to make the transition to blue carbon and seaweed farming is also founded on the overall idea that the potential of carbon sequestration and the effect on the environment are to be estimated appropriately (Barros et al., 2024; Pessarrodona et al., 2023a).

Social acceptance and public perception challenges: The efficiency of new carbon removal technologies depends on the fact that people should learn about them and be ready to adopt them (Low et al., 2024). The policies should involve the stakeholders, besides being receptive to the benefits and possible hazards of algae sequestration projects. The growing amount of literature on carbon capture and utilization (CCU) with the help of microalgae means that the policy should be demonstrated in detail to reach the net-zero emission goal (D. Li et al., 2023). Table 2 provides a summary of the main global, regional, and national policy tools intended to assist algal CO_2 sequestration, their mechanisms, challenges, and area applicability.

Table 2. Summary of Key Policy Instruments Supporting Algal CO_2 Sequestration Across Different Countries

Policy Instrument Category	Specific Instrument/ Mechanism	Description and Impact	Applicable Regions/Countries (Examples)	Key Considerations/ Challenges	Reference
International Frameworks	Paris Agreement Article 6	Establishes international carbon markets and cooperative approaches to achieve NDCs, fostering investment in carbon removal technologies like algal sequestration.	Global (UNFCCC signatories)	Requires robust MRV to prevent double counting.	(Cook et al., 2024; Kuyper et al., 2018)
Regional Regulations	EU Emissions Trading System (ETS)	While currently limited, potential for inclusion of carbon removal credits (including blue carbon) into the ETS framework is being explored.	European Union	Need for clear certification standards for emerging blue carbon options like seaweed.	(Bleuel & Müller, 2024; van den Burg et al., 2023)
National Policies	Carbon Credit Markets	Incentivizes CO_2 reduction and sequestration by assigning monetary value to carbon removal; facilitates investment in algal CCU projects.	Global (e.g., India considering national carbon market)	Requires national methodologies for quantification and verification.	(Santos et al., 2024; Singh et al., 2014)
National Policies	Climate Policy & GHG Reduction Targets	National strategies to meet climate commitments; algae-based solutions are seen as promising for emissions reduction and carbon sink enhancement.	European Union, various countries with climate change commitments	Integration with existing climate policies; addressing technical and economic barriers.	(Prasad et al., 2021; Zieliński et al., 2023)
Incentives & Funding	Research & Development (R&D) Funding	Direct funding for studies on optimizing algal growth, CO_2 fixation efficiency, and biomass valorization.	Global	Addressing low biomass productivity and high harvesting costs for scale-up.	(Li et al., 2023; Sun et al., 2023)
Incentives & Funding	Tax Credits/ Subsidies	Financial incentives to reduce the cost of deploying algal carbon capture infrastructure and operations.	Various countries implementing green technology incentives	Ensuring economic viability and competitive advantage against conventional methods.	(Terjanika et al., 2023)
Collaborative Models	Public-Private Partnerships (PPPs)	Joint ventures between government bodies and private entities to pool resources and expertise for technology development and commercialization.	Global (e.g., partnerships for decarbonization solutions)	Facilitating knowledge transfer, risk sharing, and market creation.	(Green, 2024)
Environmental Safeguards	Environmental Impact Assessments (EIAs)	Regulatory requirements to evaluate potential ecological and socio-economic impacts of large-scale algal cultivation projects.	National regulatory bodies	Ensuring sustainable deployment and minimizing negative externalities.	(Wang et al., 2023)

Policy Implementation in the Global South: Implementation of algal CO_2 sequestration in the Global South is influenced by the larger trends in the policy of climate mitigation, blue carbon policy, and the design of carbon markets. Many nations are starting to realize the benefits of operationalizing coastal blue carbon ecosystems—mangroves, seagrasses, and, increasingly, macroalgae forests—within national climate action plans and NDCs, but macroalgal carbon is still poorly integrated into formal rules on accounting and crediting (Marquardt et al., 2023; Merk et al., 2022; Pessarrodona et al., 2023b; Vanderklift et al., 2022). Experience in blue carbon projects in Kenya, India, Vietnam, and Madagascar reveals that successful implementation requires strong institutions, clear tenure, reliable MRV (measurement, reporting, and verification), co-benefits to the local community, and long-term finance, which is often through carbon or ecosystem-service payments (Bennett et al., 2024; Merk et al., 2022; Wylie et al., 2016). Meanwhile, institutionalization of mitigation in the Global South is hampered by competing priorities on development, lack of regulatory capacity, and worries over carbon colonialism and low-quality offsets (Marquardt et al., 2023; Sovacool, 2023). To apply nature-based CDR solutions like macroalgae, researchers emphasize the necessity of locally grounded governance, the need to pay attention to justice and equity, and schemes reflecting a value of the broad range of ecosystem services, rather than solely carbon (Pessarrodona et al., 2023b; Sovacool, 2023; Vanderklift et al., 2022). The risks of expanding algal carbon credits include reproducing inequities, making the environment weak, and failing to create a sustainable mitigation of climate change.

5. ECONOMIC ANALYSIS OF ALGAL CO_2 SEQUESTRATION

One of the potentially viable solutions to climate change is algal CO_2 sequestration, which integrates carbon capture together with a potential by-product (Choi et al., 2022; Ighalo et al., 2022). The cost analysis of this technology, market potential, and investment environment are also other areas that should be prioritized so as to present a more detailed economic analysis of this kind of technology (Behera & Paramasivan, 2021).

5.1. Cost-benefit analysis

The key elements that define the economic feasibility of the algal CO_2 sequestration projects are their operating and capital costs. Photobioreactors (PBRs) and open ponds are microalgae culture systems with a high initial investment (CAPEX) and operating expenses (OPEX) (Drabold et al., 2020; Llamas et al., 2021). The research found that the prices of biomass production of microalgae can be between 0.41 and 13.06 EUR/kg, and this depends on the conditions and system of cultivation (Llamas et al., 2021). The key cost drivers are considered to be such aspects as the energy used to light, mix, and provide nourishment and nutrients (Drabold et al., 2020). An example of a techno-economic analysis of hydrothermal carbonization with an algal bioreactor revealed that the cost of nutrients is one of the major factors of the economic constraint (Drabold et al., 2020).

Funding of algal CO_2 sequestration initiatives is multi-level, and it is mostly associated with carbon credits and selling value-added by-products (Behera & Paramasivan, 2021; J. Ma et al., 2025). The algae may be applicable in the process of converting CO_2 into biomass that can further be utilized to produce very many products like biofuels, nutritional supplements, animal feed, and biochemicals (J. Ma et al., 2025). To illustrate this, the integration of microalgae cultivation into biomethane facilities can produce approximately 250 t y^{-1} of microalgal biomass with the target of attaining the emerging market of growing biostimulants (Rossi et al., 2024). A hierarchical production system that enhances the economic profits is the synthesis of docosahexaenoic acid (DHA) and the subsequent biogas production using microalgae residues (Xu et al., 2024). Carbon credits also provide another financial incentive because CO_2 sequestration can be sold in carbon markets (Plastina et al., 2024). Recovery of carbon biofixation alone was estimated to be 0.03–0.1 €/kg CO_2, which is potentially recovered (Llamas et al., 2021).

The comparison of algal CO_2 sequestration and other carbon reduction methods shows that they have unique benefits and issues. The conventional industrial methods of carbon capture are associated with the economic and environmental infeasibility problems (Behera & Paramasivan, 2021). The use of microalgae systems is frequently mentioned because of their high photosynthetic efficiency, high growth rates, low resource consumption, and eco-friendliness, which is why they can be a promising solution to sustainability (D. Li et al., 2023; Udaypal et al., 2024). Nevertheless, scalability and cost-effectiveness of its use on a large scale in industries are still a challenge (Souza et al., 2024). Other carbon removal methods, like the Direct Air Capture (DACCS), are at the kiloton scale but require high capital and energy requirements (Cole et al., 2023). Although a vital part of carbon capture, utilization, and storage (CCUS), geological CO_2 storage is mainly based on the traditional oil and gas reservoir collaborative flooding technology, with offshore storage development being one of the developmental trends considered more secure and less harmful to the environment (Lin et al., 2024). Table 3 shows the comparison of the cost-effectiveness of algal CO_2 sequestration with other carbon reduction technologies and the benefits and difficulties of scaling up each of the methods.

Table 3. Cost Comparison of Algal CO_2 Sequestration with Other Carbon Reduction Technologies

Technology	Cost Range ($/tonne CO2)	Key Advantages	Key Challenges	Reference
Algal CO_2 Sequestration	Variable	Produces valuable by-products, eco-friendly	High capital/operational costs, scalability	(Llamas et al., 2021)
Conventional Chemical/ Physical CCS	High	Proven efficacy for point sources	High energy demand, unclear long-term environmental impacts	(Ramanna et al., 2025)
Direct Air Capture (DAC)	High	Removes CO_2 directly from atmosphere	Very high capital & energy costs, nascent stage	(Cole et al., 2023)
Reforestation/Afforestation	Low-Moderate	Nature-based, co-benefits (biodiversity)	Land availability, permanence, slow sequestration rate	(Plastina et al., 2024)
Blue Carbon Ecosystems	Moderate	High sequestration in soils, long-term storage	Limited geographic scope, vulnerability to climate change	(Perera et al., 2025)

5.2. Market potential and growth projections

The algal carbon credits and products market is expected to expand as the world moves toward the reduction of CO_2 emissions and the creation of renewable supply chains (Choi et al., 2022). The amount of carbon credits issued regarding forestry and land use was increased between 2006 and 2015 with changing priorities, which is a dynamic development in voluntary carbon offset markets (Fujii et al., 2024). The CO_2 sequestration through microalgae has attracted much attention as an encouraging method of carbon dioxide reduction (Ren et al., 2023). It has been estimated that microalgae cultivated on a large scale on an area of about 13 million acres of land can reach a CO_2 fixing capacity of some 50 million tons per year and also produce over 3 million tons of biomass (J. Ma et al., 2025).

Some of the factors affecting market development in the algal sequestration industry are the development of new technologies in strain development, the optimization of culture systems, and metabolic engineering to improve CO_2 fixation and biomass valorization (J. Ma et al., 2025; Tanaka et al., 2021). Carbon pricing mechanisms and incentives for green technologies are also important as policy support (Yang et al., 2024). As an illustration, the European Union considers the microalgae-guided sequestration as a potential and sustainable solution in the context of its climate policy (Zieliński et al., 2023). The climate suitability, process input (e.g., wastewater, CO_2 sources), and funding opportunities are some of the factors that affect regional variations in market potential (Schipper et al., 2023). Industrial sources of CO_2, including power plants, also provide high potential for combined algal carbon capture systems (Behera & Paramasivan, 2021; Souza et al., 2024).

5.3. Investment landscape

The availability of the two-fold potential of environmental and economic benefit, carbon credits and valuable co-products, is compelling increasing numbers of companies to be part of algal CO_2 sequestration ventures (Schipper et al., 2023). To establish sustainable and profitable businesses, the companies are looking at alternative applications, including biofuels or high-value chemicals (Choi et al., 2022).

The impact on the investment environment is enormous, as the government funding programs are de-risking the technologies at the early stage and providing incentives to implement them. The economic motivation of carbon capture by algae may be made easier to attract as an example; to encourage it, a monetary penalty on the carbon emissions can be introduced by the regulations (Zhao et al., 2024). Carbon compensation mechanisms, which orient activities to reduce emissions or grow sequestration, are also necessary for organizing regional development and attracting investment (Yang et al., 2024).

Trends in venture capital and green investment are more inclined toward sustainable technologies capable of resolving climate change. Algal CO_2 sequestration follows these trends, and it offers a technologically modern solution to nature (Dasan et al., 2025). However, the economic feasibility and operative real-time of the algae-based industrial carbon capture are yet to be compromised by the erratic economics and resource practicability (Behera & Paramasivan, 2021). The necessity to reach a compromise between the direct carbon elimination and the production of other greenhouse gases, such as methane in some ecosystems, makes the issue of the investment decisions even more complicated (Chopin et al., 2024). Effective approaches to the investment can include the life-cycle assessment (LCA) and techno-economic analysis (TEA) to ensure that the benefits are real, not only in the environment but also in the economic sphere (Cole et al., 2023). Figure 2 provides estimates of the algal CO_2 sequestration market's development, as well as the regional and most important drivers that should shape its future.

Figure 2. Projected growth of algal CO_2 sequestration market, including regional breakdown and key drivers

6. ENVIRONMENTAL AND SOCIAL IMPLICATIONS

Although algal CO_2 sequestration is a promising strategy in mitigating climate change, the interaction between environmental and social outcomes of this strategy is complicated and needs to be approached carefully and with well-developed indicators of sustainability (Enríquez-de-Salamanca, 2024; Portner et al., 2021; Wilson et al., 2020). The factors discussed are the key to successful deployment and maintenance of carbon capture and utilization (CCU) algae-based solutions.

6.1. Environmental Impacts

Large-scale algae growth, particularly in carbon capture, is associated with several environmental issues related to land and water use, biodiversity, and potential ecological risk factors. Microalgae are highly efficient in photosynthesis, and they convert CO_2 into biomass at rates that are usually higher than those of terrestrial plants and can withstand high CO_2 levels (up to 15–20%) (David et al., 2025). This efficiency, however, can also be translated into huge resource demands.

Land and water use requirements for large-scale algal cultivation: Although microalgae growth may be done on non-arable land and thus avoid direct competition with food crops, the overall land area of the large-scale process may be large, especially with open raceway ponds (Walsh et al., 2016). The issue of water demand is also one of the main ones, with the evaporation of open pond systems potentially being high (Avinash et al., 2020). Nevertheless, wastewater treatment can be combined with some of these systems, providing a co-benefit through the use of nutrient-rich effluents and freshwater reduction (J. Chen et al., 2022; Ishizaki et al., 2020). The land and water use also depends on the decision between open raceway ponds and closed photobioreactors (PBRs), where the latter usually use less land per unit of biomass but are also pricier in terms of energy requirements and capital expenditure (Sarat Chandra et al., 2018). Comparative studies of these systems in certain conditions have revealed that they have different energy requirements and environmental effects (Sarat Chandra et al., 2018). As an example, a comparative life cycle analysis (LCA) of Scenedesmus dimorphus growth in a 1000-L pilot-scale raceway pond and a 5-L airlift photobioreactor in India estimated primary energy requirements, environmental effects, and net CO_2 sequestration of each system (Sarat Chandra et al., 2018).

Biodiversity impacts of algal sequestration systems: Non-native microalgae species are introduced, particularly in large-scale outdoor growing systems, which are a risk to the environment because they may cause spillage into the natural ecosystem (Gressel et al., 2013). The evaluation of these risks is difficult because the balance of species in natural planktonic communities is quite complex and changes over time, which is referred to as the paradox of the plankton (Gressel et al., 2013). Unintended release of grown algae may outcompete natural species, change aquatic food webs, or bring pathogens, resulting in the unexpected disruptive ecological impacts (Usher et al., 2014). Effective containment and use of native or genetically modified organisms that have lower ecological fitness are important mitigation measures (Gressel et al., 2013). Macroalgae farming, including cultivated kelp, also plays a role in carbon sequestration but must be properly estimated to release biomass into the environment, which will help to determine the possible rates (Dolliver & O'Connor, 2022).

Potential risks and mitigation strategies for environmental protection: In addition to species displacement, other environmental hazard is the nutrient release of cultivation plants, which may add to the eutrophication of the water bodies around them (Usher et al., 2014). Greenhouse gas (GHG) emissions may occur when pumping, mixing, and harvesting require energy of fossil origin, which will offset the carbon sequestration advantages (Béchet et al., 2017; Walsh et al., 2016). Some of the mitigation measures are energy efficiency, renewable energy sources, and a strong wastewater management system (J. Chen et al., 2022; Ishizaki et al., 2020).

Algal biomass conversion to biochar is also becoming a viable solution, which leads to carbon sequestration and better soil properties, which further reduce waste.

6.2. Social and Ethical Considerations

The social aspects of massive algal CO_2 sequestration initiatives include the creation of employment, food security, and fairness in sharing the benefits, in particular, carbon credits.

Job creation and economic development opportunities: The algal industry has the potential to grow the economy of regions by creating new employment opportunities in cultivation, processing, research and development, and other support services (Portner et al., 2021). Algae can be used to produce various high-value products, including food, feed, pharmaceuticals, and biomaterials, which can increase the economic viability and establish a strong bioeconomy (Porcelli et al., 2020; D. Zhang et al., 2022). Nevertheless, these social impacts need to be measured through a comprehensive social life cycle assessment (S-LCA) that considers the conditions of the workers, the well-being of the local community, and equitable labor practices (Tavakoli & Barkdoll, 2020).

Food security implications of large-scale algal cultivation: Algae may be a sustainable food and feed source, with high areal productivity and capability to cultivate on non-arable soils, potentially reducing the bioenergy-food tradeoffs that are typical of crops grown on land (Walsh et al., 2016). Some species of microalgae contain proteins, fatty acids, and vitamins, which is why they can be used as dietary supplements or ingredients of animal feed (Portner et al., 2021; D. Zhang et al., 2022). Algal systems can enhance land and water-use efficiency and minimize GHG emissions by producing food and fuel (Walsh et al., 2016). The two-fold benefit can be exploited in addressing the challenge of energy, water, food, and climate security (Walsh et al., 2016).

Equity in carbon credit distribution and benefit sharing: Since projects in carbon sequestration of algae can be traded on carbon credits, there is a great necessity to make sure that the gains can be shared equally. This process involves the description of open mechanisms on how to determine the amount of carbon sequestered and how to offer a fair compensation plan to the developers of the projects and the locals (Liu et al., 2020). Carbon credits have economic rewards due to biofixation of CO_2 and creation of applications on algal biomass (Liu et al., 2020). Enríquez-de-Salamanca (2024) points out that the rationale behind the integrated approach is to cover the environmental, economic, and social dimensions to ensure that CCU solutions lead to sustainable development and the creation of equal opportunities in the society.

6.3. Sustainability Assessment

The process of the algal CO_2 sequestration is rather complicated and requires deep knowledge and analysis, including the life cycle analysis and the energy balance considerations.

Life cycle analysis (LCA) of algal CO_2 sequestration systems: LCA is a significant method of evaluating the environmental impact of cradle-to-grave fabrication of products of algae (de Souza Celente et al., 2023; D. Zhang et al., 2022). This is the calculation of the energy utilized, greenhouse emissions, water usage, and other environmental effects at each step, including growing, harvesting, and processing biomass (Gaber et al., 2021; D. Zhang et al., 2022). It has been discovered that the environmental effect greatly differs based on the cultivation system (i.e., open raceway ponds or PBRs), location, design, and operating conditions (Béchet et al., 2017). The case study on the application of modified Johnson medium (MJM) to grow Dunaliella suggested that the artificial components of the media posed a massive environmental impact (de Souza Celente et al., 2023). Another application of LCA is the determination of hotspots in the production chain where extremely high gains can result in a substantial benefit to the environment (Gaber et al., 2021). Figure 4 shows a combined circular algal bioeconomy system, which depicts the interconnection between the wastewater treatment, the CO_2 sequences the generation of bioproducts, and the carbon credit system within the life-cycle sustainability framework.

Figure 3. Roadmap for future development of algal CO_2 sequestration technology and policy

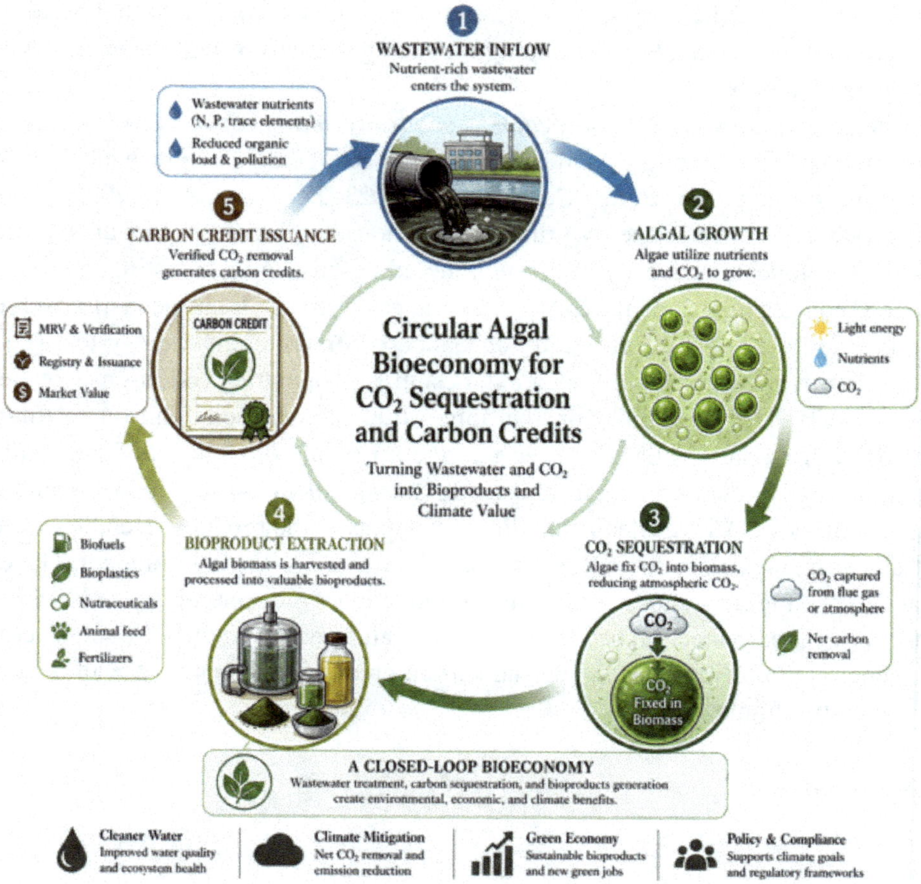

Energy balance considerations in algal cultivation and processing: Algae cultivation, including mixing, CO_2 supply, and harvesting power, can be large-scale and offsets the possible emission cuts unless regulated (Avinash et al., 2020). Harvesting is one more energy-consuming process that could be referenced (Nazari et al., 2020). The fungi-based bioflocculation technologies are under investigation to lower the cost of harvesting and energy consumption and enhance sustainability in general (Nazari et al., 2020). The power plants are able to form the synergistic effect of the flue gas (CO_2) and the algal cultivation, which means that the energy source balance is to be treated with a reserved approach to the techno-economic and the life cycle analysis (Wilson et al., 2020). A significant difference in energy consumption can be achieved by the maximization of the supply of upstream CO_2 and the biomass of downstream algae (Liu et al., 2020). Ideally, microalgae could hold up to 273.75 g/L of CO_2 within one year (Rame et al., 2023).

Long-term environmental effects of sustained algal sequestration: The environmental impact of the long-term algal sequestration and its long-term effectiveness remain uninvestigated. Algae would be a possible way out of the issues with the CCU, yet the macro performance in terms of the carbon emission reduction should be correctly understood (B. Zhao & Su, 2020). Microalgae-mediated carbon sequestration is a part of the circular carbon bioeconomy, as it transforms CO_2 into useful products, thus eliminating waste and enhancing sustainability (Leong & Chang, 2023; Z. Ma et al., 2022). Nevertheless, it needs long-term and thorough monitoring and adaptive management measures to respond to the possible changes in the ecosystem, guarantee the permanence of the carbon that has been sequestered, and avoid the unintended environmental impacts (Enríquez-de-Salamanca, 2024). Table 4 introduces an environmental impact assessment matrix, which also determines the potential environmental impact of large-scale algal cultivation and can provide mitigation strategies for various environmental factors.

Table 4. Environmental Impact Assessment Matrix for Algal CO_2 Sequestration Projects

Environmental Factor	**Potential Impact (Large-Scale Algal Cultivation)**	**Mitigation Strategies**	**Reference**
Land Use	Competition with other land uses, habitat alteration	Utilize non-arable land, integrate with existing infrastructure (e.g., industrial sites, wastewater treatment plants), use vertical farming techniques for PBRs, optimize cultivation density to maximize yield per area.	(Usher et al., 2014; Walsh et al., 2016)
Water Use	High freshwater demand, evaporation losses	Use saline or brackish water, integrate with wastewater treatment for nutrient recycling, implement water recycling technologies, use closed photobioreactors to reduce evaporation.	(Béchet et al., 2017; Chen et al., 2022)
Biodiversity	Risk of invasive species, disruption of aquatic ecosystems from spills	Cultivate native algal species, employ robust containment measures, genetic modification for reduced environmental fitness, bioremediation of spills, careful site selection.	(Gressel et al., 2013; Usher et al., 2014)
Energy Consumption	High energy inputs for mixing, CO_2 supply, harvesting, processing	Use renewable energy sources (solar, wind), optimize system design for energy efficiency (e.g., gravity-fed systems), develop less energy-intensive harvesting methods (e.g., bioflocculation), waste heat recovery.	(Avinash et al., 2020; Ishizaki et al., 2020)

continued on following page

Table 4. Continued

Environmental Factor	Potential Impact (Large-Scale Algal Cultivation)	Mitigation Strategies	Reference
Nutrient Loading	Eutrophication from nutrient discharge	Integrate with wastewater treatment to recycle nutrients, optimize nutrient delivery, implement stringent effluent treatment and recycling.	(Chen et al., 2022; Iqbal et al., 2022)
Greenhouse Gas Emissions	Emissions from energy use, N_2O from nitrogen cycling	Use renewable energy, enhance CO_2 capture efficiency, optimize nutrient management to reduce N_2O emissions, valorize biomass into stable carbon products like biochar.	(Béchet et al., 2017; Li et al., 2023)
Chemical Use	Use of fertilizers, flocculants, and processing chemicals	Optimize nutrient use, explore biological flocculants, develop sustainable extraction and processing techniques, minimize use of hazardous chemicals.	(Nazari et al., 2020)
Waste Generation	Biomass residues, processing byproducts	Develop biorefinery concepts for complete biomass utilization, convert residues into biochar, explore secondary product development	(Ma et al., 2022)

7. FUTURE DIRECTIONS AND RECOMMENDATIONS

7.1. Technological Advancements

The advancement of algal CO_2 sequestration in the future is highly dependent on technology advancement, especially in genetic engineering, integrated energy systems, and breaking commercialization barriers. Genetic engineering can be used to enhance the CO_2 fixation of algae by altering the key metabolic pathways, including ribulose-1,5-bisphosphate carboxylase/oxygenase (RuBisCO) and carbon-concentrating mechanisms (CCMs) (R. R. Kumar et al., 2022; Sadvakasova et al., 2025). Although common in photosynthesis, RuBisCO is inefficient since it combines with oxygen leading to photorespiration. Photosynthetic efficiency can be greatly improved by enhancing its CO_2 specificity or by improving CCMs, which actively carry and concentrate CO_2 around RuBisCO. For example, it has been shown that the pdxY gene, which encodes pyridoxal kinase, can be optimized to enhance carbon capture in Chlorella sorokiniana under dynamic CO_2 conditions (Liang et al., 2025). Flow improvement through native enzymes, cofactor balances, and alteration of CO_2 bio-fixation pathways using CRISPR-based tools can also be used to improve biomass production and carbon capture (Mustafa et al., 2025; Priyadharsini et al., 2022). It is also important to integrate algal systems with other renewable energy sources. Microalgae can also be grown using CO_2 extracted using industrial flue gases, which is not only a viable way of integrating carbon capture with renewable energy formation by biomass production to convert it to biofuels but also with carbon sequestration (Dasan et al., 2025; D. Li et al., 2023; Priyadharsini et al., 2022). This method will lead to a circular bioeconomy because CO_2 is transformed into useful products (Z. Ma et al., 2022; Sadvakasova et al., 2025). However, the issue of low biomass production and the high cost of harvesting these technologies makes them difficult to scale in industries (D. Li et al., 2023; J. Ma et al., 2025). The way to overcome these limitations is the creation of new bioreactor models and optimization of culture parameters to maximize CO_2 consumption and biomass (Hou et al., 2025; J. Ma et al., 2025).

7.2. Policy Recommendations

Effective policy frameworks cannot be overlooked as far as the translation of the scientific potential of algal CO_2 sequestration into the real world is concerned. Among the key suggestions is the need to streamline the process of carbon credits in so far as algal sequestration is concerned (Singh et al., 2014). The carbon credit market is evolving today, and standardized and transparent methodologies are needed to measure and certify the right quantities of carbon that have been sequestered by the algal systems (B. Zhang et al., 2025). It is essential to balance the international norms and regulations to make cross-border projects easier and to provide comparative and transparent carbon accounting (Ogunyemi, 2025). This process entails the establishment of powerful monitoring, reporting, and verification (MRV) systems that are specifically outlined with consideration of algal technologies (Ogunyemi, 2025; Singh et al., 2014). Moreover, policies should be a mix of economic motivation and environmental protection. Although financial tools, e.g., reverse auctions to remove carbon, can be used to stimulate investment, it is necessary to introduce strict environmental impact analyses to avoid unexpected ecological effects, including the possible effects on the local aquatic ecosystems or competition of resources (Ramanna et al., 2025). Environmental effects of the large-scale algal cultivation, such as possible nutrient runoff or alterations in the local biodiversity, have to be properly assessed and addressed in the long term (Ramanna et al., 2025). Figure 4 provides a roadmap for the future development of algal CO_2 sequestration technologies and the policy frameworks associated with them, indicating a clear direction for the scientific and regulatory development of the field.

Figure 4. Roadmap for future development of algal CO_2 sequestration technology and policy

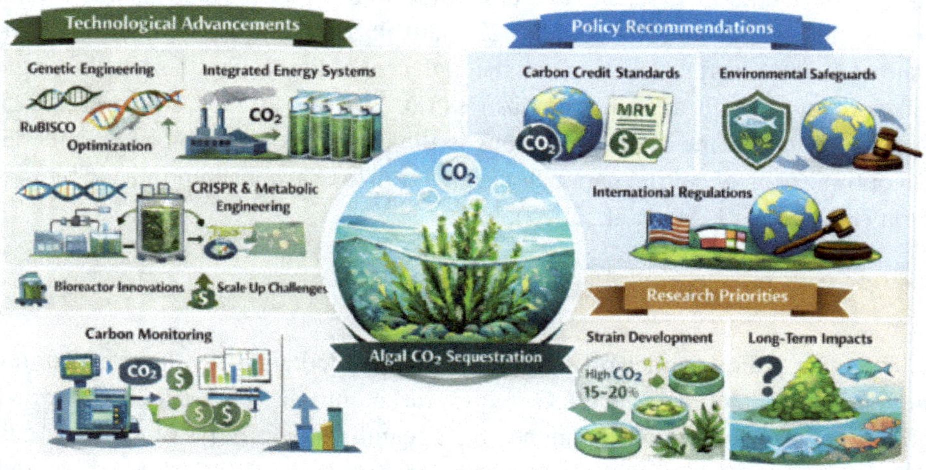

7.3. Research Priorities

Further research is required to address the existing constraints and to exploit the potential of algal CO_2 sequestration to the maximum. One of the key concerns is to improve the sequestered carbon monitoring and measurement processes. The regularity of the measurement of the CO_2 uptake and biomass carbon content is crucial to gain trust in the carbon markets and reveal a positive effect on the environment (Souza et al., 2024). It entails the planning of advanced sensors and methods of real-time observation of the CO_2 fixation rate and development of biomass (Behera et al., 2020). The second one is to streamline and choose the algal strains in CO_2 uptake (J. Ma et al., 2025). Synergistic interactions between the bacteria and the algae can also help improve the carbon fixation as well as the resilience of the system (Hu et al., 2025). Lastly, it is important to fill the gaps in knowledge regarding the long-term effects of algal sequestration. This involves determining what happens to carbon that is sequestered in the biomass, the stability of algal products in the long term, and the ecological footprint of large-scale cultivation over a long period (Liu et al., 2020; Ramanna et al., 2025). The following Table 5 describes the key research priorities in the improvement of algal CO_2 sequestration, indicating the activities and outcomes to be promoted, including increased CO_2 fixation and increased economic feasibility.

Table 5. Key Research Priorities and Potential Outcomes for Advancing Algal CO_2 Sequestration

Research Priority	Key Activities	Potential Outcomes	Reference
Genetic Engineering for Enhanced CO_2 Fixation	- Modifying RuBisCO for increased CO_2 specificity	- Development of algal strains with significantly higher CO_2 uptake rates and biomass productivity	(Barati et al., 2021)
	- Optimizing carbon-concentrating mechanisms (CCMs)	- Improved photosynthetic efficiency under diverse environmental conditions	(Barati et al., 2021; Sadvakasova et al., 2025)
	- Metabolic pathway engineering for increased carbon partitioning to desired products	- Enhanced production of biofuels, bioplastics, and other high-value compounds from algal biomass	(Dasan et al., 2025; Priyadharsini et al., 2022)
Integration with Renewable Energy Technologies	- Co-location of algal farms with industrial CO_2 emitters	- Reduced greenhouse gas emissions from industrial sources and increased production of sustainable energy	(Dasan et al., 2025; Priyadharsini et al., 2022)
	- Development of efficient energy recovery systems from algal biomass	- Creation of closed-loop systems that minimize waste and maximize resource utilization, supporting a circular bioeconomy	(Sadvakasova et al., 2025)
Scaling Up and Commercialization	- Innovative bioreactor designs for improved mass transfer and light distribution	- Cost-effective, large-scale algal cultivation systems for industrial applications	(Hou et al., 2025; Ma et al., 2025)
	- Development of low-cost harvesting and processing technologies	- Economic viability of algal products, fostering market penetration and commercial success	(Sun et al., 2023)
Streamlining Carbon Credit Processes	- Development of standardized MRV protocols for algal sequestration	- Increased credibility and transparency in carbon markets, attracting investment	(Singh et al., 2014)
	- Clear valuation mechanisms for algal carbon credits	- Fair and consistent pricing for sequestered carbon, incentivizing adoption of algal technologies	(Singh et al., 2014)
Harmonizing International Standards	- Collaboration among international bodies to establish unified regulations	- Global acceptance and interoperability of algal carbon sequestration projects, simplifying international trade and compliance	(Singh et al., 2014)
	- Policy development for environmental safeguards and sustainability criteria	- Prevention of negative ecological impacts and assurance of the long-term sustainability of algal cultivation	(Liu et al., 2020; Ramanna et al., 2025)

continued on following page

Table 5. Continued

Research Priority	Key Activities	Potential Outcomes	Reference
Improving Quantification & Monitoring	- Development of advanced sensors and real-time monitoring systems	- Accurate and precise measurement of CO_2 fixation rates and biomass production, improving data for carbon accounting	(Behera et al., 2020; Souza et al., 2024)
	- Integration of remote sensing and AI for large-scale monitoring	- Efficient and comprehensive assessment of environmental performance and progress toward climate goals	(Ren et al., 2023)
Enhancing Strain Selection & Optimization	- Screening and isolation of highly efficient native algal strains	- Identification of robust microalgae with superior CO_2 tolerance and growth characteristics	(Gao et al., 2026)
	- Research into algal-bacterial consortia for synergistic effects	- Development of more resilient and productive cultivation systems through microbial interactions	(Hu et al., 2025)
Addressing Long-Term Environmental Impacts	- Life cycle assessments (LCAs) of algal-based CCUS technologies	- Comprehensive understanding of the cradle-to-grave environmental footprint, guiding sustainable development	(Li et al., 2023; Liu et al., 2020)
	- Studies on ecological effects of large-scale algal cultivation	- Mitigation strategies for potential negative impacts on water quality, biodiversity, and ecosystem services	(Ramanna et al., 2025)

CONCLUSION

The chapter of the book provides algal CO_2 capture as a sustainable and all-encompassing solution to reducing climate change, supplementing carbon markets, and developing the circular bioeconomy. It analyzes the algae-based systems in terms of CO_2 capture using technological, economic, and policy criteria and provides a discussion of the environmental and social impacts. The microalgae and macroalgae are noted to have a high rate of photosynthesis and ability to tolerate carbon dioxide, which enables them to transform biocarbon into useful biomass. Developed as algal sequestration systems and combined with carbon credit schemes, algal sequestration systems can establish the quantifiable and verifiable linkage between environmental sustainability and economic gains. Recent developments such as the production of enhanced strains of algae, application of photobioreactors and hybrid systems have greatly increased rates of carbon fixation. Integrating algal cultivation with wastewater treatment and the production of renewable energy enhances even better resource efficiency and makes the entire process more sustainable. These combined applications demonstrate the versatility of algal CO_2 as a climate mitigation and bioeconomic solution. Nevertheless, the chapter also reveals that there are still challenges, including high costs of capital, significant energy requirements, and the lack of uniform monitoring and verification models. Further studies should focus on enhancing techno-economic efficiency, long-term and secure storage of carbon, and coming up with balanced international certification standards. Altogether, algal CO_2 sequestration is a feasible and proposed approach to the problem that connects science, policy, and sustainable development globally towards net zero.

REFERENCES

Alami, A. H., Alasad, S., Ali, M., & Alshamsi, M. (2021). Investigating algae for CO2 capture and accumulation and simultaneous production of biomass for biodiesel production. *The Science of the Total Environment*, *759*, 143529. DOI: 10.1016/j.scitotenv.2020.143529 PMID: 33229076

An, Y., Wang, D., Du, J., Wang, X., & Xiao, J. (2023). Pyrenoid: Organelle with efficient CO2-Concentrating mechanism in algae. *Journal of Plant Physiology*, *287*, 154044. DOI: 10.1016/j.jplph.2023.154044 PMID: 37392525

Arcusa, S., & Sprenkle-Hyppolite, S. (2022). Snapshot of the Carbon Dioxide Removal Certification and Standards Ecosystem (2021-2022). *Climate Policy*, *22*(9–10), 1319–1332. DOI: 10.1080/14693062.2022.2094308

Aryanpur, V., & Rogan, F. (2024). Decarbonising road freight transport: The role of zero-emission trucks and intangible costs. *Scientific Reports*, *14*(1), 1–11. DOI: 10.1038/s41598-024-52682-4 PMID: 38267587

Ashour, M., Mansour, A. T., Alkhamis, Y. A., & Elshobary, M. (2024). Usage of Chlorella and diverse microalgae for CO2 capture - towards a bioenergy revolution. *Frontiers in Bioengineering and Biotechnology*, *12*, 1387519. DOI: 10.3389/fbioe.2024.1387519 PMID: 39229458

Atikah, M. S. N., Harun, R., Ilyas, R. A., Norrrahim, M. N. F., & Knight, V. F. (2025). Algae cultivation systems integrated with photovoltaic cell: A systematic review. *Results in Engineering*, *27*, 106294. DOI: 10.1016/j.rineng.2025.106294

Avinash, A., Sasikumar, P., & Pugazhendhi, A. (2020). Analysis of the limiting factors for large scale microalgal cultivation: A promising future for renewable and sustainable biofuel industry. *Renewable & Sustainable Energy Reviews*, *134*, 110250. DOI: 10.1016/j.rser.2020.110250

Barati, B., Zeng, K., Baeyens, J., Wang, S., Addy, M., Gan, S. Y., & El-Fatah Abomohra, A. (2021). Recent progress in genetically modified microalgae for enhanced carbon dioxide sequestration. *Biomass and Bioenergy*, *145*, 105927. DOI: 10.1016/j.biombioe.2020.105927

Barros, D., Nobre, L., Bessa, J., Leite, L., Mota, C., Cunha, F., & Fangueiro, R. (2024). Bio-Inspired Eco-Composite Materials Seaweed Waste Integration for Sustainable Structural Applications. *Sustainability 2024, Vol. 16*, *16*(24). DOI: 10.3390/su162411051

Baskaran, D., Saravanan, P., Nagarajan, L., & Byun, H. S. (2024). An overview of technologies for capturing, storing, and utilizing carbon dioxide: Technology readiness, large-scale demonstration, and cost. *Chemical Engineering Journal, 491*, 151998. DOI: 10.1016/j.cej.2024.151998

Béchet, Q., Plouviez, M., Chambonnière, P., & Guieysse, B. (2017). Environmental impacts of full-scale algae cultivation. *Microalgae-Based Biofuels and Bioproducts: From Feedstock Cultivation to End-Products*, 505–525. DOI: 10.1016/B978-0-08-101023-5.00021-2

Behera, B., Aly, N., & Paramasivan, B. (2020). Theoretical Modeling of Algal Productivity and Carbon Capture Potential in Selected Places of Odisha, India. *Journal of The Institution of Engineers (India): Series A 2020 101:3, 101*(3), 503–512. DOI: 10.1007/s40030-020-00450-8

Behera, B., & Paramasivan, B. (2021). Uncertainty analysis and stochastic studies of techno-economics of algal carbon sequestration at Indian coal powered plants. *Environmental Technology & Innovation, 24*, 101897. DOI: 10.1016/j.eti.2021.101897

Bennett, M., March, A., Raguain, J., & Failler, P. (2024). Blueprint for Blue Carbon: Lessons from Seychelles for Small Island States. *Oceans 2024, Vol. 5, Pages 81-108, 5*(1), 81–108. DOI: 10.3390/oceans5010006

Bleuel, S., & Müller, C. (2024). Unlocking the potential: Expert insights on the long-term compatibility of forest carbon credits with the EU ETS. *Forest Policy and Economics, 162*, 103185. DOI: 10.1016/j.forpol.2024.103185

Boettcher, M., Winkler, M., & Michaelowa, A. (2025). Accounting and monitoring challenges for blue carbon enhancement in national climate policy targets and international carbon markets. *Carbon Management, 16*(1), 2585895. Advance online publication. DOI: 10.1080/17583004.2025.2585895

Chen, J., Dai, L., Mataya, D., Cobb, K., Chen, P., & Ruan, R. (2022). Enhanced sustainable integration of CO2 utilization and wastewater treatment using microalgae in circular economy concept. *Bioresource Technology, 366*, 128188. DOI: 10.1016/j.biortech.2022.128188 PMID: 36309175

Chen, L., & Xie, M. (2023). How do hard regimes absorb, overlap, and squeeze out soft regimes? Insights from global carbon markets. *Global Public Policy and Governance 2023 3:1, 3*(1), 60–85. DOI: 10.1007/s43508-023-00064-3

Cheong, B. C. (2025). The Paradox and Fallacy of Global Carbon Credits: A Theoretical Framework for Strengthening Climate Change Mitigation Strategies. *Anthropocene Science 2025 4:1, 4*(1), 72–83. DOI: 10.1007/s44177-025-00084-0

Choi, H., Sung, Y. J., Hong, M. E., Han, J., Min, B. K., & Sim, S. J. (2022). Reconsidering the potential of direct microalgal biomass utilization as end-products: A review. *Renewable & Sustainable Energy Reviews, 155*, 111930. DOI: 10.1016/j.rser.2021.111930

Chopin, T., Costa-Pierce, B. A., Troell, M., Hurd, C. L., Costello, M. J., Backman, S., Buschmann, A. H., Cuhel, R., Duarte, C. M., Gröndahl, F., Heasman, K., Haroun, R. J., Johansen, J., Jueterbock, A., Lench, M., Lindell, S., Pavia, H., Ricart, A. M., Sundell, K. S., & Yarish, C. (2024). Deep-ocean seaweed dumping for carbon sequestration: Questionable, risky, and not the best use of valuable biomass. *One Earth, 7*(3), 359–364. DOI: 10.1016/j.oneear.2024.01.013

Chua, E., & Palevsky, H. (2026). *Filling the monitoring gap: Aquatic ecosystem metabolism as a cost-effective, scalable tool for assessing marine carbon dioxide removal.* https://doi.org/DOI: 10.31223/X5NF49

Cole, G. M., Greene, J. M., Quinn, J. C., McDaniel, B., Kemp, L., Simmons, D., Hodges, T., Nobles, D., Weiss, T. L., McGowen, J., & McDaniel, S. (2023). Integrated techno-economic and life cycle assessment of a novel algae-based coating for direct air carbon capture and sequestration. *Journal of CO2 Utilization, 69*, 102421. DOI: 10.1016/j.jcou.2023.102421

Cook, G., Zakkour, P., Neades, S., & Dixon, T. (2024). CCS under Article 6 of the Paris Agreement. *International Journal of Greenhouse Gas Control, 134*, 104110. DOI: 10.1016/j.ijggc.2024.104110

Dahai, H., Zhihong, Y., Lin, Q., Yuhong, L., Lei, T., Jiang, L., & Liandong, Z. (2024). The application of magical microalgae in carbon sequestration and emission reduction: Removal mechanisms and potential analysis. *Renewable & Sustainable Energy Reviews, 197*, 114417. DOI: 10.1016/j.rser.2024.114417

Daneshvar, E., Wicker, R. J., Show, P. L., & Bhatnagar, A. (2022). Biologically-mediated carbon capture and utilization by microalgae towards sustainable CO2 biofixation and biomass valorization – A review. *Chemical Engineering Journal, 427*, 130884. DOI: 10.1016/j.cej.2021.130884

Dasan, Y. K., Lam, M. K., Yusup, S., Lim, J. W., Show, P. L., Tan, I. S., & Lee, K. T. (2020). Cultivation of Chlorella vulgaris using sequential-flow bubble column photobioreactor: A stress-inducing strategy for lipid accumulation and carbon dioxide fixation. *Journal of CO2 Utilization, 41*, 101226. DOI: 10.1016/j.jcou.2020.101226

Dasan, Y. K., Suparmaniam, U., Lam, M. K., Tan, I. S., Lau, S. Y., & Bridgid Lai, F. C. (2025). CO2 mitigation via seaweed and microalgae farming for sustainable biofuel production in the blue bioeconomy. *Current Opinion in Green and Sustainable Chemistry, 56*, 101053. DOI: 10.1016/j.cogsc.2025.101053

David, K. M., Ghadge, S. A., Rathod, M. K., Pandey, R., Shamkuwar, S. G., Babarao, P. Y., & Panigrahi, C. K. (2025). Microalgal Systems for Efficient CO\(_2\) Utilisation and Carbon Sequestration: A Review. *Journal of Advances in Biology & Biotechnology*, *28*(8), 1304–1317. DOI: 10.9734/jabb/2025/v28i82804

de Souza, M. F., Meers, E., & Mangini, S. (2024). The potential of microalgae for carbon capture and sequestration. *EFB Bioeconomy Journal*, *4*, 100067. DOI: 10.1016/j.bioeco.2024.100067

de Souza Celente, G., de Cassia de Souza Schneider, R., Julich, J., Rizzetti, T. M., Lobo, E. A., & Sui, Y. (2023). Life cycle assessment of microalgal cultivation medium: biomass, glycerol, and beta-carotene production by Dunaliella salina and Dunaliella tertiolecta. *The International Journal of Life Cycle Assessment 2023 29:12*, *29*(12), 2269–2282. DOI: 10.1007/s11367-023-02209-2

Do, C. V. T., Nguyen, N. T. T., Tran, T. D., Pham, M. H. T., & Pham, T. Y. T. (2021). Capability of carbon fixation in bicarbonate-based and carbon dioxide-based systems by Scenedesmus acuminatus TH04. *Biochemical Engineering Journal*, *166*, 107858. DOI: 10.1016/j.bej.2020.107858

Dolliver, J., & O'Connor, N. E. (2022). Estimating growth, loss and potential carbon sequestration of farmed kelp: A case study of Saccharina latissima at Strangford Lough, Northern Ireland. *Applied Phycology*, *3*(1), 324–339. DOI: 10.1080/26388081.2022.2081934

Drabold, E., McGaughy, K., Agner, J., Sellars, D., Johnson, R., Hajer, A. A., Reza, M. T., & Bayless, D. (2020). Challenges and process economics for algal carbon capture with novel integration: Hydrothermal carbonization. *Bioresource Technology Reports*, *12*, 100556. DOI: 10.1016/j.biteb.2020.100556

El Zrelli, R., Yacoubi, L., Wakkaf, T., Castet, S., Grégoire, M., Mansour, L., Courjault-Radé, P., & Rabaoui, L. (2021). Surface sediment enrichment with trace metals in a heavily human-impacted lagoon (Bizerte Lagoon, Southern Mediterranean Sea): Spatial distribution, ecological risk assessment, and implications for environmental protection. *Marine Pollution Bulletin*, *169*, 112512. DOI: 10.1016/j.marpolbul.2021.112512 PMID: 34052590

Elouardi, M., Mabrouki, J., & El Belghiti, M. A. (2023). Evaluation of algal absorbed CO2 sequestration techniques: a review. *Euro-Mediterranean Journal for Environmental Integration 2023 8:3*, *8*(3), 481–491. DOI: 10.1007/s41207-023-00379-x

Enríquez-de-Salamanca, Á. (2024). Environmental and social impacts of carbon sequestration. *Integrated Environmental Assessment and Management*, *20*(6), 1812–1838. DOI: 10.1002/ieam.4925 PMID: 38651985

Farooqui, A., Tripathi, G., Moheet, K., Dubey, P., Ahmad, S., Husain, A., Shamim, A., & Mahfooz, S. (2021). *Algal Biomass: Potential Renewable Feedstock for Bioenergy Production*. 85–113. DOI: 10.1007/978-981-16-1888-8_5

Fei, C., Wilson, A. T., Mangan, N. M., Wingreen, N. S., & Jonikas, M. C. (2022). Modelling the pyrenoid-based CO2-concentrating mechanism provides insights into its operating principles and a roadmap for its engineering into crops. *Nature Plants 2022 8:5, 8*(5), 583–595. DOI: 10.1038/s41477-022-01153-7

Friedlingstein, P., O'Sullivan, M., Jones, M. W., Andrew, R. M., Bakker, D. C. E., Hauck, J., Landschützer, P., Le Quéré, C., Li, H., Luijkx, I. T., Peters, G. P., Peters, W., Pongratz, J., Schwingshackl, C., Sitch, S., Canadell, J. G., Ciais, P., Aas, K., Alin, S. R., & Zeng, J. (2025). Global Carbon Budget 2025. *Earth System Science Data Discussions*, 1–139. DOI: 10.5194/essd-2025-659

Fujii, H., Webb, J., Mundree, S., Rowlings, D., Grace, P., Wilson, C., & Managi, S. (2024). Priority change and driving factors in the voluntary carbon offset market. *Cleaner Environmental Systems, 13*, 100164. DOI: 10.1016/j.cesys.2024.100164

Gaber, K., Rösch, C., & Biondi, N. (2021). Life Cycle Assessment of Total Fatty Acid (TFA) Production from Microalgae Nannochloropsis oceanica at Different Sites and Under Different Sustainability Scenarios. *BioEnergy Research 2021 15:3, 15*(3), 1595–1615. DOI: 10.1007/s12155-021-10279-z

Gao, X., Yuan, L., Yeong, H. Y., Lv, G., Zhang, X., Zhang, X., Zhang, L., & Liao, J. (2026). YEONG, H. Y., Lv, G., Zhang, X., Zhang, X., Zhang, L., & Liao, J. (2026). Physiological changes in carbon sequestration and lipid production characteristics of microalgae under high CO2 domestication. *Environmental Science and Pollution Research International, 33*(2), 1–15. DOI: 10.1007/s11356-025-37345-9 PMID: 41498902

Green, J. F. (2024). The Climate Establishment and the Paris partnerships. *Climatic Change 2024 177:6, 177*(6), 84-. DOI: 10.1007/s10584-024-03730-5

Gressel, J., van der Vlugt, C. J. B., & Bergmans, H. E. N. (2013). Environmental risks of large scale cultivation of microalgae: Mitigation of spills. *Algal Research, 2*(3), 286–298. DOI: 10.1016/j.algal.2013.04.002

Guan, Y., Rausser, G., Zilberman, D., & Choi, E. (2026). Alternative Bioeconomy Strategies for Decarbonization. *Natural Resource Management and Policy, 61*, 175–207. DOI: 10.1007/978-3-032-07112-5_11

Günther, S. A., Staake, T., Schöb, S., & Tiefenbeck, V. (2020). The behavioral response to a corporate carbon offset program: A field experiment on adverse effects and mitigation strategies. *Global Environmental Change*, *64*, 102123. DOI: 10.1016/j.gloenvcha.2020.102123

Hanson, E., Nwakile, C., & Hammed, V. O. (2025). Carbon capture, utilization, and storage (CCUS) technologies: Evaluating the effectiveness of advanced CCUS solutions for reducing CO2 emissions. *Results in Surfaces and Interfaces*, *18*, 100381. DOI: 10.1016/j.rsurfi.2024.100381

Hasnain, M., Zainab, R., Ali, F., Abideen, Z., Yong, J. W. H., El-Keblawy, A., Hashmi, S., & Radicetti, E. (2023). Utilization of microalgal-bacterial energy nexus improves CO2 sequestration and remediation of wastewater pollutants for beneficial environmental services. *Ecotoxicology and Environmental Safety*, *267*, 115646. DOI: 10.1016/j.ecoenv.2023.115646 PMID: 37939556

Hou, Y., Wang, W., Liu, Z., Yu, L., & Zhao, L. (2025). Boosting microalgae-based carbon sequestration with the artificial CO2 concentration system. *Critical Reviews in Biotechnology*, *45*(8), 1615–1633. DOI: 10.1080/07388551.2025.2498464 PMID: 40374568

Hu, L., Ye, Y., Li, Y., Tan, X., Liu, X., Zhang, T., Wang, J., Du, Z., & Ye, M. (2025). Bacteria–algae synergy in carbon sequestration: Molecular mechanisms, ecological dynamics, and biotechnological innovations. *Biotechnology Advances*, *83*, 108655. DOI: 10.1016/j.biotechadv.2025.108655 PMID: 40701356

Ighalo, J. O., Dulta, K., Kurniawan, S. B., Omoarukhe, F. O., Ewuzie, U., Eshiemogie, S. O., Ojo, A. U., & Abdullah, S. R. S. (2022). Progress in Microalgae Application for CO2 Sequestration. *Cleaner Chemical Engineering*, *3*, 100044. DOI: 10.1016/j.clce.2022.100044

Iqbal, K., Sharma, N., Takkar, S., Shukla, S., Shukla, K., Varma, A., & Mishra, A. (2022). Integrated CO2 sequestration, wastewater treatment, and biofuel production by microalgae culturing: Needs and limitations. *Integrated Environmental Technologies for Wastewater Treatment and Sustainable Development*, 217–240. DOI: 10.1016/B978-0-323-91180-1.00027-2

Ishizaki, R., Noguchi, R., Putra, A. S., Ichikawa, S., Ahamed, T., & Watanabe, M. M. (2020). Reduction in Energy Requirement and CO2 Emission for Microalgae Oil Production Using Wastewater. *Energies 2020, Vol. 13, 13*(7). DOI: 10.3390/en13071641

Islam, M. M., Alam, H., Acharjee, A., Salatul, M., & Mozumder, I. (2024). Establishment of an effective photobioreactor for growing microalgae: A review. *International Journal of Agricultural Research, Innovation and Technology*, *14*(2), 153–162. DOI: 10.3329/ijarit.v14i2.79511

Jain, S., Sharma, T., & Gupta, A. K. (2022). End-of-life management of solar PV waste in India: Situation analysis and proposed policy framework. *Renewable & Sustainable Energy Reviews*, *153*, 111774. DOI: 10.1016/j.rser.2021.111774

Jaiswal, A., Babu, V., Baishya, B., & Jaiswal, P. (2020). Biochemical Pathways Regulated by Algae to Mitigate Global Carbon Emissions: A Review. *Journal of Environmental Pathology, Toxicology and Oncology : Official Organ of the International Society for Environmental Toxicology and Cancer*, *39*(4), 317–334. DOI: 10.1615/JEnvironPatholToxicolOncol.2020034921 PMID: 33389904

Kanna Dasan, Y., Lam, M. K., Chai, Y. H., Lim, J. W., Ho, Y. C., Tan, I. S., Lau, S. Y., Show, P. L., & Lee, K. T. (2023). Unlocking the potential of microalgae bio-factories for carbon dioxide mitigation: A comprehensive exploration of recent advances, key challenges, and energy-economic insights. *Bioresource Technology*, *380*, 129094. DOI: 10.1016/j.biortech.2023.129094 PMID: 37100295

Király, É., & Borovics, A. (2024). Carbon sequestration of Hungarian forests by management system and protection status. *Trees, Forests and People*, *15*, 100511. DOI: 10.1016/j.tfp.2024.100511

Kumar, A., Sharma, A., Majumder, P., Vishal, V., & Dutta, A. (2025). Chemical and microalgal conversion of carbon dioxide into fuels and materials: a review. *Environmental Chemistry Letters 2025 23:5, 23*(5), 1209–1229. DOI: 10.1007/s10311-025-01855-2

Kumar, K., Mishra, S. K., Choi, G. G., & Yang, J. W. (2016). CO2 Sequestration Through Algal Biomass Production. *Algal Biorefinery: An Integrated Approach*, 35–57. https://doi.org/DOI: 10.1007/978-3-319-22813-6_2/SAVE-RESEARCH

Kumar, R. R., Jency, S. M., Bhavatarini, G., & Devapriya, M. J. P. (2022). CO2 sequestration: Microalgae genome analysis and its application of effective green source technology. *Plant Science Today*, *9*(2), 243–261. DOI: 10.14719/pst.1365

Kusi, P. A., McGee, D., Tabraiz, S., & Ahmed, A. (2024). Bicarbonate concentration influences carbon utilization rates and biochemical profiles of freshwater and marine microalgae. *Biotechnology Journal*, *19*(8), 2400361. DOI: 10.1002/biot.202400361 PMID: 39212191

Kuyper, J., Schroeder, H., & Linnér, B. O. (2018). The evolution of the UNFCCC. *Annual Review of Environment and Resources, 43*(Volume 43, 2018), 343–368. https://doi.org/DOI: 10.1146/ANNUREV-ENVIRON-102017-030119/CITE/REFWORKS

Leong, Y. K., & Chang, J. S. (2023). Waste stream valorization-based low-carbon bioeconomy utilizing algae as a biorefinery platform. *Renewable & Sustainable Energy Reviews, 178*, 113245. DOI: 10.1016/j.rser.2023.113245

Li, D., Dong, H., Cao, X., Wang, W., & Li, C. (2023). Enhancing photosynthetic CO2 fixation by assembling metal-organic frameworks on Chlorella pyrenoidosa. *Nature Communications 2023 14:1, 14*(1), 5337-. DOI: 10.1038/s41467-023-40839-0

Li, D., & Zhao, Q. (2023). Study of carbon fixation and carbon partitioning of evolved Chlorella sp.'s strain under different carbon dioxide conditions. *Biocatalysis and Agricultural Biotechnology, 48*, 102655. DOI: 10.1016/j.bcab.2023.102655

Li, G., & Yao, J. (2024). A Review of Algae-Based Carbon Capture, Utilization, and Storage (Algae-Based CCUS). *Gases 2024, Vol. 4, Pages 468-503, 4*(4), 468–503. DOI: 10.3390/gases4040024

Li, H., Kong, L., Ai, W., & Zhang, Y. (2025). Algal systems reimagined: exploring the potential for sustainable solutions to diverse wastewater challenges. *Blue Biotechnology 2025 2:1, 2*(1), 21-. DOI: 10.1186/s44315-025-00033-x

Li, R., Pan, J., Xing, J., & Yang, W. (2025). *The Secret of High Carbon Fixation Efficiency in Chlamydomonas and Its Potential to Boost Yields in Land Plants*. Green Carbon., DOI: 10.1016/j.greenca.2025.02.005

Liang, R. X., Hung, J. C., Diankristanti, P. A., Chen, Y. T., Chung, C. W., & Ng, I. S. (2025). Enhanced carbon capture and utilization in transgenic Chlorella sorokiniana harboring pyridoxal kinase under dynamic carbon dioxide levels. *Bioresource Technology, 425*, 132315. DOI: 10.1016/j.biortech.2025.132315 PMID: 40023335

Lin, Z., Kuang, Y., Li, W., & Zheng, Y. (2024). Research status and prospects of CO2 geological sequestration technology from onshore to offshore: A review. *Earth-Science Reviews, 258*, 104928. DOI: 10.1016/j.earscirev.2024.104928

Liu, Z., Liu, C., Han, S., & Yang, X. (2020). Optimization upstream CO2 deliverable with downstream algae deliverable in quantity and quality and its impact on energy consumption. *The Science of the Total Environment, 709*, 136197. DOI: 10.1016/j.scitotenv.2019.136197 PMID: 31887503

Llamas, B., Suárez-Rodríguez, M. C., González-López, C. V., Mora, P., & Acién, F. G. (2021). Techno-economic analysis of microalgae related processes for CO2 bio-fixation. *Algal Research, 57*, 102339. DOI: 10.1016/j.algal.2021.102339

Low, S., Fritz, L., Baum, C. M., & Sovacool, B. K. (2024). Public perceptions on carbon removal from focus groups in 22 countries. *Nature Communications 2024 15:1, 15*(1), 3453-. DOI: 10.1038/s41467-024-47853-w

Ma, J., Cui, Y., Sun, F., Zhou, R., Liu, F., Cao, K., Zhu, B., Zhang, H., Meng, C., Chen, G., & Gao, Z. (2025). Strategies for enhancing microalgal carbon sequestration: A review on strain development, culture system optimization, parameter control, and metabolic engineering. *Bioresource Technology, 435*, 132868. DOI: 10.1016/j.biortech.2025.132868 PMID: 40555329

Ma, Z., Cheah, W. Y., Ng, I. S., Chang, J. S., Zhao, M., & Show, P. L. (2022). Microalgae-based biotechnological sequestration of carbon dioxide for net zero emissions. *Trends in Biotechnology, 40*(12), 1439–1453. DOI: 10.1016/j.tibtech.2022.09.002 PMID: 36216714

Mamilla, S., Sahu, N., & Sridhar, S. (2024). Algal Biomass as Global Demand for Alternative Biofuel. *Algal Biotechnology: Applications for Industrial Development and Human Welfare*, 157–172. https://doi.org/DOI: 10.1201/9781003219194-12/ALGAL-BIOMASS-GLOBAL-DEMAND-ALTERNATIVE-BIOFUEL-SHYAMSUNDER-MAMILLA-NIVEDITA-SAHU-SRIDHAR

Marquardt, J., Fünfgeld, A., & Elsässer, J. P. (2023). Institutionalizing climate change mitigation in the Global South: Current trends and future research. *Earth System Governance, 15*, 100163. DOI: 10.1016/j.esg.2022.100163

Merk, C., Grunau, J., Riekhof, M. C., & Rickels, W. (2022). The need for local governance of global commons: The example of blue carbon ecosystems. *Ecological Economics, 201*, 107581. DOI: 10.1016/j.ecolecon.2022.107581

Mustafa, Z., Auroona, N., Sarwar, A., & Lee, E. Y. (2025). Recent advancements in microbial carbon dioxide fixation: Metabolic engineering strategies. *Green Chemistry, 28*(1), 21–36. DOI: 10.1039/D5GC03349G

Nandhini, V., & Venkataraman, R. (2025). Carbon Credit Mechanisms: A Market Based Approach to Climate Mitigation. *Archives of Current Research International, 25*(6), 1–14. DOI: 10.9734/acri/2025/v25i61246

Nazari, M. T., Freitag, J. F., Cavanhi, V. A. F., & Colla, L. M. (2020). Microalgae harvesting by fungal-assisted bioflocculation. *Reviews in Environmental Science and Bio/Technology 2020 19:2, 19*(2), 369–388. DOI: 10.1007/s11157-020-09528-y

Ngo, H. H., Vo, H. N. P., Guo, W., Lee, D. jong, & Zhang, S. (2022). Carbon dioxide fixation and phycoremediation by algae-based technologies for biofuels and biomaterials. *Biomass, Biofuels, Biochemicals: Climate Change Mitigation: Sequestration of Green House Gases*, 253–277. DOI: 10.1016/B978-0-12-823500-3.00011-X

Ogunyemi, F. M. (2025). Harmonizing global carbon accounting standards: Comparative analysis of GHG protocol, ISO 14064, and emerging regulatory frameworks for multinational corporations. *International Journal of Research in Marketing Management and Sales*, *7*(2), 425–434. DOI: 10.33545/26633329.2025.v7.i2e.313

Ojaniemi, U., Tamminen, A., Syrjänen, J., & Barth, D. (2025). CFD modeling of CO2 fixation by microalgae cultivated in a lab scale photobioreactor. *Bioresource Technology*, *415*, 131715. DOI: 10.1016/j.biortech.2024.131715 PMID: 3949054_

Otundo Richard, M. (2024). Policy Framework and Implementation Strategies for Carbon Dioxide Capture and Storage in Kenya's Coastal Region. SSRN *Electronic Journal*. https://doi.org/DOI: 10.2139/SSRN.4936151

Perera, N. S., Costa, M. D. P., Macreadie, P. I., & Wartman, M. (2025). Trends in Market-Based Blue Carbon Projects. *Sustainable Development (Bradford)*, *33*(3), 3377–3402. DOI: 10.1002/sd.3293

Pessarrodona, A., Franco-Santos, R. M., Wright, L. S., Vanderklift, M. A., Howard, J., Pidgeon, E., Wernberg, T., & Filbee-Dexter, K. (2023a). Carbon sequestration and climate change mitigation using macroalgae: A state of knowledge review. *Biological Reviews of the Cambridge Philosophical Society*, *98*(6), 1945–1971. DOI: 10.1111/brv.12990 PMID: 37437379

Pessarrodona, A., Franco-Santos, R. M., Wright, L. S., Vanderklift, M. A., Howard, J., Pidgeon, E., Wernberg, T., & Filbee-Dexter, K. (2023b). Carbon sequestration and climate change mitigation using macroalgae: A state of knowledge review. *Biological Reviews of the Cambridge Philosophical Society*, *98*(6), 1945–1971. DOI: 10.1111/brv.12990 PMID: 37437379

Plastina, A., Jo, H., & Wongpiyabovorn, O. (2024). The business case for carbon farming in the USA. *Carbon Balance and Management 2024 19:1, 19*(1), 7-. DOI: 10.1186/s13021-024-00253-5

Porcelli, R., Dotto, F., Pezzolesi, L., Marazza, D., Greggio, N., & Righi, S. (2020). Comparative life cycle assessment of microalgae cultivation for non-energy purposes using different carbon dioxide sources. *The Science of the Total Environment*, *721*, 137714. DOI: 10.1016/j.scitotenv.2020.137714 PMID: 32171140

Portner, B. W., Valente, A., & Guenther, S. (2021). Sustainability Assessment of Combined Animal Fodder and Fuel Production from Microalgal Biomass. *International Journal of Environmental Research and Public Health, 18*(21), 11351. Advance online publication. DOI: 10.3390/ijerph182111351 PMID: 34769867

Prasad, R., Gupta, S. K., Shabnam, N., Oliveira, C. Y. B., Nema, A. K., Ansari, F. A., & Bux, F. (2021). Role of Microalgae in Global CO2 Sequestration: Physiological Mechanism, Recent Development, Challenges, and Future Prospective. *Sustainability 2021, Vol. 13, 13*(23). DOI: 10.3390/su132313061

Priyadharsini, P., Nirmala, N., Dawn, S. S., Baskaran, A., SundarRajan, P., Gopinath, K. P., & Arun, J. (2022). Genetic improvement of microalgae for enhanced carbon dioxide sequestration and enriched biomass productivity: Review on CO2 bio-fixation pathways modifications. *Algal Research, 66*, 102810. DOI: 10.1016/j.algal.2022.102810

Quiroz, D., Greene, J. M., Limb, B. J., & Quinn, J. C. (2023). Global Life Cycle and Techno-Economic Assessment of Algal-Based Biofuels. *Environmental Science & Technology, 57*(31), 11541–11551. DOI: 10.1021/acs.est.3c02892 PMID: 37499260

Ramanna, L., Ansari, F. A., Rawat, I., & Bux, F. (2025). Microalgae-driven carbon sequestration and bio-fertiliser: Steps towards a sustainable future. *Chemical Engineering Journal, 519*, 164892. DOI: 10.1016/j.cej.2025.164892

Rame, R., Purwanto, P., & Sudarno, S. (2023). Sustainable energy harnessing: Microalgae as a potential biofuel source and carbon sequestration solution. *Renewable Energy Focus, 47*, 100498. DOI: 10.1016/j.ref.2023.100498

Ren, H., Zhou, D., Lu, J., Show, P. L., & Sun, F. F. (2023). Mapping the field of microalgae CO2 sequestration: a bibliometric analysis. *Environmental Science and Pollution Research 2023 30:32, 30*(32), 78030–78040. DOI: 10.1007/s11356-023-27850-0

Rios Pinto, L. F., Ferreira, G. F., & Tasic, M. (2021). Cultivation techniques. *Microalgae: Cultivation, Recovery of Compounds and Applications*, 1–33. DOI: 10.1016/B978-0-12-821218-9.00001-3

Rossi, S., Carecci, D., Marazzi, F., Di Benedetto, F., Mezzanotte, V., Parati, K., Alberti, D., Geraci, I., & Ficara, E. (2024). Integrating microalgae growth in biomethane plants: Process design, modelling, and cost evaluation. *Heliyon, 10*(1), e23240. DOI: 10.1016/j.heliyon.2023.e23240 PMID: 38163195

Sadvakasova, A. K., Balouch, H., Bauenova, M. O., Kossalbayev, B. D., Allakhverdiev, S. I., & Shabala, S. (2025). Strategies to intensify CO2 capture by microalgae for the circular bioeconomy. *Trends in Plant Science*. Advance online publication. DOI: 10.1016/j.tplants.2025.10.012 PMID: 41168026

Santos, A. A., Santos, A. A., & Garbossa, L. H. P. (2024). Potencial de sequestro do dióxido de carbono na biomassa de macroalga Kappaphycus alvarezii nas fazendas marinhas de Santa Catarina. *Agropecuária Catarinense*, *37*(2), 16–18. DOI: 10.52945/rac.v37i2.1815

Sarat Chandra, T., Maneesh Kumar, M., Mukherji, S., Chauhan, V. S., Sarada, R., & Mudliar, S. N. (2018). Comparative life cycle assessment of microalgae-mediated CO2 capture in open raceway pond and airlift photobioreactor system. *Clean Technologies and Environmental Policy 2018 20:10, 20*(10), 2357–2364. DOI: 10.1007/s10098-018-1612-5

Schipper, K., Al Jabri, H. M. S. J., Wijffels, R. H., & Barbosa, M. J. (2023). Realizing algae value chains in arid environments: An Arabian Peninsula perspective. *Trends in Biotechnology*, *41*(6), 750–759. DOI: 10.1016/j.tibtech.2022.11.004 PMID: 36581482

Singh, S. K., Dixit, K., & Sundaram, S. (2014). Algal-based CO 2 Sequestration Technology and Global Scenario of Carbon Credit Market: A Review. [AJER]. *American Journal of Engineering Research*, *03*, 35–37. www.ajer.org. DOI: 10.7603/s40632-015-0035-2

Sovacool, B. K. (2023). Expanding carbon removal to the Global South: Thematic concerns on systems, justice, and climate governance. *Energy and Climate Change*, *4*, 100103. DOI: 10.1016/j.egycc.2023.100103

Srimongkol, P., Sangtanoo, P., Songserm, P., Watsuntorn, W., & Karnchanatat, A. (2022). Microalgae-based wastewater treatment for developing economic and environmental sustainability: Current status and future prospects. *Frontiers in Bioengineering and Biotechnology*, *10*, 904046. DOI: 10.3389/fbioe.2022.904046 PMID: 36159694

Sun, X., Dregulo, A. M., Zhenni, S., Karthikeyan, O. P., Harirchi, S., Salama, E. S., Li, Y., Sindhu, R., Binod, P., Pandey, A., & Awasthi, M. K. (2023). Role of Marine Algae for GHG Reduction/CO2 Sequestration: Strategies and Applications. *Sustainable Marine Food and Feed Production Technologies*, 157–166. https://doi.org/DOI: 10.1201/9781003326946-12/ROLE-MARINE-ALGAE-GHG-REDUCTION-CO2-SEQUESTRATION-XINWEI-SUN-ANDREI-MIKHAILOVICH-DREGULO-SU-ZHENNI-OBULISAMY-PARTHIBA-KARTHIKEYAN-EL-SAYED-SALAMA-YUE-LI-RAVEENDRAN-SINDHU-PARAMESWARAN-BINOD-ASHOK-PANDEY-MUKESH-KUMAR-AWASTHI

Sun, Z., Bo, C., Cao, S., & Sun, L. (2025). Enhancing CO2 Fixation in Microalgal Systems: Mechanistic Insights and Bioreactor Strategies. *Marine Drugs 2025, Vol. 23, 23*(3). DOI: 10.3390/md23030113

Tanaka, T., Maeda, Y., Suhaimi, N., Tsuneoka, C., Nonoyama, T., Yoshino, T., Kato, N., & Lauersen, K. J. (2021). Intron-mediated enhancement of transgene expression in the oleaginous diatom Fistulifera solaris towards bisabolene production. *Algal Research, 57*, 102345. DOI: 10.1016/j.algal.2021.102345

Tarafdar, A., Sowmya, G., Yogeshwari, K., Rattu, G., Negi, T., Awasthi, M. K., Hoang, A. T., & Sirohi, R. (2023). Environmental pollution mitigation through utilization of carbon dioxide by microalgae. *Environmental Pollution, 328*, 121623. DOI: 10.1016/j.envpol.2023.121623 PMID: 37072107

Tavakoli, H., & Barkdoll, B. D. (2020). Blended Lifecycle Integrated Social System Method. *International Journal of Environmental Research 2020 14:6, 14*(6), 727–749. DOI: 10.1007/s41742-020-00284-z

Terjanika, V., Pubule, J., Blumberga, D., & Zarins, E. (2023). Policy instruments for CO2 valorisation support. *E-Prime - Advances in Electrical Engineering. Electronics and Energy, 4*, 100181. DOI: 10.1016/j.prime.2023.100181

Tripathi, S., Choudhary, S., Meena, A., & Poluri, K. M. (2023). Carbon capture, storage, and usage with microalgae: a review. *Environmental Chemistry Letters 2023 21:4, 21*(4), 2085–2128. DOI: 10.1007/s10311-023-01609-y

Udaypal, G., Goswami, R. K., Mehariya, S., & Verma, P. (2024). Advances in microalgae-based carbon sequestration: Current status and future perspectives. *Environmental Research, 249*, 118397. DOI: 10.1016/j.envres.2024.118397 PMID: 38309563

Usher, P. K., Ross, A. B., Camargo-Valero, M. A., Tomlin, A. S., & Gale, W. F. (2014). An overview of the potential environmental impacts of large-scale microalgae cultivation. *Biofuels, 5*(3), 331–349. DOI: 10.1080/17597269.2014.913925

van den Burg, S., Koch, S., Raoult, J., Tsvetkov, B., & Selnes, T. (2023). Time to Include Blue Carbon from Seaweed in Voluntary Carbon Removal Certification. *EuroChoices (Uckfield), 22*(2), 13–19. DOI: 10.1111/1746-692X.12398

Vanderklift, M. A., Herr, D., Lovelock, C. E., Murdiyarso, D., Raw, J. L., & Steven, A. D. L. (2022). A Guide to International Climate Mitigation Policy and Finance Frameworks Relevant to the Protection and Restoration of Blue Carbon Ecosystems. *Frontiers in Marine Science, 9*, 872064. DOI: 10.3389/fmars.2022.872064

Viswanaathan, S., Perumal, P. K., & Sundaram, S. (2022). Integrated Approach for Carbon Sequestration and Wastewater Treatment Using Algal–Bacterial Consortia: Opportunities and Challenges. *Sustainability 2022, Vol. 14, 14*(3). DOI: 10.3390/su14031075

Walsh, M. J., Gerber Van Doren, L., Sills, D. L., Archibald, I., Beal, C. M., Lei, X. G., Huntley, M. E., Johnson, Z., & Greene, C. H. (2016). Algal food and fuel coproduction can mitigate greenhouse gas emissions while improving land and water-use efficiency. *Environmental Research Letters, 11*(11), 114006. DOI: 10.1088/1748-9326/11/11/114006

Wang, L., Zhang, F., Shi, X., Zeng, C., Ahmad, I., Wang, G., Thapa, S., & Xu, X. (2023). Water resources system vulnerability in high mountain areas under climate change. *Journal of Cleaner Production, 403*, 136789. DOI: 10.1016/j.jclepro.2023.136789

Wilson, M. H., Shea, A., Groppo, J., Crofcheck, C., Quiroz, D., Quinn, J. C., & Crocker, M. (2020). Algae-Based Beneficial Re-use of Carbon Emissions Using a Novel Photobioreactor: a Techno-Economic and Life Cycle Analysis. *BioEnergy Research 2020 14:1, 14*(1), 292–302. DOI: 10.1007/s12155-020-10178-9

Workman, M., Heap, R., Mackie, E., & Connon, I. (2023). Decision making for net zero policy design and climate action: considerations for improving translation at the research-policy interface: a UK Carbon Dioxide Removal case study. *Frontiers in Climate, 5*, 1288001. DOI: 10.3389/fclim.2023.1288001

Wylie, L., Sutton-Grier, A. E., & Moore, A. (2016). Keys to successful blue carbon projects: Lessons learned from global case studies. *Marine Policy, 65*, 76–84. DOI: 10.1016/j.marpol.2015.12.020

Xie, W., Aryanpur, V., Deane, P., & Daly, H. E. (2025). Negative emissions technologies in energy system models and mitigation scenarios - a systematic review. *Applied Energy, 380*, 125064. DOI: 10.1016/j.apenergy.2024.125064

Xie, Y., Khoo, K. S., Chew, K. W., Devadas, V. V., Phang, S. J., Lim, H. R., Rajendran, S., & Show, P. L. (2022). Advancement of renewable energy technologies via artificial and microalgae photosynthesis. *Bioresource Technology, 363*, 127830. DOI: 10.1016/j.biortech.2022.127830 PMID: 36029982

Xu, P., Shao, S., Qian, J., Li, J., Xu, R., Liu, J., & Zhou, W. (2024). Scale-up of microalgal systems for decarbonization and bioproducts: Challenges and opportunities. *Bioresource Technology, 398*, 130528. Advance online publication. DOI: 10.1016/j.biortech.2024.130528 PMID: 38437968

Yan, L., Xue, S., Cha, J., Wen, X., Wang, B., Shi, J., Li, P., Zhang, Y., & Xie, T. (2024). Effect of excess CO2 on semi-continuous microalgae systems: Carbon biofixation. *Environmental Technology & Innovation*, *35*, 103645. DOI: 10.1016/j.eti.2024.103645

Yang, R., Jin, X., Zhou, H., Ren, F., Zhang, X., Ma, Z., Yao, L., & Zhang, H. (2024). Regional Differences in Carbon Budgets and Inter-Regional Compensation Zoning: A Case Study of Chongqing, China. *Land 2024, Vol. 13, 13*(9). DOI: 10.3390/land13091495

Zeng, J., Wang, Z., & Chen, G. (2021). Biological characteristics of energy conversion in carbon fixation by microalgae. *Renewable & Sustainable Energy Reviews*, *152*, 111661. DOI: 10.1016/j.rser.2021.111661

Zhang, B., Gong, M., & Dong, C. (2025). Microalgae photosynthetic carbon sequestration technology, financing subsidies and carbon trading mechanism: An evolutionary game mechanism modeling study from a biomechanical perspective. *Molecular & Cellular Biomechanics*, *22*(1), 1005–1005. DOI: 10.62617/mcb1005

Zhang, D., An, S., Yao, R., Fu, W., Han, Y., Du, M., Chen, Z., Lei, A., & Wang, J. (2022). Life cycle assessment of auto-tropically cultivated economic microalgae for final products such as food, total fatty acids, and bio-oil. *Frontiers in Marine Science*, *9*, 990635. DOI: 10.3389/fmars.2022.990635

Zhao, B., & Su, Y. (2020). Macro assessment of microalgae-based CO2 sequestration: Environmental and energy effects. *Algal Research*, *51*, 102066. DOI: 10.1016/j.algal.2020.102066

Zhao, S., Chen, H., Jia, C., Wang, Y., Xin, C., & Jiang, X. (2024). Comprehensive Evaluation and Scheme Optimization for Power Transmission and Transformation Projects with the Adoption of Carbon Reduction Technologies. *Energies 2024, Vol. 17, 17*(3). DOI: 10.3390/en17030598

Zieliński, M., Dębowski, M., Kazimierowicz, J., & Świca, I. (2023). Microalgal Carbon Dioxide (CO2) Capture and Utilization from the European Union Perspective. *Energies 2023, Vol. 16, 16*(3). DOI: 10.3390/en16031446

Compilation of References

Abatzoglou, N., & Boivin, S. A. (2009). A review of biogas purification processes. *Biofuels, Bioproducts & Biorefining : Biofpr*, *3*(1), 42–71. DOI: 10.1002/bbb.117

Abbasi, I. A., Đức, A. N., Nam, S., & Yeom, I. T. (2025). Efficient removal of polystyrene nano plastics from aqueous solution by chitosan-coated black tea magnetic nanocomposite: A green synthesis, influencing factors and mechanisms. *Journal of Water Process Engineering*, *77*, 108650. DOI: 10.1016/j.jwpe.2025.108650

Abdelfattah, A., Ali, S. S., Ramadan, H., El-Aswar, E. I., Eltawab, R., Ho, S. H., Elsamahy, T., Li, S., El-Sheekh, M. M., Schagerl, M., Kornaros, M., & Sun, J. (2023). Microalgae-Based Wastewater Treatment: Mechanisms, Challenges, Recent Advances, and Future Prospects. *Environmental Science and Ecotechnology*, *13*, 100205. DOI: 10.1016/j.ese.2022.100205 PMID: 36247722

Abdel-Raouf, N., Al-Homaidan, A. A., & Ibraheem, I. (2012). Microalgae and wastewater treatment. *Saudi Journal of Biological Sciences*, *19*(3), 257–275. DOI: 10.1016/j.sjbs.2012.04.005 PMID: 24936135

Abinandan, S., Subashchandrabose, S. R., Venkateswarlu, K., & Megharaj, M. (2018). Nutrient removal and biomass production: Advances in microalgal biotechnology for wastewater treatment. *Critical Reviews in Biotechnology*, *38*(8), 1244–1260. DOI: 10.1080/07388551.2018.1472066 PMID: 29768936

Abreu, A. P., Martins, R., & Nunes, J. J. B. (2023). Emerging applications of Chlorella sp. and Spirulina (Arthrospira) sp. *10*(8), 955.

Aceituno-Caicedo, D. F. (2025). *Stress Tested: Impact of Salinity on Cyanobacteria Surface Chemistry and Mineral Precipitation-Implications for Biotechnological Applications* (Doctoral dissertation, University of Toronto (Canada)).

Achal, V., & Mukherjee, A. (2015). A review of microbial precipitation for sustainable construction. *Construction & Building Materials*, *93*, 1224–1235. DOI: 10.1016/j.conbuildmat.2015.04.051

Acién Fernández, F. G. (2016). Wastewater Treatment Using Microalgae: How Realistic a Contribution Might It Be to Significant Urban Wastewater Treatment? *Applied Microbiology and Biotechnology, 100*(21), 9013–9022. DOI: 10.1007/s00253-016-7835-7 PMID: 27645298

Acién, F. G., Gómez-Serrano, C., Morillas, A., Arbib, Z., Sanchez, A., Nordio, R., Rodríguez, E., Guzmán, J. L., & Fernández-Sevilla, J. M. (2023). Wastewater treatment by microalgae-based processes. In *IWA Publishing eBooks* (p. 77). UWA Publishing., DOI: 10.2166/9781789063547_0077

Acién, F., Fernández, J., Magán, J., & Molina, E. (2012). Production cost of a real microalgae production plant and strategies to reduce it. *Biotechnology Advances, 30*(6), 1344–1353. DOI: 10.1016/j.biotechadv.2012.02.005 PMID: 22361647

Addissouky, T. A. (2024). Precision medicine for personalized cholecystitis care: Integrating molecular diagnostics and biotherapeutics. *Bulletin of the National Research Center, 48*(1), 89. DOI: 10.1186/s42269-024-01244-9

Agarwal, N. (2025). *Diatoms: basic and applied research*. Educohack Press.

Ahmad Kamal, A. H., Mohd Hamidi, N. F., Zakaria, M. F., Ahmad, A., Harun, M. R., Chandra Segaran, T., & Jusoh, M. (2024). Genetically engineered microalgae for enhanced bioactive compounds. *Discover Applied Sciences, 6*(9), 482. DOI: 10.1007/s42452-024-06116-5

Ahmad, Maqbool, David Plischke, Petar Zuanović, and Zoran Zorić. 2025. "Interactive Effects of Salinity Variation and Nitrogen Deficiency on Fatty Acids Composition in Nannochloropsis." *Journal of Applied Phycology 2025 37:5* 37(5):3487–99. .DOI: 10.1007/s10811-025-03624-8

Ahmad, A., Rambabu, K., Hasan, S. W., Show, P. L., & Banat, F. (2024). Biohydrogen production through dark fermentation: Recent trends and advances in transition to a circular bioeconomy. *International Journal of Hydrogen Energy, 52*, 335–357. DOI: 10.1016/j.ijhydene.2023.05.161

Ahmed, A. T. A., Seol, J. H., Seok, J. H., Jana, A., Meena, A., Cho, S., Sree, V. G., Park, Y., Lee, S. U., & Im, H. (2025). Boosting the electrocatalytic performance of CuCo2S4 via surface-state engineering for ampere current water electrolysis applications. *Applied Surface Science, 680*, 161353. DOI: 10.1016/j.apsusc.2024.161353

Ahmed, T., Luo, J., Noman, M., Ijaz, M., Wang, X., Masood, H. A., Manzoor, N., Wang, Y., & Li, B. (2023). Microbe-mediated nanoparticle intervention for the management of plant diseases. *Crop Health, 1*(1), 3. Advance online publication. DOI: 10.1007/s44297-023-00006-9 PMID: 41652775

Aizawa, K., & Miyachi, S. (1986). Carbonic anhydrase and CO2 concentrating mechanisms in microalgae and cyanobacteria. *FEMS Microbiology Letters*, *2*(3), 215–233. DOI: 10.1111/j.1574-6968.1986.tb01860.x

Akash, S., Sivaprakash, B., Rajamohan, N., & Vo, D. V. N. (2023). Biotechnology to convert carbon dioxide into biogas, bioethanol, bioplastic and succinic acid using algae, bacteria and yeast: A review. *Environmental Chemistry Letters*, *21*(3), 1477–1497. DOI: 10.1007/s10311-023-01569-3

Akhlaghi, N., & Najafpour-Darzi, G. (2020). A comprehensive review on biological hydrogen production. *International Journal of Hydrogen Energy*, *45*(43), 22492–22512. DOI: 10.1016/j.ijhydene.2020.06.182

Akinbuja, O., Orta, S. V., & Boodhoo, K. (2025). Life cycle assessment of microalgae-assisted microbial fuel cells. *The International Journal of Life Cycle Assessment*, *30*(10), 2133–2144. DOI: 10.1007/s11367-025-02518-8

Akram, F., Fatima, T., Ibrar, R., & ul Haq, I. (2024). Biohydrogen: Production, promising progressions and challenges of a green carbon-free energy. *Sustainable Energy Technologies and Assessments*, *69*, 103893. DOI: 10.1016/j.seta.2024.103893

Akram, S., Guan, G., Hu, B., Saddique, M. A. B., Xi, Y., Luo, X., & Ren, M. (2026). Microalgae-based strategies for cadmium remediation: Insights, challenges, and future directions. *Critical Reviews in Biotechnology*, *46*(2), 318–337. DOI: 10.1080/07388551.2025.2585681 PMID: 41260883

Alaka, S. (2025). Integrating machine learning and IoT technologies for smart water quality monitoring: Methods, challenges, and future directions.

Alami, A. H., Alasad, S., Ali, M., & Alshamsi, M. (2021). Investigating algae for CO2 capture and accumulation and simultaneous production of biomass for biodiesel production. *The Science of the Total Environment*, *759*, 143529. DOI: 10.1016/j.scitotenv.2020.143529 PMID: 33229076

Alamiery, A. (2024). Advancements in materials for hydrogen production: A review of cutting-edge technologies. *ChemPhysMater*, *3*(1), 64–73. DOI: 10.1016/j.chphma.2023.09.002

Alarcon, H. (2023). *Growth of Purple Sulfur Bacteria Allochromatium Vinosum on Solid Phase Metal Sulfides as Sulfur and Electron Sources* (Doctoral dissertation, The University of Texas at El Paso).

Alazhari, M., Sharma, T., Heath, A., Cooper, R., & Paine, K. (2018). Application of expanded perlite encapsulated bacteria and growth media for self-healing concrete. *Construction & Building Materials*, *160*, 610–619. DOI: 10.1016/j.conbuildmat.2017.11.086

Albalawi, F. (2025). Hydrogen Energy Review: Challenges, Innovations, and Pathways to a Sustainable Future.

Al-Dailami, A., Ahmad, I., Abdullah, N., Koji, I., & Yuzir, A. (2022, April). Feasibility and viability of procuring biohydrogen from microalgae: An emerging and sustainable energy resource technology. [). IOP Publishing.]. *Journal of Physics: Conference Series*, *2259*(1), 012014. DOI: 10.1088/1742-6596/2259/1/012014

Aledhari, M., & Rahouti, M. (2024). Gene and RNA editing: methods, enabling technologies, applications, and future directions. *arXiv preprint arXiv:2409.09057.*

Alengaram, U. J., Ibrahim, S., Phang, S. M., Vello, V., Jun, H. K., & Alnahhal, A. M. (2021). Evaluation of crack healing potential of cement mortar incorporated with blue-green microalgae. *Journal of Building Engineering*, *44*, 102958. DOI: 10.1016/j.jobe.2021.102958

Alengaram, U. J., Ibrahim, S., Vello, V., & Phang, S. M. (2024). Investigation on the enhancement of crack restoration properties in cement incorporated with Arthrospira platensis cultured in modified medium. *Environmental Science and Pollution Research International*, *31*(17), 25538–25558. DOI: 10.1007/s11356-024-32784-2 PMID: 38478311

Alewell, Christine, Bruno Ringeval, Cristiano Ballabio, David A. Robinson, Panos Panagos, and Pasquale Borrelli. 2020. "Global Phosphorus Shortage Will Be Aggravated by Soil Erosion." *Nature Communications 2020 11:1* 11(1):4546-. .DOI: 10.1038/s41467-020-18326-7

Algaifi, H. A., Bakar, S. A., Sam, A. R. M., Ismail, M., Abidin, A. R. Z., Shahir, S., & Altowayti, W. A. H. (2020). Insight into the role of microbial calcium carbonate and the factors involved in self-healing concrete. *Construction & Building Materials*, *254*, 119258. DOI: 10.1016/j.conbuildmat.2020.119258

Ali, M. S., & Almurshedi, A. D. (2024). Bio-cementations: A Review on Enzyme and Microbially Induced Calcite Precipitation Mechanisms and Applications in Geotechnical Engineering. *Journal of Engineering*, *30*(12), 167–188. DOI: 10.31026/j.eng.2024.12.11

Ali, M., Shabbir, K., Ali, S., Mohsin, M., Kumar, A., Aziz, M., Zubair, M., & Sultan, H. M. (2024). A New Era of Discovery: How Artificial Intelligence has Revolutionized the Biotechnology. *Nepal Journal of Biotechnology*, *12*(1), 1–11. DOI: 10.54796/njb.v12i1.312

Alizadeh, S. M., Khalili, Y., & Ahmadi, M. (2024). Comprehensive review of carbon capture and storage integration in hydrogen production: Opportunities, challenges, and future perspectives. *Energies*, *17*(21), 5330. DOI: 10.3390/en17215330

Almomani, F., Abdelbar, A., & Ghanimeh, S. (2023). A review of the recent advancement of bioconversion of carbon dioxide to added value products: A state of the art. *Sustainability (Basel)*, *15*(13), 10438. DOI: 10.3390/su151310438

Alzate, M. E., Muñoz, R., Rogalla, F., Fdz-Polanco, F., & Perez-Elvira, S. I. (2012). Biochemical methane potential of microalgae: Influence of substrate to inoculum ratio, biomass concentration and pretreatment. *Bioresource Technology*, *123*, 488–494. DOI: 10.1016/j.biortech.2012.06.113 PMID: 22940359

Aminian-Dehkordi, J., Rahimi, S., Golzar-Ahmadi, M., Singh, A., Lopez, J., Ledesma-Amaro, R., & Mijakovic, I. (2023). Synthetic biology tools for environmental protection. *Biotechnology Advances*, *68*, 108239. DOI: 10.1016/j.biotechadv.2023.108239

Amin, M., Tahir, F., Akbar, I., Alessa, A. H., Alsaigh, A. A., Liu, C. G., Xin, F., Chi, Z., Syafiuddin, A., Mehmood, M. A., & Boopathy, R. (2025). Towards Environmental Sustainability: Employing Adaptive Laboratory Evolution to Develop Elite Algae Strains for Industrial and Environmental Applications. *Current Pollution Reports*, *11*(1), 14. DOI: 10.1007/s40726-025-00346-w

Amjith, L. R., & Bavanish, B. (2022). A review on biomass and wind as renewable energy for sustainable environment. *Chemosphere*, *293*, 133579. DOI: 10.1016/j.chemosphere.2022.133579

Anabtawi, H. M., Ikhlaq, A., Kumar, S., Rafique, S., & Hassan, A. A. (2025). Addressing Challenges for Eco-Friendly and Sustainable Wastewater Treatment Solutions Using Extremophile Microorganisms. *Sustainability (Basel)*, *17*(6), 2339. DOI: 10.3390/su17062339

Ananthakumar, B., Nanjundeshwara, N., Lakshmaiah, A., & Raghavan, R. (2025). Nanocarriers for Delivering Nucleic Acids in Plants: Mechanisms and Applications. *Nanocarriers in Plant Science and Agriculture*, 207–234. DOI: 10.4018/979-8-3693-9869-2.ch008

Ananthi, V., Ramesh, U., Balaji, P., Kumar, P., Govarthanan, M., & Arun, A. (2024). A review on the impact of various factors on biohydrogen production. *International Journal of Hydrogen Energy*, *52*, 33–45. DOI: 10.1016/j.ijhydene.2022.08.046

An, B., Wang, Y., Huang, Y., Wang, X., Liu, Y., Xun, D., Church, G. M., Dai, Z., Yi, X., Tang, T.-C., & Zhong, C. (2022). Engineered living materials for sustainability. *Chemical Reviews*, *123*(5), 2349–2419. DOI: 10.1021/acs.chemrev.2c00512

Angmo, S., Srivastava, A., Saxena, A. K. R., & Tiwari, A. (2026). Diatoms nanofactories: Synthesis, properties, and biomedical applications. *Systems Microbiology and Biomanufacturing*, *6*(2), 30. DOI: 10.1007/s43393-025-00427-3

An, H., Hu, W., Huang, S., Huang, S., Li, R., Liang, Y., Shao, J., Song, Y., Wang, Z., Yuan, C., Zhang, C., Zhang, H., Zhuang, W., & Li, X. (2026). Ai flow: Perspectives, scenarios, and approaches. *Vicinagearth*, *3*(1), 1. DOI: 10.1007/s44336-025-00031-y

Antonini, C., Treyer, K., Moioli, E., Bauer, C., Schildhauer, T. J., & Mazzotti, M. (2021). Hydrogen from wood gasification with CCS–a techno-environmental analysis of production and use as transport fuel. *Sustainable Energy & Fuels*, *5*(10), 2602–2621. DOI: 10.1039/D0SE01637C

Anvar, S. A. A., Nowruzi, B., & Afshari, G. (2023). A Review of the Application of Nanoparticles Biosynthesized by Microalgae and Cyanobacteria in Medical and Veterinary Sciences [Review of A Review of the Application of Nanoparticles Biosynthesized by Microalgae and Cyanobacteria in Medical and Veterinary Sciences]. *Iranian Journal of Veterinary Medicine*, *17*(1), 1–18. DOI: 10.32598/IJVM.17.1.1005309

An, Y., Wang, D., Du, J., Wang, X., & Xiao, J. (2023). Pyrenoid: Organelle with efficient CO2-Concentrating mechanism in algae. *Journal of Plant Physiology*, *287*, 154044. DOI: 10.1016/j.jplph.2023.154044 PMID: 37392525

Aragón, D., García-Merino, B., Barquín, C., Bringas, E., Rivero, M. J., & Ortíz, I. (2024). Advanced green capture of microplastics from different water matrices by surface-modified magnetic nanoparticles. *Separation and Purification Technology*, *354*, 128813. DOI: 10.1016/j.seppur.2024.128813

Arashiro, L. T., Boto-Ordóñez, M., Van Hulle, S. W., Ferrer, I., Garfí, M., & Rousseau, D. P. (2020). Natural pigments from microalgae grown in industrial wastewater. *Bioresource Technology*, *303*, 122894. DOI: 10.1016/j.biortech.2020.122894 PMID: 32032937

Arashiro, L. T., Josa, I., Ferrer, I., Van Hulle, S. W., Rousseau, D. P., & Garfí, M. (2022). Life cycle assessment of microalgae systems for wastewater treatment and bioproducts recovery: Natural pigments, biofertilizer and biogas. *The Science of the Total Environment*, *847*, 157615. DOI: 10.1016/j.scitotenv.2022.157615 PMID: 35901897

Arcusa, S., & Sprenkle-Hyppolite, S. (2022). Snapshot of the Carbon Dioxide Removal Certification and Standards Ecosystem (2021-2022). *Climate Policy*, *22*(9–10), 1319–1332. DOI: 10.1080/14693062.2022.2094308

Armah, E. K., Chetty, M., Adedeji, J., Kukwa, D. T., Mutsvene, B., Shabangu, K. P., & Bakare, B. F. (2020). Emerging Trends in Wastewater Treatment Technologies: The Current Perspective. In *IntechOpen eBooks*. IntechOpen. DOI: 10.5772/intechopen.93898

Artamonova, E., Vasskog, T., & Eilertsen, H. C. (2017). Lipid content and fatty acid composition of Porosira glacialis and Attheya longicornis in response to carbon dioxide (CO2) aeration. *PLoS One*, *12*(5), e0177703. DOI: 10.1371/journal pone.0177703 PMID: 28542308

Arteaga-Castrejón, A. A., Agarwal, V., & Khandual, S. (2024). Microalgae as a potential natural source for the green synthesis of nanoparticles. *Chemical Communications (Cambridge)*, *60*(29), 3874–3890. DOI: 10.1039/D3CC05767D PMID: 38529840

Artemyev, V., Gubaeva, A., Paremskaia, A. I., Dzhioeva, A. A., Deviatkin, A., Feoktistova, S. G., Mityaeva, O., & Volchkov, P. Y. (2024). Synthetic promoters in gene therapy: Design approaches, features and applications. *Cells*, *13*(23), 1963. DOI: 10.3390/cells13231963 PMID: 39682712

Aryanpur, V., & Rogan, F. (2024). Decarbonising road freight transport: The role of zero-emission trucks and intangible costs. *Scientific Reports*, *14*(1), 1–11. DOI: 10.1038/s41598-024-52682-4 PMID: 38267587

Asghar, A., Sairash, S., Hussain, N., Baqar, Z., Sumrin, A., & Bilal, M. (2022). Current challenges of biomass refinery and prospects of emerging technologies for sustainable bioproducts and bioeconomy. *Biofuels, Bioproducts & Biorefining : Biofpr*, *16*(6), 1478–1494. DOI: 10.1002/bbb.2403

Ashitha, S., & George, S. C. (2022). Splitting of water: biological and non-biological approaches. In *Handbook of Biofuels* (pp. 453–469). Academic Press. DOI: 10.1016/B978-0-12-822810-4.00023-3

Ashour, M., Mansour, A. T., Alkhamis, Y. A., Elshobary, M. J. F. i. B., & Biotechnology. (2024). Usage of Chlorella and diverse microalgae for CO2 capture-towards a bioenergy revolution. *12*, 1387519.

Ashour, M., Mansour, A. T., Alkhamis, Y. A., & Elshobary, M. (2024). Usage of Chlorella and Diverse Microalgae for CO2 Capture - towards a Bioenergy Revolution. *Frontiers in Bioengineering and Biotechnology*, *12*, 1387519. DOI: 10.3389/fbioe.2024.1387519 PMID: 39229458

Asimakis, E., Shehata, A. A., Eisenreich, W., Acheuk, F., Lasram, S., Basiouni, S., Emekci, M., Ntougias, S., Taner, G., May-Simera, H., Yilmaz, M., & Tsiamis, G. (2022). Algae and their metabolites as potential bio-pesticides. *Microorganisms*, *10*(2), 307. DOI: 10.3390/microorganisms10020307 PMID: 35208762

Asrul, M. A. M., Atan, M. F., Yun, H. A. H., & Lai, J. C. H. (2022). A review of advanced optimization strategies for fermentative biohydrogen production processes. *International Journal of Hydrogen Energy*, *47*(38), 16785–16804. DOI: 10.1016/j.ijhydene.2022.03.197

Assis, L. R., Ferreira, J., Assemany, P. P., Teixeira, J. S., Castro, J. S., Pereira, H. A., & Calijuri, M. L. (2023). Environmental benefits of a hybrid system for algal biomass production, harvesting and nutrient recovery under a life-cycle assessment. *Algal Research*, *73*, 103163. 10.1016/j.algal.2023.103163. DOI: 10.1016/j.algal.2023.103163

Atikah, M. S. N., Harun, R., Ilyas, R. A., Norrrahim, M. N. F., & Knight, V. F. (2025). Algae cultivation systems integrated with photovoltaic cell: A systematic review. *Results in Engineering*, *27*, 106294. DOI: 10.1016/j.rineng.2025.106294

Avinash, A., Sasikumar, P., & Pugazhendhi, A. (2020). Analysis of the limiting factors for large scale microalgal cultivation: A promising future for renewable and sustainable biofuel industry. *Renewable & Sustainable Energy Reviews*, *134*, 110250. DOI: 10.1016/j.rser.2020.110250

Avrahami, Y., Koplovitz, G., & Frada, M. J. (2025). Diatom community succession and bloom variability as a function of winter-mixing depth in the subtropical Gulf of Aqaba, Red Sea. *Marine Ecology Progress Series*, *760*, 39–54. Advance online publication. DOI: 10.3354/meps14848

Ayodele, D. T., Ogunbiyi, O. D., Akamo, D. O., Otun, K. O., Akinpelu, D. A., Adegoke, J. A., Fapojuwo, D. P., & Oladoye, P. O. (2023). Factors affecting biohydrogen production: Overview and perspectives. *International Journal of Hydrogen Energy*, *48*(71), 27513–27539. DOI: 10.1016/j.ijhydene.2023.04.001

Ayyadurai, P., & Ragavendran, C. (2025). Green nanotechnologies for removing cosmetic pollutants: Approach towards preserving and protecting aquatic health. *Nanotechnology for Environmental Engineering*, *10*(2), 21. Advance online publication. DOI: 10.1007/s41204-025-00416-w

Azizan, A., & Jalil, R. (2024). 9 Biophotolysis Process. *Hydrogen Production from Renewable Resources and Wastes*, 201.

Babenko, I., Friedrich, B. M., & Kröger, N. (2022). Structure and morphogenesis of the frustule. In *The molecular life of diatoms* (pp. 287–312). Springer International Publishing. DOI: 10.1007/978-3-030-92499-7_11

Babu, S. S., Gondi, R., Vincent, G. S., JohnSamuel, G. C., & Jeyakumar, R. B. (2022). Microalgae biomass and lipids as feedstock for biofuels: Sustainable biotechnology strategies. *Sustainability (Basel)*, *14*(22), 15070. DOI: 10.3390/su142215070

Badger, M. R., Price, G. D., Long, B. M., & Woodger, F. J. (2006). The environmental plasticity and ecological genomics of the cyanobacterial CO_2 concentrating mechanism. *Journal of Experimental Botany*, *57*(2), 249–265. DOI: 10.1093/jxb/erj286 PMID: 16216846

Baeyens, J., Zhang, H., Nie, J., Appels, L., Dewil, R., Ansart, R., & Deng, Y. (2020). Reviewing the potential of bio-hydrogen production by fermentation. *Renewable & Sustainable Energy Reviews*, *131*, 110023. DOI: 10.1016/j.rser.2020.110023

Bajhaiya, A. K., Dean, A. P., Zeef, L. A. H., Webster, R. E., & Pittman, J. K. (2016). PSR1 Is a Global Transcriptional Regulator of Phosphorus Deficiency Responses and Carbon Storage Metabolism in Chlamydomonas Reinhardtii. *Plant Physiology*, *170*(3), 1216–1234. DOI: 10.1104/pp.15.01907 PMID: 26704642

Balabanova, L., Averianova, L., Marchenok, M., Son, O., & Tekutyeva, L. (2021). Microbial and genetic resources for cobalamin (vitamin B12) biosynthesis: From ecosystems to industrial biotechnology. *International Journal of Molecular Sciences*, *22*(9), 4522. DOI: 10.3390/ijms22094522 PMID: 33926061

Balamurugan, S., Li, D. W., Wang, X., & Li, H. Y. (2025). Unleashing the potential of biotechnological strategies for the sustainable production of microalgal polysaccharides. *Critical Reviews in Food Science and Nutrition*, *65*(31), 1–18. DOI: 10.1080/10408398.2025.2475240 PMID: 40084718

Balan, V., Bals, B., Chundawat, S. P., Marshall, D., & Dale, B. E. (2009). Lignocellulosic biomass pretreatment using AFEX. *Methods in Molecular Biology (Clifton, N.J.)*, *581*, 61–77. DOI: 10.1007/978-1-60761-214-8_5 PMID: 19768616

Ball, S., Han, P. G., James, M., Myers, A., Keeling, P., Mouille, G., Buléon, A., Colonna, P., & Preiss, J. (1996). From Glycogen to Amylopectin: A Model for the Biogenesis of the Plant Starch Granule. *Cell*, *86*(3), 349–352. DOI: 10.1016/S0092-8674(00)80107-5 PMID: 8756717

Balouch, H., Sadvakasova, A. K., Kossalbayev, B. D., Bauenova, M. O., Zaletova, D. E., Kumarbekuly, S., & Kirbayeva, D. K. (2025). Phyto-Algal Consortia as a Complementary System for Wastewater Treatment and Biorefinery. *Plants*, *14*(19), 3069. DOI: 10.3390/plants14193069 PMID: 41095212

Baragi, L. V., Khandeparker, L., & Anil, A. C. (2015). Influence of elevated temperature and pCO2 on the marine periphytic diatom Navicula distans and its associated organisms in culture. *Hydrobiologia*, *762*(1), 127–142. DOI: 10.1007/s10750-015-2343-9

Barati, B., Zeng, K., Baeyens, J., Wang, S., Addy, M., Gan, S. Y., & Abomohra, A. E. F. (2021). Recent progress in genetically modified microalgae for enhanced carbon dioxide sequestration. *Biomass and Bioenergy*, *145*, 105927. DOI: 10.1016/j.biombioe.2020.105927

Barber, L. (2025). *Swedish Algae Factory raises €3 million to expand algae production and boost beauty partnerships*. Cosmetics Business. https://cosmeticsbusiness .com/swedish-algae-factory-raises-%E2%82%AC3million-expand-algica-production-boost-beauty-partnerships

Barros, D., Nobre, L., Bessa, J., Leite, L., Mota, C., Cunha, F., & Fangueiro, R. (2024). Bio-Inspired Eco-Composite Materials Seaweed Waste Integration for Sustainable Structural Applications. *Sustainability 2024, Vol. 16*, *16*(24). DOI: 10.3390/su162411051

Barua, P. J. M. B. (2025). Diatoms as Microbial Architects of the Silica Cycle. *Ecological Significance, Biogenic Bioactives, and Emerging Biotechnological Applications.*, *8*(1), 1–12.

Baskaran, D., Saravanan, P., Nagarajan, L., & Byun, H. S. (2024). An overview of technologies for capturing, storing, and utilizing carbon dioxide: Technology readiness, large-scale demonstration, and cost. *Chemical Engineering Journal*, *491*, 151998. DOI: 10.1016/j.cej.2024.151998

Battin, T. J., Luyssaert, S., Kaplan, L. A., Aufdenkampe, A. K., Richter, A., & Tranvik, L. J. (2009). The boundless carbon cycle. *Nature Geoscience*, *2*(9), 598–600. DOI: 10.1038/ngeo618

Bayramoğlu, K., & Bayramoğlu, T. (2025). Integrated Modeling of Steam Methane Reforming and Carbon Capture for Blue Hydrogen Production. *Hydrogen*, *6*(4), 94. DOI: 10.3390/hydrogen6040094

Beardall, J., Raven, J. A., & Giordano, M. (2009). Environmental regulation of photosynthesis in microalgae. *Journal of Experimental Botany*, *60*(10), 2491–2507.

Béchet, Q., Plouviez, M., Chambonnière, P., & Guieysse, B. (2017). Environmental impacts of full-scale algae cultivation. *Microalgae-Based Biofuels and Bioproducts: From Feedstock Cultivation to End-Products*, 505–525. DOI: 10.1016/B978-0-08-101023-5.00021-2

Behera, B., Aly, N., & Paramasivan, B. (2020). Theoretical Modeling of Algal Productivity and Carbon Capture Potential in Selected Places of Odisha, India. *Journal of The Institution of Engineers (India): Series A 2020 101:3, 101*(3), 503–512. DOI: 10.1007/s40030-020-00450-8

Behera, B., & Paramasivan, B. (2021). Uncertainty analysis and stochastic studies of techno-economics of algal carbon sequestration at Indian coal powered plants. *Environmental Technology & Innovation*, *24*, 101897. DOI: 10.1016/j.eti.2021.101897

Behrenfeld, M. J., Halsey, K. H., Boss, E., Karp-Boss, L., Milligan, A. J., & Peers, G. (2021). Thoughts on the evolution and ecological niche of diatoms. *Ecological Monographs*, *91*(3), e01457. DOI: 10.1002/ecm.1457

Bello, M., Modu, A. K., Boryo, D. E. A., & Ranganathan, P. (2024). Sustainable algal biorefinery: A review on current perspective on technical maturity, supply infrastructure, business and industrial opportunities. *Journal of Environmental Management*, *370*, 122208. DOI: 10.1016/j.jenvman.2024.122208 PMID: 39243640

Bennett, M., March, A., Raguain, J., & Failler, P. (2024). Blueprint for Blue Carbon: Lessons from Seychelles for Small Island States. *Oceans 2024, Vol. 5, Pages 81-108, 5*(1), 81–108. DOI: 10.3390/oceans5010006

Beschkov, V., Parvanova-Mancheva, T., & Vasileva, E. (2023). Experimental Study of Bio-Hydrogen Production by Clostridium beijerinckii from Different Substrates. *Energies*, *16*(6), 2747. DOI: 10.3390/en16062747

Bhagat, S. P., Wagh, B. D., Satpute, S. M., Gunjal, M. B., Kasar, S. K., & Roylawar, P. B. (2025). Fourth-Generation Biofuels: Genetic Engineering for Sustainable Energy. In *Recent Trends in Lignocellulosic Biofuels and Bioenergy: Advancements and Sustainability Assessment* (pp. 419–438). Springer Nature Singapore. DOI: 10.1007/978-981-96-4636-4_15

Bilal, M., Niu, D., & Wang, Z. (2025). Biotechnological Strategies and Perspectives for Food Waste Treatment: The Role of Lactic Acid and Microbial Biomass. *Waste and Biomass Valorization*, *16*(2), 547–569. DOI: 10.1007/s12649-024-02705-y

Bin Abu Sofian, A. D. A., Lim, H. R., Manickam, S., Ang, W. L., & Show, P. L. (2024). Towards a sustainable circular economy: Algae-based bioplastics and the role of internet-of-things and machine learning. *ChemBioEng Reviews*, *11*(1), 39–59. DOI: 10.1002/cben.202300028

Blanken, W., Janssen, M., Cuaresma, M., Libor, Z., Bhaiji, T., & Wijffels, R. H. (2014). Biofilm growth of *Chlorella sorokiniana* in a rotating biological contactor-based photobioreactor. *Biotechnology and Bioengineering*, *111*(12), 2436–2445. DOI: 10.1002/bit.25301 PMID: 24895246

Bleuel, S., & Müller, C. (2024). Unlocking the potential: Expert insights on the long-term compatibility of forest carbon credits with the EU ETS. *Forest Policy and Economics*, *162*, 103185. DOI: 10.1016/j.forpol.2024.103185

Boelen, P., van de Poll, W. H., van der Strate, H. J., Neven, I. A., Beardall, J., & Buma, A. G. (2011). Neither elevated nor reduced CO2 affects the photophysiological performance of the marine Antarctic diatom Chaetoceros brevis. *Journal of Experimental Marine Biology and Ecology*, *406*(1-2), 38–45. DOI: 10.1016/j.jembe.2011.06.012

Boettcher, M., Winkler, M., & Michaelowa, A. (2025). Accounting and monitoring challenges for blue carbon enhancement in national climate policy targets and international carbon markets. *Carbon Management*, *16*(1), 2585895. Advance online publication. DOI: 10.1080/17583004.2025.2585895

Bohutskyi, P., & Bouwer, E. (2013). Biogas production from algae and cyanobacteria through anaerobic digestion: A review, analysis, and research needs. In Lee, J. W. (Ed.), *Advanced Biofuels and Bioproducts* (pp. 873–975). Springer. DOI: 10.1007/978-1-4614-3348-4_36

Bongirwar, R., & Shukla, P. (2023). Metabolic sink engineering in cyanobacteria: Perspectives and applications. *Bioresource Technology*, *379*, 128974. DOI: 10.1016/j.biortech.2023.128974 PMID: 36990331

Boonchai, R., Kaewsuk, J., & Seo, G. (2015). Effect of Nutrient Starvation on Nutrient Uptake and Extracellular Polymeric Substance for Microalgae Cultivation and Separation. *Desalination and Water Treatment*, *55*(2), 360–367. DOI: 10.1080/19443994.2014.939501

Borowitzka, M. A. (1999). Commercial production of microalgae: Ponds, tanks, tubes and fermenters. *Journal of Biotechnology*, *70*(1–3), 313–321. DOI: 10.1016/S0168-1656(99)00083-8

Botelho, D. F., Dias, B. H., de Oliveira, L. W., Soares, T. A., Rezende, I., & Sousa, T (2021). Innovative business models as drivers for prosumers integration-Enablers and barriers. *Renewable & Sustainable Energy Reviews*, *144*, 111057. DOI: 10.1016/j.rser.2021.111057

Botte, P., d'Ippolito, G., Gallo, C., Sardo, A., & Fontana, A. J. J. o. A. P. (2018b). Combined exploitation of CO2 and nutrient replenishment for increasing biomass and lipid productivity of the marine diatoms Thalassiosira weissflogii and Cyclotella cryptica. *30*(1), 243-251.

Botte, P., d'Ippolito, G., Gallo, C., Sardo, A., & Fontana, A. (2018a). Combined exploitation of CO2 and nutrient replenishment for increasing biomass and lipid productivity of the marine diatoms Thalassiosira weissflogii and Cyclotella cryptica. *Journal of Applied Phycology*, *30*(1), 243–251. DOI: 10.1007/s10811-017-1221-4

Bouazzara, H., & Chaibi, R. (2025). Diatoms and nanotechnology, current trends in nanomaterials applications and environmental monitoring. *Environmental Monitoring and Assessment*, *197*(7), 793. DOI: 10.1007/s10661-025-14227-8 PMID: 40550980

Bozieva, A. M., Khasimov, M. K., Voloshin, R. A., Sinetova, M. A., Kupriyanova, E. V., Zharmukhamedov, S. K., Dunikov, D. O., Tsygankov, A. A., Tomo, T., & Allakhverdiev, S. I. (2023). New cyanobacterial strains for biohydrogen production. *International Journal of Hydrogen Energy*, *48*(21), 7569–7581. DOI: 10.1016/j.ijhydene.2022.11.198

Bradley, J. M. (2025). Transporting to the matrix: A pyrenoid-localized phosphate transporter required for optimal photoautotrophic growth in algae.

Braga, M. Q., Lorentz, J. F., Pereira, A. S. A. de P., Magalhães, I. B., Silva, T. A., Avelar, N. V., Henriques, B. S., de Castro, A. P. M., Assemany, P. P., & Calijuri, M. L. (2026). Application of nanoparticles to microalgae-based wastewater treatment: Mechanisms, bioremediation potential and technological frontiers. *Journal of Environmental Management*, *401*, 128827. DOI: 10.1016/j.jenvman.2026.128827 PMID: 41650870

Branco-Vieira, M., San Martin, S., Agurto, C., Freitas, M. A. V., Martins, A. A., Mata, T. M., & Caetano, N. S. (2020). Biotechnological potential of Phaeodactylum tricornutum for biorefinery processes. *Fuel*, *268*, 117357. https://doi.org/https://doi.org/10.1016/j.fuel.2020.117357. DOI: 10.1016/j.fuel.2020.117357

Brányiková, I., Maršálková, B., Doucha, J., Brányik, T., Bišová, K., Zachleder, V., & Vítová, M. (2011). Microalgae-Novel Highly Efficient Starch Producers. *Biotechnology and Bioengineering, 108*(4), 766–776. DOI: 10.1002/bit.23016 PMID: 21404251

Briat, C., & Khammash, M. (2018). Perfect adaptation and optimal equilibrium productivity in a simple microbial biofuel metabolic pathway using dynamic integral control. *ACS Synthetic Biology, 7*(2), 419–431. DOI: 10.1021/acssynbio.7b00188 PMID: 29343065

Buckel, W. (2021). Energy conservation in fermentations of anaerobic bacteria. *Frontiers in Microbiology, 12*, 703525. DOI: 10.3389/fmicb.2021.703525

Bueno, G., Deniz, O., Pedraza, A., Ruiz-Santaquiteria, J., Salido, J., Cristóbal, G., Borrego-Ramos, M., & Blanco, S. (2017). Automated diatom classification (Part A): Handcrafted feature approaches. *Applied Sciences (Basel, Switzerland), 7*(8), 753. DOI: 10.3390/app7080753

Buono, S., Colucci, A., Angelini, A., Langellotti, A. L., Massa, M., Martello, A., Fogliano, V., & Dibenedetto, A. (2016). Productivity and biochemical composition of Tetradesmus obliquus and Phaeodactylum tricornutum: Effects of different cultivation approaches. *Journal of Applied Phycology, 28*(6), 3179–3192. DOI: 10.1007/s10811-016-0876-6

Busch, F. A. (2020). Photorespiration in the context of Rubisco biochemistry, CO2 diffusion and metabolism. *The Plant Journal : For Cell and Molecular Biology, 101*(4), 919–939. DOI: 10.1111/tpj.14674 PMID: 31910295

Butean, A., Cutean, I., Barbero, R., Enriquez, J., & Matei, A. (2025). A review of artificial intelligence applications for biorefineries and bioprocessing: From data-driven processes to optimization strategies and real-time control. *Processes (Basel, Switzerland), 13*(8), 2544. DOI: 10.3390/pr13082544

Byalahunasishetter, B. I., Chowdareddy, S., & Raghavan, R. (2026). Cancer Risk Prediction Based on Gut Microbiota Analysis. *IGI Global Scientific Publishing*, 307–342. DOI: 10.4018/979-8-3373-8172-5.ch009

Cai, W.-L., Yu, S.-Y., & Hu, Y.-H. (2025). Synergistic mechanisms of DGAT and PDAT in shaping triacylglycerol diversity: evolutionary insights and metabolic engineering strategies [Review]. *Frontiers in Plant Science, Volume 16 - 2025*. DOI: 10.3389/fpls.2025.1598815

Cai, T., Park, S. Y., & Li, Y. (2013). Nutrient recovery from wastewater streams by microalgae: Status and prospects. *Renewable & Sustainable Energy Reviews, 19*, 360–369. DOI: 10.1016/j.rser.2012.11.030

Calijuri, M. L., Do Couto, E. A., Assemany, P. P., Ribeiro, V. J., Lorentz, J. F., Castro, J. S., de Assis, L. R., Oliveira, A. P. S., Pereira, A., Marangon, B. B., Magalhães, I. B., Silva, T. A., Ferreira, J., Braga, M. Q., da Gama, R. C. N., & Henriques, B. S. (2025). Microalgae-Based Wastewater Treatment and Biomass Valorization: Insights, Challenges, and Opportunities from 15 Years of Research. *ACS Omega, 10*(42), 49273–49299. DOI: 10.1021/acsomega.5c04364 PMID: 41179135

Cannon, G. C., Heinhorst, S., & Kerfeld, C. A. (2010). Carboxysomal carbonic anhydrases: Structure and role in microbial CO2 fixation. *Biochimica et Biophysica Acta (BBA)-. Biochimica et Biophysica Acta. Proteins and Proteomics, 1804*(2), 382–392. DOI: 10.1016/j.bbapap.2009.09.026 PMID: 19818881

Canzoneri, M., Horner, M., & Banerjee, P. (2024). Digital twins in biomanufacturing. In *Bioconvergence Revolution: Challenges and Opportunities in Research Productivity* (pp. 59–81). Springer Nature Switzerland. DOI: 10.1007/978-3-031-61593-1_4

Cao, K., Cui, Y., Sun, F., Zhang, H., Fan, J., Ge, B., Cao, Y., Wang, X., Zhu, X., Wei, Z., Yao, Q., Ma, J., Wang, Y., Meng, C., & Gao, Z. (2023). Metabolic engineering and synthetic biology strategies for producing high-value natural pigments in Microalgae. *Biotechnology Advances, 68*, 108236. DOI: 10.1016/j.biotechadv.2023.108236 PMID: 37586543

Cardiff, R. A., Carothers, J. M., Zalatan, J. G., & Sauro, H. M. (2024). Systems-level modeling for CRISPR-based metabolic engineering. *ACS Synthetic Biology, 13*(9), 2643–2652. DOI: 10.1021/acssynbio.4c00053 PMID: 39119666

Cardoso, R., Borges, I., Vieira, J., Duarte, S. O., & Monteiro, G. A. (2023). Interactions between clay minerals, bacteria growth and urease activity on biocementation of soils. *Applied Clay Science, 240*, 106972. DOI: 10.1016/j.clay.2023.106972

Carvalho, A. P., Meireles, L. A., & Malcata, F. X. (2006). Microalgal reactors: A review of enclosed system designs and performances. *Biotechnology Progress, 22*(6), 1490–1506. DOI: 10.1002/bp060065r PMID: 17137294

Carver, S. M., Hulatt, C. J., Thomas, D. N., & Tuovinen, O. H. (2011). Thermophilic, anaerobic co-digestion of microalgal biomass and cellulose for H_2 production. *Biodegradation, 22*(4), 805–814. DOI: 10.1007/s10532-010-9419-z PMID: 20878208

Catherall, E., Musial, S., Atkinson, N., Walker, C. E., Mackinder, L. C., & McCormick, A. J. (2025). From algae to plants: Understanding pyrenoid-based CO2-concentrating mechanisms. *Trends in Biochemical Sciences*, *50*(1), 33–45. DOI: 10.1016/j.tibs.2024.10.010 PMID: 39592300

Cazzonelli, C. I. (2011). Carotenoids in Nature: Insights from Plants and Beyond. *Functional Plant Biology : FPB*, *38*(11), 833–847. DOI: 10.1071/FP11192 PMID: 32480941

Cen, S. Y., Li, D. W., Xue, L. H., Huang, D., Balamurugan, S., Wan, J. L., Jian, W. Z., Wei, D. Y., & Hong, Y. L. (2022). Crucial Carotenogenic Genes Elevate Hyper-accumulation of Both Fucoxanthin and β-Carotene in Phaeodactylum Tricornutum. *Algal Research*, *64*, 102691. DOI: 10.1016/j.algal.2022.102691

Cha, M., Park, M. S., & Kim, S. J. (2025). Biological Routes for Biohydrogen Production: A Clean and Carbon-Free Fuel. *Biotechnology Journal, 20*(7), e70074.

Chakraborty, S., Talukdar, A., Dey, S., & Bhattacharya, S. (2025). Role of fungi, bacteria and microalgae in bioremediation of emerging pollutants with special reference to pesticides, heavy metals and pharmaceuticals. *Discover Environment*, *3*(1), 91. DOI: 10.1007/s44274-025-00217-7

Cha, M., Park, M. S., & Kim, S. J. (2025). Biological Routes for Biohydrogen Production: A Clean and Carbon-Free Fuel. *Biotechnology Journal*, *20*(7), e70074. DOI: 10.1002/biot.70074

Chandra, S., Bhattacharya, S., Pandit, S., Banerjee, S., Roy, A., & Rai, A. K. (2023). Cyanobacteria as an eco-friendly bioresource for EPS production with crack healing capacity in concrete. *Biocatalysis and Agricultural Biotechnology*, *54*, 102909. DOI: 10.1016/j.bcab.2023.102909

Chandrasekhar, K., Kumar, A. N., Raj, T., Kumar, G., & Kim, S. H. (2021). Bioelectrochemical system-mediated waste valorization. *Systems Microbiology and Biomanufacturing*, *1*(4), 432–443. DOI: 10.1007/s43393-021-00039-7

Chandrasekhar, T., Reddy, P. C. O., Swapna, B., Veeranjaneya Reddy, L., Anuprasanna, V., Dakshayani, L., Ramachandra Reddy, P., & Reddy, M. C. (2025). Algae: The game-changers in biohydrogen sector. *Critical Reviews in Biotechnology*, *45*(3), 509–529. DOI: 10.1080/07388551.2024.2387176

Chaudhary, R., Nawaz, K., Khan, A. K., Hano, C., Abbasi, B. H., & Anjum, S. (2020). An overview of the algae-mediated biosynthesis of nanoparticles and their biomedical applications. *Biomolecules*, *10*(11), 1498. DOI: 10.3390/biom10111498 PMID: 33143289

Chaudry, S., Hurtado-McCormick, V., Cheng, K. Y., Willis, A., Speight, R., & Kaksonen, A. H. (2025). Microalgae to bioplastics–Routes and challenges. *Cleaner Engineering and Technology*, *25*, 100922. DOI: 10.1016/j.clet.2025.100922

Chayaamor-Heil, N., Perricone, V., Gruber, P., & Guéna, F. (2023). Bioinspired, biobased and living material designs: A review of recent research in architecture and construction. *Bioinspiration & Biomimetics*, *18*(4), 041001. DOI: 10.1088/1748-3190/acd82e

Cheenakula, D., Hoffstadt, K., Krafft, S., Reinecke, D., Klose, H., Kuperjans, I., & Grömping, M. (2023). Anaerobic digestion of algal–bacterial biomass of an Algal Turf Scrubber system. *Biomass Conversion and Biorefinery*, *13*(15), 13605–13619. DOI: 10.1007/s13399-022-03236-z

Chen, L., & Xie, M. (2023). How do hard regimes absorb, overlap, and squeeze out soft regimes? Insights from global carbon markets. *Global Public Policy and Governance 2023 3:1, 3*(1), 60–85. DOI: 10.1007/s43508-023-00064-3

Chen, B., Wan, C., Mehmood, M. A., Chang, J. S., Bai, F., & Zhao, X. (2017). Manipulating Environmental Stresses and Stress Tolerance of Microalgae for Enhanced Production of Lipids and Value-Added Products–A Review. *Bioresource Technology*, *244*, 1198–1206. DOI: 10.1016/j.biortech.2017.05.170 PMID: 28601395

Chen, H., Ma, H., Yu, L., Zou, S., Li, Y., Lu, Y., Yang, M., Ge, F., Yuan, L., Zeng, X., Zhang, C.-C., Yoon, K., Han, D., & Hu, Q. (2025). Functional characterization of the plastidial enzyme GPAT in microalgae and its potential role in lipid biosynthesis. *Plant Physiology*, *199*(1), kiaf395. Advance online publication. DOI: 10.1093/plphys/kiaf395 PMID: 40892446

Chen, J., Dai, L., Mataya, D., Cobb, K., Chen, P., & Ruan, R. (2022). Enhanced sustainable integration of CO2 utilization and wastewater treatment using microalgae in circular economy concept. *Bioresource Technology*, *366*, 128188. DOI: 10.1016/j.biortech.2022.128188 PMID: 36309175

Chen, J., Liu, D., Wang, Y., Wang, S., & Huang, R. (2023). Effects of selective RNA processing and stabilization enzymes on carbon sequestration by photosynthesis of Synechococcus sp. PCC7002. *Frontiers in Marine Science*, *10*, 1223060. DOI: 10.3389/fmars.2023.1223060

Chen, T., Hojka, M., Davey, P., Sun, Y., Zhou, F., Lawson, T., Nixon, P. J., Lin, Y., & Liu, L. N. (2025). Engineering Rubisco condensation in chloroplasts to manipulate plant photosynthesis. *Plant Biotechnology Journal*, *23*(6), 2140–2149. DOI: 10.1111/pbi.70047 PMID: 40087764

Chen, W. H., Biswas, P. P., Lee, K. T., Chang, J. S., Lin, K. Y. A., Petrissans, M., & Petrissans, A. (2025). Integrating comparative life cycle assessment and applications of torrefaction and pyrolysis: A comprehensive review of net carbon and environmental credits. *Chemical Engineering Journal*, •••, 172106.

Chen, X., & Gao, K. (2003). Effect of CO2 concentrations on the activity of photosynthetic CO2 fixation and extracelluar carbonic anhydrase in the marine diatom Skeletonema costatum. *Chinese Science Bulletin*, *48*(23), 2616–2620. DOI: 10.1360/03wc0084

Cheong, B. C. (2025). The Paradox and Fallacy of Global Carbon Credits: A Theoretical Framework for Strengthening Climate Change Mitigation Strategies. *Anthropocene Science 2025 4:1*, *4*(1), 72–83. DOI: 10.1007/s44177-025-00084-0

Chew, K. W., Yap, J. Y., Show, P. L., Suan, N. H., Juan, J. C., Ling, T. C., Lee, D. J., & Chang, J. S. (2017). Microalgae Biorefinery: High Value Products Perspectives. *Bioresource Technology*, *229*, 53–62. DOI: 10.1016/j.biortech.2017.01.006 PMID: 28107722

Chinnasamy, S., Bhatnagar, A., Hunt, R. W., & Das, K. C. (2010). Microalgae cultivation in a wastewater dominated by carpet mill effluents for biofuel applications. *Bioresource Technology*, *101*(9), 3097–3105. DOI: 10.1016/j.biortech.2009.12.026 PMID: 20053551

Chisti, Y. (2007). Biodiesel from microalgae. *Biotechnology Advances*, *25*(3), 294–306. DOI: 10.1016/j.biotechadv.2007.02.001 PMID: 17350212

Choi, A. H. (2022). Biomaterials and bioceramics—part 1: traditional, natural, and nano. In *Innovative Bioceramics in Translational Medicine I: Fundamental Research* (pp. 1–45). Springer Singapore. DOI: 10.1007/978-981-16-7435-8_1

Choi, H. I., Hwang, S. W., Kim, J., Park, B., Jin, E., Choi, I. G., & Sim, S. J. (2021). Augmented CO2 tolerance by expressing a single H+-pump enables microalgal valorization of industrial flue gas. *Nature Communications*, *12*(1), 6049. DOI: 10.1038/s41467-021-26325-5 PMID: 34663809

Choi, H., Sung, Y. J., Hong, M. E., Han, J., Min, B. K., & Sim, S. J. (2022). Reconsidering the potential of direct microalgal biomass utilization as end-products: A review. *Renewable & Sustainable Energy Reviews*, *155*, 111930. DOI: 10.1016/j.rser.2021.111930

Choi, K. R., Jang, W. D., Yang, D., Cho, J. S., Park, D., & Lee, S. Y. (2019). Systems metabolic engineering strategies: Integrating systems and synthetic biology with metabolic engineering. *Trends in Biotechnology, 37*(8), 817–837. DOI: 10.1016/j.tibtech.2019.01.003 PMID: 30737009

Chopin, T., Costa-Pierce, B. A., Troell, M., Hurd, C. L., Costello, M. J., Backman, S., Buschmann, A. H., Cubel, R., Duarte, C. M., Gröndahl, F., Heasman, K., Haroun, R. J., Johansen, J., Jueterbock, A., Lench, M., Lindell, S., Pavia, H., Ricart, A. M., Sundell, K. S., & Yarish, C. (2024). Deep-ocean seaweed dumping for carbon sequestration: Questionable, risky, and not the best use of valuable biomass. *One Earth, 7*(3), 359–364. DOI: 10.1016/j.oneear.2024.01.013

Chu, Wan Loy, and Siew Moi Phang. 2019. "Biosorption of Heavy Metals and Dyes from Industrial Effluents by Microalgae." *Microalgae Biotechnology for Development of Biofuel and Wastewater Treatment* 599–634. doi:.DOI: 10.1007/978-981-13-2264-8_23/TABLES/2

Chua, E., & Palevsky, H. (2026). *Filling the monitoring gap: Aquatic ecosystem metabolism as a cost-effective, scalable tool for assessing marine carbon dioxide removal*. https://doi.org/DOI: 10.31223/X5NF49

Chubukov, V., Gerosa, L., Kochanowski, K., & Sauer, U. (2014). Coordination of microbial metabolism. *Nature Reviews. Microbiology, 12*(5), 327–340. DOI: 10.1038/nrmicro3238 PMID: 24658329

Chunzhuk, E. A., Grigorenko, A. V., Chernova, N. I., Kiseleva, S. V., Ryndin, K. G., Popel, O. S., Malaniy, S. Y., Slavkina, O. V., de Farias Neves, F., Leng, L., Kumar, V., & Vlaskin, M. S. (2023). Direct study of CO2 capture efficiency during microalgae Arthrospira platensis cultivation at high CO2 concentrations. *Energies, 16*(2), 822. DOI: 10.3390/en16020822

Chu, Y., Jiao, X., Li, D., Yu, J., Wang, C., & He, M. (2025). Integration of dairy manure wastewater treatment and biofuel production by microalgae-bacteria consortia in a biofilm reactor. *Algal Research, 86*, 103912. DOI: 10.1016/j.algal.2025.103912

Chu, Y., Zhang, C., Wang, R., Chen, X., Ren, N., & Ho, S. H. (2022). Biotransformation of Sulfamethoxazole by Microalgae: Removal Efficiency, Pathways, and Mechanisms. *Water Research, 221*, 118834. DOI: 10.1016/j.watres.2022.118834 PMID: 35839594

Cichoński, J., & Chrzanowski, G. (2022). Microalgae as a source of valuable phenolic compounds and carotenoids. *Molecules (Basel, Switzerland), 27*(24), 8852. DOI: 10.3390/molecules27248852 PMID: 36557985

Cole, G. M., Greene, J. M., Quinn, J. C., McDaniel, B., Kemp, L., Simmons, D., Hodges, T., Nobles, D., Weiss, T. L., McGowen, J., & McDaniel, S. (2023). Integrated techno-economic and life cycle assessment of a novel algae-based coating for direct air carbon capture and sequestration. *Journal of CO2 Utilization, 69*, 102421. DOI: 10.1016/j.jcou.2023.102421

Cong, P., Du, R., Gao, H., & Chen, Z. (2024). Comparison and assessment of carbon dioxide emissions between alkali-activated materials and OPC cement concrete. [English Edition]. *Journal of Traffic and Transportation Engineering, 11*(5), 918–938. DOI: 10.1016/j.jtte.2023.07.011

Converti, A., Casazza, A. A., Ortiz, E. Y., Perego, P., & Del Borghi, M. (2009). Effect of temperature and nitrogen concentration on the growth and lipid content of Nannochloropsis oculata and Chlorella vulgaris. *Chemical Engineering and Processing, 48*(6), 1146–1151. DOI: 10.1016/j.cep.2009.03.006

Cook, G., Zakkour, P., Neades, S., & Dixon, T. (2024). CCS under Article 6 of the Paris Agreement. *International Journal of Greenhouse Gas Control, 134*, 104110. DOI: 10.1016/j.ijggc.2024.104110

Coşgun, A., Günay, M. E., & Yıldırım, R. (2023). Machine learning for algal biofuels: A critical review and perspective for the future. *Green Chemistry, 25*(9), 3354–3373. DOI: 10.1039/D3GC00389B

Cozma, P., Ro ca, M., Minu , M., & Gavrilescu, M. (2025). Phytoremediation: A sustainable and promising bio-based approach to heavy metal pollution management. *The Science of the Total Environment, 1001*, 180458. DOI: 10.1016/j.scitotenv.2025.180458 PMID: 40992187

Cristóbal, G., Blanco, S., & Bueno, G. (Eds.). (2020). *Modern trends in diatom identification: fundamentals and applications*. Springer Nature. DOI: 10.1007/978-3-030-39212-3

Cuellar-Bermudez, S. P., Aguilar-Hernandez, I., Cardenas-Chavez, D. L., Ornelas-Soto, N., Romero-Ogawa, M. A., & Parra-Saldivar, R. (2015). Extraction and Purification of High-Value Metabolites from Microalgae: Essential Lipids, Astaxanthin and Phycobiliproteins. *Microbial Biotechnology, 8*(2), 190–209. DOI: 10.1111/1751-7915.12167 PMID: 25223877

Cuellar-Bermudez, S. P., Garcia-Perez, J. S., Rittmann, B. E., & Parra-Saldivar, R. (2015). Photosynthetic bioenergy utilizing CO_2: An approach on flue gases utilization for microalgae cultivation. *Journal of Cleaner Production, 98*, 53–65. DOI: 10.1016/j.jclepro.2014.03.034

Cvjetinovic, J., Luchkin, S. Y., Statnik, E. S., Davidovich, N. A., Somov, P. A., Salimon, A. I., Korsunsky, A. M., & Gorin, D. A. (2023). Revealing the static and dynamic nanomechanical properties of diatom frustules—Nature's glass lace. *Scientific Reports*, *13*(1), 5518. DOI: 10.1038/s41598-023-31487-x PMID: 37015973

da Trindade, L. G., Bürck, M., de Souza, E. L., Zanchet, L., Assis, M., & Braga, A. R. C. (2025). Diatomaceous biosilica: A multifunctional resource for biomedicine and sustainable applications. *Ceramics*, *8*(2), 62. DOI: 10.3390/ceramics8020062

Dahai, H., Zhihong, Y., Lin, Q., Yuhong, L., Lei, T., Jiang, L., & Liandong, Z. (2024). The application of magical microalgae in carbon sequestration and emission reduction: Removal mechanisms and potential analysis. *Renewable & Sustainable Energy Reviews*, *197*, 114417. DOI: 10.1016/j.rser.2024.114417

Daneshvar, E., Wicker, R. J., Show, P. L., & Bhatnagar, A. (2022). Biologically-mediated carbon capture and utilization by microalgae towards sustainable CO2 biofixation and biomass valorization – A review. *Chemical Engineering Journal 427*, 130884. DOI: 10.1016/j.cej.2021.130884

Danouche, Mohammed, Naïma El Ghatchouli, and Hicham Arroussi. 2022. "Overview of the Management of Heavy Metals Toxicity by Microalgae." *Journal of Applied Phycology 2021 34:1* 34(1):475–88. .DOI: 10.1007/s10811-021-02668-w

Danquah, M. K., Ang, L., Uduman, N., Moheimani, N., & Forde, G. M. (2009). Dewatering of microalgal culture for biodiesel production: Exploring polymer flocculation and tangential flow filtration. *Journal of Chemical Technology and Biotechnology*, *84*(7), 1078–1083. DOI: 10.1002/jctb.2137

Dasan, Y. K., Lam, M. K., Yusup, S., Lim, J. W., Show, P. L., Tan, I. S., & Lee, K. T. (2020). Cultivation of Chlorella vulgaris using sequential-flow bubble column photobioreactor: A stress-inducing strategy for lipid accumulation and carbon dioxide fixation. *Journal of CO2 Utilization*, *41*, 101226. DOI: 10.1016/j.jcou.2020.101226

Dasan, Y. K., Suparmaniam, U., Lam, M. K., Tan, I. S., Lau, S. Y., & Bridgid Lai, F. C. (2025). CO2 mitigation via seaweed and microalgae farming for sustainable biofuel production in the blue bioeconomy. *Current Opinion in Green and Sustainable Chemistry*, *56*, 101053. DOI: 10.1016/j.cogsc.2025.101053

Das, D., & Veziroglu, T. N. (2008). Advances in biological hydrogen production processes. *International Journal of Hydrogen Energy*, *33*(21), 6046–6057. DOI: 10.1016/j.ijhydene.2008.07.098

Dash, S. K., Chakraborty, S., & Elangovan, D. (2023). A brief review of hydrogen production methods and their challenges. *Energies*, *16*(3), 1141. DOI: 10.3390/en16031141

Das, P., Quadir, M. A., Thaher, M., Alghasal, G., & Aljabri, H. (2019). Microalgal nutrients recycling from the primary effluent of municipal wastewater and use of the produced biomass as bio-fertilizer. *International Journal of Environmental Science and Technology*, *16*(7), 3355–3364. DOI: 10.1007/s13762-018-1867-8

David, K. M., Ghadge, S. A., Rathod, M. K., Pandey, R., Shamkuwar, S. G., Babarao, P. Y., & Panigrahi, C. K. (2025). Microalgal Systems for Efficient CO\(_2\) Utilisation and Carbon Sequestration: A Review. *Journal of Advances in Biology & Biotechnology*, *28*(8), 1304–1317. DOI: 10.9734/jabb/2025/v28i82804

Dayanath, W. M. S., Ilankoon, I. M. S. K., Ali, S. A. S., Chong, M. N., & Foo, S. C. (2026). Microalgae-Based Direct Air Capture Systems for Carbon Dioxide Emission Reductions in Developing Megacities: A Techno-Economic and Policy Analysis. *Journal of Environmental Chemical Engineering*, *14*(3), 122763. DOI: 10.1016/j.jece.2026.122763

de Oliveira, K. S., Volsi, B., Telles, T. S., Mendes, A. D. R., Yunes, J. S., & Andrade, D. S. (2025). Co-inoculation with Rhizobium, Azospirillum, and microalgae increases common bean yield and profitability. *Agronomy Journal*, *117*(1), e21719. DOI: 10.1002/agj2.21719

De Porcellinis, A. J., Nørgaard, H., Brey, L. M. F., Erstad, S. M., Jones, P. R., Heazlewood, J. L., & Sakuragi, Y. (2018). Overexpression of bifunctional fructose-1, 6-bisphosphatase/sedoheptulose-1, 7-bisphosphatase leads to enhanced photosynthesis and global reprogramming of carbon metabolism in Synechococcus sp. PCC 7002. *Metabolic Engineering*, *47*, 170–183. DOI: 10.1016/j.ymben.2018.03.001 PMID: 29510212

de Souza Celente, G., de Cassia de Souza Schneider, R., Julich, J., Rizzetti, T. M., Lobo, E. A., & Sui, Y. (2023). Life cycle assessment of microalgal cultivation medium: biomass, glycerol, and beta-carotene production by Dunaliella salina and Dunaliella tertiolecta. *The International Journal of Life Cycle Assessment 2023 29:12*, *29*(12), 2269–2282. DOI: 10.1007/s11367-023-02209-2

de Souza, M. F., Meers, E., & Mangini, S. (2024). The potential of microalgae for carbon capture and sequestration. *EFB Bioeconomy Journal*, *4*, 100067. DOI: 10.1016/j.bioeco.2024.100067

Dębowski, M., Kazimierowicz, J., & Zieliński, M. (2025). Multi-Sensing Monitoring of the Microalgae Biomass Cultivation Systems for Biofuels and Added Value Products Synthesis—Challenges and Opportunities. *Applied Sciences (Basel, Switzerland)*, *15*(13), 7324. DOI: 10.3390/app15137324

Demirkaya, C., & De la Hoz Siegler, H. (2024). Dark Fermentation of Microalgae and Cyanobacteria for Hydrogen Production. In *Biotechnological Processes for Green Energy, and High Value Bioproducts by Microalgae, and Cyanobacteria Cultures* (pp. 99-111). Cham: Springer International Publishing.

Derco, Ján, Patrícia Guľašová, Maša Legan, Ronald Zakhar, and Andreja Žgajnar Gotvajn. 2024. "Sustainability Strategies in Municipal Wastewater Treatment." *Sustainability 2024, Vol. 16, Page 9038* 16(20):9038. .DOI: 10.3390/su16209038

Dhaouadi, F., Awwad, F., Diamond, A., & Desgagné-Penix, I. (2020). Diatoms' breakthroughs in biotechnology: Phaeodactylum tricornutum as a model for producing high-added value molecules. *American Journal of Plant Sciences*, *11*(10), 1632–1670. DOI: 10.4236/ajps.2020.1110118

Diaz, R. J., & Rosenberg, R. (2008). Spreading Dead Zones and Consequences for Marine Ecosystems. *Science*, *321*(5891), 926–929. DOI: 10.1126/science.1156401 PMID: 18703733

Dineshkumar, R., Kumaravel, R., Gopalsamy, J., Sikder, M. N. A., & Sampathkumar, P. (2018). Microalgae as bio-fertilizers for rice growth and seed yield productivity. *Waste and Biomass Valorization*, *9*(5), 793–800. DOI: 10.1007/s12649-017-9873-5

Dineshkumar, R., Subramanian, J., Gopalsamy, J., Jayasingam, P., Arumugam, A., Kannadasan, S., & Sampathkumar, P. (2019). The impact of using microalgae as biofertilizer in maize (Zea mays L.). *Waste and Biomass Valorization*, *10*(5), 1101–1110. DOI: 10.1007/s12649-017-0123-7

Do, C. V. T., Nguyen, N. T. T., Tran, T. D., Pham, M. H. T., & Pham, T. Y. T. (2021). Capability of carbon fixation in bicarbonate-based and carbon dioxide-based systems by Scenedesmus acuminatus TH04. *Biochemical Engineering Journal*, *166*, 107858. DOI: 10.1016/j.bej.2020.107858

Dolliver, J., & O'Connor, N. E. (2022). Estimating growth, loss and potential carbon sequestration of farmed kelp: A case study of Saccharina latissima at Strangford Lough, Northern Ireland. *Applied Phycology*, *3*(1), 324–339. DOI: 10.1080/26388081.2022.2081934

Doppler, P., & Spadiut, O. (2021). Introduction to autotrophic cultivation of microalgae in photobioreactors. In Kourist, R., & Schmidt, S. (Eds.), *The Autotrophic Biorefinery* (pp. 113–130). DOI: 10.1515/9783110550603-005

D, R. G., v P, H., K P, R. A. H., & Bhide, A.VP. (2024). Attention-driven LSTM and GRU deep learning techniques for precise water quality prediction in smart aquaculture. *Aquaculture International*, *32*(6), 8455–8478. DOI: 10.1007/s10499-024-01574-5

Drabold, E., McGaughy, K., Agner, J., Sellars, D., Johnson, R., Hajer, A. A., Reza, M. T., & Bayless, D. (2020). Challenges and process economics for algal carbon capture with novel integration: Hydrothermal carbonization. *Bioresource Technology Reports*, *12*, 100556. DOI: 10.1016/j.biteb.2020.100556

Driver, T., Bajhaiya, A., & Pittman, J. K. (2014). Potential of bioenergy production from microalgae. *Current Sustainable/Renewable Energy Reports*, *1*(3), 94–103. DOI: 10.1007/s40518-014-0011-8

Ebadi, A. G., Naeem, M. Y., & Selamoglu, Z. (2025). Biohybrid Systems for Waste Valorization: Synergizing Synthetic Biology and Green Engineering to Enable Circular Resource Recovery. *International Journal of Environmental Science and Green Technology*, *1*(2), 25–37.

Ehimen, E. A., Holm-Nielsen, J. B., Poulsen, M., & Boelsmand, J. E. (2013). Influence of different pretreatment routes on the anaerobic digestion of a filamentous algae. *Renewable Energy*, *50*, 476–480. DOI: 10.1016/j.renene.2012.06.064

Eissler, Y., Anabalon, J., Celis-Plá, P. S., Kieft, B., Junier, P., Rivera, P., Cruces, F., Ascencio, E., Dorador, C., & Molina, V. (2025). High nutrient requirement on the photosynthetic performance of a diatom species of the genus Nitzschia (Bacillariophyta) isolated from the high-altitude wetland, Salar de Huasco, Chile. *Phycological Research*, *73*(2), 134–147. DOI: 10.1111/pre.12581

El Zrelli, R., Yacoubi, L., Wakkaf, T., Castet, S., Grégoire, M., Mansour, L., Courjault-Radé, P., & Rabaoui, L. (2021). Surface sediment enrichment with trace metals in a heavily human-impacted lagoon (Bizerte Lagoon, Southern Mediterranean Sea): Spatial distribution, ecological risk assessment, and implications for environmental protection. *Marine Pollution Bulletin*, *169*, 112512. DOI: 10.1016/j.marpolbul.2021.112512 PMID: 34052590

Elgendy, I. M., Elkaliny, N. E., Saleh, H. M., Darwish, G. O., Almostafa, M. M., Metwally, K., Yahya, G., & Mahmoud, Y. A. G. (2025). Bacteria-powered self-healing concrete: Breakthroughs, challenges, and future prospects. *Journal of Industrial Microbiology & Biotechnology*, *52*, kuae051. DOI: 10.1093/jimb/kuae051 PMID: 39673695

El-Hosari, D. G., Mokhtar, F. A., Khalaf, H. A., Ibrahim, A. R., Mohamed, R. M., & Makhlof, M. E. (2024). Minimum-Run resolution IV design for optimized bio removal of Fe2+ Using enteromorpha intestinalis aqueous extract and its extract-Coated silver nanoparticles. *Plants*, *14*(1), 40. DOI: 10.3390/plants14010040 PMID: 39795301

Elouardi, M., Mabrouki, J., & El Belghiti, M. A. (2023). Evaluation of algal absorbed CO2 sequestration techniques: a review. *Euro-Mediterranean Journal for Environmental Integration 2023 8:3*, *8*(3), 481–491. DOI: 10.1007/s41207-023-00379-x

El-Sheekh, Mostafa M., Hamdy R. Galal, Amal Sh H. Mousa, and Abla A. M. Farghl. 2023. "Coupling Wastewater Treatment, Biomass, Lipids, and Biodiesel Production of Some Green Microalgae." *Environmental Science and Pollution Research 2023 30:12* 30(12):35492–504. .DOI: 10.1007/s11356-023-25628-y

El-Sheekh, M. M., El-Kassas, H. Y., & Ali, S. S. (2025). Microalgae-based bioremediation of refractory pollutants: An approach towards environmental sustainability. *Microbial Cell Factories*, *24*(1), 19. DOI: 10.1186/s12934-024-02638-0 PMID: 39810167

Engel, B. D., Schaffer, M., & Cuellar, L. K.. (2015). Native architecture of the Chlamydomonas chloroplast revealed by in situ cryo-electron tomography. *eLife*, *4*, e04889. DOI: 10.7554/eLife.04889 PMID: 25584625

Ennaceri, H., Mkpuma, V. O., Amo-Duodu, G., Menkiti, M. C., Ishika, T., Vadiveloo, A., & Moheimani, N. R. (2025). Microalgae as sustainable bio-factories for nanoparticle biosynthesis: Progress and challenges. *Bioresource Technology Reports*, *30*, 102089. DOI: 10.1016/j.biteb.2025.102089

Enríquez-de-Salamanca, Á. (2024). Environmental and social impacts of carbon sequestration. *Integrated Environmental Assessment and Management*, *20*(6), 1812–1838. DOI: 10.1002/ieam.4925 PMID: 38651985

Espie, G. S., & Kimber, M. S. (2011). Carboxysomes: Cyanobacterial RubisCO comes in small packages. *Photosynthesis Research*, *109*(1), 7–20. DOI: 10.1007/s11120-011-9656-y PMID: 21556873

Espitia-Nery, M. E., Corredor-Pulido, D. E., Castaño-Oliveros, P. A., Rodríguez-Medina, J. A., Ordoñez-Bello, Q. Y., & Pérez-Fuentes, M. S. (2019). Mechanisms of encapsulation of bacteria in self-healing concrete. *Dyna*, *86*(210), 17–22. DOI: 10.15446/dyna.v86n210.75343

Fabris, M., Abbriano, R. M., Pernice, M., Sutherland, D. L., Commault, A. S., Hall, C. C., Labeeuw, L., McCauley, J. I., Kuzhiuparambil, U., Ray, P., Kahlke, T., & Ralph, P. J. (2020). Emerging technologies in algal biotechnology: Toward the establishment of a sustainable, algae-based bioeconomy. *Frontiers in Plant Science*, *11*, 279. DOI: 10.3389/fpls.2020.00279 PMID: 32256509

Fakhimi, N., Gonzalez-Ballester, D., Fernández, E., Galván, A., & Dubini, A. (2020). Algae-bacteria consortia as a strategy to enhance H2 production. *Cells*, *9*(6), 1353. DOI: 10.3390/cells9061353

Faktorova, D., Nisbet, R. E. R., Fernandez Robledo, J. A., Casacuberta, E., Sudek, L., Allen, A. E., Ares, M.Jr, Aresté, C., Balestreri, C., Barbrook, A. C., Beardslee, P., Bender, S., Booth, D. S., Bouget, F.-Y., Bowler, C., Breglia, S. A., Brownlee, C., Burger, G., Cerutti, H., & Lukeš, J. (2020). Genetic tool development in marine protists: Emerging model organisms for experimental cell biology. *Nature Methods*, *17*(5), 481–494. https://doi.org/10. DOI: 10.1038/s41592-020-0796-x PMID: 32251396

Falkowski, P. G., Barber, R. T., & Smetacek, V. (1998). Biogeochemical controls and feedbacks on ocean primary production. *Science*, *281*(5374), 200–206. DOI: 10.1126/science.281.5374.200 PMID: 9660741

Farooqui, A., Tripathi, G., Moheet, K., Dubey, P., Ahmad, S., Husain, A., Shamim, A., & Mahfooz, S. (2021). *Algal Biomass: Potential Renewable Feedstock for Bioenergy Production*. 85–113. DOI: 10.1007/978-981-16-1888-8_5

Faruque, M. O., Hossain, M. M., Farooq, W., & Razzak, S. A. (2023). Phototrophic bioremediation of municipal tertiary wastewater coupling with lipid biosynthesis using Scenedesmus dimorphus: Effect of nitrogen to phosphorous ratio with/without CO2 supplementation. *Sustainability (Basel)*, *15*(2), 1409. DOI: 10.3390/su15021409

Fazal, T., Wang, Y., Zhang, Y., Mushtaq, A., Hou, S., & Zhang, C. (2026). Enhancing ocean alkalinity and CO2 sequestration via microalgae-driven carbonate precipitation and biomimetic catalysis. *Journal of Environmental Management*, *398*, 128457. DOI: 10.1016/j.jenvman.2025.128457 PMID: 41494387

Fei, C., Wilson, A. T., Mangan, N. M., Wingreen, N. S., & Jonikas, M. C. (2022). Modelling the pyrenoid-based CO2-concentrating mechanism provides insights into its operating principles and a roadmap for its engineering into crops. *Nature Plants 2022 8:5, 8*(5), 583–595. DOI: 10.1038/s41477-022-01153-7

Feng, J., Ma, D., Gao, S., Liao, Y., Feng, J., Xu, S., Wang, X., & Chen, K. (2023). Recent advances in engineering heterotrophic microorganisms for reinforcing CO2 fixation based on Calvin–Benson–Bassham cycle. *ACS Sustainable Chemistry & Engineering, 11*(26), 9509–9522. DOI: 10.1021/acssuschemeng.2c06627

Ferguson, R. M., O'Gorman, E. J., McElroy, D. J., McKew, B. A., Coleman, R. A., Emmerson, M. C., & Dumbrell, A. J. (2021). The ecological impacts of multiple environmental stressors on coastal biofilm bacteria. *Global Change Biology, 27*(13), 3166–3178. DOI: 10.1111/gcb.15626 PMID: 33797829

Fernández-Marín, B., Roach, T., Verhoeven, A., & García-Plazaola, J. I. (2021). Shedding Light on the Dark Side of Xanthophyll Cycles. *The New Phytologist, 230*(4), 1336–1344. DOI: 10.1111/nph.17191 PMID: 33452715

Fernández, P. M., Buscaglia, L., Jordaan, D., Barbosa, M., & Janssen, M. (2024). Model-driven design and validation of glucose supply control in a tubular photobioreactor operated under oxygen-balanced mixotrophy. *Chemical Engineering Journal, 499*, 155718. DOI: 10.1016/j.cej.2024.155718

Ferreira, J., Quintão Braga, M., Carvalho Nogueira da Gama, R., Barbosa Magalhães, I., Barros Marangon, B., de Siqueira Castro, J., Ferreira Lorentz, J., Silva Henriques, B., Saleme Aona de Paula Pereira, A., Peixoto Assemany, P., & Lúcia Calijuri, M. (2024). Carotenoids from Wastewater-Grown Microalgae Biomass: Life Cycle Assessment and Techno-Economical Analysis. *Journal of Cleaner Production, 434*, 140526. Advance online publication. DOI: 10.1016/j.jclepro.2023.140526

Fetisoa, M. R., Acheampong, A., Bondzie-Quaye, P., Bashir, M. A., & Huang, Q. (2024). Synergistic strategies for optimizing microalgae-based wastewater treatment: A comprehensive review. *Journal of Water Process Engineering, 68*, 106409. DOI: 10.1016/j.jwpe.2024.106409

Fettke, J., & Fernie, A. R. (2015). Intracellular and cell-to-apoplast compartmentation of carbohydrate metabolism. *Trends in Plant Science, 20*(8), 490–497. DOI: 10.1016/j.tplants.2015.04.012 PMID: 26008154

Field, C. B., Behrenfeld, M. J., Randerson, J. T., & Falkowski, P. J. s. (1998). Primary production of the biosphere: integrating terrestrial and oceanic components. *281*(5374), 237-240.

Field, C. B., Behrenfeld, M. J., Randerson, J. T., & Falkowski, P. (1998). Primary production of the biosphere: Integrating terrestrial and oceanic components. *Science*, *281*(5374), 237–240. DOI: 10.1126/science.281.5374.237 PMID: 9657713

Field, J. L. (2021). Revisiting "Additional Carbon": Tracking atmosphere–ecosystem carbon exchange to establish mitigation and negative emissions from bio-based systems. *Frontiers in Climate*, *3*, 603239. DOI: 10.3389/fclim.2021.603239

Firth, L. B., Bone, J., Bartholomew, A., Bishop, M. J., Bugnot, A., Bulleri, F., Chee, S.-Y., Claassens, L., Dafforn, K. A., & Fairchild, T. P. (2024). Coastal greening of grey infrastructure: An update on the state of the art. Proceedings of the Institution of Civil Engineers-Maritime Engineering, Flori, S., Mikus, F., Flaum, E., Moog, K., Guessoum, S., Beavis, T., Zwahlen, S. M., Romero-Brey, I., Oorshot, V., Olivetta, M., Steele-Ogus, M., Yeh, E., Dudin, O., Schwab, Y., Dey, G., & Vincent, F. (2025). Diatom ultrastructural diversity across controlled and natural environments. *Current Biology : CB*, *35*(23), 5709–5720.e5704. https://doi.org/https://doi.org/10.1016/j.cub.2025.10.024

Forero-Pineda, C. (2006). The impact of stronger intellectual property rights on science and technology in developing countries. *Research Policy*, *35*(6), 808–824. DOI: 10.1016/j.respol.2006.04.003

Fortunato, A. E., Jaubert, M., Enomoto, G., Bouly, J. P., Raniello, R., Thaler, M., Malviya, S., Bernardes, J. S., Rappaport, F., Gentili, B., Huysman, M. J. J., Carbone, A., Bowler, C., d'Alcalà, M. R., Ikeuchi, M., & Falciatore, A. (2016). Diatom phytochromes reveal the existence of far-red- light-based sensing in the ocean. *The Plant Cell*, *28*(3), 616–628. DOI: 10.1105/tpc.15.00928 PMID: 26941092

Friedlingstein, P., O'Sullivan, M., Jones, M. W., Andrew, R. M., Bakker, D. C. E., Hauck, J., Landschützer, P., Le Quéré, C., Li, H., Luijkx, I. T., Peters, G. P., Peters, W., Pongratz, J., Schwingshackl, C., Sitch, S., Canadell, J. G., Ciais, P., Aas, K., Alin, S. R., & Zeng, J. (2025). Global Carbon Budget 2025. *Earth System Science Data Discussions*, 1–139. DOI: 10.5194/essd-2025-659

Frigon, J. C., Matteau-Lebrun, F., Abdou, R. H., McGinn, P. J., O'Leary, S. J. B., & Guiot, S. R. (2013). Screening microalgae strains for their productivity in methane following anaerobic digestion. *Applied Energy*, *108*, 100–107. DOI: 10.1016/j.apenergy.2013.02.051

Frongia, F., Cramarossa, M., Arru, L., & Forti, L. (2023). Integrated wastewater treatment and CO2 capture by microalgae-based system. *Journal of Biotech Research*, *14*, 227–233.

Fujii, H., Webb, J., Mundree, S., Rowlings, D., Grace, P., Wilson, C., & Managi, S. (2024). Priority change and driving factors in the voluntary carbon offset market. *Cleaner Environmental Systems*, *13*, 100164. DOI: 10.1016/j.cesys.2024.100164

Fuqua, C., Parsek, M. R., & Greenberg, E. P. (2001). Regulation of gene expression by cell-to-cell communication: Acyl-homoserine lactone quorum sensing. *Annual Review of Genetics*, *35*(1), 439–468. DOI: 10.1146/annurev.genet.35.102401.090913 PMID: 11700290

Fu, X., Ye, W., Wang, C., Zhu, P., & Tian, J. (2025). A Review on the Improvement Technology of Plant-Based Bio-Cement (PBBC). *Geomicrobiology Journal*, *42*(8), 1–15. DOI: 10.1080/01490451.2025.2516468

Gaber, K., Rösch, C., & Biondi, N. (2021). Life Cycle Assessment of Total Fatty Acid (TFA) Production from Microalgae Nannochloropsis oceanica at Different Sites and Under Different Sustainability Scenarios. *BioEnergy Research 2021 15:3*, *15*(3), 1595–1615. DOI: 10.1007/s12155-021-10279-z

Ganesan, A., Nawkarkar, P., Ali, S., Kajlaand, S., & Kumar, S. (2022). Genetic engineering of microalgae for enhanced photosynthetic efficiency, CO2 fixation, and fuel-based products.

Ganguly, A., Sun, P., Liu, X., Delgado, H. E., Sun, L., & Elgowainy, A. (2025). Techno-economic and life cycle analysis of bio-hydrogen production using bio-based waste streams through the integration of dark fermentation and microbial electrolysis. *Green Chemistry*, *27*(21), 6213–6231. DOI: 10.1039/D4GC05020G

Gao, B., Chen, A., Zhang, W., Li, A., & Zhang, C. (2017). Co-production of lipids, eicosapentaenoic acid, fucoxanthin, and chrysolaminarin by Phaeodactylum tricornutum cultured in a flat-plate photobioreactor under varying nitrogen conditions. *Journal of Ocean University of China*, *16*(5), 916–924. DOI: 10.1007/s11802-017-3174-2

Gao, X., Yuan, L., Yeong, H. Y., Lv, G., Zhang, X., Zhang, X., Zhang, L., & Liao, J. (2026). YEONG, H. Y., Lv, G., Zhang, X., Zhang, X., Zhang, L., & Liao, J. (2026). Physiological changes in carbon sequestration and lipid production characteristics of microalgae under high CO2 domestication. *Environmental Science and Pollution Research International*, *33*(2), 1–15. DOI: 10.1007/s11356-025-37345-9 PMID: 41498902

Gao, Y. X., Li, X., Xiao, Y. F., Jun, R. Z., & Zhong, X. Z. (2022). Wastewater Treatment Plants as Reservoirs and Sources for Antibiotic Resistance Genes: A Review on Occurrence, Transmission and Removal. *Journal of Water Process Engineering*, *46*, 102539. DOI: 10.1016/j.jwpe.2021.102539

García Sánchez, J., Berenguel, M., Rodríguez, F., Fernández Sevilla, J., Brindley Alias, C., & Acién Fernández, F. (2003). Minimization of carbon losses in pilot-scale outdoor photobioreactors by model-based predictive control. *Biotechnology and Bioengineering*, *84*(5), 533–543. DOI: 10.1002/bit.10819 PMID: 14574687

García-Abad, L., López-Rosales, L., María del Carmen, C.-G., Fernández-García, M., García-Camacho, F., & Molina-Grima, E. (2022). Influence of Abiotic Conditions on the Biofouling Formation of Flagellated Microalgae Culture. *Biofouling*, *38*(5), 507–520. DOI: 10.1080/08927014.2022.2089564 PMID: 35729852

Gautam, R., Nayak, J. K., Talapatra, K. N., & Ghosh, U. K. (2023). Assessment of different organic substrates for bio-electricity and bio-hydrogen generation in an integrated bio-electrochemical system. *Materials Today: Proceedings*, *80*, 2255–2259. DOI: 10.1016/j.matpr.2021.06.223

Gayathri, K., Saranraj, P., Manigandan, M., Lokeshwari, B., Yuvalakshmi, L., & Monisha, S. (2025). Applications of nanosciences and nanotechnology in life sciences. *American-Eurasian Journal of Scientific Research*, *20*(3), 91–97.

Gayathri, R., Mahboob, S., Govindarajan, M., Al-Ghanim, K. A., Ahmed, Z., Al-Mulhm, N., Vodovnik, M., & Vijayalakshmi, S. (2021). A review on biological carbon sequestration: A sustainable solution for a cleaner air environment, less pollution and lower health risks. *Journal of King Saud University. Science*, *33*(2), 101282. DOI: 10.1016/j.jksus.2020.101282

Gebru, K. A., Kidanemariam, T. G., & Gebretinsae, H. K. (2021). Bio-cement production using microbially induced calcite precipitation (MICP) method: A review. *Chemical Engineering Science*, *238*, 116610. DOI: 10.1016/j.ces.2021.116610

Geng, Y., Shaukat, A., Azhar, W., Raza, Q. U. A., Tahir, A., Abideen, M. Z. U., Zia, M. A. B., Bashir, M. A., & Rehim, A. (2025). Microalgal biorefineries: A systematic review of technological trade-offs and innovation pathways. *Biotechnology for Biofuels and Bioproducts*, *18*(1), 93. DOI: 10.1186/s13068-025-02694-7 PMID: 40817213

Gerotto, C., Norici, A., & Giordano, M. (2020). Toward Enhanced Fixation of CO2 in Aquatic Biomass: Focus on Microalgae. *Frontiers in Energy Research*, *8*, 550930. DOI: 10.3389/fenrg.2020.00213

Ghirardi, M. L., Zhang, L., Lee, J. W., Flynn, T., Seibert, M., Greenbaum, E., & Melis, A. (2000). Microalgae: A green source of renewable H_2. *Trends in Biotechnology*, *18*(12), 506–511. DOI: 10.1016/S0167-7799(00)01511-0

Ghobara, M. M., Gordon, R., & Reissig, L. (2021a). The mesopores of Raphid pennate diatoms: toward natural controllable anisotropic mesoporous silica microparticles. In *Diatom morphogenesis* (pp. 383–409). Wiley., DOI: 10.1002/9781119488170.ch16

Ghobara, M., El-Sheekh, M., Hamed, A. F., Abdelhamid, M. A., & Pack, S. P. (2023). Diatom nanostructured biosilica. In *Value-Added Products from Algae: Phycochemical Production and Applications* (pp. 461–492). Springer International Publishing.

Gholami, P., Khataee, A., & Bhatnagar, A. (2020). Environmentally superior cleaning of diatom frustules using sono-Fenton process: Facile fabrication of nanoporous silica with homogeneous morphology and controlled size. *Ultrasonics Sonochemistry*, *64*, 105044. DOI: 10.1016/j.ultsonch.2020.105044 PMID: 32146334

Giordano, M., Beardall, J., & Raven, J. A. (2005). CO_2 concentrating mechanisms in algae: Mechanisms, environmental modulation, and evolution. *Annual Review of Plant Biology*, *56*(1), 99–131. DOI: 10.1146/annurev.arplant.56.032604.144052 PMID: 15862091

Global Monitoring Laboratory. 2025. https://gml.noaa.gov/ccgg/trends/global.html

Golewski, G. L. (2023). The phenomenon of cracking in cement concretes and reinforced concrete structures: The mechanism of cracks formation, causes of their initiation, types and places of occurrence, and methods of detection—a review. *Buildings (Basel, Switzerland)*, *13*(3), 765. DOI: 10.3390/buildings13030765

Golueke, C. G., Oswald, W. J., & Gotaas, H. B. (1957). Anaerobic digestion of algae. *Applied Microbiology*, *5*(1), 47–55. DOI: 10.1128/am.5.1.47-55.1957 PMID: 13403639

Gomez-Casati, D. F., Barchiesi, J., & Busi, M. V. (2022). Mitochondria and Chloroplasts Function in Microalgae Energy Production. *PeerJ*, *10*, e14576. DOI: 10.7717/peerj.14576 PMID: 36545385

Gonçalves, J., Freitas, J., Fernandes, I., & Silva, P. (2023). Microalgae as biofertilizers: A sustainable way to improve soil fertility and plant growth. *Sustainability (Basel)*, *15*(16), 12413. DOI: 10.3390/su151612413

Gonzalez, D. I., & Ynalvez, R. A. (2023). Comparison of the effects of nitrogen-, sulfur- and combined nitrogen- and sulfur-deprivations on cell growth, lipid bodies and gene expressions in Chlamydomonas reinhardtii cc5373-sta6. *BMC Biotechnology*, *23*(1), 35. DOI: 10.1186/s12896-023-00808-3 PMID: 37684579

Goold, H. D., Moseley, J. L., & Lauersen, K. J. (2024). The synthetic future of algal genomes. *Cell Genomics*, *4*(3), 100505. DOI: 10.1016/j.xgen.2024.100505 PMID: 38395701

Goold, H. D., Wright, P., & Hailstones, D. (2018). Emerging opportunities for synthetic biology in agriculture. *Genes*, *9*(7), 341. DOI: 10.3390/genes9070341 PMID: 29986428

Gorelova, Olga, Olga Baulina, Tatiana Ismagulova, Kamilya Kokabi, Elena Lobakova, Irina Selyakh, Larisa Semenova, Olga Chivkunova, Olga Karpova, Pavel Scherbakov, Inna Khozin-Goldberg, and Alexei Solovchenko. 2018. "Stress-Induced Changes in the Ultrastructure of the Photosynthetic Apparatus of Green Microalgae." *Protoplasma 2018 256:1* 256(1):261–77. .DOI: 10.1007/s00709-018-1294-1

Gotovtsev, P., Sergeeva, Y., Sukhinov, D., Petrova, M., Vishnevskaya, M., Shestakova, A., & Vasilov, R. (2025). Internet of things (IoT) use for remote monitoring of algal bioreactors. In *Algal Bioreactors* (pp. 205–214). Elsevier Science Ltd. DOI: 10.1016/B978-0-443-14058-7.00044-0

Götz, W., Tobiasch, E., Witzleben, S., & Schulze, M. (2019). Effects of silicon compounds on biomineralization, osteogenesis, and hard tissue formation. *Pharmaceutics*, *11*(3), 117. DOI: 10.3390/pharmaceutics11030117 PMID: 30871062

Grama, S. B., Liu, Z., & Li, J. (2022). Emerging trends in genetic engineering of microalgae for commercial applications. *Marine Drugs*, *20*(5), 285. DOI: 10.3390/md20050285 PMID: 35621936

Green, J. F. (2024). The Climate Establishment and the Paris partnerships. *Climatic Change 2024 177:6*, *177*(6), 84-. DOI: 10.1007/s10584-024-03730-5

Greene, J. M., Quiroz, D., Compton, S., Lammers, P. J., & Quinn, J. C. (2021). A validated thermal and biological model for predicting algal productivity in large scale outdoor cultivation systems. *Algal Research*, *54*, 102224. DOI: 10.1016/j.algal.2021.102224

Gressel, J., van der Vlugt, C. J. B., & Bergmans, H. E. N. (2013). Environmental risks of large scale cultivation of microalgae: Mitigation of spills. *Algal Research*, *2*(3), 286–298. DOI: 10.1016/j.algal.2013.04.002

Griffiths, M. J., & Harrison, S. T. L. (2009). Lipid productivity as a key characteristic for choosing algal species for biodiesel production. *Journal of Applied Phycology*, *21*(5), 493–507. DOI: 10.1007/s10811-008-9392-7

Grypioti, E., Richard, H., Kryovrysanaki, N., Jaubert, M., Falciatore, A., Verret, F., & Kalantidis, K. (2024). Dicer-dependent heterochromatic small RNAs in the model diatom species Phaeodactylum tricornutum. *The New Phytologist*, *241*(2), 811–826. DOI: 10.1111/nph.19429 PMID: 38044751

Guan, Y., Rausser, G., Zilberman, D., & Choi, E. (2026). Alternative Bioeconomy Strategies for Decarbonization. *Natural Resource Management and Policy*, *61*, 175–207. DOI: 10.1007/978-3-032-07112-5_11

Guiry, M. D. (2024). How many species of algae are there? A reprise. Four kingdoms, 14 phyla, 63 classes and still growing. *Journal of Phycology*, *60*(2), 214–228. DOI: 10.1111/jpy.13431 PMID: 38245909

Güler, N. U., & Yumurtaci, Z. (2025). Techno-economic evaluation of hydrogen production via biomass gasification: The role of carbon capture and storage in steam and oxygen gasification routes. *International Journal of Hydrogen Energy*, *188*, 152134. DOI: 10.1016/j.ijhydene.2025.152134

Günther, S. A., Staake, T., Schöb, S., & Tiefenbeck, V. (2020). The behavioral response to a corporate carbon offset program: A field experiment on adverse effects and mitigation strategies. *Global Environmental Change*, *64*, 102123. DOI: 10.1016/j.gloenvcha.2020.102123

Guo, B., Lu, X., Jiang, X., Shen, X. L., Wei, Z., & Zhang, Y. (2025). Artificial Intelligence in Advancing Algal Bioactive Ingredients: Production, Characterization, and Application. *Foods*, *14*(10), 1783. DOI: 10.3390/foods14101783 PMID: 40428562

Guo, W., Weng, Y., Ma, W., Chang, C., Gao, Y., Huang, X., & Zhang, F. (2024). Improving lipid content in the diatom Phaeodactylum tricornutum by the knockdown of the enoyl-CoA hydratase using CRISPR interference. *Current Issues in Molecular Biology*, *46*(10), 10923–10933. DOI: 10.3390/cimb46100649 PMID: 39451529

Gupta, Riya, Niranjan Mishra, Gayatri Singh, Sandhya Mishra, Neelu Lodhiyal, and Blue Biotechnology. 2024. "Microalgae Cultivation and Value-Based Products from Wastewater: Insights and Applications." *Blue Biotechnology 2024 1:1* 1(1):20-. .DOI: 10.1186/s44315-024-00019-1

Gupta, Y., & Philip, L. (2026). Mechanistic insights and resistance profiling of chlorophyta–proteobacteria–cyanobacteria dominated consortia in removing antibiotic mixtures and nutrients from synthetic and real municipal wastewater. *Journal of Environmental Chemical Engineering*, *14*(1), 121004. DOI: 10.1016/j.jece.2025.121004

Gurau, S., Imran, M., & Ray, R. L. (2025). Algae: A cutting-edge solution for enhancing soil health and accelerating carbon sequestration–A review. *Environmental Technology & Innovation*, *37*, 103980. DOI: 10.1016/j.eti.2024.103980

Gu, X., Shi, Y., Luo, C., & Cheng, J. (2024). Establishment of Saccharomyces cerevisiae as a cell factory for efficient de novo production of monogalactosyldiacylglycerol. *Biotechnology for Biofuels and Bioproducts*, *17*(1), 111. DOI: 10.1186/s13068-024-02560-y PMID: 39129014

Habib, M. A., Abdulrahman, G. A., Alquaity, A. B., & Qasem, N. A. (2024). Hydrogen combustion, production, and applications: A review. *Alexandria Engineering Journal*, *100*, 182–207. DOI: 10.1016/j.aej.2024.05.030

Hakkoum, Z., Minaoui, F., Chabili, A., Douma, M., Mouhri, K., & Loudiki, M. (2025). Biofertilizing effect of soil Cyanobacterium Anabaena cylindrica–based formulations on wheat growth, physiology, and soil fertility. *Agriculture*, *15*(2), 189. DOI: 10.3390/agriculture15020189

Hamano, R., Shoumura, S., Takeda, Y., Yamazaki, T., Hirayama, K., Hanada, Y., Mayama, S., Takemura, M., Lin, H.-J., Umemura, K. J. M., & Microanalysis. (2021). Sinking of four species of living diatom cells directly observed by a "tumbled" optical microscope. *27*(5), 1154-1160.

Hamed, A. (2023). A brief and general overview on diatoms and their applications. *Egyptian journal of phycology*, *24*(1), 1-53.

Hamida, R. S., Ali, M. A., Alkhateeb, M. A., Alfassam, H. E., Momenah, M. A., & Bin-Meferij, M. M. (2022). Algal-Derived Synthesis of Silver Nanoparticles Using the Unicellular ulvophyte sp. MBIC10591: Optimisation, Characterisation, and Biological Activities. *Molecules (Basel, Switzerland)*, *28*(1), 279. DOI: 10.3390/molecules28010279 PMID: 36615473

Hamilton, M. L., Warwick, J., Terry, A., Allen, M. J., Napier, J. A., & Sayanova, O. (2015). Towards the Industrial Production of Omega-3 Long Chain Polyunsaturated Fatty Acids from a Genetically Modified Diatom Phaeodactylum tricornutum. *PLoS One*, *10*(12), e0144054. DOI: 10.1371/journal.pone.0144054 PMID: 26658738

Hanson, E., Nwakile, C., & Hammed, V. O. (2025). Carbon capture, utilization, and storage (CCUS) technologies: Evaluating the effectiveness of advanced CCUS solutions for reducing CO2 emissions. *Results in Surfaces and Interfaces*, *18*, 100381. DOI: 10.1016/j.rsurfi.2024.100381

Hasan, M. N., Altaf, M. M., Khan, N. A., Khan, A. H., Khan, A. A., Ahmed, S., Kumar, P. S., Naushad, M., Rajapaksha, A. U., Iqbal, J., Tirth, V., & Islam, S. (2021). Recent Technologies for Nutrient Removal and Recovery from Wastewaters: A Review. *Chemosphere*, *277*, 130328. DOI: 10.1016/j.chemosphere.2021.130328 PMID: 33794428

Hase, R., Oikawa, H., Sasao, C., Morita, M., & Watanabe, Y. (2000). Photosynthetic production of microalgal biomass in a raceway system under greenhouse conditions in Sendai City. *Journal of Bioscience and Bioengineering*, *89*(2), 157–163. DOI: 10.1016/S1389-1723(00)88730-7 PMID: 16232719

Hashimoto, Koji. 2019. "Global Temperature and Atmospheric Carbon Dioxide Concentration." 5–17. .DOI: 10.1007/978-981-13-8584-1_3

Hasnain, M., Zainab, R., Ali, F., Abideen, Z., Yong, J. W. H., El-Keblawy, A., Hashmi, S., & Radicetti, E. (2023). Utilization of microalgal-bacterial energy nexus improves CO2 sequestration and remediation of wastewater pollutants for beneficial environmental services. *Ecotoxicology and Environmental Safety*, *267*, 115646. DOI: 10.1016/j.ecoenv.2023.115646 PMID: 37939556

Hassan, N. S., Jalil, A. A., Vo, D. V. N., & Nabgan, W. (2023). An overview on the efficiency of biohydrogen production from cellulose. *Biomass Conversion and Biorefinery, 13*(10), 8485-8507.

Hassan, S., Meenatchi, R., Pachillu, K., Bansal, S., Brindangnanam, P., Arockiaraj, J., Kiran, G. S., & Selvin, J. (2022). Identification and characterization of the novel bioactive compounds from microalgae and cyanobacteria for pharmaceutical and nutraceutical applications. *Journal of Basic Microbiology*, *62*(9), 999–1029. DOI: 10.1002/jobm.202100477 PMID: 35014044

Hattab, M. A., & Ghaly, A. (2015). Microalgae Harvesting Methods for Industrial Production of Biodiesel:Critical Review and Comparative Analysis. *Journal of Fundamentals of Renewable Energy and Applications*, *05*(02). Advance online publication. DOI: 10.4172/2090-4541.1000154

Hawrot-Paw, M., & Sąsiadek, M. (2023). Optimization of microalgal biomass production in vertical tubular photobioreactors. *Energies*, *16*(5), 2429. DOI: 10.3390/en16052429

Hazarika, A., Hazarika, I., Gogoi, M., Bora, S. S., Borah, R. R., Goutam, P. J., & Saikia, N. (2018). Use of a plant based polymeric material as a low cost chemical admixture in cement mortar and concrete preparations. *Journal of Building Engineering*, *15*, 194–202. DOI: 10.1016/j.jobe.2017.11.017

HD. A., & Vijayan P. U, H. (2025). Bio-green Evolution: The Journey of Microalgae Products from Lab to Market—Considerations and Recommendations—Part II. In *Microalgae Horizons: Fundamentals, Innovations, and Industrial Applications* (pp. 277-299). Cham: Springer Nature Switzerland.

He, L. N., Rogers, R. D., Su, D., Tundo, P., & Zhang, Z. C. (2016). *Green chemistry and sustainable technology* (Doctoral dissertation, Dalian Institute of Chemical Physics, Chinese Academy of Sciences).

He, A., Dean, J. M., & Lodhi, I. J. (2021). Peroxisomes as Cellular Adaptors to Metabolic and Environmental Stress. *Trends in Cell Biology*, *31*(8), 656–670. DOI: 10.1016/j.tcb.2021.02.005 PMID: 33674166

Heidarnezhad, S., Wang, S., & Rahbar, N. (2024). Carbonic Anhydrase as a Sustainable Corrosion Inhibitor for Concrete. *ACS Applied Engineering Materials*, *3*(1), 128–141. DOI: 10.1021/acsaenm.4c00643

Heintze, C., Formanek, P., Pohl, D., Hauptstein, J., Rellinghaus, B., & Kröger, N. (2020). An intimate view into the silica deposition vesicles of diatoms. *BMC Materials*, *2*(1), 11. DOI: 10.1186/s42833-020-00017-8

Hemashenpagam, N., Selvajeyanthi, S., & Gunasheela, N. (2024). Effect of Substrate on the Efficiency of Microbial Electrolysis Cells. In *Microbial Electrolysis Cell Technology: Fundamental to Sustainable Applications* (pp. 75–91). Springer Nature Singapore. DOI: 10.1007/978-981-97-3356-9_5

Hepburn, Cameron, Ella Adlen, John Beddington, Emily A. Carter, Sabine Fuss, Niall Mac Dowell, Jan C. Minx, Pete Smith, and Charlotte K. Williams. 2019. "The Technological and Economic Prospects for CO2 Utilization and Removal." *Nature 2019 575:7781* 575(7781):87–97. .DOI: 10.1038/s41586-019-1681-6

He, T., Liu, Z., Son, H., Gundersen, T., & Lin, W. (2023). Comparative analysis of cryogenic distillation and chemical absorption for carbon capture in integrated natural gas liquefaction processes. *Journal of Cleaner Production*, *383*, 135264. DOI: 10.1016/j.jclepro.2022.135264

Hickland, P. R., Geisler, K., Newsad, S., Pasquina, M. L., Faessler, A. C., Mendoza-Ochoa, G. I., Holzer, A., Mehrshahi, P., & Smith, A. G. (2025). A new tool for engineering Phaeodactylum tricornutum: The METE promoter drives both high expression and B12-tuneable regulation of transgenes. *The Plant Journal : For Cell and Molecular Biology*, *124*(2), e70210. DOI: 10.1111/tpj.70210 PMID: 41116758

Hildebrand, M., Manandhar-Shrestha, K., & Abbriano, R. (2017). Effects of Chrysolaminarin Synthase Knockdown in the Diatom Thalassiosira Pseudonana: Implications of Reduced Carbohydrate Storage Relative to Green Algae. *Algal Research*, *23*, 66–77. DOI: 10.1016/j.algal.2017.01.010

Hitam, C. N. C., & Jalil, A. A. (2023). A review on biohydrogen production through photo-fermentation of lignocellulosic biomass. *Biomass Conversion and Biorefinery*, *13*(10), 8465–8483. DOI: 10.1007/s13399-020-01140-y

Homburg, S. V., Kruse, O., & Patel, A. V. (2019). Growth and photosynthetic activity of Chlamydomonas reinhardtii entrapped in lens-shaped silica hydrogels. *Journal of Biotechnology*, *302*, 58–66. DOI: 10.1016/j.jbiotec.2019.06.009

Hooshmand Zaferani, S., Sams, M. W., Shi, X. L., Mehrabian, N., Ghomashchi, R., & Chen, Z. G. (2022). Applications of thermoelectric generators to improve catalytic-assisted hydrogen production efficiency: Future directions. *Energy & Fuels*, *36*(15), 8096–8106. DOI: 10.1021/acs.energyfuels.2c01754

Hopkinson, B. M., Dupont, C. L., & Matsuda, Y. (2016). The physiology and genetics of CO2 concentrating mechanisms in model diatoms. *Current Opinion in Plant Biology*, *31*, 51–57. DOI: 10.1016/j.pbi.2016.03.013 PMID: 27055267

Hoque, M. M., Iannelli, V., Padula, F., Radice, R. P., Saha, B. K., Martelli, G., Scopa, A., & Drosos, M. J. B. (2025). Microalgae: Green engines for achieving carbon sequestration, circular economy, and environmental sustainability—A review based on last ten years of research. *12*(9), 909.

Ho, S. H., Chen, C. Y., Lee, D. J., & Chang, J. S. (2011). Perspectives on microalgal CO_2-emission mitigation systems. *Biotechnology Advances*, *29*(2), 189–198. DOI: 10.1016/j.biotechadv.2010.11.001 PMID: 21094248

Hou, C., Zhao, J., Huang, B., Zhou, X., & Zhang, Y. (2024). Microalgae-based technologies for carbon neutralization and pollutant remediation: A comprehensive and systematic review. *Resources, Conservation and Recycling*, *202*, 107323. DOI: 10.1016/j.resconrec.2023.107323

Hou, Y., Wang, W., Liu, Z., Yu, L., & Zhao, L. (2025). Boosting microalgae-based carbon sequestration with the artificial CO2 concentration system. *Critical Reviews in Biotechnology*, *45*(8), 1615–1633. DOI: 10.1080/07388551.2025.2498464 PMID: 40374568

Hsiao, T. Y., & Blanch, H. W. (2006). Physiological studies of eicosapentaenoic acid production in the marine microalga *Glossomastix chrysoplasta*. *Biotechnology and Bioengineering*, *93*(3), 465–475. DOI: 10.1002/bit.20761 PMID: 16304671

Hu, B., Li, Y., Zhu, S., Zhang, H., Jing, Y., Jiang, D., He, C., & Zhang, Z. (2020). Evaluation of biohydrogen yield potential and electron balance in the photo-fermentation process with different initial pH from starch agricultural leftover. *Bioresource Technology*, *305*, 122900. DOI: 10.1016/j.biortech.2020.122900

Hu, D., Liu, H., Yang, C., & Hu, E. (2008). The design and optimization for light-algae bioreactor controller based on Artificial Neural Network-Model Predictive Control. *Acta Astronautica*, *63*(7-10), 1067–1075. DOI: 10.1016/j.actaastro.2008.02.008

Huerlimann, R., & Heimann, K. (2013). Comprehensive Guide to Acetyl-Carboxylases in Algae. *Critical Reviews in Biotechnology*, *33*(1), 49–65. DOI: 10.3109/07388551.2012.668671 PMID: 22524446

Hughes, A. D., Kelly, M. S., Black, K. D., & Stanley, M. S. (2012). Biogas from macroalgae: Is it time to revisit the idea? *Biotechnology for Biofuels*, *5*(1), 86. DOI: 10.1186/1754-6834-5-86 PMID: 23186536

Hu, H., & Gao, K. (2008). Impacts of CO2 enrichment on growth and photosynthesis in freshwater and marine diatoms. *Chinese Journal of Oceanology and Limnology*, *26*(4), 407–414. DOI: 10.1007/s00343-008-0407-7

Hu, L., Li, Z., Kong, L., Wei, J., Chang, J., & Shi, W. (2024). Reassessing the greenhouse effect of biogenic carbon emissions in constructed wetlands. *Journal of Environmental Management*, *354*, 120263. DOI: 10.1016/j.jenvman.2024.120263 PMID: 38387360

Hu, L., Ye, Y., Li, Y., Tan, X., Liu, X., Zhang, T., Wang, J., Du, Z., & Ye, M. (2025). Bacteria–algae synergy in carbon sequestration: Molecular mechanisms, ecological dynamics, and biotechnological innovations. *Biotechnology Advances*, *83*, 108655. DOI: 10.1016/j.biotechadv.2025.108655 PMID: 40701356

Hundleby, P. A., Sack, M., & Twyman, R. M. (2018). Biosafety, risk assessment, and regulation of molecular farming. *Molecular Pharming: Applications, Challenges, and Emerging Areas*, 327-351.

Hussein, H. A., Nazir, M. S., Azra, N., Qamar, Z., Seeni, A., Tengku Din, T. A. D. A. A., & Abdullah, M. A. (2022). Novel drug and gene delivery system and imaging agent based on marine diatom biosilica nanoparticles. *Marine Drugs*, *20*(8), 480. DOI: 10.3390/md20080480 PMID: 36005484

Ighalo, J. O., Dulta, K., Kurniawan, S. B., Omoarukhe, F. O., Ewuzie, U., Eshiemogie, S. O., Ojo, A. U., & Abdullah, S. R. S. (2022). Progress in microalgae application for CO2 sequestration. *Cleaner Chemical Engineering*, *3*, 100044. DOI: 10.1016/j.clce.2022.100044

Ihnken, S., Roberts, S., & Beardall, J. (2011). Differential responses of growth and photosynthesis in the marine diatom Chaetoceros muelleri to CO2 and light availability. *Phycologia*, *50*(2), 182–193. DOI: 10.2216/10-11.1

Ikeda, T. (2021). Bacterial biosilicification: A new insight into the global silicon cycle. *Bioscience, Biotechnology, and Biochemistry*, *85*(6), 1324–1331. DOI: 10.1093/bbb/zbab069 PMID: 33877302

Indhur, R., Amoah, I. D., Bux, F., & Kumari, S. (2023). Nanomaterials for Microplastic Removal from Wastewater: Current State of the Art Nanomaterials and Future Prospects. *ACS ES&T Water*, *3*(12), 3741–3754. DOI: 10.1021/acsestwater.3c00203

Insights, S. (2023). *Global diatomite market size, share, and COVID-19 impact analysis, by type (calcined, flux-calcined, and natural), by application (filter aids, cementitious materials, fillers, absorbents, and others), and by region (North America, Europe, Asia-Pacific, Latin America, Middle East, and Africa), analysis and forecast 2023 – 2033*. https://www.sphericalinsights.com/reports/diatomite-market

Intasian, P., Prakinee, K., Phintha, A., Trisrivirat, D., Weeranoppanant, N., Wongnate, T., & Chaiyen, P. (2021). Enzymes, in vivo biocatalysis, and metabolic engineering for enabling a circular economy and sustainability. *Chemical Reviews*, *121*(17), 10367–10451. DOI: 10.1021/acs.chemrev.1c00121

Inuwa, A. B., A. Pervez, and R. Nazir. 2023. "Microalgae-Based Wastewater Treatment System: Current State, Antibiotic Resistant Bacteria and Antibiotic Resistance Genes Reduction Potentials." *International Journal of Environmental Science and Technology 2023 20:12* 20(12):14053–72. .DOI: 10.1007/s13762-023-05069-3

IPCC. 2022. "IPCC AR6 WGIII: CDR Factsheet."

Iqbal, K., Sharma, N., Takkar, S., Shukla, S., Shukla, K., Varma, A., & Mishra, A. (2022). Integrated CO2 sequestration, wastewater treatment, and biofuel production by microalgae culturing: Needs and limitations. *Integrated Environmental Technologies for Wastewater Treatment and Sustainable Development*, 217–240. DOI: 10.1016/B978-0-323-91180-1.00027-2

Iqbal, D. M., Wong, L. S., & Kong, S. Y. (2021). Bio-cementation in construction materials: A review. *Materials (Basel)*, *14*(9), 2175. DOI: 10.3390/ma14092175 PMID: 33922871

Ishizaki, R., Noguchi, R., Putra, A. S., Ichikawa, S., Ahamed, T., & Watanabe, M. M. (2020). Reduction in Energy Requirement and CO2 Emission for Microalgae Oil Production Using Wastewater. *Energies 2020, Vol. 13, 13*(7). DOI: 10.3390/en13071641

Islam, M. M., Alam, H., Acharjee, A., Salatul, M., & Mozumder, I. (2024). Establishment of an effective photobioreactor for growing microalgae: A review. *International Journal of Agricultural Research, Innovation and Technology*, *14*(2), 153–162. DOI: 10.3329/ijarit.v14i2.79511

Islam, M. S., Senaha, I., Rahman, M. M., Yoda, Y., & Saha, B. B. (2022). Mathematical modelling and statistical optimization of fast cultivation of Agardhiella subulata: Response surface methodology. *Energy Nexus*, *7*, 100115. DOI: 10.1016/j.nexus.2022.100115

Ivanenko, A. A., Laikova, A. A., Zhuravleva, E. A., Shekhurdina, S. V., Vishnyakova, A. V., Kovalev, A. A., Kovalev, D. A., Trchounian, K. A., & Litti, Y. V. (2024). Biological production of hydrogen: From basic principles to the latest advances in process improvement. *International Journal of Hydrogen Energy*, *55*, 740–755. DOI: 10.1016/j.ijhydene.2023.11.179

Ivaškė, A., Gribniak, V., Jakubovskis, R., & Urbonavičius, J. (2023). Bacterial viability in self-healing concrete: A case study of non-ureolytic Bacillus species. *Microorganisms*, *11*(10), 2402. DOI: 10.3390/microorganisms11102402 PMID: 37894059

Jabri, H. A., Das, P., Khan, S., & Thaher, M. (2020). Treatment of Wastewaters by Microalgae and the Potential Applications of the Produced Biomass—A Review. *Water (Basel)*, *13*(1), 27. DOI: 10.3390/w13010027

Jagdeo, K. R. (2023). Physical methods for synthesis of nanoparticles. In *Nanochemistry* (pp. 66–76). CRC Press. DOI: 10.1201/9781003081944-4

Jain, S., & Kumar, S. (2026). Advancing bioethanol: Exploring feedstock diversity, production pathways, and environmental implications. *RSC Sustainability*, *4*(3), 1129–1159. Advance online publication. DOI: 10.1039/D5SU00152H

Jain, S., Sharma, T., & Gupta, A. K. (2022). End-of-life management of solar PV waste in India: Situation analysis and proposed policy framework. *Renewable & Sustainable Energy Reviews*, *153*, 111774. DOI: 10.1016/j.rser.2021.111774

Jaiswal, A., Babu, V., Baishya, B., & Jaiswal, P. (2020). Biochemical Pathways Regulated by Algae to Mitigate Global Carbon Emissions: A Review. *Journal of Environmental Pathology, Toxicology and Oncology : Official Organ of the International Society for Environmental Toxicology and Cancer*, *39*(4), 317–334. DOI: 10.1615/JEnvironPatholToxicolOncol.2020034921 PMID: 33389904

Jaiswal, D. K., Verma, J. P., Belwal, T., Pereira, A. P. D. A., & Ade, A. B. (2023). Microbial co-cultures: A new era of synthetic biology and metabolic engineering. *Frontiers in Microbiology*, *14*, 1235565. DOI: 10.3389/fmicb.2023.1235565 PMID: 37426012

Jalil, A., Ahmadi, H., Ndayisenga, F., Khan, S., Ahmad, A., Wang, X., & Yu, Z. (2025). Integrating dark fermentation and electrohydrogenesis for enhanced biohydrogen production from food waste. *Sustainable Energy & Fuels*, *9*(20), 5432–5457. DOI: 10.1039/D5SE00571J

Jangama, R. (2026). *Hybrid Coating Systems: Bioactive Molecules and Nanoparticles for Cancer Therapy*. IGI Global Scientific Publishing., DOI: 10.4018/979-8-3373-4104-0.ch003

Jaramillo, H. Y., Vasco-Echeverri, O. H., Moreno-Pacheco, L. A., & García-León, R. A. (2023). Biomaterials in concrete for engineering applications: A bibliometric review. *Infrastructures*, *8*(11), 161. DOI: 10.3390/infrastructures8110161

Jiang, K., Bai, R., Gao, T., Lu, P., Zhang, J., Zhang, S., Xu, F., Wang, S., & Zhao, H. (2023). Optimization of hydrogen production in Enterobacter aerogenes by Complex I peripheral fragments destruction and maeA overexpression. *Microbial Cell Factories*, *22*(1), 137. DOI: 10.1186/s12934-023-02155-6

Jiang, Q. (2022). *Synthetic engineering of biological carbon-fixing organelles shells*. The University of Liverpool.

Ji, M., & Wang, J. (2021). Review and comparison of various hydrogen production methods based on costs and life cycle impact assessment indicators. *International Journal of Hydrogen Energy*, *46*(78), 38612–38635. DOI: 10.1016/j.ijhydene.2021.09.142

Jimenez, R., Markou, G., Tayibi, S., Barakat, A., Chapsal, C., & Monlau, F. (2020). Production of microalgal slow-release fertilizer by valorizing liquid agricultural digestate: Growth experiments with tomatoes. *Applied Sciences (Basel, Switzerland)*, *10*(11), 3890. DOI: 10.3390/app10113890

Jin, K., Chen, G., Yang, Y., Zhang, Z., & Lu, T. (2023). Strategies for manipulating Rubisco and creating photorespiratory bypass to boost C3 photosynthesis: Prospects on modern crop improvement. *Plant, Cell & Environment*, *46*(2), 363–378. DOI: 10.1111/pce.14500 PMID: 36444099

Jones, E. R., Van Vliet, M. T. H., Qadir, M., & Bierkens, M. F. P. (2021). Country-Level and Gridded Estimates of Wastewater Production, Collection, Treatment and Reuse. *Earth System Science Data*, *13*(2), 237–254. DOI: 10.5194/essd-13-237-2021

Kabato, W. S., Hailegnaw, N., Ermias, T., Samuel, A., De Silva, G., & Molnár, Z. (2025). Microalgae-Based Strategies for Soil Health and Crop Productivity: Mechanisms, Challenges, and Pathways to Climate-Resilient Agriculture. *Agronomy (Basel), 15*(11), 2669. Advance online publication. DOI: 10.3390/agronomy15112669

Kaewpintong, K., Shotipruk, A., Powtongsook, S., & Pavasant, P. (2007). Photoautotrophic high-density cultivation of vegetative cells of Haematococcus pluvialis in airlift bioreactor. *Bioresource Technology, 98*(2), 288–295. https://doi.org/https://doi.org/10.1016/j.biortech.2006.01.011. DOI: 10.1016/j.biortech.2006.01.011 PMID: 16516464

Kahlert, M., Bailet, B., Chonova, T., Karjalainen, S. M., Schneider, S. C., & Tapolczai, K. (2021). Same same, but different: The response of diatoms to environmental gradients in Fennoscandian streams and lakes–barcodes, traits and microscope data compared. *Ecological Indicators, 130*, 108088. DOI: 10.1016/j.ecolind.2021.108088

Kaliyaperumal, G., Nagabhooshanam, N., Rathore, S., Kulshreshta, A., Sheela, C. S., Beulah, D., Maranan, R., Srinivasan, R., & Sathiyamurthy, S. (2025). Converging pathways in biomass-to-biohydrogen conversion: A review on hybrid systems, nanotechnological interventions, and sustainable integration frameworks. *Environmental Progress & Sustainable Energy*, •••, 70295. DOI: 10.1002/ep.70295

Kamennaya, N. A., Ajo-Franklin, C. M., Northen, T., & Jansson, C. (2012). Cyanobacteria as biocatalysts for carbonate mineralization. *Minerals (Basel), 2*(4), 338–364. DOI: 10.3390/min2040338

Kandasamy, S., Shah, M. P., Subbiah, K., & Manickam, N. (Eds.). (2025). *Microbial Niche Nexus Sustaining Environmental Biological Wastewater and Water-Energy-Environment Nexus*. Springer Nature. DOI: 10.1007/978-3-031-62660-9

Kanna Dasan, Y., Lam, M. K., Chai, Y. H., Lim, J. W., Ho, Y. C., Tan, I. S., Lau, S. Y., Show, P. L., & Lee, K. T. (2023). Unlocking the potential of microalgae bio-factories for carbon dioxide mitigation: A comprehensive exploration of recent advances, key challenges, and energy-economic insights. *Bioresource Technology, 380*, 129094. DOI: 10.1016/j.biortech.2023.129094 PMID: 37100295

Kao, C. Y., Chiu, S. Y., Huang, T. T., Dai, L., Wang, G. H., Tseng, C. P., Chen, C. H., & Lin, C. S. (2012). A mutant strain of microalga *Chlorella sp.* for the carbon dioxide capture from biogas. *Biomass and Bioenergy, 36*, 132–140. DOI: 10.1016/j.biombioe.2011.10.046

Kapdan, I. K., & Kargi, F. (2006). Bio-hydrogen production from waste materials. *International Journal of Hydrogen Energy, 31*(12), 1664–1673.

Kapoor, S., Singh, M., Paul, S., Kar, S., Bagchi, T., Chavali, M., & Chandrasekhar, K. (2023). Sustainable bio-applications of diatom silica as nanoarchitectonic material. *Algae refinery*, 161-186.

Karabulut, G., Purkiewicz, A., & Gökşen, G. (2024). Recent developments and challenges in algal protein and peptide extraction strategies, functional and technological properties, bioaccessibility, and commercial applications. *Comprehensive Reviews in Food Science and Food Safety*, *23*(3), e13372. Advance online publication. DOI 10.1111/1541-4337.13372 PMID: 38795380

Kareya, M. S., Mariam, I., Shaikh, K. M., Nesamma, A. A., & Jutur, P. P. (2020) Photosynthetic carbon partitioning and metabolic regulation in response to very-low and high CO2 in Microchloropsis gaditana NIES 2587. *Frontiers in Plant Science*, *11*, 981. DOI: 10.3389/fpls.2020.00981 PMID: 32719702

Karimi, K., & Ghaemi, A. (2025). A comprehensive review of the physicochemical properties and performance of novel carbon-based adsorbents for CO2 capture. *Environmental Science and Pollution Research International*, *32*(34), 20248–20287. DOI: 10.1007/s11356-025-36803-8 PMID: 40836200

Kariyawasam, T., Helvig, C., Petkovich, M., & Vriens, B. (2024). Pharmaceutical Removal from Wastewater by Introducing Cytochrome P450s into Microalgae. *Microbial Biotechnology*, *17*(6), e14515. Advance online publication. DOI: 10.1111/1751-7915.14515 PMID: 38925623

Karoui, H., Patwal, P. S., Pavan Kumar, B. V. V. S., & Martin, N. (2022). Chemical communication in artificial cells: Basic concepts, design and challenges. *Frontiers in Molecular Biosciences*, *9*, 880525. DOI: 10.3389/fmolb.2022.880525 PMID: 35720123

Kar, S., Sanderson, H., Roy, K., Benfenati, E., & Leszczynski, J. (2021). Green chemistry in the synthesis of pharmaceuticals. *Chemical Reviews*, *122*(3), 3637–3710. DOI: 10.1021/acs.chemrev.1c00631

Kashyap, M., Chakraborty, S., Kumari, A., Rai, A., Varjani, S., & Vinayak, V. (2023). Strategies and challenges to enhance commercial viability of algal biorefineries for biofuel production. *Bioresource Technology*, *387*, 129551. DOI: 10.1016/j.biortech.2023.129551 PMID: 37506948

Kavitha, S., Subbulakshmi, P., Rajesh Banu, J., Gobi, M., & Yeom, I. T. (2017). Enhancement of biogas production from microalgal biomass through cellulolytic bacterial pretreatment. *Bioresource Technology*, *233*, 34–43. DOI: 10.1016/j.biortech.2017.02.081 PMID: 28258994

Kazi, T. A., & Biswas, S. R. (2021). CRISPR/dCas system as the modulator of gene expression. *Progress in Molecular Biology and Translational Science*, *178*, 99–122. DOI: 10.1016/bs.pmbts.2020.12.002 PMID: 33685602

Khalili, Y., Yasemi, S., Abdi, M., Ghasemi Ertian, M., Mohammadi, M., & Bagheri, M. (2025). A Review of Integrated Carbon Capture and Hydrogen Storage: AI-Driven Optimization for Efficiency and Scalability. *Sustainability (Basel)*, *17*(13), 5754. DOI: 10.3390/su17135754

Khan, S., Das, P., Abdul Quadir, M., Thaher, M. I., Mahata, C., Sayadi, S., & Al-Jabri, H. (2023). Microalgal feedstock for biofuel production: recent advances, challenges, and future perspective. *Fermentation*, *9*(3), 281.

Khan, A. H., Bahar, A. N., Islam, A., & Wahid, K. (2026). A Review of Battery Storage Technology in Renewable Energy Utilization. *Arabian Journal for Science and Engineering*, *51*(8), 1–23. DOI: 10.1007/s13369-025-11038-4

Khan, A. Z., Bilal, M., Mehmood, S., Sharma, A., & Iqbal, H. M. (2019). State-of-the-art genetic modalities to engineer cyanobacteria for sustainable biosynthesis of biofuel and fine-chemicals to meet bio–economy challenges. *Life (Basel, Switzerland)*, *9*(3), 54. DOI: 10.3390/life9030054 PMID: 31252652

Khan, M. I., Shin, J. H., & Kim, J. D. (2018). The promising future of microalgae: Current status, challenges, and optimization of a sustainable and renewable industry for biofuels, feed, and other products. *Microbial Cell Factories*, *17*(1), 36. DOI: 10.1186/s12934-018-0879-x PMID: 29506528

Khan, S. A., Sharma, G. K., Malla, F. A., Kumar, A., & Gupta, N. (2019). Microalgae based biofertilizers: A biorefinery approach to phycoremediate wastewater and harvest biodiesel and manure. *Journal of Cleaner Production*, *211*, 1412–1419. DOI: 10.1016/j.jclepro.2018.11.281

Khatoon, H., Yusoff, F., Banerjee, S., Shariff, M., & Bujang, J. S. (2007). Formation of periphyton biofilm and subsequent biofouling on different substrates in nutrient enriched brackishwater shrimp ponds. *Aquaculture (Amsterdam, Netherlands)*, *273*(4), 470–477. DOI: 10.1016/j.aquaculture.2007.10.040

Khosravitabar, F., & Spetea, C. (2024). Evaluating technical parameters for microalgae immobilization to optimize green hydrogen photoproduction: A comparative review. *International Journal of Hydrogen Energy*, *67*, 925–932. DOI: 10.1016/j.ijhydene.2024.01.299

Kiai, M. S., Aslfattahi, N., Karakoc, H., Baydogan, N., Samylingam, L., Kadirgama, K., Kok, C. K., & Schmirlerova, H. (2026). A Comprehensive Review, Future Possibilities and the Impact of Sustainable Battery Technologies for Practical Applications. *Korean Journal of Chemical Engineering*, *43*(5), 1–31. DOI: 10.1007/s11814-025-00642-7

Kichouh-Aiadi, S., H. Kawakita, J. J. Gallardo-Rodríguez, A. Sánchez-Mirón, and F. García-Camacho. 2025. "Sequential Recovery of Phycobiliproteins and Polysaccharides from Spirulina Using a Novel Two-Step Process Integrating Adsorption and Cross-Flow Dialysis." *Journal of Applied Phycology 2025 37:5* 37(5):3581–93 .DOI: 10.1007/s10811-025-03606-w

Kichouh-Aiadi, S., Gallardo-Rodríguez, J. J., Cerón-García, M. C., López-Rosales, L., García-Camacho, F., & Sánchez-Mirón, A. (2024). Exploring the Potential of Epigenetic Chemicals to Increase Metabolite Production in the Dinoflagellate Microalga Amphidinium Carterae. *Journal of Applied Phycology*, *36*(3), 1169–1179. Advance online publication. DOI: 10.1007/s10811-024-03189-y

Kichouh-Aiadi, S., López-Rosales, L., Gallardo-Rodríguez, J. J., Cerón-García, M. C., Sánchez-Mirón, A., & García-Camacho, F. (2025). Effects of Hormones on the Growth and Metabolite Production of Amphidinium Carterae under Carbon Sufficient and Carbon Limited Conditions. *Algal Research*, *85*, 103810. DOI: 10.1016/j.algal.2024.103810

Kim, D. S., Moreno-Cabezuelo, J. Á., Schulz, E. N., Lea-Smith, D. J., & Sagaram, U. S. (2024). Recent advances in engineering fast-growing cyanobacterial species for enhanced CO2 fixation. *Frontiers in Climate*, *6*, 1412232. DOI: 10.3389/fclim.2024.1412232

Kim, H., Hillson, N. J., Cho, B. K., Sung, B. H., Lee, D. H., Kim, D. M., Oh, M.-K., Chang, M. W., Jin, Y.-S., Rosser, S. J., Vegh, P., Fragkoudis, R., Le Feuvre, R., Scrutton, N. S., Storch, M., Seong, W., Freemont, P. S., & Lee, S. G. (2025). Abstraction hierarchy to define biofoundry workflows and operations for interoperable synthetic biology research and applications. *Nature Communications*, *16*(1), 6056. DOI: 10.1038/s41467-025-61263-6 PMID: 40593727

Kim, J., & Kang, C. M. (2015). Increased anaerobic production of methane by co-digestion of sludge with microalgal biomass and food waste leachate. *Bioresource Technology*, *189*, 409–412. DOI: 10.1016/j.biortech.2015.04.028 PMID: 25911192

Kim, S. H., Kumar, G., Chen, W. H., & Khanal, S. K. (2021). Renewable hydrogen production from biomass and wastes (ReBioH2-2020). *Bioresource Technology*, *331*, 125024. DOI: 10.1016/j.biortech.2021.125024

Király, É., & Borovics, A. (2024). Carbon sequestration of Hungarian forests by management system and protection status. *Trees, Forests and People*, *15*, 100511. DOI: 10.1016/j.tfp.2024.100511

Kirbayeva, D. K., Balouch, H., Xu, T., Bauenova, M. O., Zaletova, D. E., & Kossalbayev, B. D. (2025). The role of microalgae in circular carbon capture systems. *Functional Plant Biology : FPB*, *52*(9), FP25195. DOI: 10.1071/FP25195 PMID: 40977150

Kirbayeva, D. K., Shayakhmetova, A. Y., Kossalbayev, B. D., Sadvakasova, A. K., & Bauenova, M. O. (2025). Harnessing MicroRNAs and CRISPR to enhance biofuel production in microalgae. *International Journal of Hydrogen Energy*, *157*, 150399. DOI: 10.1016/j.ijhydene.2025.150399

Kirubakaran, D., Wahid, J. B. A., Karmegam, N., Jeevika, R., Sellapillai, L., Rajkumar, M., & SenthilKumar, K. J. (2026). A comprehensive review on the green synthesis of nanoparticles: Advancements in biomedical and environmental applications. *Biomedical Materials & Devices*, *4*(1), 388–413. DOI: 10.1007/s44174-025-00295-4

Kivaisi, A. K. (2001). The potential for constructed wetlands for wastewater treatment and reuse in developing countries: A review. *Ecological Engineering*, *16*(4), 545–560. DOI: 10.1016/S0925-8574(00)00113-0

Knox, S. W. (2026). *Machine learning: a concise introduction*. John Wiley & Sons. DOI: 10.1002/9781394325313

Koller, M., & Braunegg, G. (2018). Advanced Approaches to Produce Polyhydroxyalkanoate (PHA) Biopolyesters in a Sustainable and Economic Fashion. *The EuroBiotech Journal*, *2*(2), 89–103. DOI: 10.2478/ebtj-2018-0013

Kourkoumpas, D. S., Bon, A., Sagani, A., Atsonios, K., Grammelis, P., Karellas, S., & Kakaras, E. (2024). Life cycle assessment of novel thermochemical–biochemical biomass-to-liquid pathways for sustainable aviation and maritime fuel production. *Bioresource Technology*, *393*, 130115. DOI: 10.1016/j.biortech.2023.130115 PMID: 38013031

Krajewska, B. (2018). Urease-aided calcium carbonate mineralization for engineering applications: A review. *Journal of Advanced Research*, *13*, 59–67. DOI: 10.1016/j.jare.2017.10.009 PMID: 30094083

Krishnamoorthy, S., Kuppam, C., & Mamilla, R. C. R. (2024). Evaluation of carbon capture methodologies, mechanisms, and improvements for sustainable carbon dioxide mitigation using microalgae. *Industrial Biotechnology (New Rochelle, N.Y.)*, *20*(5), 186–203. DOI: 10.1089/ind.2024.0015

Krysenko, S. (2023). Impact of nitrogen-containing compounds on secondary metabolism in Streptomyces spp.—A source of metabolic engineering strategies. *SynBio*, *1*(3), 204–225. DOI: 10.3390/synbio1030015

Kulichová, J., & Urbánková, P. (2020). Symmetric and Asymmetric Components of Shape Variation in the Diatom Genus Frustulia (Bacillariophyta). *Symmetry*, *12*(10), 1626. DOI: 10.3390/sym12101626

Kumar, A., Sharma, A., Majumder, P., Vishal, V., & Dutta, A. (2025). Chemical and microalgal conversion of carbon dioxide into fuels and materials: a review. *Environmental Chemistry Letters 2025 23:5*, *23*(5), 1209–1229. DOI: 10.1007/s10311-025-01855-2

Kumar, K., Mishra, S. K., Choi, G. G., & Yang, J. W. (2016). CO2 Sequestration Through Algal Biomass Production. *Algal Biorefinery: An Integrated Approach*, 35–57. https://doi.org/DOI: 10.1007/978-3-319-22813-6_2/SAVE-RESEARCH

Kumar, A., Pandey, S. S., Kumar, D., & Tripathi, B. N. (2023). Genetic manipulation of photosynthesis to enhance crop productivity under changing environmental conditions. *Photosynthesis Research*, *155*(1), 1–21. DOI: 10.1007/s11120-022-00977-w PMID: 36319887

Kumar, G., Shekh, A., Jakhu, S., Sharma, Y., Kapoor, R., & Sharma, T. R. (2020). Bioengineering of microalgae: Recent advances, perspectives, and regulatory challenges for industrial application. *Frontiers in Bioengineering and Biotechnology*, *8*, 914. DOI: 10.3389/fbioe.2020.00914 PMID: 33014997

Kumar, K., Anand, A., & Moholkar, V. S. (2024). Molecular hydrogen (H2) metabolism in microbes: a special focus on biohydrogen production. In *Biohydrogen-advances and processes* (pp. 25–58). Springer Nature Switzerland. DOI: 10.1007/978-3-031-49818-3_2

Kumar, K., Banerjee, D., & Das, D. (2014). Carbon Dioxide Sequestration from Industrial Flue Gas by Chlorella Sorokiniana. *Bioresource Technology*, *152*, 225–233. DOI: 10.1016/j.biortech.2013.10.098 PMID: 24292202

Kumar, K., Dasgupta, C. N., Nayak, B., Lindblad, P., & Das, D. (2011). Development of suitable photobioreactors for CO2 sequestration addressing global warming using green algae and cyanobacteria. *Bioresource Technology*, *102*(8), 4945–4953. DOI: 10.1016/j.biortech.2011.01.054 PMID: 21334885

Kumar, L., & Bharadvaja, N. (2024). A review on microalgae biofuel and biorefinery: Challenges and way forward. *Energy Sources. Part A, Recovery, Utilization, and Environmental Effects*, *46*(1), 16599–16622. DOI: 10.1080/15567036.2020.1836084

Kumar, N., Kar, S., & Shukla, P. (2022). Role of regulatory pathways and multi-omics approaches for carbon capture and mitigation in cyanobacteria. *Bioresource Technology*, *366*, 128104. DOI: 10.1016/j.biortech.2022.128104 PMID: 36257524

Kumar, P. V., Gobikannan, K., Nandakumar, S., & Manavaalan, G. (2025). Enhancing hydrogen production and wastewater treatment through an integrated bio-electrochemical system combining dark fermentation and galvanic cells. *Biomass and Bioenergy*, *201*, 108127. DOI: 10.1016/j.biombioe.2025.108127

Kumar, R. R., Jency, S. M., Bhavatarini, G., & Devapriya, M. J. P. (2022). CO2 sequestration: Microalgae genome analysis and its application of effective green source technology. *Plant Science Today*, *9*(2), 243–261. DOI: 10.14719/pst.1365

Kumar, R., Karmakar, R., Sierra, B. G., & Andjelković, S. (2026). The Future of Respiratory Bioinformatics. In *Next-Generation Bioinformatics for Pulmonary Disease Research* (pp. 373–412). IGI Global Scientific Publishing.

Kumar, R., Singh, S., & Singh, O. V. (2008). Bioconversion of lignocellulosic biomass: Biochemical and molecular perspectives. *Journal of Industrial Microbiology & Biotechnology*, *35*(5), 377–391. DOI: 10.1007/s10295-008-0327-8 PMID: 18338189

Kumar, V., & Verma, P. (2025). Advances in microbial biotechnology for sustainable wastewater reclamation: Recent trends and future prospects. *World Journal of Microbiology & Biotechnology*, *41*(12), 478. DOI: 10.1007/s11274-025-04674-3 PMID: 41307588

Kundu, P., Dutta, N., & Bhattacharya, S. (2024). Application of microalgae in wastewater treatment with special reference to emerging contaminants: A step towards sustainability. *Frontiers in Analytical Science*, *4*, 1513153. Advance online publication. DOI: 10.3389/frans.2024.1513153

Kuppan, P., Sudharsanam, A., Venkateswarlu, K., & Megharaj, M. (2023). Solar technology–closed loop synergy facilitates low-carbon circular bioeconomy in microalgal wastewater treatment. *NPJ Clean Water*, *6*(1), 43. DOI: 10.1038/s41545-023-00256-8

Kupriyanova, E. V., Pronina, N. A., & Los, D. A. J. P. (2023). Adapting from low to high: an update to CO2-concentrating mechanisms of cyanobacteria and microalgae. *12*(7), 1569.

Kupriyanova, E. V., Pronina, N. A., & Los, D. A. (2023). Adapting from low to high: An update to CO2-concentrating mechanisms of cyanobacteria and microalgae. *Plants*, *12*(7), 1569. DOI: 10.3390/plants12071569 PMID: 37050194

Kushwaha, O. S., Uthayakumar, H., & Kumaresan, K. (2023). Modeling of carbon dioxide fixation by microalgae using hybrid artificial intelligence (AI) and fuzzy logic (FL) methods and optimization by genetic algorithm (GA). *Environmental Science and Pollution Research International*, *30*(10), 24927–24948. DOI: 10.1007/s11356-022-19683-0 PMID: 35349067

Kusi, P. A., McGee, D., Tabraiz, S., & Ahmed, A. (2024). Bicarbonate concentration influences carbon utilization rates and biochemical profiles of freshwater and marine microalgae. *Biotechnology Journal*, *19*(8), 2400361. DOI: 10.1002/biot.202400361 PMID: 39212191

Kuyper, J., Schroeder, H., & Linnér, B. O. (2018). The evolution of the UNFCCC. *Annual Review of Environment and Resources*, *43*(Volume 43, 2018), 343–368. https://doi.org/DOI: 10.1146/ANNUREV-ENVIRON-102017-030119/CITE/REFWORKS

Lakkakula, J., Kalra, P., Mungaji, H., Suprasanna, P., & Kadam, U. S. (2025). Microalgae-Derived Metal Nanostructures: Biosynthesis, Characterization, and Applications. *Chemistry & Biodiversity*, *23*(1), e03229. Advance online publication. DOI: 10.1002/cbdv.202503229 PMID: 41264401

Lak, S. Z., Rezaei, J., & Rahimpour, M. R. (2024). Health and pollution challenges of fossil fuels utilization. *Encycl. Renew. Energy Sustain. Environ*, *8*, 155.

Lanigan, T. M., Kopera, H. C., & Saunders, T. L. (2020). Principles of genetic engineering. *Genes*, *11*(3), 291. DOI: 10.3390/genes11030291 PMID: 32164255

Lan, K., Ou, L., Park, S., Kelley, S. S., Nepal, P., Kwon, H., Cai, H., & Yao, Y. (2021). Dynamic life-cycle carbon analysis for fast pyrolysis biofuel produced from pine residues: Implications of carbon temporal effects. *Biotechnology for Biofuels*, *14*(1), 191. DOI: 10.1186/s13068-021-02027-4 PMID: 34587989

Lapierre, F. M., Mattaliano, P., Raith, D., Castillo-Cota, M., Bermeitinger, J., & Huber, R. (2025). Multi-cycle high-throughput growth media optimization using batch Bayesian optimization. *Journal of Chemical Technology and Biotechnology*, *100*(8), 1571–1583. DOI: 10.1002/jctb.7860

Lardon, L., Hélias, A., Sialve, B., Steyer, J. P., & Bernard, O. (2009). Life-Cycle Assessment of Biodiesel Production from Microalgae. *Environmental Science & Technology*, *43*(17), 6475–6481. DOI: 10.1021/es900705j PMID: 19764204

Lastella, G., Testa, C., Cornacchia, G., Notornicola, M., Voltasio, F., & Sharma, V. K. (2002). Anaerobic digestion of semi-solid organic waste: Biogas production and its purification. *Energy Conversion and Management*, *43*(1), 63–75. DOI: 10.1016/S0196-8904(01)00011-5

Launay, H., Avilan, L., Gérard, C., Parsiegla, G., Receveur-Brechot, V., Gontero, B., & Carriere, F. (2023). Location of the Photosynthetic Carbon Metabolism in Microcompartments and Separated Phases in Microalgal Cells. *FEBS Letters*, *597*(23), 2853–2878. DOI: 10.1002/1873-3468.14754 PMID: 37827572

Launay, H., Huang, W., Maberly, S. C., & Gontero, B. (2020). Regulation of carbon metabolism by environmental conditions: A perspective from diatoms and other chromalveolates. *Frontiers in Plant Science*, *11*, 1033. DOI: 10.3389/fpls.2020.01033 PMID: 32765548

Laxa, R. A. (2021). Construction of Dichotomous Taxonomic Keys for San Francisco Bay Planktonic Diatoms.

La, Y. H., Lee, K. S., Kim, T. W., & Song, J. Y. (2022). A bacteria-based carbon sequestration and waste recycling system. *Scientific Reports*, *12*(1), 10236. DOI: 10.1038/s41598-022-14239-1 PMID: 35764782

Lebeau, T., & Robert, J. M. (2003). Diatom cultivation and biotechnologically relevant products. Part I: Cultivation at various scales. *Applied Microbiology and Biotechnology*, *60*(6), 612–623. DOI: 10.1007/s00253-002-1176-4 PMID: 12664139

Leech, D., Previtera, D. A., Zhang, Y., Botella, J. R., & Crisp, P. A. (2025). Precision plant epigenome editing: What, how, and why. *Trends in Plant Science*. PMID: 40915965

Lee, J., Lim, K., Kim, A., Mok, Y. G., Chung, E., Cho, S. I., Lee, J. M., & Kim, J.-S. (2023). Prime editing with genuine Cas9 nickases minimizes unwanted indels. *Nature Communications*, *14*(1), 1786. DOI: 10.1038/s41467-023-37507-8 PMID: 36997524

Lee, M., Lin, Y. L., Chiueh, P. T., & Den, W. (2020). Environmental and energy assessment of biomass residues to biochar as fuel: A brief review with recommendations for future bioenergy systems. *Journal of Cleaner Production*, *251*, 119714. DOI: 10.1016/j.jclepro.2019.119714

Lee, Y. S., & Park, W. (2018). Current challenges and future directions for bacterial self-healing concrete. *Applied Microbiology and Biotechnology*, *102*(7), 3059–3070. DOI: 10.1007/s00253-018-8830-y PMID: 29487987

Leong, Y. K., & Chang, J. S. (2023). Waste stream valorization-based low-carbon bioeconomy utilizing algae as a biorefinery platform. *Renewable & Sustainable Energy Reviews*, *178*, 113245. DOI: 10.1016/j.rser.2023.113245

Li, D., Dong, H., Cao, X., Wang, W., & Li, C. (2023). Enhancing photosynthetic CO2 fixation by assembling metal-organic frameworks on Chlorella pyrenoidosa *Nature Communications 2023 14:1, 14*(1), 5337-. DOI: 10.1038/s41467-023-40839-0

Li, G., & Yao, J. (2024). A Review of Algae-Based Carbon Capture, Utilization, and Storage (Algae-Based CCUS). *Gases 2024, Vol. 4, Pages 468-503, 4*(4), 468–503. DOI: 10.3390/gases4040024

Li, H., Kong, L., Ai, W., & Zhang, Y. (2025). Algal systems reimagined: exploring the potential for sustainable solutions to diverse wastewater challenges. *Blue Biotechnology 2025 2:1, 2*(1), 21-. DOI: 10.1186/s44315-025-00033-x

Li, Y., Han, D., Yoon, K., Zhu, S., Sommerfeld, M., & Hu, Q. 2013. "Molecular and Cellular Mechanisms for Lipid Synthesis and Accumulation in Microalgae: Biotechnological Implications." *Handbook of Microalgal Culture: Applied Phycology and Biotechnology: Second Edition* 545–65. doi:DOI: 10.1002/9781118567166. CH28;JOURNAL:JOURNAL:BOOKS;WGROUP:STRING:PUBLICATION

Liang, R. X., Hung, J. C., Diankristanti, P. A., Chen, Y. T., Chung, C. W., & Ng, I. S. (2025). Enhanced carbon capture and utilization in transgenic Chlorella sorokiniana harboring pyridoxal kinase under dynamic carbon dioxide levels. *Bioresource Technology, 425*, 132315. DOI: 10.1016/j.biortech.2025.132315 PMID: 40023335

Liang, Y., Zheng, Z., Hua, R., & Luo, X. (2011). A preliminary study of simultaneous lime treatment and dry digestion of smooth cord grass for biogas production. *Chemical Engineering Journal, 174*(1), 175–181. DOI: 10.1016/j.cej.2011.08.071

Li, C., Pradhan, P., Chen, G., Kropp, J. P., & Schellnhuber, H. J. (2025). Carbon footprint of the construction sector is projected to double by 2050 globally. *Communications Earth & Environment, 6*(1), 831. DOI: 10.1038/s43247-025-02840-x PMID: 41163641

Li, D. W., Shi, Y. C., Yu, H. L., Balamurugan, S., Xin, Y. Z., Alimujiang, A., Wei, D. Y., Jie, S. L., & Hong, Y. L. (2016). A Type 2 Diacylglycerol Acyltransferase Accelerates the Triacylglycerol Biosynthesis in Heterokont Oleaginous Microalga Nannochloropsis Oceanica. *Journal of Biotechnology, 229*, 65–71. DOI: 10.1016/j.jbiotec.2016.05.005 PMID: 27164260

Li, D., & Zhao, Q. (2023). Study of carbon fixation and carbon partitioning of evolved Chlorella sp.'s strain under different carbon dioxide conditions. *Biocatalysis and Agricultural Biotechnology, 48*, 102655. DOI: 10.1016/j.bcab.2023.102655

Li, G., Xiao, W., Yang, T., & Lyu, T. (2023). Optimization and process effect for microalgae carbon dioxide fixation technology applications based on carbon capture: A comprehensive review. *C*, *9*(1), 35.

Li, H., Chen, L., Zhang, F., & Cai, Z. (2025). Graph-learning-based machine learning improves prediction and cultivation of commercial-grade marine microalgae Porphyridium. *Bioresource Technology*, *416*, 131728. DOI: 10.1016/j.biortech.2024.131728 PMID: 39521188

Li, J. (2026). Policy Gradient Algorithms. In *Artificial Intelligence for Drug Design* (pp. 169–177). Springer Nature Singapore. DOI: 10.1007/978-981-95-2525-6_16

Li, J., Gao, M., Liu, B., Fan, Y., Wei, J., Zhang, Y., Li, F., Zhang, N., & Hu, Z. (2025). Seasonal Dynamics of Phytoplankton Community Structure and Environmental Drivers in the Coastal Waters of the Leizhou Peninsula, China. *Diversity*, *17*(12), 867. DOI: 10.3390/d17120867

Li, J., Han, D., Wang, D., Ning, K., Jia, J., Wei, L., Jing, X., Huang, S., Chen, J., Li, Y., Hu, Q., & Xu, J. (2014). Choreography of Transcriptomes and Lipidomes of Nannochloropsis Reveals the Mechanisms of Oil Synthesis in Microalgae. *The Plant Cell*, *26*(4), 1645–1665. DOI: 10.1105/tpc.113.121418 PMID: 24692423

Li, L., Gao, K., Yang, M., Zheng, Q., Zhang, M., & Deng, X. (2023). Challenges and potential solutions of microalgae-based systems for wastewater treatment and resource recovery. *Frontiers in Bioengineering and Biotechnology*, *11*, 1210228. DOI: 10.3389/fbioe.2023.1210228 PMID: 37342510

Lima, P. J. M., da Silva, R. M., Neto, C. A. C. G., Gomes e Silva, N. C., Souza, J. E. D. S., Nunes, Y. L., & Sousa dos Santos, J. C. (2022). An overview on the conversion of glycerol to value-added industrial products via chemical and biochemical routes. *Biotechnology and Applied Biochemistry*, *69*(6), 2794–2818. DOI: 10.1002/bab.2098

Lim, L. W. K. (2025). CRISPR-Cas applications in extremophiles: Thermophiles, psychrophiles, halophiles, acidophiles, alkaliphiles, piezophiles, xerophiles, radiophiles and metallophiles. *Total Environment Microbiology*, *1*(3), 100029. DOI: 10.1016/j.temicr.2025.100029

Lin, Z., Kuang, Y., Li, W., & Zheng, Y. (2024). Research status and prospects of CO2 geological sequestration technology from onshore to offshore: A review. *Earth-Science Reviews*, *258*, 104928. DOI: 10.1016/j.earscirev.2024.104928

Li, Q., Gluch, J., Liao, Z., Posseckardt, J., Clausner, A., Łępicka, M., Grądzka-Dahlke, M., & Zschech, E. (2021). Morphology and mechanical properties of fossil diatom frustules from genera of Ellerbeckia and Melosira. *Nanomaterials (Basel, Switzerland), 11*(6), 1615. DOI: 10.3390/nano11061615 PMID: 34202999

Li, R., Pan, J., Xing, J., & Yang, W. (2025). *The Secret of High Carbon Fixation Efficiency in Chlamydomonas and Its Potential to Boost Yields in Land Plants* Green Carbon., DOI: 10.1016/j.greenca.2025.02.005

Li, S., Tabatabaei, M., Li, F., & Ho, S. H. (2024). A review of green biohydrogen production using anoxygenic photosynthetic bacteria for hydrogen economy: Challenges and opportunities. *International Journal of Hydrogen Energy, 54*, 218–238. DOI: 10.1016/j.ijhydene.2022.11.014

Li, S., Zhao, S., Yan, S., Qiu, Y., Song, C., Li, Y., & Kitamura, Y. (2019). Food Processing Wastewater Purification by Microalgae Cultivation Associated with High Value-Added Compounds Production — A Review. *Chinese Journal of Chemical Engineering, 27*(12), 2845–2856. DOI: 10.1016/j.cjche.2019.03.028

Li, S., & Zhu, L. (2023). Copper regulates degradation of typical antibiotics by microalgal-fungal consortium in simulated swine wastewater: Insights into metabolic routes and dissolved organic matters. *Water Research, 245*, 120654. DOI: 10.1016/j.watres.2023.120654 PMID: 37778083

Liu, J., Wu, X., Yao, M., Xiao, W., & Zha, J. (2020). Chassis engineering for microbial production of chemicals: From natural microbes to synthetic organisms. *Current Opinion in Biotechnology, 66*, 105–112. DOI: 10.1016/j.copbio.2020.06.013 PMID: 32738762

Liu, X., Wang, Y., Liu, H., Zhang, Y., Zhou, Q., Wen, X., Guo, W., & Zhang, Z. (2024). A Systematic Review on Aquaculture Wastewater: Pollutants, Impacts, and Treatment Technology. *Environmental Research, 262*(Pt 1), 119793. Advance online publication. DOI: 10.1016/j.envres.2024.119793 PMID: 39147181

Liu, Y., Cruz-Morales, P., Zargar, A., Belcher, M. S., Pang, B., Englund, E., Dan, Q., Yin, K., & Keasling, J. D. (2021). Biofuels for a sustainable future. *Cell, 184*(6), 1636–1647. DOI: 10.1016/j.cell.2021.01.052 PMID: 33639085

Liu, Y., Wang, Y., Wen, X., Shimizu, K., Lei, Z., Kobayashi, M., Zhang, Z., Sumi, I., Yao, Y., & Mogi, Y. (2018). Enhanced bioconversion of hydrogen and carbon dioxide to methane using a micro-nano sparger system: Mass balance and energy consumption. *RSC Advances, 8*(47), 26488–26496. DOI: 10.1039/C8RA02924E

Liu, Z., Liu, C., Han, S., & Yang, X. (2020). Optimization upstream CO2 deliverable with downstream algae deliverable in quantity and quality and its impact on energy consumption. *The Science of the Total Environment*, *709*, 136197. DOI: 10.1016/j.scitotenv.2019.136197 PMID: 31887503

Li, X., He, X., Ren, K., Dong, H., & Lian, B. (2023). Mechanisms of carbonate precipitation induced by two model bacteria. *Chemical Geology*, *628*, 121461. DOI: 10.1016/j.chemgeo.2023.121461

Li, X., Sun, X., Gao, L., Xu, J., & Gao, G. (2021). Effects of periodical dehydration on biomass yield and biochemical composition of the edible red alga Pyropia yezoensis grown at different salinities. *Algal Research*, *56*, 102315. DOI: 10.1016/j.algal.2021.102315

Li, Y., Deng, L., Walker, E. J. L., Karas, B. J., & Mock, T. (2025). Genetic engineering in diatoms: Advances and prospects. *The Plant Journal : For Cell and Molecular Biology*, *121*(6), e70102. DOI: 10.1111/tpj.70102 PMID: 40089910

Li, Y., Horsman, M., Wu, N., Lan, C. Q., & Dubois-Calero, N. (2008). Biofuels from microalgae. *Biotechnology Progress*, *24*(4), 815–820. DOI: 10.1021/bp070371k PMID: 18335954

Li, Y., Park, S. Y., & Zhu, J. (2011). Solid-state anaerobic digestion for methane production from organic waste. *Renewable & Sustainable Energy Reviews*, *15*(1), 821–826. DOI: 10.1016/j.rser.2010.07.042

Li, Y., Yi, F. C., Chen, P., Min, M., Zhou, W., Martinez, B., Zhu, J., & Ruan, R. (2011). Characterization of a Microalga Chlorella Sp. Well Adapted to Highly Concentrated Municipal Wastewater for Nutrient Removal and Biodiesel Production. *Bioresource Technology*, *102*(8), 5138–5144. DOI: 10.1016/j.biortech.2011.01.091 PMID: 21353532

Li, Z., Liu, A., Sun, C., Li, H., Kong, Z., & Zhai, H. (2024). Biomineralization process of CaCO3 precipitation induced by Bacillus mucilaginous and its potential application in microbial self-healing concrete. *Applied Biochemistry and Biotechnology*, *196*(4), 1896–1920. DOI: 10.1007/s12010-023-04634-3 PMID: 37440115

Lizárraga, G. L. L. (2016). Considerations for photobioreactor design and operation for mass cultivation of microalgae. *Algae: Organisms for Imminent Biotechnology*, 55.

Llamas, B., Suarez-Rodriguez, M. C., Gonzalez-Lopez, C. V., Mora, P., & Acien, F. G. (2021). Techno-economic analysis of microalgae related processes for CO2 bio-fixation. *Algal Research*, *57*, 102339. DOI: 10.1016/j.algal.2021.102339

Long, B. M., Badger, M. R., Whitney, S. M., & Price, G. D. (2007). Analysis of Rubisco biogenesis in the chloroplast reveals alternative pathways for protein targeting. *Proceedings of the National Academy of Sciences of the United States of America, 104*(6), 2178–2183.

Long, B., Fischer, B., Zeng, Y., Amerigian, Z., Li, Q., Bryant, H., Li, M., Dai, S. Y., & Yuan, J. S. (2022). Machine learning-informed and synthetic biology-enabled semi-continuous algal cultivation to unleash renewable fuel productivity. *Nature Communications, 13*(1), 541. DOI: 10.1038/s41467-021-27665-y PMID: 35087023

López i Losada, R., Rosenbaum, R. K., Brady, M. V., Wilhelmsson, F., & Hedlund, K.i Losada. (2024). Agent-Based Life Cycle Assessment enables joint economic-environmental analysis of policy to support agricultural biomass for biofuels. *The Science of the Total Environment, 916*, 170264. DOI: 10.1016/j.scitotenv.2024.170264 PMID: 38253104

López-Pacheco, I. Y., Castillo-Vacas, E. I., Castañeda-Hernández, L., Gradiz-Menjivar, A., Rodas-Zuluaga, L. I., Castillo-Zacarías, C., Sosa-Hernández, J. E., Barceló, D., Iqbal, H. M. N., & Parra-Saldívar, R. (2021). CO2 biocapture by Scenedesmus sp. grown in industrial wastewater. *The Science of the Total Environment, 790*, 148222. DOI: 10.1016/j.scitotenv.2021.148222 PMID: 34380253

Loria, M. H., Griffin, J. S., Wells, G. F., & Rhoads, K. R. (2023). Effects of Feast-Famine Nutrient Regimes on Wastewater Algal Biofuel Communities. *PLoS One, 18*(1), e0279943. Advance online publication. DOI: 10.1371/journal.pone.0279943 PMID: 36598899

Low, S., Fritz, L., Baum, C. M., & Sovacool, B. K. (2024). Public perceptions on carbon removal from focus groups in 22 countries. *Nature Communications 2024 15:1, 15*(1), 3453-. DOI: 10.1038/s41467-024-47853-w

Luís, A. T., Hlúbiková, D., Vaché, V., Choquet, P., Hoffmann, L., & Ector, L. (2017). Atomic force microscopy (AFM) application to diatom study: Review and perspectives. *Journal of Applied Phycology, 29*(6), 2989–3001. DOI: 10.1007/s10811-017-1177-4

Luo, M., Qian, C. X., & Li, R. Y. (2015). Factors affecting crack repairing capacity of bacteria-based self-healing concrete. *Construction & Building Materials, 87*, 1–7. DOI: 10.1016/j.conbuildmat.2015.03.117

Lu, Q. (2025). A State-of-the-Art Review of Microalgae-Based Food Processing Wastewater Treatment: Progress, Problems, and Prospects. *Water (Basel), 17*(4), 536. DOI: 10.3390/w17040536

Lv, H., Cui, X., Wahid, F., Xia, F., Zhong, C., & Jia, S. (2016). Analysis of the Physiological and Molecular Responses of Dunaliella Salina to Macronutrient Deprivation. *PLoS One*, *11*(3), e0152226. DOI: 10.1371/journal.pone.0152226 PMID: 27023397

Lv, J., Guo, B., Feng, J., Liu, Q., Nan, F., Liu, X., & Xie, S. (2019). Integration of Wastewater Treatment and Flocculation for Harvesting Biomass for Lipid Production by a Newly Isolated Self-Flocculating Microalga Scenedesmus Rubescens SX. *Journal of Cleaner Production*, *240*, 118211. DOI: 10.1016/j.jclepro.2019.118211

Madadi Avargani, V., Habibzadeh, M., Abdlla Maarof, H., Zendehboudi, S., & Duan, X. (2025). Harnessing renewable energy for hydrogen production: Advances, challenges, and opportunities. *Industrial & Engineering Chemistry Research*, *64*(25), 12368–12418. DOI: 10.1021/acs.iecr.5c00061

Madhumithra, E., Elakkiya, K., & Sumathi, A. (2025). Enhancing Self-Healing Concrete Using Sodium Alginate Beads and Citrullus lanatus Seed Powder for Bacterial Encapsulation. *Arabian Journal for Science and Engineering*, •••, 1–21.

Maeda, T., Sanchez-Torres, V., & Wood, T. K. (2007). Metabolic engineering to enhance bacterial hydrogen production. *Applied Microbiology and Biotechnology*, *73*(1), 130–136.

Mahalakshmi, K., P, D. A., J, S. P., Moger, S. S., & Raghavan, R. (2026). Spatial Multi-Omics in Cancers: Integrating Molecular Landscapes for Precision Oncology. *Spatially Resolved Single-Cell and Multiple Omics in Human Diseases*, 217–248. DOI: 10.4018/979-8-3373-4849-0.ch007

Mahlangu, D., Mphahlele, K., Paola, F. D., & Mthombeni, N. H. (2024). Microalgae-Mediated Biosorption for Effective Heavy Metals Removal from Wastewater: A Review. *Water (Basel)*, *16*(5), 718. DOI: 10.3390/w16050718

Ma, J., Cui, Y., Sun, F., Zhou, R., Liu, F., Cao, K., Zhu, B., Zhang, H., Meng, C., Chen, G., & Gao, Z. (2025). Strategies for enhancing microalgal carbon sequestration: A review on strain development, culture system optimization, parameter control, and metabolic engineering. *Bioresource Technology*, *435*, 132868. DOI: 10.1016/j.biortech.2025.132868 PMID: 40555329

Maliha, A., & Abu-Hijleh, B. (2023). A review on the current status and post-pandemic prospects of third-generation biofuels. *Energy systems, 14*(4), 1185-1216.

Malik, H. A.. (2026). *"Development of algal industry 4.0 for efficient cultivation, resource recovery, and bioproducts production." Algae Biotechnology*. Elsevier.

Malik, S., Shahid, A., Betenbaugh, M. J., Liu, C.-G., & Mehmood, M. A. (2022). A novel wastewater-derived cascading algal biorefinery route for complete valorization of the biomass to biodiesel and value-added bioproducts. *Energy Conversion and Management*, *256*, 115360. DOI: 10.1016/j.enconman.2022.115360

Malode, S. J., Pandiaraj, S., Alodhayb, A., & Shetti, N. P. (2025). Review on Sustainable Biohydrogen Production from Biomass: Recent Developments and Prospects toward a Promising Source. *Energy & Fuels*, *39*(17), 7967–7989. DOI: 10.1021/acs.energyfuels.5c00315

Mamilla, S., Sahu, N., & Sridhar, S. (2024). Algal Biomass as Global Demand for Alternative Biofuel. *Algal Biotechnology: Applications for Industrial Development and Human Welfare*, 157–172. https://doi.org/DOI: 10.1201/9781003219194-12/ALGAL-BIOMASS-GLOBAL-DEMAND-ALTERNATIVE-BIOFUEL-SHYAMSUNDER-MAMILLA-NIVEDITA-SAHU-SRIDHAR

Ma, N., Zhao, W., Wang, W., Li, X., & Zhou, H. (2024). Large scale of green hydrogen storage: Opportunities and challenges. *International Journal of Hydrogen Energy*, *50*, 379–396. DOI 10.1016/j.ijhydene.2023.09.021

Manaia, C. M., Rocha, J., Scaccia, N., Marano, R., Radu, E., Biancullo, F., Cerqueira, F., Fortunato, G., Iakovides, I. C., Zammit, I., Kampouris, I., Vaz-Moreira, I., & Nunes, O. C. (2018). Antibiotic Resistance in Wastewater Treatment Plants: Tackling the Black Box. *Environment International*, *115*, 312–324. DOI: 10.1016/j.envint.2018.03.044 PMID: 29626693

Manoharan, L., Suparmaniam, U., Lam, M. K., Chai, Y. H., Loy, A. C. M., Tan, I. S., Foo, H. C. Y., & Lim, S. (2025). Advances in attached growth microalgae cultivation for third-generation biofuel production: Opportunities, challenges, and future prospect. *Brazilian Journal of Chemical Engineering*, *43*(2), 1–36. DOI: 10.1007/s43153-025-00567-y

Mansoori, G. A., Agyarko, L. B., Estevez, L. A., Fallahi, B., Gladyshev, G., Santos, R. G. D., . . . Yen, J. (2021). Fuels of the future for renewable energy sources (ammonia, biofuels, hydrogen). *arXiv preprint arXiv:2102.00439.*

Mantovani, M., Collina, E., Lasagni, M., Marazzi, F., & Mezzanotte, V. (2022). Production of microalgal-based carbon encapsulated iron nanoparticles (ME-nFe) to remove heavy metals in wastewater. *Environmental Science and Pollution Research International*, *30*(3), 6730–6745. DOI: 10.1007/s11356-022-22506-x PMID: 36008581

Maran, F. N., Türk, İ., Akkoyun, B., Meligy, M. T. F., Lubiçeva, E., İkballı, D., & Şen, F. (2023). Nanotechnology and its versatile applications in medicine, environment, energy, textiles, food industry. *International Journal of Boron Science and Nanotechnology*, (1), 18–49.

Marquardt, J., Fünfgeld, A., & Elsässer, J. P. (2023). Institutionalizing climate change mitigation in the Global South: Current trends and future research. *Earth System Governance*, *15*, 100163. DOI: 10.1016/j.esg.2022.100163

Mata, T. M., Martins, A. A., & Caetano, N. S. (2010). Microalgae for biodiesel production and other applications: A review. *Renewable & Sustainable Energy Reviews*, *14*(1), 217–232. DOI: 10.1016/j.rser.2009.07.020

Mathew, M. M., Khatana, K., Vats, V., Dhanker, R., Kumar, R., Dahms, H. U., & Hwang, J. S. (2022). Biological approaches integrating algae and bacteria for the degradation of wastewater contaminants—A review. *Frontiers in Microbiology*, *12*, 801051. DOI: 10.3389/fmicb.2021.801051 PMID: 35185825

Mathimani, T., Le, T. T. H., Salmen, S. H., Chinnathambi, A., & Jhanani, G. K. (2024). A first-of-its-kind study on founding concatenated trio fuels production platform.... *Fuel*, *361*, 130639. DOI: 10.1016/j.fuel.2023.130639

Mathur, M., Choudhary, P., Sujathan, S., Naaz, F., Trenado-Yuste, C., & Malik, A. (2024). A biologically driven model for rural wastewater management: Feasibility and efficiency of algal-bacterial biofilm reactors for combined treatment and algae farming. *Frontiers in Water*, *6*, 1430900. DOI: 10.3389/frwa.2024.1430900

Matinha-Cardoso, J., Santos, T., Pereira, H., Varela, J., Tamagnini, P., & Mota, R. (2023). Pilot scale production of Crocosphaera chwakensis CCY0110 and evaluation of its biomass nutritional potential. *Algal Research*, *69*, 102939. DOI: 10.1016/j.algal.2022.102939

Matthijs, M., Fabris, M., Obata, T., Foubert, I., Franco-Zorrilla, J. M., Solano, R., Fernie, A. R., Vyverman, W., & Goossens, A. (2017). The transcription factor bZIP14 regulates the TCA cycle in the diatom Phaeodactylum tricornutum. *The EMBO Journal*, *36*(11), 1559–1576. DOI: 10.15252/embj.201696392 PMID: 28420744

Matzko, R., & Konur, S. (2024). Technologies for design-build-test-learn automation and computational modelling across the synthetic biology workflow: A review. *Network Modeling and Analysis in Health Informatics and Bioinformatics*, *13*(1), 22. DOI: 10.1007/s13721-024-00455-4

Maurya, A. K., Rathi, S., Saksena, Y. P., Awasthi, A., Kaushik, B., & Sheetal. (2025). Uptake, Translocation and Consequences of Application of Si/SiO 2 Nanoparticles (SiNPs) in Plants. *Silicon Nanoparticles: Synthesis, Uptake, and Applications* 213-261.

Ma, Z., Cheah, W. Y., Ng, I. S., Chang, J. S., Zhao, M., & Show, P. L. (2022). Microalgae-based biotechnological sequestration of carbon dioxide for net zero emissions. *Trends in Biotechnology*, *40*(12), 1439–1453. DOI: 10.1016/j.tibtech.2022.09.002 PMID: 36216714

McLaughlin, H., Littlefield, A. A., Menefee, M., Kinzer, A., Hull, T., Sovacool, B. K., Bazilian, M. D., Kim, J., & Griffiths, S. (2023). Carbon Capture Utilization and Storage in Review: Sociotechnical Implications for a Carbon Reliant World. *Renewable & Sustainable Energy Reviews*, *177*, 113215. DOI: 10.1016/j.rser.2023.113215

Medlin, L. K. J. P. (2016). Evolution of the diatoms: major steps in their evolution and a review of the supporting molecular and morphological evidence. *55*(1), 79-103.

Medlin, L. K. (2009). Diatoms (Bacillariophyta). *Timetree Life*, *1*, 127–130. DOI: 10.1093/oso/9780199535033.003.0010

Mei, R., Yang, H., Guo, C., Hong, Z., Hu, Z., Wu, Y., Huang, D., & Wang, C. (2025). Effects of High Light Exposure and Heterologous Expression of β-Carotene Ketolase on the Metabolism of Carotenoids in Chlamydomonas Reinhardtii. *Frontiers in Bioengineering and Biotechnology*, *13*, 1533661. Advance online publication. DOI: 10.3389/fbioe.2025.1533661 PMID: 40144396

Meiser, A., Schmid-Staiger, U., & Trösch, W. (2004). Optimization of eicosapentaenoic acid production byPhaeodactylum tricornutumin the flat panel airlift (FPA) reactor. *Journal of Applied Phycology*, *16*(3), 215–225. DOI: 10.1023/B:JAPH.0000048507.95878.b5

Mekuye, B., & Abera, B. (2023). Nanomaterials: An overview of synthesis, classification, characterization, and applications. *Nano Select*, *4*(8), 486–501. DOI: 10.1002/nano.202300038

Melis, A., Zhang, L., Forestier, M., Ghirardi, M. L., & Seibert, M. (2000). Sustained photobiological hydrogen gas production upon reversible inactivation of oxygen evolution in the green alga Chlamydomonas reinhardtii. *Plant Physiology*, *122*(1), 127–135. DOI: 10.1104/pp.122.1.127

Meloni, M., Gurrieri, L., Fermani, S., Velie, L., Sparla, F., Crozet, P., Henri, J., & Zaffagnini, M. (2023). Ribulose-1, 5-bisphosphate regeneration in the Calvin-Benson-Bassham cycle: Focus on the last three enzymatic steps that allow the formation of Rubisco substrate. *Frontiers in Plant Science*, *14*, 1130430. DOI: 10.3389/fpls.2023.1130430 PMID: 36875598

Mérida, L. G. R., & Padrón, R. A. R. (2023). Application of microalgae in wastewater: Opportunity for sustainable development. *Frontiers in Environmental Science*, *11*, 1238640. Advance online publication. DOI: 10.3389/fenvs.2023.1238640

Merk, C., Grunau, J., Riekhof, M. C., & Rickels, W. (2022). The need for local governance of global commons: The example of blue carbon ecosystems. *Ecological Economics*, *201*, 107581. DOI: 10.1016/j.ecolecon.2022.107581

Meyer, M. T., Griffiths, H., & Olofsson, L. (2017). The pyrenoid: A microcompartment for carbon-concentrating mechanisms in algae. *Plant Physiology*, *173*(1), 1–10. PMID: 28049854

Michael, M., Vanness, V., Henry, H., Tjandra, T. M., Kristian, K., Nasution, J. A., Masytha, M., Sidabutar, R., Trisakti, B., Irvan, I., Dalimunthe, N. F., Al Fath, M. T., Nabilah, Y., Kobayashi, T., Daimon, H., & Takriff, M. S. (2026). Microalgae for sustainable biodiesel and omega-3: A comprehensive review of production, processing, and implementation. *Renewable & Sustainable Energy Reviews*, *226*, 116327. DOI: 10.1016/j.rser.2025.116327

Milledge, J. J., Nielsen, B. V., Maneein, S., & Harvey, P. J. (2019). A brief review of anaerobic digestion of algae for bioenergy. *Energies*, *12*(6), 1166. DOI: 10.3390/en12061166

Miller, R., Wu, G., Deshpande, R. R., Vieler, A., Gartner, K., Li, X., Moellering, E. R., Zauner, S., Cornish, A. J., Liu, B., Bullard, B., Sears, B. B., Min, H. K., Hegg, E. L., Shachar-Hill, Y., Shiu, S. H., & Benning, C. (2010). Changes in Transcript Abundance in Chlamydomonas Reinhardtii Following Nitrogen Deprivation Predict Diversion of Metabolism. *Plant Physiology*, *154*(4), 1737–1752. DOI: 10.1104/pp.110.165159 PMID: 20935180

Min, K. H., Kim, D. H., Youn, S., & Pack, S. P. (2024). Biomimetic diatom biosilica and its potential for biomedical applications and prospects: A review. *International Journal of Molecular Sciences*, *25*(4), 2023. DOI: 10.3390/ijms25042023 PMID: 38396701

Mipeshwaree Devi, A., Khedashwori Devi, K., Premi Devi, P., Lakshmipriyari Devi, M., & Das, S. (2023). Metabolic engineering of plant secondary metabolites Prospects and its technological challenges. *Frontiers in Plant Science*, *14*, 1171154 DOI: 10.3389/fpls.2023.1171154 PMID: 37251773

Miranda, A. M., Hernandez-Tenorio, F., Ocampo, D., Vargas, G. J., & Sáez, A. A. (2022). Trends on CO2 capture with microalgae: A bibliometric analysis. *Molecules (Basel, Switzerland)*, *27*(15), 4669. DOI: 10.3390/molecules27154669 PMID: 35897845

Mitchell, H. J., & Bartsch, D. (2020). Regulation of GM organisms for invasive species control. *Frontiers in Bioengineering and Biotechnology*, *7*, 454. DOI: 10.3389/fbioe.2019.00454 PMID: 32039172

Mitra, S., Paliya, S., & Mandpe, A. (2024). Microbial engineering in biofuel production—A global outlook, advances, and roadmap. In *Emerging sustainable technologies for biofuel production* (pp. 547–593). Springer Nature Switzerland. DOI: 10.1007/978-3-031-52167-6_22

Miyake, M., Miyake, J., & Asada, Y. (2013). Photosynthetic hydrogen production and its potential for practical application. *Photosynthesis Research*, *116*(2–3), 409–420.

Moghri, P., Sabahi, H., Rezayan, A. H., & Akbari, B. (2025). A green approach: Utilizing untreated pine pollen grains as a natural biosorbent for microplastic removal from water systems. *Journal of Contaminant Hydrology*, *104675*, 104675. Advance online publication. DOI: 10.1016/j.jconhyd.2025.104675 PMID: 40680663

Mohapatra, A., Guliani, E., Nandal, V., Solanki, M., Pathak, V. V., Banga, S., & Sharma, D. K. (2025). Recent innovations in sustainable algal biofuel production: A comprehensive review. *Brazilian Journal of Chemical Engineering*, *43*(2), 549–577. Advance online publication. DOI: 10.1007/s43153-025-00598-5

Mohsenpour, S. F., Hennige, S., Willoughby, N., Adeloye, A. J., & Gutiérrez, T. (2020). Integrating micro-algae into wastewater treatment: A review. *The Science of the Total Environment*, *752*, 142168. DOI: 10.1016/j.scitotenv.2020.142168 PMID: 33207512

Mokhtar, N., Johari, M. A. M., Tajarudin, H. A., Al-Gheethi, A. A., & Algaifi, H. A. (2021). A sustainable enhancement of bio-cement using immobilised Bacillus sphaericus: Optimization, microstructural properties, and techno-economic analysis for a cleaner production of bio-cementitious mortars. *Journal of Cleaner Production*, *318*, 128470. DOI: 10.1016/j.jclepro.2021.128470

Molina Grima, E., Belarbi, E. H., Acién Fernández, F. G., Robles Medina, A., & Chisti, Y. (2003). Recovery of microalgal biomass and metabolites: Process options and economics. *Biotechnology Advances*, *20*(7–8), 491–515. DOI: 10.1016/S0734-9750(02)00050-2 PMID: 14550018

Molino, Antonio, Sanjeet Mehariya, Giuseppe Di Sanzo, Vincenzo Larocca, Maria Martino, Gian Paolo Leone, Tiziana Marino, Simeone Chianese, Roberto Balducchi, and Dino Musmarra. 2020. "Recent Developments in Supercritical Fluid Extraction of Bioactive Compounds from Microalgae: Role of Key Parameters, Technological Achievements and Challenges." *Journal of CO2 Utilization* 36:196–209. .DOI: 10.1016/j.jcou.2019.11.014

Moñino Fernández, P., López Morales, M., de Winter, A., van den End, F., Janssen, M., & Barbosa, M. (2025). Pilot-scale oxygen-balanced mixotrophic cultivation of Galdieria sulphuraria. *ACS Sustainable Chemistry & Engineering*, *13*(5), 2132–2140. DOI: 10.1021/acssuschemeng.4c09186 PMID: 39950107

Monlau, F., Suarez-Alvarez, S., Lallement, A., Vaca-Medina, G., Giacinti, G., Munarriz, M., Urreta, I., Raynaud, C., Ferrer, C., & Castañón, S. (2021). A Cascade Biorefinery for the Valorization of Microalgal Biomass: Biodiesel, Biogas, Fertilizers and High Valuable Compounds. *Algal Research*, *59*, 102433. DOI: 10.1016/j.algal.2021.102433

Montaño López, J., Duran, L., & Avalos, J. L. (2022). Physiological limitations and opportunities in microbial metabolic engineering. *Nature Reviews. Microbiology*, *20*(1), 35–48. DOI: 10.1038/s41579-021-00600-0 PMID: 34341566

Montgomery, L. F. R., & Bochmann, G. (2014). *Pretreatment of feedstock for enhanced biogas production*. IEA Bioenergy.

Moosburner, M., Allen, A. E., & Daboussi, F. (2022) Genetic engineering in marine diatoms: current practices and emerging technologies. In: Falcia- tore, A. & Mock, T. (Eds.) The molecular life of diatoms. London: Springer Press, pp. 743–773. DOI: 10.1007/978-3-030-92499-7_25

Moraes, L. C., Figueiredo, R. C., Ribeiro-Andrade, R., Pontes-Silva, A. V., Arantes, M. L., Giani, A., & Figueredo, C. C. (2021). High diversity of microalgae as a tool for the synthesis of different silver nanoparticles: A species-specific green synthesis. *Colloid and Interface Science Communications*, *42*, 100420. DOI: 10.1016/j.colcom.2021.100420

Morales, M., Sánchez, L., & Revah, S. (2018). The Impact of Environmental Factors on Carbon Dioxide Fixation by Microalgae. *FEMS Microbiology Letters*, *365*(3). Advance online publication. DOI: 10.1093/femsle/fnx262 PMID: 29228188

Morgan, S. H., Mahmoud, M. A., Mottaleb, S. A., El-Bahbohy, R. M., & Mahmoud-Aly, M. (2023). A sustainable nanotechnology producing high-quality remediated sewage wastewater used for microalgal protein-rich biomass and biodiesel production. *Biomass Conversion and Biorefinery*, *14*(24), 31021–31037. DOI: 10.1007/s13399-023-05016-9

Moroi, K., Yamamoto, T., & Kurita, T. (2025). Double-strand break-free and transgene-free genome editing in the microalga Nannochloropsis oceanica using removable vectors containing the CRISPR base editing system. *Scientific Reports*, *15*(1), 42431. DOI: 10.1038/s41598-025-26657-y PMID: 41309901

Morrow, R., Bridgens, B., & Mackenzie, L. (Eds.). (2023). *Bioprotopia: Designing the built environment with living organisms*. De Gruyter. DOI: 10.1515/9783035625806

Msanne, J., Di Xu, A. R. K., Casas-Mollano, J. A., Awada, T., Cahoon, E. B., & Cerutti, H. (2012). Metabolic and Gene Expression Changes Triggered by Nitrogen Deprivation in the Photoautotrophically Grown Microalgae Chlamydomonas Reinhardtii and Coccomyxa Sp. C-169. *Phytochemistry*, *75*, 50–59. DOI: 10.1016/j.phytochem.2011.12.007 PMID: 22226037

Muga, H. E., & Mihelcic, J. R. (2008). Sustainability of wastewater treatment technologies. *Journal of Environmental Management*, *88*(3), 437–447. DOI: 10.1016/j.jenvman.2007.03.008 PMID: 17467148

Mukherjee, A., Sarkar, D., & Sasmal, S. (2021). A Review of Green Synthesis of Metal Nanoparticles Using Algae. *Frontiers in Microbiology*, *12*, 693899. DOI: 10.3389/fmicb.2021.693899 PMID: 34512571

Mulbry, W., Kondrad, S., Pizarro, C., & Kebede-Westhead, E. (2008). Treatment of dairy manure effluent using freshwater algae: Algal productivity and recovery of manure nutrients using pilot-scale algal turf scrubbers. *Bioresource Technology*, *99*(17), 8137–8142. DOI: 10.1016/j.biortech.2008.03.073 PMID: 18487042

Müller, M. M., Arndt, K. M., & Hoffmann, S. A. (2025). Genetic circuits in synthetic biology: Broadening the toolbox of regulatory devices. *Frontiers in Synthetic Biology*, *3*, 1548572. DOI: 10.3389/fsybi.2025.1548572

Musenich, L. (2024). Design of Multifunctional Bioinspired Structures.

Mussgnug, J. H., Klassen, V., Schlüter, A., & Kruse, O. (2010). Microalgae as substrates for fermentative biogas production in a combined biorefinery concept. *Journal of Biotechnology*, *150*(1), 51–56. DOI: 10.1016/j.jbiotec.2010.07.030 PMID: 20691224

Mustafa, Z., Auroona, N., Sarwar, A., & Lee, E. Y. (2025). Recent advancements in microbial carbon dioxide fixation: Metabolic engineering strategies. *Green Chemistry, 28*(1), 21–36. DOI: 10.1039/D5GC03349G

Mutitu, K. D., Munyao, M. O., Wachira, M. J., Mwirichia, R., Thiong'o, K. J., & Marangu, M. J. (2019). Effects of biocementation on some properties of cement-based materials incorporating Bacillus species bacteria–a review. *Journal of Sustainable Cement-Based Materials, 8*(5), 309–325. DOI: 10.1080/21650373.2019.1640141

Mwandira, W., Mavroulidou, M., Gunn, M. J., Purchase, D., Garelick, H., & Garelick, J. (2023). Concurrent carbon capture and biocementation through the carbonic anhydrase (CA) activity of microorganisms-a review and outlook. *Environmental Processes, 10*(4), 56. DOI: 10.1007/s40710-023-00667-2

Naduthodi, M. I. S., Claassens, N. J., D'Adamo, S., van der Oost, J., & Barbosa, M. J. (2021). Synthetic biology approaches to enhance microalgal productivity. *Trends in Biotechnology, 39*(10), 1019–1036. DOI: 10.1016/j.tibtech.2020.12.010 PMID: 33541719

Nagappan, S., Devendran, S., Tsai, P. C., Jayaraman, H., Alagarsamy, V., Pugazhendhi, A., & Ponnusamy, V. K. (2020). Metabolomics Integrated with Transcriptomics and Proteomics: Evaluation of Systems Reaction to Nitrogen Deficiency Stress in Microalgae. *Process Biochemistry (Barking, London, England), 91*, 1–14. DOI: 10.1016/j.procbio.2019.11.027

Nagar, S., Antony, R., & Thamban, M. (2021). Extracellular Polymeric Substances in Antarctic Environments: A Review of Their Ecological Roles and Impact on Glacier Biogeochemical Cycles. *Polar Science, 30*, 100686. DOI: 10.1016/j.polar.2021.100686

Nagi, G. K., Minhas, A. K., Gaur, S., Jain, P., & Mandal, S. (2021). Integration of algal biofuels with bioremediation coupled industrial commodities towards cost-effectiveness. *Frontiers in Energy Research, 9*, 735141. DOI: 10.3389/fenrg.2021.735141

Nair, A., & Chakraborty, S. (2020). Synergistic effects between autotrophy and heterotrophy in optimization of mixotrophic cultivation of Chlorella sorokiniana in bubble-column photobioreactors. *Algal Research, 46*, 101799. DOI: 10.1016/j.algal.2020.101799

Najeeb, M. I., Anjum, A. A., Ahmad, M. U. D., Ali, T., Maqbool, A., & Manzoor, R. (2023). Optimization of physicochemical factors for microalgal biomass using indigenously isolated microalgae. *JAPS: Journal of Animal & Plant Sciences, 33*(6).

Nanda, S., Reddy, S. N., Mitra, S. K., & Kozinski, J. A. (2016). The progressive routes for carbon capture and sequestration. *Energy Science & Engineering*, *4*(2), 99–122. DOI: 10.1002/ese3.117

Nandhini, V., & Venkataraman, R. (2025). Carbon Credit Mechanisms: A Market Based Approach to Climate Mitigation. *Archives of Current Research International*, *25*(6), 1–14. DOI: 10.9734/acri/2025/v25i61246

Nanjundeshwara, N., Lakshmaiah, A., & Raghavan, R. (2025). Aflatoxigenic Fungi. *Aflatoxigenic Fungi*, 225–239. DOI: 10.1201/9781003487043-29

Narayanan, Ishvarya, Ricky Rajamanickam, Rishu Kumari Singh, Pardeep Singh, Ashish A. Prabhu, Ashutosh Pandey, Nirakar Pradhan, Y. V. Nancharaiah, Vikram Kumar, Rangabhashiyam Selvasembian, and Joyabrata Mal. 2025. "Microalgae-Based Remediation of Pharmaceutical Contaminants: Emerging Strategies and Technological Synergies." *Discover Applied Sciences 2025 7:11* 7(11):1288-. .DOI: 10 1007/s42452-025-07226-4

Narwal, N., Katyal, D., & Bathi, J. R. (2025). Green synthesis of Iron Oxide Nanoparticles using Momordica charantia: Kinetics of removal of heavy metal and microplastic pollutants. *Regional Studies in Marine Science*, *104189*, 104189. Advance online publication. DOI: 10.1016/j.rsma.2025.104189

Nath, K., & Das, D. (2004). Improvement of fermentative hydrogen production: Various approaches. *Biotechnology Letters*, *26*(1), 13–19.

Nayak, S. K., Bhattacharyya, P., Padhy, S. R., Das, A., Parida, S. P., Rath, M., & Nayak, A. (2025). Harnessing the potential of microalgae for carbon sequestration to achieve net-zero emissions. *Mitigation and Adaptation Strategies for Global Change*, *30*(7), 58. DOI: 10.1007/s11027-025-10249-2

Nayeem, M. A., Saduwale, S., Vempada, S. R., & Rao, M. S. (2025, March). Bio-inspired Self-Healing Concrete using Algae-Based Healing Agents for Crack Repair. In *International Conference on Advanced Materials, Manufacturing and Sustainable Development (ICAMMSD 2024)* (pp. 1002-1016). Atlantis Press. DOI: 10.2991/978-94-6463-662-8_78

Nazari, M. T., Freitag, J. F., Cavanhi, V. A. F., & Colla, L. M. (2020). Microalgae harvesting by fungal-assisted bioflocculation. *Reviews in Environmental Science and Bio/Technology 2020 19:2*, *19*(2), 369–388. DOI: 10.1007/s11157-020-09528-y

Ndaba, J., Cotiyane-Pondo, P., Human, L., Puccinelli, E., Pieterse, P., Pattrick, P., & Porri, F. (2025). Diatom colonisation and biofilm metal bioaccumulation: Can Indigenous Knowledge Systems aid the ecological engineering of urban coastlines? *Ecological Engineering, 219*, 107696. DOI: 10.1016/j.ecoleng.2025.107696

Ndubuisi, O. G. N. OBIORAH, C. A., Ugah, T. A., Agbakhamen, C. O., Ali, S. E., & Nesiama, O. (2025). *Integrated Water Resource Management in a Changing Climate: Assessing Adaptive Strategies, Technologies, and Policies for Resilient Water System*. DOI: 10.5281/zenodo.15769652

Negoro, M., Shioji, N., Miyamoto, K., & Micira, Y. (1991). Growth of microalgae in high CO2 gas and effects of SOx and NOx. *Applied Biochemistry and Biotechnology, 28*(1), 877–886. DOI: 10.1007/BF02922657 PMID: 1929389

Nenadović, T., Šarčević, T., Čižmek, H., Godrijan, J., Marić Pfannkuchen, D., Pfannkuchen, M., & Ljubešić, Z. (2015). Development of periphytic diatoms on different artificial substrates in the Eastern Adriatic Sea. *Acta Botanica Croatica, 74*(2), 377–392. DOI: 10.1515/botcro-2015-0026

Ng, I. S., Keskin, B. B., & Tan, S. I. (2020). A critical review of genome editing and synthetic biology applications in metabolic engineering of microalgae and cyanobacteria. *Biotechnology Journal, 15*(8), 1900228. DOI: 10.1002/biot.201900228 PMID: 32080963

Ngo, H. H., Vo, H. N. P., Guo, W., Lee, D. jong, & Zhang, S. (2022). Carbon dioxide fixation and phycoremediation by algae-based technologies for biofuels and biomaterials. *Biomass, Biofuels, Biochemicals: Climate Change Mitigation: Sequestration of Green House Gases*, 253–277. DOI: 10.1016/B978-0-12-823500-3.00011-X

Nguyen, D. D., Shin, M. C., Han, J. H., & Kim, G. M. (2025). Artificial Intelligence–Assisted Optimization of Microfluidic and Microbial Fuel Cells: An Updated Review. *International Journal of Precision Engineering and Manufacturing, 26*(9), 2389–2409. DOI: 10.1007/s12541-025-01340-4

Niekerk, L. A., Gokul, A., Basson, G., Badiwe, M., Nkomo, M., Klein, A., & Keyster, M. (2024). Heavy Metal Stress and Mitogen Activated Kinase Transcription Factors in Plants: Exploring Heavy Metal-ROS Influences on Plant Signalling Pathways. *Plant, Cell & Environment, 47*(8), 2793–2810. DOI: 10.1111/pce.14926 PMID: 38650576

Niescioruk, M. J., Bandrow, P., Szufa, S., Wozniak, M., & Siczek, K. (2025). Biomass-Based Hydrogen Extraction and Accompanying Hazards—Review. *Molecules, 30*(3).

Nievergelt, A. P. (2025). Genome editing in the green alga Chlamydomonas: Past, present practice and future prospects. *The Plant Journal : For Cell and Molecular Biology*, *122*(1), e70140. DOI: 10.1111/tpj.70140 PMID: 40186543

Nigam, H., Jain, R., Malik, A., & Singh, V. (2022). Comparative Life-Cycle assessment of microalgal biomass production in conventional growth media versus newly developed nanoemulsion media. *Bioresource Technology*, *352*, 127069. DOI: 10.1016/j.biortech.2022.127069 PMID: 35367603

Nigishi, M., Shimakawa, G., Yamagishi, K., Amano, R., Ito, S., Tsuji, Y., Nagasato, C., & Matsuda, Y. (2024). Low-CO2-inducible bestrophins outside the pyrenoid sustain high photosynthetic efficacy in diatoms. *Plant Physiology*, *195*(2), 1432–1445. DOI: 10.1093/plphys/kiae137 PMID: 38478576

Ni, Y., & Sun, Z. (2009). Recent progress on industrial fermentative production of acetone–butanol–ethanol by *Clostridium acetobutylicum* in China. *Applied Microbiology and Biotechnology*, *83*(3), 415–423. DOI: 10.1007/s00253-009-2003-y PMID: 19430776

Niyogi, K. K. (1999). Photoprotection revisited: Genetic and molecular approaches. *Annual Review of Plant Biology*, *50*(1), 333–359. DOI: 10.1146/annurev.arplant.50.1.333 PMID: 15012213

Nneoma, U. C., Chukwudi, O. F., Nnenna, U. J., & Paul-Chima, U. O. (2025). Hybrid biofactories: Integrating microalgae and engineered microbiomes for enhanced biofuel production in circular carbon systems. *Frontiers in Energy Research*, *13*, 1654079. DOI: 10.3389/fenrg.2025.1654079

Nobre, M. L. F., Tavares, D., Fraga, C., Oliveira, B., Dias, M., Mesquita, S., Oliveira, C. M., & Pires, J. C. M. (2024). Techno-Economic Analysis of a Circular Microalgal Approach for Enhanced Wastewater Treatment and Resource Recovery in Northern Portugal. *Journal of Cleaner Production*, *434*, 140389. Advance online publication. DOI: 10.1016/j.jclepro.2023.140389

Noreen, S., Saleem, M. H., Ali, B., Khan, K. A., Hafeez, A., & Javed, M. A. (2024). Advancing Plant Resilience Against Microplastics and Metals Through Nanotechnology. *BioNanoScience*, *14*(2), 2065–2079. DOI: 10.1007/s12668-024-01301-7

Novoveská, L., Nielsen, S. L., Eroldoğan, O. T., Haznedaroglu, B. Z., Rinkevich, B., Fazi, S., Robbens, J., Vasquez, M., & Einarsson, H. (2023). Overview and challenges of large-scale cultivation of photosynthetic microalgae and cyanobacteria. *Marine Drugs*, *21*(8), 445. DOI: 10.3390/md21080445 PMID: 37623726

Nowicka, B. (2022). Heavy Metal-Induced Stress in Eukaryotic Algae-Mechanisms of Heavy Metal Toxicity and Tolerance with Particular Emphasis on Oxidative Stress in Exposed Cells and the Role of Antioxidant Response. *Environmental Science and Pollution Research International*, *29*(12), 16860–16911. DOI: 10.1007/s11356-021-18419-w PMID: 35006558

Nugroho, Y. T., Hardianto, A., Syihab, A. B. M. I., Sidquni, S. A., Nurjannah, I., Adinisa, L., & Subroto, T. J. T. S. (2025). Carbon Capture at the Point Source: CO2 Gas Bubbling vs. *Bicarbonate Supplementation in Microalgal Cultivation.*, *22*(11), 10991–10991.

Nur, M. M. A. Mahreni, Murni, S. W., Setyoningrum, T. M., Hadi, F., Widayati, T. W., Jaya, D., Sulistyawati, R. R. E., Puspitaningrum, D. A., Dewi, R. N., Hadiyanto, & Hasanuzzaman, M. (2025). Innovative strategies for utilizing microalgae as dual-purpose biofertilizers and phycoremediators in agroecosystems. *Biotechnology Reports, 45*, e00870. 10.1016/j.btre.2024.e00870

Nur, M. M. A., & Dewi, R. N. (2024). Opportunities and challenges of microalgae in biocement production and self-repair mechanisms. *Biocatalysis and Agricultural Biotechnology*, *56*, 103048. DOI: 10.1016/j.bcab.2024.103048

Nwali, O. I., Oladunjoye, M. A., & Alao, O. A. (2024). A review of atmospheric carbon dioxide sequestration pathways; processes and current status in Nigeria. *Carbon Capture Science & Technology*, *12*, 100208. DOI: 10.1016/j.ccst.2024.100208

Nymark, M., Sharma, A. K., Sparstad, T., Bones, A. M., & Winge, P. (2016). A CRISPR/Cas9 system adapted for gene editing in marine algae. *Scientific Reports*, *6*(1), 24951. DOI: 10.1038/srep24951 PMID: 27108533

Nzayisenga, Jean Claude, Xavier Farge, Sophia Leticia Groll, and Anita Sellstedt. 2020. "Effects of Light Intensity on Growth and Lipid Production in Microalgae Grown in Wastewater." *Biotechnology for Biofuels 2020 13:1* 13(1):4-. .DOI: 10.1186/s13068-019-1646-x

Nzereogu, P. U., Omah, A. D., Ezema, F. I., Iwuoha, E. I., & Nwanya, A. C. (2026). Silica derived from rice husk waste as anode material for lithium-ion battery: A comprehensive study. *Next Materials*, *11*, 101615. DOI: 10.1016/j.nxmate.2026.101615

Obidi, P. O., & Bayless, D. J. (2025). The synergy of artificial intelligence and algal systems for sustainable wastewater treatment and carbon sequestration. *Environmental Technology Reviews*, *14*(1), 1040–1067. DOI: 10.1080/21622515.2025.2581330

Ogunyemi, F. M. (2025). Harmonizing global carbon accounting standards: Comparative analysis of GHG protocol, ISO 14064, and emerging regulatory frameworks for multinational corporations. *International Journal of Research in Marketing Management and Sales*, *7*(2), 425–434. DOI: 10.33545/26633329.2025.v7.i2e.313

Ojaniemi, U., Tamminen, A., Syrjänen, J., & Barth, D. (2025). CFD modeling of CO2 fixation by microalgae cultivated in a lab scale photobioreactor. *Bioresource Technology*, *415*, 131715. DOI: 10.1016/j.biortech.2024.131715 PMID: 39490541

Olabi, A. G., Alami, A. H., Alasad, S., Aljaghoub, H., Sayed, E. T., Shehata, N., Rezk, H., & Abdelkareem, M. A. (2022). Emerging technologies for enhancing microalgae biofuel production: Recent progress, barriers, and limitations. *Fermentation (Basel, Switzerland)*, *8*(11), 649. DOI: 10.3390/fermentation8110649

Oliver, N., Avramov, A. P., Nürnberg, D. J., Dau, H., & Burnap, R. L. (2022). From manganese oxidation to water oxidation: Assembly and evolution of the water-splitting complex in photosystem II. *Photosynthesis Research*, *152*(2), 107–133. DOI: 10.1007/s11120-022-00912-z

Olofsson, M., Lindehoff, E., Frick, B., Svensson, F., & Legrand, C. (2015). Baltic Sea microalgae transform cement flue gas into valuable biomass. *Algal Research*, *11*, 227–233. https://doi.org/https://doi.org/10.1016/j.algal.2015.07.001. DOI: 10.1016/j.algal.2015.07.001

Onuaguluchi, O., & Banthia, N. (2016). Plant-based natural fibre reinforced cement composites: A review. *Cement and Concrete Composites*, *68*, 96–108. DOI: 10.1016/j.cemconcomp.2016.02.014

Oostlander, P. C., van Houcke, J., Wijffels, R. H., & Barbosa, M. J. (2020). Microalgae production cost in aquaculture hatcheries. *Aquaculture (Amsterdam, Netherlands)*, *525*, 735310. DOI: 10.1016/j.aquaculture.2020.735310

Ördög, V., Stirk, W., Takács, G., Pőthe, P., Illés, A., Bojtor, C., Széles, A., Tóth, B., Van Staden, J., & Nagy, J. (2021). Plant biostimulating effects of the cyanobacterium Nostoc piscinale on maize (Zea mays L.) in field experiments. *South African Journal of Botany*, *140*, 153–160. DOI: 10.1016/j.sajb.2021.03.026

Orr, F. M.Jr. (2018). Carbon capture, utilization, and storage: An update. *SPE Journal*, *23*(06), 2444–2455. DOI: 10.2118/194190-PA

Osman, Mohamed E. H., Atef M. Abo-Shady, Saly F. Gheda, Samy M. Desoki, and Mostafa E. Elshobary. 2023. "Unlocking the Potential of Microalgae Cultivated on Wastewater Combined with Salinity Stress to Improve Biodiesel Production." *Environmental Science and Pollution Research 2023 30:53* 30(53):114610–24. .DOI: 10.1007/s11356-023-30370-6

Osman, A. I., Fawzy, S., Farghali, M., El-Azazy, M., Elgarahy, A. M., Fahim, R. A., Maksoud, M. I. A. A., Ajlan, A. A., Yousry, M., Saleem, Y., & Rooney, D. W. (2022). Biochar for agronomy, animal farming, anaerobic digestion, composting, water treatment, soil remediation, construction, energy storage, and carbon sequestration: A review. *Environmental Chemistry Letters*, *20*(4), 2385–2485. DOI: 10.1007/s10311-022-01424-x PMID: 35571983

Oswald, W. J. (2003). My sixty years in applied algology. *Journal of Applied Phycology*, *15*(2-3), 99–106. DOI: 10.1023/A:1023871903434

Otundo Richard, M. (2024). Policy Framework and Implementation Strategies for Carbon Dioxide Capture and Storage in Kenya's Coastal Region. SSRN *Electronic Journal*. https://doi.org/DOI: 10.2139/SSRN.4936151

Pachouri, V., Chandramauli, A., Singh, R., Gehlot, A., Priyadarshi, N., & Twala, B. (2024). Removal of contaminants by chlorella species: An effort towards sustainable remediation. *Discover Sustainability*, *5*(1), 19. Advance online publication. DOI: 10.1007/s43621-024-00199-1

Pallavi, V. N., Chowdareddy, S., Byalahunasishetter, B. I., & Raghavan, R. (2026). CRISPR-Cas9 Technology for Precision Cancer Medicine: Target Discovery, Disease Modeling, and Therapeutics. *CRISPR-Cas9 Technology for Precision Medicine*, 35–72. DOI: 10.4018/979-8-3373-4857-5.ch002

Panbehkar Bisheh, M., & Amini Rad, H. (2023). Optimization of the culture of Chlorella sorokiniana PA. 91 by RSM: effect of temperature, light intensity, and MgAC-NPs. Environmental Science and Pollution Research, 30(17), 50896-50919.

Panday, D., Bhusal, N., Das, S., & Ghalehgolabbehbahani, A. (2024). Rooted in nature: The rise, challenges, and potential of organic farming and fertilizers in agroecosystems. *Sustainability (Basel)*, *16*(4), 1530. DOI: 10.3390/su16041530

Paniagua-Michel, J., Olmos-Soto, J., & Ruiz, M. A. (2012). Pathways of Carotenoid Biosynthesis in Bacteria and Microalgae. *Methods in Molecular Biology (Clifton, N.J.)*, *892*, 1–12. DOI: 10.1007/978-1-61779-879-5_1 PMID: 22623294

Pan, M., Lyu, T., Zhan, L., Matamoros, V., Angelidaki, I., Cooper, M., & Pan, G. (2021). Mitigating antibiotic pollution using cyanobacteria: Removal efficiency, pathways and metabolism. *Water Research, 190*, 116735. DOI: 10.1016/j.watres.2020.116735 PMID: 33352526

Pant, S., Goel, M., Sahoo, N. K., Yuan, Q., & Rout, P. R. (2025). Microalgae-based treatment for wastewater management and valorisation. *Current Opinion in Environmental Science & Health, 48*, 100662. DOI: 10.1016/j.coesh.2025.100662

Panwar, V., & Dutta, T. (2019). Diatom biogenic silica as a felicitous platform for biochemical engineering: Expanding frontiers. *ACS Applied Bio Materials, 2*(6), 2295–2316. DOI: 10.1021/acsabm.9b00050 PMID: 35030724

Papachristou, I., S. Akaberi, A. Silve, E. Navarro-López, R. Wüstner, K. Leber, N. Nazarova, G. Müller, and W. Frey. 2021. "Analysis of the Lipid Extraction Performance in a Cascade Process for Scenedesmus Almeriensis Biorefinery." *Biotechnology for Biofuels 2021 14:1* 14(1):20-. .DOI: 10.1186/s13068-020-01870-1

Parakh, S. K., Tian, Z., Wong, J. Z. E., & Tong, Y. W. (2023). From microalgae to bioenergy: Recent advances in biochemical conversion processes. *Fermentation (Basel, Switzerland), 9*(6), 529. DOI: 10.3390/fermentation9060529

Parikh, H. (2021). Diatom Biosilica as a source of Nanomaterials. [IJARESM]. *International Journal of All Research Education and Scientific Methods, 9*(11).

Parikh, H. S., Dave, G., Singh, P. K., & Tiwari, A. (2025). Exploring the significance of xenic and axenic cultures for diatoms: Methodological insights, scope, and biotechnological implications. *Systems Microbiology and Biomanufacturing, 5*(3), 969–989. DOI: 10.1007/s43393-025-00363-2

Parker, L., Ward, K., Pilarski, T., Price, J., Derkach, P., Correa, M., Miller, R., Benites, V., Athanasiadis, D., Doherty, B., Edy, L., Alnozaili, G., Reyes, N., Wittenberg, J., Eliares, G., Destaillats, F., Rakitsky, W., & Franklin, S. (2025). Development and Production of High-Oleic Palm Oil Alternative by Fermentation of Microalgae. *Fermentation (Basel, Switzerland), 11*(4), 207. DOI: 10.3390/fermentation11040207

Park, J. B. K., Craggs, R. J., & Shilton, A. N. (2011). Wastewater treatment high rate algal ponds for biofuel production. *Bioresource Technology, 102*(1), 35–42. DOI: 10.1016/j.biortech.2010.06.158 PMID: 20674341

Parmar, P., Kumar, R., Neha, Y., & Srivatsan, V. (2023). Microalgae as next generation plant growth additives: Functions, applications, challenges and circular bioeconomy based solutions. *Frontiers in Plant Science, 14*, 1073546. DOI: 10.3389/fpls.2023.1073546 PMID: 37063190

Parthiban, N., & Thiagarajan, K. (2026). Biodegradation of phthalate esters by microalgae: A comprehensive review. *The Science of the Total Environment, 1012*, 181248. DOI: 10.1016/j.scitotenv.2025.181248 PMID: 41447808

Passos, F., Hernandez-Marine, M., Garcia, J., & Ferrer, I. (2014). Long-term anaerobic digestion of microalgae grown in HRAP for wastewater treatment: Effect of microwave pretreatment. *Water Research, 49*, 351–359. DOI: 10.1016/j.watres.2013.10.013 PMID: 24183557

Pataro, I. M., Gil, J. D., Guzman, J. L., Berenguel, M., & Lemos, J. M. (2023). A learning-based model predictive strategy for pH control in raceway photobioreactors with freshwater and wastewater cultivation media. *Control Engineering Practice, 138*, 105619. DOI: 10.1016/j.conengprac.2023.105619

Patel, A. K., Singhania, R. R., Albarico, F. P. J. B., Pandey, A., Chen, C. W., & Dong, C. D. (2022). Organic wastes bioremediation and its changing prospects. *The Science of the Total Environment, 824*, 153889. DOI: 10.1016/j.scitotenv.2022.153889

Patil, D., Mishra, A., Tyagi, G., Reddy, P. M., Lahoti, M., Pable, A., Jadhav, U. U., & Sachs, S. G. (2025). Comparative evaluation of bacterial strains and novel SCM-based biocarriers for self-healing and strength enhancement in mortar. *Clean Technologies and Environmental Policy, 27*(12), 7611–7631. DOI: 10.1007/s10098-025-03312-8

Pawar, S. S., & Parekar, S. R. (2018). Bacteria based self-healing concrete. *International Research Journal of Engineering and Technology, 5*(3), 1001–1004.

Penfold, C. M., Miyan, M., Reeves, T., & Grierson, I. T. (1995). Biological farming for sustainable agricultural production. *Australian Journal of Experimental Agriculture, 35*(7), 849–856. DOI: 10.1071/EA9950849

Peng, M., Liu, P., Peng, R., Jiang, X., Li, S., & Jiang, M. (2025). Effects of nitrogen, phosphorus, iron, and silicon on the growth, total lipid content, and fatty acid composition of the diatom Corethron sp. *Journal of Applied Phycology*, •••, 1–14.

Peng, S., & Colosi, L. M. (2016). Anaerobic digestion of algae biomass to produce energy during wastewater treatment. *Water Environment Research : A Research Publication of the Water Environment Federation, 88*(1), 29–39. DOI: 10.2175/106143014X14062131179195 PMID: 26803024

Perera, N. S., Costa, M. D. P., Macreadie, P. I., & Wartman, M. (2025). Trends in Market-Based Blue Carbon Projects. *Sustainable Development (Bradford), 33*(3), 3377–3402. DOI: 10.1002/sd.3293

Pérez, O. F. A., Arrieta, V. S., Ospina, J. H. G., Herrera, S. H., Rojas, C. F. R., & Navarro, A. M. S. (2024). Carbon dioxide emissions from traditional and modified concrete. A review. *Environmental Development*, *52*, 101036. DOI: 10.1016/j.envdev.2024.101036

Pessarrodona, A., Franco-Santos, R. M., Wright, L. S., Vanderklift, M. A., Howard, J., Pidgeon, E., Wernberg, T., & Filbee-Dexter, K. (2023a). Carbon sequestration and climate change mitigation using macroalgae: A state of knowledge review. *Biological Reviews of the Cambridge Philosophical Society*, *98*(6), 1945–1971. DOI: 10.1111/brv.12990 PMID: 37437379

Petrucciani, A., Moretti, P., Ortore, M. G., & Norici, A. (2023). Integrative effects of morphology, silicification, and light on diatom vertical movements. *Frontiers in Plant Science*, *14*, 1143998. DOI: 10.3389/fpls.2023.1143998 PMID: 37056507

Phillips, A. J., Gerlach, R., Lauchnor, E., Mitchell, A. C., Cunningham, A. B., & Spangler, L. (2013). Engineered applications of ureolytic biomineralization: A review. *Biofouling*, *29*(6), 715–733. DOI: 10.1080/08927014.2013.796550 PMID: 23802871

Phogat, P., Sharma, S., & Rawat, S. (2026). *Next-Generation Hydrogen Economy: Production, Storage, and Fuel Cell Technologies*. John Wiley & Sons.

Phong, W. N., Regmi, R., Adhikari, K. C., Bhattarai, R. R., & Show, P. L. (2023). Bioactive Compounds from Diatoms. In *Diatoms Biotechnology* (pp. 20–37). CRC Press. DOI: 10.1201/9781003436553-2

Piiparinen, J., Barth, D., Eriksen, N. T., Teir, S., Spilling, K., & Wiebe, M. G. (2018). Microalgal CO2 capture at extreme pH values. *Algal Research*, *32*, 321–328. DOI: 10.1016/j.algal.2018.04.021

Pivac, I., Šimunović, J., Barbir, F., & Nižetić, S. (2024). Reduction of greenhouse gases emissions by use of hydrogen produced in a refinery by water electrolysis. *Energy*, *296*, 131157. DOI: 10.1016/j.energy.2024.131157

Piyatilleke, S., Thevarajah, B., Nimarshana, P., & Ariyadasa, T. U. (2025). Microalgal biofuels: Challenges and prospective in the framework of circular bioeconomy. *Energy Nexus*, *17*, 100338. DOI: 10.1016/j.nexus.2024.100338

Plastina, A., Jo, H., & Wongpiyabovorn, O. (2024). The business case for carbon farming in the USA. *Carbon Balance and Management 2024 19:1, 19*(1), 7-. DOI: 10.1186/s13021-024-00253-5

Polat, E., & Altınbaş, M. (2023). A review on microalgal growth stress parameters for sustainable biofuel production. *Clean Technologies and Environmental Policy*, *25*(8), 2469–2487. DOI: 10.1007/s10098-023-02535-x

Poliner, E., Farré, E. M., & Benning, C. (2018). Advanced genetic tools enable synthetic biology in the oleaginous microalgae Nannochloropsis sp. *Plant Cell Reports*, *37*(10), 1383–1399. DOI: 10.1007/s00299-018-2270-0 PMID: 29511798

Popa, M. D., Simionov, I.-A., Petrea, S. M., Georgescu, P.-L., Ifrim, G. A., & Iticescu, C. (2025). Efficiency of microalgae employment in nutrient removal (nitrogen and phosphorous) from municipal wastewater. *Water (Basel)*, *17*(2), 260. DOI: 10.3390/w17020260

Porcelli, R., Dotto, F., Pezzolesi, L., Marazza, D., Greggio, N., & Righi, S. (2020). Comparative life cycle assessment of microalgae cultivation for non-energy purposes using different carbon dioxide sources. *The Science of the Total Environment*, *721*, 137714. DOI: 10.1016/j.scitotenv.2020.137714 PMID: 32171140

Porras Reyes, L., Havlik, I., & Beutel, S. (2024). Software sensors in the monitoring of microalgae cultivations. *Reviews in Environmental Science and Biotechnology*, *23*(1), 67–92. DOI: 10.1007/s11157-023-09679-8

Portner, B. W., Valente, A., & Guenther, S. (2021). Sustainability Assessment of Combined Animal Fodder and Fuel Production from Microalgal Biomass. *International Journal of Environmental Research and Public Health*, *18*(21), 11351. Advance online publication. DOI: 10.3390/ijerph182111351 PMID: 34769867

Posadas, E., del Mar Morales, M., Gomez, C., Acién, F. G., & Muñoz, R. (2015). Influence of PH and CO2 Source on the Performance of Microalgae-Based Secondary Domestic Wastewater Treatment in Outdoors Pilot Raceways. *Chemical Engineering Journal*, *265*, 239–248. DOI: 10.1016/j.cej.2014.12.059

Poschenrieder, C., Fernández, J. A., Rubio, L., Pérez, L., Terés, J., & Barceló, J. (2018). Transport and use of bicarbonate in plants: Current knowledge and challenges ahead. *International Journal of Molecular Sciences*, *19*(5), 1352. DOI: 10.3390/ijms19051352 PMID: 29751549

Posten, C. (2009). Design principles of photo-bioreactors for cultivation of microalgae. *Engineering in Life Sciences*, *9*(3), 165–177. DOI: 10.1002/elsc.200900003

Poulsen, N., & Kröger, N. (2023). Thalassiosira pseudonana (Cyclotella nana) (Hustedt) Hasle et Heimdal (Bacillariophyceae): A genetically tractable model organism for studying diatom biology, including biological silica formation. *Journal of Phycology*, *59*(5), 809–817. DOI: 10.1111/jpy.13362 PMID: 37424141

Prajapati, A. K., Mahajan, A., Jadhav, S. M., & Kumar, K. (2026). Fourth-generation (4G) biodiesel: Paving the way for a greener and sustainable energy future in emerging economies. *Renewable & Sustainable Energy Reviews*, *225*, 116103. DOI: 10.1016/j.rser.2025.116103

Prajapati, S. K., Malik, A., & Vijay, V. K. (2014). Comparative evaluation of biomass production and bioenergy generation potential of Chlorella spp. through anaerobic digestion. *Applied Energy*, *114*, 790–797. DOI: 10.1016/j.apenergy.2013.08.021

Prasad, R., Gupta, S. K., Shabnam, N., Oliveira, C. Y. B., Nema, A. K., Ansari, F. A., & Bux, F. (2021). Role of microalgae in global CO2 sequestration: Physiological mechanism, recent development, challenges, and future prospective. *Sustainability (Basel)*, *13*(23), 13061. DOI: 10.3390/su132313061

Premarathna, K. S. D., Lau, S. Y., Chiong, T., Show, P. L., Vithanage, M., & Lam, M. K. (2024). Greening up the fight against emerging contaminants: Algae-based nanoparticles for water remediation. *Clean Technologies and Environmental Policy*. Advance online publication. DOI: 10.1007/s10098-024-02862-7

Price, G. D., Badger, M. R., Woodger, F. J., & Long, B. M. (2008). Advances in understanding the cyanobacterial CO2-concentrating-mechanism (CCM): Functional components, Ci transporters, diversity, genetic regulation and prospects for engineering into plants. *Journal of Experimental Botany*, *59*(7), 1441–1461. DOI: 10.1093/jxb/erm112 PMID: 17578868

Priyadharsini, P., Nirmala, N., Dawn, S. S., Baskaran, A., SundarRajan, P., Gopinath, K. P., & Arun, J. (2022). Genetic improvement of microalgae for enhanced carbon dioxide sequestration and enriched biomass productivity: Review on CO2 bio-fixation pathways modifications. *Algal Research*, *66*, 102810. DOI: 10.1016/j.algal.2022.102810

Priyanka, S. J., Moger, S. S., Byalahunasishetter, B. I., & Raghavan, R. (2026). The One Health Framework and Antimicrobial Resistance: Role of Herbal Medicine. *Herbal Medicine Approaches to Antimicrobial Resistance and One Health*, 1–36. DOI: 10.4018/979-8-3373-7810-7.ch001

Pulz, O. (2001). Photobioreactors: Production systems for phototrophic microorganisms. *Applied Microbiology and Biotechnology*, *57*(3), 287–293. DOI: 10.1007/s002530100702 PMID: 11759675

Qadir, M., Drechsel, P., Cisneros, B. J., Kim, Y., Pramanik, A., Mehta, P., & Olaniyan, O. (2020). Global and Regional Potential of Wastewater as a Water, Nutrient and Energy Source. *Natural Resources Forum*, *44*(1), 40–51. DOI: 10.1111/1477-8947.12187

Qian, C., Zheng, T., Zhang, X., & Su, Y. (2021). Application of microbial self-healing concrete: Case study. *Construction & Building Materials, 290*, 123226. DOI: 10.1016/j.conbuildmat.2021.123226

Qian, Z., Yanqiu, S., Lin, G., Hongmei, D., Lihan, Z., Shu, Y., & Qi, L. (2024). Sulfur source promotes the biosorption and bioprecipitation of Cd in purple non-sulfur bacteria. *International Biodeterioration & Biodegradation, 188*, 105742. DOI: 10.1016/j.ibiod.2024.105742

Qin, K., Ye, X., Luo, S., Fernie, A. R., & Zhang, Y. (2025). Engineering carbon assimilation in plants. *Journal of Integrative Plant Biology*, *67*(4), 926–948. DOI: 10.1111/jipb.13825 PMID: 39783795

Qiu, S., Yang, A., & Zeng, H. (2023). Flux balance analysis-based metabolic modeling of microbial secondary metabolism: Current status and outlook. *PLoS Computational Biology*, *19*(8), e1011391. DOI: 10.1371/journal.pcbi.1011391 PMID: 37619239

Quinn, J. C., & Davis, R. (2015). The potentials and challenges of algae based biofuels: A review of the techno-economic, life cycle, and resource assessment modeling. *Bioresource Technology*, *184*, 444–452. DOI: 10.1016/j.biortech.2014.10.075 PMID: 25453439

Quiroz, D., Greene, J. M., Limb, B. J., & Quinn, J. C. (2023). Global Life Cycle and Techno-Economic Assessment of Algal-Based Biofuels. *Environmental Science & Technology*, *57*(31), 11541–11551. DOI: 10.1021/acs.est.3c02892 PMID: 37499260

Rabiee, N., Khatami, M., Jamalipour Soufi, G., Fatahi, Y., Iravani, S., Varma, R. S. J. A. b. s., & engineering. (2021). Diatoms with invaluable applications in nanotechnology, biotechnology, and biomedicine: recent advances. *7*(7), 3053-3068.

Rabiee, N., Khatami, M., Jamalipour Soufi, G., Fatahi, Y., Iravani, S., & Varma, R. S. (2021). Diatoms with invaluable applications in nanotechnology, biotechnology, and biomedicine: Recent advances. *ACS Biomaterials Science & Engineering*, *7*(7), 3053–3068. DOI: 10.1021/acsbiomaterials.1c00475 PMID: 34152742

Rafique, M. S., Rafique, M., Tahir, M. B., Hajra, S., Nawaz, T., & Shafiq, F. (2020). Synthesis methods of nanostructures. In *Nanotechnology and photocatalysis for environmental applications* (pp. 45–56). Elsevier. DOI: 10.1016/B978-0-12-821192-2.00003-6

Rahul, S., & Arumugam, A. (2026). Current Advancements in Algal Feedstock Cultivation and Processing Aiming 3G Biorefinery. In *Biorefineries in Circular Bioeconomy: Reactor Configurations Analysis* (pp. 57–80). Springer Nature Singapore.

Rai, I., Achankunju, J., Gordon, R., & Vinayak, V. (2025). Terraforming Mars with Microalgae, Especially Diatoms. *Diatom Cultivation for Biofuel, Food and High-Value Products*, 203-230.

Raines, C. A., Cavanagh, A. P., & Simkin, A. J. (2022). Improving carbon fixation. In *Photosynthesis in action* (pp. 175–192). Academic Press. DOI: 10.1016/B978-0-12-823781-6.00009-5

Rai, S. K., Kim, G., & Song, H. (2025). Algae to biofuels: Catalytic strategies and sustainable technologies for green energy conversion. *Catalysts*, *15*(9), 806. DOI: 10.3390/catal15090806

Rajadesingu, S., Palani, N., Mendonce, K. C., Vijayakumar, P., Monisha, P., & Ayyadurai, S. (2024). State-of-the-art review on advancements of eco-friendly bacterial-infused self-healing concrete for sustainable constructions. *Journal of Building Engineering*, *91*, 109669. DOI: 10.1016/j.jobe.2024.109669

Rajesh, A., Hariny, V. S., & Sumathi, A. (2024). Performance of Bacillus tropicus on Mechanical, Durable and Crack Remediation Properties in Sustainable Vermiculite Concrete. *Journal of Bionic Engineering*, *21*(4), 1987–1999. DOI: 10.1007/s42235-024-00546-y

Rajesh, A., & Sumathi, A. (2024). Improvement on strength, durability, and crack closure behavior of bacteria concrete under marine soil exposures. *Journal of Testing and Evaluation*, *52*(2), 1269–1279. DOI: 10.1520/JTE20230403

Rajesh, A., & Sumathi, A. (2024, January). Strength and self-healing behavior of bacteria biocomposite concrete in soil exposure condition. [). Elsevier.]. *Structures (Oxford)*, *59*, 105673. DOI: 10.1016/j.istruc.2023.105673

Rajpoot, A. S., Shende, V., Chelladurai, H. M., Dwivedi, G., Verma, T. N., & Choudhary, T. (2025). Exploring the potential and progress of microalgae biomass production, harvesting, and pre-treatment for sustainable biofuel production: A comprehensive review. *Environment, Development and Sustainability*, •••, 1–51. DOI: 10.1007/s10668-025-05984-0

Ramadan, R., Ghanem, H., Khatib, J. M., & ElKordi, A. M. (2025). Effect of Plant-based natural fibers on the mechanical properties and volume change of cement paste. *International Journal of Building Pathology and Adaptation*, *43*(4), 693–711. DOI: 10.1108/IJBPA-11-2023-0166

Ramakrishnan, P., Sundaram, T., Lahiri, D., Nag, M., & Bhattacharya, D. (2025). Genetic Engineering and Modulation of Metabolic Pathways. In *Introduction to Metabolic Engineering and Application* (pp. 295–330). Springer Nature Switzerland. DOI: 10.1007/978-3-031-93189-5_12

Ramanna, L., Ansari, F. A., Rawat, I., & Bux, F. (2025). Microalgae-driven carbon sequestration and bio-fertiliser: Steps towards a sustainable future. *Chemical Engineering Journal, 519*, 164892. DOI: 10.1016/j.cej.2025.164892

Rame, R., Purwanto, P., & Sudarno, S. (2023). Sustainable energy harnessing: Microalgae as a potential biofuel source and carbon sequestration solution. *Renewable Energy Focus, 47*, 100498. DOI: 10.1016/j.ref.2023.100498

Ranganathan, P., & Savithri, S. (2019). Techno-Economic Analysis of Microalgae-Based Liquid Fuels Production from Wastewater via Hydrothermal Liquefaction and Hydroprocessing. *Bioresource Technology, 284*, 256–265. DOI: 10.1016/j.biortech.2019.03.087 PMID: 30947140

Ranjbar, S., & Malcata, F. X. (2022). Is genetic engineering a route to enhance microalgae-mediated bioremediation of heavy metal-containing effluents? *Molecules (Basel, Switzerland), 27*(5), 1473. DOI: 10.3390/molecules27051473 PMID: 35268582

Ras, M., Lardon, L., Bruno, S., Bernet, N., & Steyer, J.-P. (2011). Experimental study on a coupled process of production and anaerobic digestion of *Chlorella vulgaris. Bioresource Technology, 102*(1), 200–206. DOI: 10.1016/j.biortech.2010.06.146 PMID: 20678925

Rasmussen, R. S., & Morrissey, M. T. (2007). Marine biotechnology for production of food ingredients. In Taylor, S. L. (Ed.), *Advances in Food and Nutrition Research, 52, 237–292*. Academic Press. DOI: 10.1016/S1043-4526(06)52005-4

Ratledge, C., & Wynn, J. P. (2002). The Biochemistry and Molecular Biology of Lipid Accumulation in Oleaginous Microorganisms. *Advances in Applied Microbiology, 51*, 1–52. DOI: 10.1016/S0065-2164(02)51000-5 PMID: 12236054

Raut, N., Al Saadi, A., Ghasemi, M., & Vaidya, R. (2025). Enhancing Biohydrogen Yield and Slaughterhouse Wastewater Treatment Efficiency through Microalgae and Bacterial Synergy. *Periodica Polytechnica. Chemical Engineering, 69*(2), 313–328. DOI: 10.3311/PPch.38818

Raven, J. A., & Falkowski, P. G. (1999). Oceanic sinks for atmospheric CO_2. *Plant, Cell & Environment, 22*(6), 741–755. DOI: 10.1046/j.1365-3040.1999.00419.x

Raven, J. A., Giordano, M., Beardall, J., & Maberly, S. C. (2012). Algal and aquatic plant carbon concentrating mechanisms in relation to environmental change. *Photosynthesis Research, 109*(1–3), 281–296. PMID: 21327536

Raven, J. A., Giordano, M., Beardall, J., & Maberly, S. C. (2012). Algal evolution in relation to atmospheric CO2: Carboxylases, carbon-concentrating mechanisms and carbon oxidation cycles. *Philosophical Transactions of the Royal Society of London. Series B, Biological Sciences*, *367*(1588), 493–507. DOI: 10.1098/rstb.2011.0212 PMID: 22232762

Ravichandran, M., Kumar, T. T. A., & Dineshkumar, R. (2024). Carbon dioxide capture, sequestration, and utilization models for carbon management and transformation. *Environmental Science and Pollution Research International*, *31*(44), 55895–55916. DOI: 10.1007/s11356-024-34861-y PMID: 39256334

Rawindran, H., Khoo, K. S., Satpati, G. G., Maity, S., Chandran, K., Lim, J. W., Tong, W.-Y., Setiabudi, H. D., & Yunus, N. M. (2025). Composition of carbohydrate, protein and lipid derived from microalgae using thermally pretreated solid waste. *Journal of the Science of Food and Agriculture*, *105*(9), 4672–4679. DOI: 10.1002/jsfa.14038

Rayamajhi, V., Hussain, M., Shin, H., & Jung, S. (2025). Recent advances in the application of artificial intelligence in microalgal cultivation. *Processes (Basel, Switzerland)*, *13*(12), 3764. DOI: 10.3390/pr13123764

Ray, S., Abraham, J., Jordan, N., Lindsay, M., & Chauhan, N. (2022). Synthetic, photosynthetic, and chemical strategies to enhance carbon dioxide fixation. *C*, *8*(1), 18.

Razaviarani, V., Arab, G., Lerdwanawattana, N., & Gadia, Y. (2023). Algal biomass dual roles in phycoremediation of wastewater and production of bioenergy and value-added products. *International Journal of Environmental Science and Technology*, *20*(7), 8199–8216. DOI: 10.1007/s13762-022-04696-6

Razzak, S. A., Bahar, K., Islam, K. O., Haniffa, A. K., Faruque, M. O., Hossain, S. Z., & Hossain, M. M. (2024). Microalgae cultivation in photobioreactors: Sustainable solutions for a greener future. *Green Chemical Engineering*, *5*(4), 418–439. DOI: 10.1016/j.gce.2023.10.004

Razzak, S. A., Hossain, M. M., Lucky, R. A., Bassi, A. S., & de Lasa, H. (2013). Integrated CO_2 capture, wastewater treatment and biofuel production by microalgae culturing. *Renewable & Sustainable Energy Reviews*, *27*, 622–653. DOI: 10.1016/j.rser.2013.05.063

Ream, M., & Prather, K. L. (2024). Engineered autonomous dynamic regulation of metabolic flux. *Nature Reviews Bioengineering*, *2*(3), 233–243. DOI: 10.1038/s44222-023-00140-7

Rees, T. J. J. o. E. B. (1984). Sodium dependent photosynthetic oxygen evolution in a marine diatom. *35*(3), 332-337.

Reid, A., Buchanan, F., Julius, M., & Walsh, P. J. (2021). A review on diatom biosilicification and their adaptive ability to uptake other metals into their frustules for potential application in bone repair. *Journal of Materials Chemistry. B*, *9*(34), 6728–6737. DOI: 10.1039/D1TB00322D PMID: 34346480

Ren, H., Zhou, D., Lu, J., Show, P. L., & Sun, F. F. (2023). Mapping the field of microalgae CO2 sequestration: a bibliometric analysis. *Environmental Science and Pollution Research 2023 30:32, 30*(32), 78030–78040. DOI: 10.1007/s11356-023-27850-0

Renuka, N., Prasanna, R., Sood, A., Ahluwalia, A. S., Bansal, R., Babu, S., Singh, R., Shivay, Y. S., & Nain, L. (2016). Exploring the efficacy of wastewater-grown microalgal biomass as a biofertilizer for wheat. *Environmental Science and Pollution Research International*, *23*(7), 6608–6620. DOI: 10.1007/s11356-015-5884-6 PMID: 26638970

Rinku, Angmo, S., Srivastava, A., Merina, K., Sindhwal, S., & Tiwari, A. (2026). Diatoms Nanotechnology: Prospects and Challenges for Sustainable Environment. *Advanced Nanotechnology Applications for Environmental Sustainability*, 41-70.

Rios Pinto, L. F., Ferreira, G. F., & Tasic, M. (2021). Cultivation techniques. *Microalgae: Cultivation, Recovery of Compounds and Applications*, 1–33. DOI: 10.1016/B978-0-12-821218-9.00001-3

Rockström, J., Steffen, W., Noone, K., Persson, Å., Chapin III, F. S., Lambin, E., ... & Foley, J. (2009). Planetary boundaries: exploring the safe operating space for humanity. Ecology and society, 14(2).

Roell, M. S., & Zurbriggen, M. D. (2020). The impact of synthetic biology for future agriculture and nutrition. *Current Opinion in Biotechnology*, *61*, 102–109. DOI: 10.1016/j.copbio.2019.10.004 PMID: 31812911

Rogato, A., & De Tommasi, E. (2020). Physical, chemical, and genetic techniques for diatom frustule modification: Applications in nanotechnology. *Applied Sciences (Basel, Switzerland)*, *10*(23), 8738. DOI: 10.3390/app10238738

Ronga, D., Biazzi, E., Parati, K., Carminati, D., Carminati, E., & Tava, A. (2019). Microalgal Biostimulants and Biofertilisers in Crop Productions. *Agronomy (Basel)*, *9*(4), 192. https://www.mdpi.com/2073-4395/9/4/192. DOI: 10.3390/agronomy9040192

Rosero-Chasoy, G., Rodríguez-Jasso, R. M., Aguilar, C. N., Buitrón, G., Chairez, I., & Ruiz, H. A. (2021). Microbial co-culturing strategies for the production high value compounds, a reliable framework towards sustainable biorefinery implementation–an overview. *Bioresource Technology*, *321*, 124458. DOI: 10.1016/j.biortech.2020.124458 PMID: 33338739

Rossi, S. (2025). Bioeconomy analysis of microalgal products and process: the role of microalgae technology in the blue economy. In *Microalgae as Promising Source of Commercial Bioproducts: Advances and Emerging Technologies in Bioactive Compounds, Biofuels, Food and Biorefineries* (pp. 343–352). Springer Nature Switzerland. DOI: 10.1007/978-3-031-86433-9_23

Rossi, S., Carecci, D., Marazzi, F., Di Benedetto, F., Mezzanotte, V., Parati, K., Alberti, D., Geraci, I., & Ficara, E. (2024). Integrating microalgae growth in biomethane plants: Process design, modelling, and cost evaluation. *Heliyon*, *10*(1), e23240. DOI: 10.1016/j.heliyon.2023.e23240 PMID: 38163195

Round, F. E., Crawford, R. M., & Mann, D. G. (1990). *Diatoms: biology and morphology of the genera*. Cambridge University Press.

Rubin-Blum, M., Dubilier, N., & Kleiner, M. (2019). Genetic evidence for two carbon fixation pathways (the Calvin-Benson-Bassham cycle and the reverse tricarboxylic acid cycle) in symbiotic and free-living bacteria. *MSphere*, *4*(1), 10–1128. DOI: 10.1128/mSphere.00394-18 PMID: 30602523

Ruiz, J., Olivieri, G., de Vree, J., Bosma, R., Willems, P., Reith, J. H., Eppink, M. H. M., Kleinegris, D. M. M., Wijffels, R. H., & Barbosa, M. J. (2016). Towards industrial products from microalgae. *Energy & Environmental Science*, *9*(10), 3036–3043. DOI: 10.1039/C6EE01493C

Rusznyák, A., Akob, D. M., Nietzsche, S., Eusterhues, K., Totsche, K. U., Neu, T. R., Frosch, T., Popp, J., Keiner, R., Geletneky, J., Katzschmann, L., Schulze, E.-D., & Küsel, K. (2012). Calcite biomineralization by bacterial isolates from the recently discovered pristine karstic Herrenberg cave. *Applied and Environmental Microbiology*, *78*(4), 1157–1167. DOI: 10.1128/AEM.06568-11 PMID: 22179248

Saccardo, A., Bezzo, F., & Sforza, E. (2022). Microalgae growth in ultra-thin steady-state continuous photobioreactors: Assessing self-shading effects. *Frontiers in Bioengineering and Biotechnology*, *10*, 977429. DOI: 10.3389/fbioe.2022.977429 PMID: 36032730

Sadvakasova, A. K., Balouch, H., Bauenova, M. O., Kossalbayev, B. D., Allakhverdiev, S. I., & Shabala, S. (2025). Strategies to intensify CO2 capture by microalgae for the circular bioeconomy. *Trends in Plant Science*. PMID: 41168026

Salamone, A. L., Robicheau, B. M., & Walker, A. K. (2016). Fungal diversity of marine biofilms on artificial reefs in the north-central Gulf of Mexico. *Botanica Marina*, *59*(5), 291–305. DOI: 10.1515/bot-2016-0032

Salar, R. K., Gahlawat, S. K., Siwach, P., & Duhan, J. S. (Eds.). (2013). *Biotechnology: prospects and applications* (p. 315). Springer. DOI: 10.1007/978-81-322-1683-4

Salleh, S., Mohammad, M., Cheah, W., Palaniveloo, K., Fadhullah, W., Khare, V., Zhongwei, H., & Ishak, S. D. (2024). The Biology of Diatom and Potential Usage for Value-Added Products. In Riti Thapar, K., Mohd, R., & Norli, I. (Eds.), *Marine Biomass* (pp. 341–366). De Gruyter., https://doi.org/doi:10.1515/9783111353951-015 DOI: 10.1515/9783111353951-015

Santos, A. A., Santos, A. A., & Garbossa, L. H. P. (2024). Potencial de sequestro do dióxido de carbono na biomassa de macroalga Kappaphycus alvarezii nas fazendas marinhas de Santa Catarina. *Agropecuária Catarinense*, *37*(2), 16–18. DOI: 10.52945/rac.v37i2.1815

Saoud, H. A. A., Sprynskyy, M., Pashaei, R., Kawalec, M., Pomastowski, P., & Buszewski, B. (2022). Diatom biosilica: Source, physical-chemical characterization, modification, and application. *Journal of Separation Science*, *45*(17), 3362–3376. DOI: 10.1002/jssc.202100981 PMID: 35652201

Saranraj, P., Singh, S., Manigandan, M., Kesavardhini, K., Lokeshwari, B., & Gayathri, K. (2026). Gold Nanoparticles: Green Synthesis, Characterization, and Biomedical Applications. In Plant-Associated Medicinal Nano-Chemistry (pp. 37-49). CRC Press.

Saranraj, P., Singh, S., Manigandan, M., Nisha, R., Lokeshwari, B., & Gayathri, K. (2025). Pharmacological Effects of Medicinal Herbs: Recent Development and Progress. In Pharmacology of Plants and Plant Derived Biologically Active Molecules (pp. 30-45). CRC Press.

Saranya, S., Thamanna, L., & Chellapandi, P. (2024). Unveiling the potential of systems biology in biotechnology and biomedical research. *Systems Microbiology and Biomanufacturing*, *4*(4), 1217–1238. DOI: 10.1007/s43393-024-00286-4

Sarat Chandra, T., Maneesh Kumar, M., Mukherji, S., Chauhan, V. S., Sarada, R., & Mudliar, S. N. (2018). Comparative life cycle assessment of microalgae-mediated CO2 capture in open raceway pond and airlift photobioreactor system. *Clean Technologies and Environmental Policy 2018 20:10, 20*(10), 2357–2364. DOI: 10.1007/s10098-018-1612-5

Sarkar, O., Santhosh, J., Dhar, A., & Mohan, S. V. (2021). Green hythane production from food waste: Integration of dark-fermentation and methanogenic process towards biogas up-gradation. *International Journal of Hydrogen Energy, 46*(36), 18832–18843. DOI: 10.1016/j.ijhydene.2021.03.053

Sarker, N. K., & Kaparaju, P. (2024). Microalgal bioeconomy: A green economy approach towards achieving sustainable development goals. *Sustainability (Basel), 16*(24), 11218. DOI: 10.3390/su162411218

Sarker, N. K., & Kaparaju, P. (2025). Advances in algae-based bioplastics: From strain engineering and fermentation to commercialization and sustainability. *Fermentation (Basel, Switzerland), 11*(10), 574. DOI: 10.3390/fermentation11100574

Sarmah, M., & Panda, M. R. (2026). Systematic review on proliferation of microalgae and carbon sequestration potential for sustainability. *World Journal of Microbiology & Biotechnology, 42*(3), 100. DOI: 10.1007/s11274-026-04849-6 PMID: 41731188

Sarma, U., Hoque, M. E., Thekkangil, A., Venkatarayappa, N., & Rajagopal, S. (2024). Microalgae in removing heavy metals from wastewater–An advanced green technology for urban wastewater treatment. *Journal of Hazardous Materials Advances, 15*, 100444. DOI: 10.1016/j.hazadv.2024.100444

Sarwan, J., Nair, M. V., Sharma, S., Uddin, N., & Bose, K. J. C. (2024). Omics technology approaches for the generation of biofuels. In *Biomass Energy for Sustainable Development* (pp. 165–191). CRC Press. DOI: 10.1201/9781003406501-9

Satrio, R. D., Fendiyanto, M. H., & Miftahudin, M. (2024). Tools and techniques used at global scale through genomics, transcriptomics, proteomics, and metabolomics to investigate plant stress responses at the molecular level. In *Molecular dynamics of plant stress and its management* (pp. 555–607). Springer Nature Singapore. DOI: 10.1007/978-981-97-1699-9_25

Saxena, A., Prakash Gupta, J., Tiwary, J. K., Kumar, A., Sharma, S., Pandey, G., Biswas, S., & Raghav Chaturvedi, K. (2024). Innovative pathways in carbon capture: Advancements and strategic approaches for effective carbon capture, utilization, and storage. *Sustainability (Basel), 16*(22), 10132. DOI: 10.3390/su162210132

Schipper, K., Al Jabri, H. M. S. J., Wijffels, R. H., & Barbosa, M. J. (2023). Realizing algae value chains in arid environments: An Arabian Peninsula perspective. *Trends in Biotechnology*, *41*(6), 750–759. DOI: 10.1016/j.tibtech.2022.11.004 PMID: 36581482

Schreiber, C., Schiedung, H., Harrison, L., Briese, C., Ackermann, B., Kant, J., Schrey, S. D., Hofmann, D., Singh, D., & Ebenhöh, O. J. J. o. a. p. (2018). Evaluating potential of green alga Chlorella vulgaris to accumulate phosphorus and to fertilize nutrient-poor soil substrates for crop plants. *30*(5), 2827-2836.

Sciuto, G., Russo, N., Randazzo, C. L., & Caggia, C. (2025). Valorization of Second Cheese Whey Through Microalgae-Based Treatments: Advantages, Limits, and Opportunities. BioTech, 14(4), 79.

Semple, A., & Pandhal, J. (2025). Engineering the phycosphere: Fundamental concepts and tools for the bottom-up design of microalgal-bacterial consortia. *Applied Phycology*, *6*(1), 21–51. DOI: 10.1080/26388081.2024.2435040

Serhane, A., Raad, M., Raad, R., & Susilo, W. (2019). Programmable logic controllers based systems (PLC-BS): Vulnerabilities and threats. *SN Applied Sciences*, *1*(8), 924. DOI: 10.1007/s42452-019-0860-2

Serik, A., Kuspanov, Z., & Daulbayev, C. (2026). Cost-effective strategies and technologies for green hydrogen production. *Renewable & Sustainable Energy Reviews*, *226*, 116242. DOI: 10.1016/j.rser.2025.116242

Sethi, D., Butler, T. O., Shuhaili, F., & Vaidyanathan, S. J. B. (2020). Diatoms for carbon sequestration and bio-based manufacturing. *9*(8), 217.

Shaaban, S., Hammad, N., Elnemr, A., & Shaaban, I. G. (2023, June). Efficiency of bacteria-based self-healing mechanism in concrete. [). Trans Tech Publications Ltd.]. *Materials Science Forum*, *1089*, 135–143. DOI: 10.4028/p-tc6w54

Shaban, M., Sadeq, A. M., Rajhi, W., & Hajlaoui, K. (2026). Multi-objective optimization of a biomass-fired supercritical CO2 Brayton–electrolysis system for high-efficiency hydrogen, power, and cogenerative thermal services. Journal of CO2 Utilization, 103(10328), 1.

Shabbirahmed, A. M., Jacob, A., Dey, P., Somu, P., & Haldar, D. (2025). Biomass as eco-friendly adsorbents for the removal of emerging pollutants from wastewater: A review. *Discover Applied Sciences*, *7*(7), 771. DOI: 10.1007/s42452-025-07463-7

Shah, M. M. R., Liang, Y., Cheng, J. J., & Daroch, M. (2016). Astaxanthin-producing green microalga Haematococcus pluvialis: from single cell to high value commercial products. Frontiers in plant science, 7, 531

Shang, Z., Qi, F., Mu, R., Ren, Y., Ma, G., & Yin, M. (2022). Enhanced potential of a Selenastrum capricornutum mutant for CO2 fixation and biofuels production. *Environmental Progress & Sustainable Energy, 41*(5), e13838. DOI: 10.1002/ep.13838

Shanmugam, S., Hari, A., Pandey, A., Mathimani, T., Felix, L., & Pugazhendhi, A. (2020). Comprehensive review on the application of inorganic and organic nanoparticles for enhancing biohydrogen production. *Fuel, 270*, 117453. DOI: 10.1016/j.fuel.2020.117453

Sharma, P. K., Goud, V. V., Yamamoto, Y., & Sahoo, L. (2021). Efficient Agrobacterium tumefaciens-mediated stable genetic transformation of green microalgae, Chlorella sorokiniana. *3 Biotech, 11*(4), 196.

Sharma, A., & Bhushan, K. (2026). A comprehensive survey on IoT security: Challenges, security issues, and countermeasures. *Computer Science Review, 59*, 100839. DOI: 10.1016/j.cosrev.2025.100839

Sharma, A., Hussain Mehdi, S. E., Shahzad, S., Hussain, F., Pandey, S., Kang, W., & Oh, S.-E. (2026). Microalgal biorefinery: Innovations in sustainable biofuel production. *Green Chemistry, 28*(4), 1352–1871. DOI: 10.1039/D5GC02940F

Sharma, P., Jain, A., Bora, B. J., Balakrishnan, D., Show, P. L., Ramaraj, R., Ağbulut, Ü., & Khoo, K. S. (2023). Application of modern approaches to the synthesis of biohydrogen from organic waste. *International Journal of Hydrogen Energy, 48*(55), 21189–21213. DOI: 10.1016/j.ijhydene.2023.03.029

Sheik, A. G., Kumar, A., Ansari, F. A., Raj, V., Peleato, N. M., Patan, A. K., Kumari, S., & Bux, F. (2024). Reinvigorating algal cultivation for biomass production with digital twin technology-a smart sustainable infrastructure. *Algal Research, 84*, 103779. DOI: 10.1016/j.algal.2024.103779

Shelake, R. M., Khalid, M. A., & Kim, J.-Y. (2025). Microalgal bioengineering for futuristic applications in synthetic and space biology. *Biotechnology Advances, 84*, 108665. DOI: 10.1016/j.biotechadv.2025.108665 PMID: 40738416

Shen, J., Zheng, X., Liu, M., Xu, K., He, L., & Lin, Z. (2024). Targeted cultivation of diatoms in mariculture wastewater by nutrient regulation and UV-C irradiation [Original Research]. *Frontiers in Microbiology, Volume 15 - 2024*. DOI: 10.3389/fmicb.2024.1371855

Shen, M., Kong, F., Tong, L., Luo, Y., Yin, S., Liu, C., Zhang, P., Wang, L., Chu, P. K., & Ding, Y. (2022). Carbon capture and storage (CCS): Development path based on carbon neutrality and economic policy. *Carbon Neutrality, 1*(1), 37. DOI: 10.1007/s43979-022-00039-z

Shen, Y., Yuan, W., Pei, Z., Wu, Q., & Mao, E. (2009). Microalgae mass production methods. *Transactions of the ASABE, 52*, 1275–1287. DOI: 10.13031/2013.27771

Shera, S. S., & Banik, R. M. (2022). Algal nanoparticles: synthesis and characterization. In *Bioprospecting Algae for Nanosized Materials* (pp. 25–69). Springer International Publishing.

Shetty, P., Gitau, M. M., & Maróti, G. (2019). Salinity stress responses and adaptation mechanisms in eukaryotic green microalgae. Cells, 8(12), 1657

Shevela, D., Kern, J. F., Govindjee, G., & Messinger, J. (2023). Solar energy conversion by photosystem II: Principles and structures. *Photosynthesis Research, 156*(3), 279–307. DOI: 10.1007/s11120-022-00991-y

Shilky, Patra, S., Harshvardhan, A., Kumar, A., & Saikia, P. (2023). Role of microbes in controlling the geochemical composition of aquatic ecosystems. *Hydrogeochemistry of Aquatic Ecosystems*, 265-281.

Shilton, A., Mara, D., Craggs, R., & Powell, N. (2008). Solar-powered aeration and disinfection, anaerobic co-digestion, biological CO_2 scrubbing and biofuel production: The energy and carbon management opportunities of waste stabilization ponds. *Water Science and Technology, 58*(1), 253–258. DOI: 10.2166/wst.2008.666 PMID: 18653962

Shi, S., Qi, N., & Nielsen, J. (2022). Microbial production of chemicals driven by CRISPR-Cas systems. *Current Opinion in Biotechnology, 73*, 34–42. DOI: 10.1016/j.copbio.2021.07.002 PMID: 34303184

Siddique, M. N. I., & Wahid, Z. B. A. (2018). Enhanced methane yield by codigestion of sewage sludge with microalgae and catering waste leachate. *Water Environment Research : A Research Publication of the Water Environment Federation, 90*(9), 835–839. DOI: 10.2175/106143017X15131012153031 PMID: 29618407

Sidhu, N., Goyal, S., & Reddy, M. S. (2022). Biomineralization of cyanobacteria Synechocystis pevalekii improves the durability properties of cement mortar. *AMB Express, 12*(1), 59. DOI: 10.1186/s13568-022-01403-z PMID: 35587839

Singaravelu, G., Arockiamary, J. S., Kumar, V. G., & Govindaraju, K. (2007). A novel extracellular synthesis of monodisperse gold nanoparticles using marine alga, Sargassum wightii Greville. *Colloids and Surfaces. B, Biointerfaces, 57*(1), 97–101. DOI: 10.1016/j.colsurfb.2007.01.010 PMID: 17350236

Singh Chauhan, D., Sahoo, L., & Mohanty, K. (2022). Maximize microalgal carbon dioxide utilization and lipid productivity by using toxic flue gas compounds as nutrient source. *Bioresource Technology*, *348*, 126784. DOI: 10.1016/j.biortech.2022.126784 PMID: 35104656

Singh, A. K., Santos-Merino, M., Sakkos, J. K., Walker, B. J., & Ducat, D. C. (2022). Rubisco regulation in response to altered carbon status in the cyanobacterium Synechococcus elongatus PCC 7942. *Plant Physiology*, *189*(2), 874–888. DOI: 10.1093/plphys/kiac065 PMID: 35201348

Singhal, R., Pal, R., & Dutta, S. (2023). Chloroplast engineering: Fundamental insights and its application in amelioration of environmental stress. *Applied Biochemistry and Biotechnology*, *195*(4), 2463–2482. DOI: 10.1007/s12010-022-03930-8 PMID: 35484466

Singh, J., & Gu, S. (2010). Commercialization potential of microalgae for biofuels production and CO_2 mitigation. *Renewable & Sustainable Energy Reviews*, *14*(9), 2596–2610. DOI: 10.1016/j.rser.2010.06.014

Singh, S. K., Badgujar, G., Reddy, V. R., Fleisher, D. H., & Bunce, J. A. (2013). Carbon dioxide diffusion across stomata and mesophyll and photo-biochemical processes as affected by growth CO2 and phosphorus nutrition in cotton. *Journal of Plant Physiology*, *170*(9), 801–813. DOI: 10.1016/j.jplph.2013.01.001 PMID: 23384758

Singh, S. K., Dixit, K., & Sundaram, S. (2014). Algal-based CO 2 Sequestration Technology and Global Scenario of Carbon Credit Market: A Review. [AJER]. *American Journal of Engineering Research*, *03*, 35–37. www.ajer.org. DOI: 10.7603/s40632-015-0035-2

Singh, S. K., Sundaram, S., Sinha, S., Rahman, M. A., & Kapur, S. (2016). Recent Advances in CO2 Uptake and Fixation Mechanism of Cyanobacteria and Microalgae. *Critical Reviews in Environmental Science and Technology*, *46*(16), 1297–1323. DOI: 10.1080/10643389.2016.1217911

Singh, U. B., & Ahluwalia, A. S. (2013). Microalgae: A promising tool for carbon sequestration. *Mitigation and Adaptation Strategies for Global Change*, *18*(1), 73–95. DOI: 10.1007/s11027-012-9393-3

Singh, V., Aishwarya, V. M., Sriprasath, V. J., Pranavi, M., & Singh, T. (2025). Green hydrogen value chain challenges and global readiness for a sustainable energy future. *iScience*, *28*(7). Advance online publication. DOI: 10.1016/j.isci.2025.112900

Singla, P. (2021). Plant synthetic biology: A paradigm shift targeting stress mitigation, reduction of ecological footprints and sustainable transformation in agriculture. In *Plant Stress Biology: Strategies and Trends* (pp. 435–489). Springer Singapore.

Sivaramakrishnan, R., Kumar, S., Hosseinzadeh, H., & Balan, V. (2026). Harnessing microalgae from wastewater for biocrude via hydrothermal Liquefaction: A sustainable pathway to biofuels and biochemicals. *Renewable & Sustainable Energy Reviews*, *231*, 116759. DOI: 10.1016/j.rser.2026.116759

Slade, R., & Bauen, A. (2013). Micro-algae cultivation for biofuels: Cost, energy balance, environmental impacts and future prospects. *Biomass and Bioenergy*, *53*, 29–38. DOI: 10.1016/j.biombioe.2012.12.019

Smith, A. M., & Stitt, M. (2007). Coordination of carbon supply and plant growth. *Plant, Cell & Environment*, *30*(9), 1126–1149. DOI: 10.1111/j.1365-3040.2007.01708.x PMID: 17661751

Smitha, M. P., Suji, D., Shanthi, M., & Adesina, A. (2022). Application of bacterial biomass in biocementation process to enhance the mechanical and durability properties of concrete. *Cleaner Materials*, *3*, 100050. DOI: 10.1016/j.clema.2022.100050

Soda, P. R. K., Mogal, A., Chakravarthy, K., Thota, N., Bandaru, N., Shukla, S. K., & Mini, K. M. (2024). Performance assessment of sustainable biocement mortar incorporated with bacteria-encapsulated cement-coated alginate beads. *Construction & Building Materials*, *411*, 134198. DOI: 10.1016/j.conbuildmat.2023.134198

Solomon, W., Mutum, L., Janda, T., & Molnár, Z. (2023). Potential benefit of microalgae and their interaction with bacteria to sustainable crop production. *Plant Growth Regulation*, *101*(1), 53–65. DOI: 10.1007/s10725-023-01019-8

Song, S., Ginige, M. P., Cheng, K. Y., Peacock, C. S., & Kaksonen, A. H. (2024). Ultrasonication-induced metabolic pathway shifts and reduced electron carrier washout with biomass enhanced hydrogen yield in a continuous stirred tank reactor. *Chemical Engineering Journal*, *493*, 152594. DOI: 10.1016/j.cej.2024.152594

Song, Y., Hu, Z., Yang, X., An, Y., & Lu, Y. (2025). Synergistic microalgae-duckweed systems for enhanced aquaculture wastewater treatment, biomass recovery, and CO2 sequestration: A novel approach for sustainable resource recovery. *Environmental Research*, *274*, 121271. DOI: 10.1016/j.envres.2025.121271 PMID: 40043930

Soon, C. F., Jamia'an, M., Mohamed Sunar, N., Arifin, S. N. H., Tay, K. G., Heng, C., See, C. H., Mat Nayan, N. H., & Tee, K. S. (2025). Smart sensing and anomaly detection for microalgae culture based on LoRaWAN sensors and LSTM autoencoder. *Aquaculture International*, *33*(6), 438. DOI: 10.1007/s10499-025-02104-7

Sosnowski, P., Wieczorek, A., & Ledakowicz, S. (2003). Anaerobic co-digestion of sewage sludge and organic fraction of municipal solid wastes. *Advances in Environmental Research*, *7*(3), 609–616. DOI: 10.1016/S1093-0191(02)00049-7

Sovacool, B. K. (2023). Expanding carbon removal to the Global South: Thematic concerns on systems, justice, and climate governance. *Energy and Climate Change*, *4*, 100103. DOI: 10.1016/j.egycc.2023.100103

Spaulding, S. A., Potapova, M. G., Bishop, I. W., Lee, S. S., Gasperak, T. S., Jovanoska, E., Furey, P. C., & Edlund, M. B. (2021). Diatoms.org: Supporting taxonomists, connecting communities. *Diatom Research*, *36*(4), 291–304. DOI: 10.1080/0269249X.2021.2006790 PMID: 35958044

Sridevi, V., Hamzah, H. T., Terapalli, A., Varma, J. V., Rao, C. S., & Ramesh, P. (2026). Net-Zero Emission and Microalgae Biotechnology Carbon Dioxide Sequestration. *Microalgae Biotechnology and Sustainable Development*, 131-155.

Srimongkol, P., Sangtanoo, P., Songserm, P., Watsuntorn, W., & Karnchanatat, A. (2022). Microalgae-Based Wastewater Treatment for Developing Economic and Environmental Sustainability: Current Status and Future Prospects. *Frontiers in Bioengineering and Biotechnology*, *10*, 904046. DOI: 10.3389/fbioe.2022.904046 PMID: 36159694

Srinivas, M. K., Alengaram, U. J., Ibrahim, S., Vello, V., & Phang, S. M. (2024). Feasibility study on the use of microalgae as an external crack healing agent for cement mortar rehabilitation. *Journal of Sustainable Cement-Based Materials*, *13*(1), 17–32. DOI: 10.1080/21650373.2023.2249889

Stadnichuk, I. N., Krasilnikov, P. M., & Zlenko, D. V. (2015). Cyanobacterial phycobilisomes and phycobiliproteins. *Microbiology*, *84*(2), 101–111. DOI: 10.1134/S0026261715020150 PMID: 26263619

Stephens, E., Ross, I. L., King, Z., Mussgnug, J. H., Kruse, O., Posten, C., Borowitzka, M. A., & Hankamer, B. (2010). An economic and technical evaluation of microalgal biofuels. *Nature Biotechnology*, *28*(2), 126–128. DOI: 10.1038/nbt0210-126 PMID: 20139944

Stephens, S., Mahadevan, R., & Allen, D. G. (2021). Engineering photosynthetic bioprocesses for sustainable chemical production: A review. *Frontiers in Bioengineering and Biotechnology*, *8*, 610723. DOI: 10.3389/fbioe.2020.610723 PMID: 33490053

Stiles, W. A., Styles, D., Chapman, S. P., Esteves, S., Bywater, A., Melville, L., Silkina, A., Lupatsch, I., Grünewald, C. F., & Lovitt, R. (2018). Using microalgae in the circular economy to valorise anaerobic digestate: Challenges and opportunities. *Bioresource Technology*, *267*, 732–742. DOI: 10.1016/j.biortech.2018.07.100 PMID: 30076074

Storrs, K., Lyhne, I., & Drustrup, R. (2023). A comprehensive framework for feasibility of CCUS deployment: A meta-review of literature on factors impacting CCUS deployment. *International Journal of Greenhouse Gas Control*, *125*, 103878. DOI: 10.1016/j.ijggc.2023.103878

Subhalakshmi, U., Saranraj, P., Kanshiya, S., Vithiya, G., Lokeshwari, B., & Manigandan, M. (2026). Diagnostic challenges of Parkinson disease in under-resourced settings. In *Evolution and innovation in Parkinson's disease therapies in under-resourced settings* (pp. 37–64). IGI Global Scientific Publishing. DOI: 10.4018/979-8-3373-7437-6.ch002

Subhash, G. V., Rajvanshi, M., Kumar, G. R. K., Sagaram, U. S., Prasad, V., Govindachary, S., & Dasgupta, S. (2022). Challenges in microalgal biofuel production: A perspective on techno economic feasibility under biorefinery stratagem. *Bioresource Technology*, *343*, 126155. DOI: 10.1016/j.biortech.2021.126155 PMID: 34673195

Sukačová, K., Vícha, D., & Dušek, J. (2020). Perspectives on microalgal biofilm systems with respect to integration into wastewater treatment technologies and phosphorus scarcity. *Water (Basel)*, *12*(8), 2245. DOI: 10.3390/w12082245

Sumathi, Y., Dong, C., Singhania, R. R., Chen, C., Baskar, G., & Patel, A. K. (2024). Advancements in Nano-Enhanced microalgae bioprocessing. *Bioresource Technology*, *401*, 130749. DOI: 10.1016/j.biortech.2024.130749 PMID: 38679239

Sun, X., Dregulo, A. M., Zhenni, S., Karthikeyan, O. P., Harirchi, S., Salama, E. S., Li, Y., Sindhu, R., Binod, P., Pandey, A., & Awasthi, M. K. (2023). Role of Marine Algae for GHG Reduction/CO2 Sequestration: Strategies and Applications. *Sustainable Marine Food and Feed Production Technologies*, 157–166. https://doi.org/DOI: 10.1201/9781003326946-12/ROLE-MARINE-ALGAE-GHG-REDUCTION-CO2-SEQUESTRATION-XINWEI-SUN-ANDREI-MIKHAILOVICH-DREGULO-SU-ZHENNI-OBULISAMY-PARTHIBA-KARTHIKEYAN-EL-SAYED-SALAMA-YUE-LI-RAVEENDRAN-SINDHU-PARAMESWARAN-BINOD-ASHOK-PANDEY-MUKESH-KUMAR-AWASTHI

Sung, C. H., Hao, T., Fang, H., Nguyen, A. T., Perricone, V., Yu, H., Huang, W., Sarmiento, E., Ornelas, A. F. D., Lublin, D., Wehling, R., Farajollahi, S., Arakaki, A., Nepal, D., Lord, N. P., & Kisailus, D. (2025). Biological and Biologically Inspired Functional Nanostructures: Insights into Structural, Optical, Thermal, and Sensing Applications. *Advanced Materials*, *37*(51), e09281. DOI: 10.1002/adma.202509281 PMID: 40981639

Sun, H., Zhao, W., Mao, X., Li, Y., Wu, T., & Chen, F. (2018). High-value biomass from microalgae production platforms: Strategies and progress based on carbon metabolism and energy conversion. *Biotechnology for Biofuels*, *11*(1), 227. DOI: 10.1186/s13068-018-1225-6 PMID: 30151055

Sun, T. H., Cheng, Q. L., Yuan, Y. H., Wen, K. W., Zhi, G. Z., & Lu, S. (2010). Coordinated Regulation of Gene Expression for Carotenoid Metabolism in Chlamydomonas Reinhardtii. *Journal of Integrative Plant Biology*, *52*(10), 868–878. DOI: 10.1111/j.1744-7909.2010.00993.x PMID: 20883439

Sun, X., Zhang, M., Liu, J., Hui, G., Chen, X., & Feng, C. (2024). The art of exploring diatom biosilica biomaterials: From biofabrication perspective. *Advanced Science (Weinheim, Baden-Württemberg, Germany)*, *11*(6), 2304695. DOI: 10.1002/advs.202304695 PMID: 38044309

Sun, Z., Bo, C., Cao, S., & Sun, L. (2025). Enhancing CO(2) Fixation in Microalgal Systems: Mechanistic Insights and Bioreactor Strategies. *Marine Drugs*, *23*(3), 113. Advance online publication. DOI: 10.3390/md23030113 PMID: 40137299

Suparmaniam, U., Lam, M. K., Lim, J. W., Yusup, S., Tan, I. S., Lau, S. Y., Kodgire, P., & Kachhwaha, S. S. (2023). Influence of Environmental Stress on Microalgae Growth and Lipid Profile: A Systematic Review. *Phytochemistry Reviews*, *22*(4), 879–901. DOI: 10.1007/s11101-022-09810-7

Swami, S., Lallawmkimi, M. C., Singh, V., Yadav, K. K., Alagendran, S., Koushal, S., & Lairenjam, G. (2025). Harnessing edaphic microalgae for soil carbon sequestration and climate-smart agriculture. *Archives of Current Research International*, *25*(3), 155–171. DOI: 10.9734/acri/2025/v25i31105

Tabassum, H., Sailaja, A., Afreen, H., & Wani, M. (2023). Sewage Treatment Using Nanoparticles. In *IntechOpen eBooks*. IntechOpen., DOI: 10.5772/intechopen.109407

Tamagnini, P., Axelsson, R., Lindberg, P., Oxelfelt, F., Wunschiers, R., & Lindblad, P. (2002). Hydrogenases and hydrogen metabolism of cyanobacteria. *FEMS Microbiology Reviews*, *26*(1), 67–81.

Tamburro, J., & Boyle, N. R. (2025). Charting the state of GEMs in microalgae: Progress, challenges, and innovations. *Frontiers in Plant Science*, *16*, 1614397. DOI: 10.3389/fpls.2025.1614397 PMID: 40584862

Tamilarasan, K., Kavitha, S., Banu, J. R., Arulazhagan, P., & Yeom, I. T. (2017). Energy-efficient methane production from macroalgal biomass through chemo-disperser liquefaction. *Bioresource Technology*, *228*, 156–163. DOI: 10.1016/j.biortech.2016.12.102 PMID: 28064132

Tan, Kenneth Wei Min, Huixin Lin, Hui Shen, and Yuan Kun Lee. 2016. "Nitrogen-Induced Metabolic Changes and Molecular Determinants of Carbon Allocation in Dunaliella Tertiolecta." *Scientific Reports 2016 6:1* 6(1):1–13. .DOI: 10.1038/srep37235

Tanaka, T., Maeda, Y., Suhaimi, N., Tsuneoka, C., Nonoyama, T., Yoshino, T., Kato, N., & Lauersen, K. J. (2021). Intron-mediated enhancement of transgene expression in the oleaginous diatom Fistulifera solaris towards bisabolene production. *Algal Research*, *57*, 102345. DOI: 10.1016/j.algal.2021.102345

Tang, D. Y. Y., Khoo, K. S., & Chew, K. W.. (2020). CO_2 biofixation by microalgae and its potential contribution to climate change mitigation. *Bioresource Technology*, *299*, 122635.

Tang, D. Y. Y., Khoo, K. S., Chew, K. W., Tao, Y., Ho, S. H., & Show, P. L. (2020). Potential utilization of bioproducts from microalgae for the quality enhancement of natural products. *Bioresource Technology*, *304*, 122997. DOI: 10.1016/j.biortech.2020.122997 PMID: 32094007

Tang, D., Li, X., Zhang, L., Xiao, P., Nie, Y., Qiu, F., Cheng, Z., Li, W., & Zhao, Y. (2025). Reactive Oxygen Species-Mediated Signal Transduction and Utilization Strategies in Microalgae. *Bioresource Technology*, *418*, 132004. DOI: 10.1016/j.biortech.2024.132004 PMID: 39710205

Tang, T., Chen, Y., Liu, M., Du, Y., & Tan, Y. (2022). Effect of pH on the performance of hydrogen production by dark fermentation coupled denitrification. *Environmental Research*, *208*, 112663. DOI: 10.1016/j.envres.2021.112663

Tao, F., Xiao, B., Qi, Q., Cheng, J., & Ji, P. (2022). Digital twin modeling. *Journal of Manufacturing Systems*, *64*, 372–389. DOI: 10.1016/j.jmsy.2022.06.015

Tarafdar, A., Sowmya, G., Yogeshwari, K., Rattu, G., Negi, T., Awasthi, M. K., Hoang, A. T., & Sirohi, R. (2023). Environmental pollution mitigation through utilization of carbon dioxide by microalgae. *Environmental Pollution*, *328*, 121623. DOI: 10.1016/j.envpol.2023.121623 PMID: 37072107

Tavakoli, H., & Barkdoll, B. D. (2020). Blended Lifecycle Integrated Social System Method. *International Journal of Environmental Research 2020 14:6, 14*(6), 727–749. DOI: 10.1007/s41742-020-00284-z

Tazon, A. W., Mérindol, N., Fantino, E., Bouhadada, A., Awwad, F., Meddeb-Mouelhi, F., & Desgagné-Penix, I. (2026). Engineering bidirectional chloroplast promoters for tunable co-expression of multiple genes in microalgae (Chlamydomonas reinhardtii). *Communications Biology, 9*(1), 202. Advance online publication. DOI: 10.1038/s42003-025-09478-7 PMID: 41484447

Terjanika, V., Pubule, J., Blumberga, D., & Zarins, E. (2023). Policy instruments for CO2 valorisation support. *E-Prime - Advances in Electrical Engineering. Electronics and Energy, 4*, 100181. DOI: 10.1016/j.prime.2023.100181

Terracciano, M., De Stefano, L., & Rea, I. (2018). Diatoms Green Nanotechnolcgy for Biosilica-Based Drug Delivery Systems. *Pharmaceutics, 10*(4), 242. Advance online publication. DOI: 10.3390/pharmaceutics10040242 PMID: 30463290

Thajuddin, F., Rasheed, A. S., Elumalai, A., Palanivel, P., Nooruddin, T., & Dharumadurai, D. (2026). Harnessing microalgae and cyanobacteria for sustainable bioenergy Production: Integration of molecular engineering and AI-driven approaches. *Renewable Energy, 256*, 123961. DOI: 10.1016/j.renene.2025.123961

Thajuddin, N., & Subramanian, G. (2005). Cyanobacterial biodiversity and potential applications in biotechnology. *Current Science, 89*(1), 11.

Thakur, N., Nigam, M., Mann, N. A., Gupta, S., Hussain, C. M., Shukla, S. K., Shah, A. A., Casini, R., Elansary, H. O., & Khan, S. A. (2023). Host-mediated gene engineering and microbiome-based technology optimization for sustainable agriculture and environment. *Functional & Integrative Genomics, 23*(1), 57. DOI: 10.1007/s10142-023-00982-9 PMID: 36752963

The, S. V., Snyder, R., & Tegeder, M. (2021). Targeting nitrogen metabolism and transport processes to improve plant nitrogen use efficiency. *Frontiers in Plant Science, 11*, 628366. DOI: 10.3389/fpls.2020.628366 PMID: 33732269

Thiviyanathan, V. A., Lim, H. R., Tham, P. E., & Show, P. L. (2022). How far has the development for industrial Internet of Things (IoT) in microalgae? In *Microalgae for Environmental Biotechnology* (pp. 145–174). CRC Press. DOI: 10.1201/9781003202196-4

Thomas, D. M., Mechery, J., & Paulose, S. V. (2016). Carbon dioxide capture strategies from flue gas using microalgae: A review. *Environmental Science and Pollution Research International, 23*(17), 16926–16940. DOI: 10.1007/s11356-016-7158-3

Thoré, E. S., Muylaert, K., Bertram, M. G., & Brodin, T. (2023). Microalgae. *Current Biology : CB*, *33*(3), R91–R95. DOI: 10.1016/j.cub.2022.12.032 PMID: 36750029

Tien, C.-J., & Chen, C. S. (2013). Patterns of metal accumulation by natural river biofilms during their growth and seasonal succession. *Archives of Environmental Contamination and Toxicology*, *64*(4), 605–616. DOI: 10.1007/s00244-012-9856-2 PMID: 23247559

Titirici, M. M. (2013). Green carbon. *Sustainable Carbon Materials from Hydrothermal Processes*, 1-36.

Tiwari, A., Marella, T. K., & Saxena, A. (2022). Diatom biorefinery: from carbon mitigation to high-value products. In *Biomass, biofuels, biochemicals* (pp. 401–420). Elsevier. DOI: 10.1016/B978-0-12-823500-3.00018-2

Tiwari, A., Saini, G. K., Parikh, H. K. R., & Saxena, A. (2026). Diatom-Derived Biochemicals: An In-Depth Analysis of Polysaccharides, Extraction Methodologies, and Diverse Applications. *Applied Biochemistry and Biotechnology*, *198*(1), 1–28. DOI: 10.1007/s12010-025-05414-x PMID: 41182542

Tiwari, M. D., Dhengare, S. W., Yenurkar, G. K., Bhagat, R. M., Pande, P. B., & S, A. (2026). Machine learning and adaptive control for optimizing MSW-algal biomass anaerobic co-digestion. *Scientific Reports*, *16*(1), 17884. DOI: 10.1038/s41598-026-48288-7 PMID: 41998065

Tonelli, F. M. P., Silva, C. S., Delgado, V. M. S., Tonelli, F. C. P., Tonelli, F. C. P., & Tonelli, F. C. P. (2023). Algae-based green AgNPs, AuNPs, and FeNPs as potential nanoremediators. *Green Processing and Synthesis*, *12*(1), 20230008. Advance online publication. DOI: 10.1515/gps-2023-0008

Torres-Momber, S., Morales-Rivera, F. N., Jacuinde-Ruíz, J. C., & González-Hernández, J. C. (2026). Extracellular Enzyme Activity and Growth of Chlorella sorokiniana in Response to CO2, Light Color, and Light Intensity in a Photobioreactor. *Current Microbiology*, *83*(2), 116. DOI: 10.1007/s00284-025-04710-3 PMID: 41483157

Traller, J. C., Cokus, S. J., Lopez, D. A., Gaidarenko, O., Smith, S. R., McCrow, J. P., Gallaher, S. D., Podell, S., Thompson, M., Cook, O., Morselli, M., Jaroszewicz, A., Allen, E. E., Allen, A. E., Merchant, S. S., Pellegrini, M., & Hildebrand, M. (2016). Genome and methylome of the oleaginous diatom Cyclotella cryptica reveal genetic flexibility toward a high lipid phenotype. *Biotechnology for Biofuels*, *9*(1), 258. DOI: 10.1186/s13068-016-0670-3 PMID: 27933100

Tredici, M. R., Rodolfi, L., Biondi, N., Bassi, N., & Sampietro, G. (2016). Techno-economic analysis of microalgal biomass production in a 1-ha Green Wall Panel (GWP®) plant. *Algal Research, 19*, 253–263. DOI: 10.1016/j.algal.2016.09.005

Trickett, C. A., Helal, A., Al-Maythalony, B. A., Yamani, Z. H., Cordova, K. E., & Yaghi, O. M. (2017). The chemistry of metal–organic frameworks for CO2 capture, regeneration and conversion. *Nature Reviews. Materials, 2*(8), 1–16. DOI: 10.1038/natrevmats.2017.45

Tripathi, S., Choudhary, S., Meena, A., & Poluri, K. M. (2023). Carbon capture, storage, and usage with microalgae: a review. *Environmental Chemistry Letters 2023 21:4, 21*(4), 2085–2128. DOI: 10.1007/s10311-023-01609-y

Tripathi, D. M., & Tripathi, S. (2024). Microbial advancements in dark fermentative biohydrogen production: applications and innovations. In *Emerging trends and techniques in biofuel production from agricultural waste* (pp. 57–80). Springer Nature Singapore. DOI: 10.1007/978-981-99-8244-8_4

Tsouko, E., Pilafidis, S., Kourmentza, K., Gomes, H. I., Sarris, G., Koralli, P., ... & Sarris, D. (2024). A sustainable bioprocess to produce bacterial cellulose (BC) using waste streams from wine distilleries and the biodiesel industry: evaluation of BC for adsorption of phenolic compounds, dyes and metals. *Biotechnology for Biofuels and Bioproducts, 17*(1), 40.

Tuck, B., Watkin, E., Somers, A., & Machuca, L. (2022). A critical review of marine biofilms on metallic materials. *npj Materials degradation, 6*(1), 25.

Turmel, M., Pombert, J.-F., Kosanovic, C., Julian, A. T., Otis, C., & Lemieux, C. (2025). Comparative Genomics of Chloropicon primus and Chloropicon roscoffensis Provide Insights into the Evolutionary Dynamics and Ecological Success of These Tiny Green Algae in Marine Environments. *Genome Biology and Evolution, 17*(7), evaf140. Advance online publication. DOI: 10.1093/gbe/evaf140 PMID: 40644408

Udaypal, G. R. K., Mehariya, S., & Verma, P. (2024). Advances in microalgae-based carbon sequestration: Current status and future perspectives. *Environmental Research, 249*, 118397. 10.1016/j.envres.2024.118397

Udaypal, G., Goswami, R. K., Mehariya, S., & Verma, P. (2024). Advances in microalgae-based carbon sequestration: Current status and future perspectives. *Environmental Research, 249*, 118397. DOI: 10.1016/j.envres.2024.118397 PMID: 38309563

Uduman, N., Qi, Y., Danquah, M. K., Forde, G. M., & Hoadley, A. (2010). Dewatering of microalgal cultures: A major bottleneck to algae-based fuels. *Journal of Renewable and Sustainable Energy*, *2*(1), 012701. DOI: 10.1063/1.3294480

Ueno, Y., Otsuka, S., & Morimoto, M. (1998). Hydrogen production from industrial wastewater by anaerobic microflora in chemostat culture. *Biotechnology Letters*, *20*(3), 195–199.

Ugwu, C. U., Aoyagi, H., & Uchiyama, H. (2008). Photobioreactors for mass cultivation of algae. *Bioresource Technology*, *99*(10), 4021–4028. DOI: 10.1016/j.biortech.2007.01.046 PMID: 17379512

Ullah, F., Ali, S., Siraj, M., Akhtar, M. S., & Zaman, W. (2025). Plant Microbiomes Alleviate Abiotic Stress-Associated Damage in Crops and Enhance Climate-Resilient Agriculture. *Plants*, *14*(12), 1890. DOI: 10.3390/plants14121890 PMID: 40573878

Ummalyma, S. B., Maharajan, T., Sreelekshmy, B. R., & Eswaran, K. (2025). Carbon Sequestration by Microalgae and Engineering of Microalgal Cells by CRISPR Technology for Efficient CO2 Capture for Carbon Neutral Bioproducts. *ACS ES&T Water*, *5*(9), 4969–4984. DOI: 10.1021/acsestwater.5c00517

Ummalyma, S. B., Sirohi, R., Udayan, A., Yadav, P., Raj, A., Sim, S. J., & Pandey, A. (2023). Sustainable microalgal biomass production in food industry wastewater for low-cost biorefinery products: A review. *Phytochemistry Reviews*, *22*(4), 969–991. DOI: 10.1007/s11101-022-09814-3 PMID: 35431709

United Nations Department of Economic and Social Affairs. Population Division. World population prospects 2022: summary of results. In: UN DESA/POP/2022/TR/NO. 3. New York, NY, USA: Un

Usher, P. K., Ross, A. B., Camargo-Valero, M. A., Tomlin, A. S., & Gale, W. F. (2014). An overview of the potential environmental impacts of large-scale microalgae cultivation. *Biofuels*, *5*(3), 331–349. DOI: 10.1080/17597269.2014.913925

Usman, M., Saidu, M., Usman, G., Agboola, P. O., Onwuchekwa, C., Onwuchekwa, N., & Oniye, M. M. (2025). Eco-Friendly Nanomaterials for Remediation of Microplastics and Heavy Metals in Wetland Ecosystems. *Deleted Journal*, *20*(1), 54–86. DOI: 10.62154/ajesre.2025.020.01014

Uzair, B., Liaqat, A., Iqbal, H., Menaa, B., Razzaq, A., Thiripuranathar, G., Rana, N. F., & Menaa, F. (2020). Green and Cost-Effective Synthesis of Metallic Nanoparticles by Algae: Safe Methods for Translational Medicine. *Bioengineering (Basel, Switzerland)*, *7*(4), 129. DOI: 10.3390/bioengineering7040129 PMID: 33081248

Vaishnav, S., Saini, T., Chauhan, A., Gaur, G. K., Tiwari, R., Dutt, T., & Tarafdar, A. (2023). Livestock and Poultry Farm Wastewater Treatment and Its Valorization for Generating Value-Added Products: Recent Updates and Way Forward. *Bioresource Technology*, *382*, 129170. DOI: 10.1016/j.biortech.2023.129170 PMID: 37196748

van den Berg, N. I., Machado, D., Santos, S., Rocha, I., Chacón, J., Harcombe, W., Mitri, S., & Patil, K. R. (2022). Ecological modelling approaches for predicting emergent properties in microbial communities. *Nature Ecology & Evolution*, *6*(7), 855–865. DOI: 10.1038/s41559-022-01746-7 PMID: 35577982

van den Burg, S., Koch, S., Raoult, J., Tsvetkov, B., & Selnes, T. (2023). Time to Include Blue Carbon from Seaweed in Voluntary Carbon Removal Certification. *EuroChoices (Uckfield)*, *22*(2), 13–19. DOI: 10.1111/1746-692X.12398

Vanderklift, M. A., Herr, D., Lovelock, C. E., Murdiyarso, D., Raw, J. L., & Steven, A. D. L. (2022). A Guide to International Climate Mitigation Policy and Finance Frameworks Relevant to the Protection and Restoration of Blue Carbon Ecosystems. *Frontiers in Marine Science*, *9*, 872064. DOI: 10.3389/fmars.2022.872064

Vandermeulen, H., & Gordin, H. (1990). Ammonium uptake using *Ulva* (Chlorophyta) in intensive fishpond systems: Mass culture and treatment of effluent. *Journal of Applied Phycology*, *2*(4), 363–374. DOI: 10.1007/BF02180927

Vavitsas, K., Crozet, P., Vinde, M. H., Davies, F., Lemaire, S. D., & Vickers, C. E. (2019). The synthetic biology toolkit for photosynthetic microorganisms. *Plant Physiology*, *181*(1), 14–27. DOI: 10.1104/pp.19.00345 PMID: 31262955

Vázquez-Romero, B., Villar-Navarro, E., Perales, J. A., Garrido-Pérez, C., & Ruiz, J. (2024). Techno-Economic Analysis of Using Microalgae to Treat Streams from Fish RAS Farming and Replace Fish Meal: A Case Study. *Journal of Water Process Engineering*, *59*, 104904. Advance online publication. DOI: 10.1016/j.jwpe.2024.104904

Veale, C. G. L., Woodland, J. G., Wicht, K. J., & Chibale, K. (2025). Chemistry as a Catalyst for Transforming the Health and Wealth of South Africa. *Angewandte Chemie*, *137*(12), e202419681. Advance online publication. DOI: 10.1002/ange.202419681 PMID: 39887802

Veeramuthu, A., & Ngamcharussrivichai, C. (2020). Potential of microalgal biodiesel: challenges and applications. *Renewable Energy-Technologies and Applications*, 167-180.

Velásquez-Orta, S. B., Yáñez-Noguez, I., Ramírez, I. M., & Ledesma, M. T. O. (2024). Pilot-scale microalgae cultivation and wastewater treatment using high-rate ponds: A meta-analysis. *Environmental Science and Pollution Research International, 31*(34), 46994–47021. DOI: 10.1007/s11356-024-34000-7 PMID: 38985422

Venkatesan, L. S., & Sathishkumar, P. (2024). Recent advances in the removal of emerging pollutant tetracycline using green nanomaterials: An eco-friendly mitigation approach. *International Journal of Environmental Science and Technology, 22*(6), 5129–5142. DOI: 10.1007/s13762-024-06163-w

Venkatrayalu, G., Choudhary, S., & Poluri, K. M. (2026). AI-Assisted Microalgal Bioprocessing: Innovations, Challenges, and Future Prospects in Sustainable Manufacturing. *Industrial Microbiology and Biotechnology: Trends in Microbial Biotechnology*, 269-307.

Verma, K., Dogra, R., & Rathour, N. S. (2025). Ecosystem dynamics: Exploring types, components and the forces shaping their transformation. *Journal of Geography. Environment and Earth Science International, 29*(4), 177–191.

Vidhyavathi, R., Venkatachalam, L., Sarada, R., & Ravishankar, G. A. (2008). Regulation of Carotenoid Biosynthetic Genes Expression and Carotenoid Accumulation in the Green Alga Haematococcus Pluvialis under Nutrient Stress Conditions. *Journal of Experimental Botany, 59*(6), 1409–1418. DOI: 10.1093/jxb/ern048 PMID: 18343887

Vishwakarma, R., Selvaraju, K., & Malik, A. (2025). Critical insights into the dynamics of carbon capture for enhancement of photosynthetic efficiency in microalgae. *Environmental Science and Pollution Research International, 32*(34), 20787–20797. DOI: 10.1007/s11356-025-36837-y PMID: 40813513

Viswanaathan, S., Perumal, P. K., & Sundaram, S. (2022). Integrated approach for carbon sequestration and wastewater treatment using algal–bacterial consortia: Opportunities and challenges. *Sustainability (Basel), 14*(3), 1075. DOI: 10.3390/su14031075

Walsh, M. J., Gerber Van Doren, L., Sills, D. L., Archibald, I., Beal, C. M., Lei, X. G., Huntley, M. E., Johnson, Z., & Greene, C. H. (2016). Algal food and fuel coproduction can mitigate greenhouse gas emissions while improving land and water-use efficiency. *Environmental Research Letters, 11*(11), 114006. DOI: 10.1088/1748-9326/11/11/114006

Wang, C., Shi, H. D., You, N., Bai, Y., Jin, P., & Han, J. (2023). Pathways of Wastewater Treatment for Resource Recovery and Energy Minimization towards Carbon Neutrality and Circular Economy: Technological Opinions. *Frontiers in Environmental Chemistry*, *4*, 4. DOI: 10.3389/fenvc.2023.1255092

Wang, J. K., & Seibert, M. (2017). Prospects for commercial production of diatoms. *Biotechnology for Biofuels*, *10*(1), 16. DOI: 10.1186/s13068-017-0699-y PMID: 28115988

Wang, J., Li, J., & Wang, X.. (2022). Microalgae-based carbon capture, utilization and storage (CCUS): A critical review. *Journal of Cleaner Production*, *344*, 131068.

Wang, J., & Yin, Y. (2021). Clostridium species for fermentative hydrogen production: An overview. *International Journal of Hydrogen Energy*, *46*(70), 34599–34625. DOI: 10.1016/j.ijhydene.2021.08.052

Wang, L., Ma, X., Huang, X., Wen, H., Jin, W., Chen, W., Wang, Y., Xu, P., Yang, H., & Zhang, Y. (2026). Inhibition of photosystems and respiratory complexes enhances lipid biosynthesis via ROS in Chlorella pyrenoidosa SHOU-1002 under nitrogen limitation. *Planta*, *263*(2), 37. DOI: 10.1007/s00425-025-04909-5 PMID: 41483224

Wang, L., Sun, Y., Zhang, R., Pan, K., Li, Y., Wang, R., Zhang, L., Zhou, C., Li, J., Li, Y., Zhu, B., & Han, J. (2023). Enhancement of hemostatic properties of Cyclotella cryptica frustule through genetic manipulation. *Biotechnology for Biofuels and Bioproducts*, *16*(1), 136. DOI: 10.1186/s13068-023-02389-x PMID: 37710352

Wang, L., Yang, T., Pan, Y., Shi, L., Jin, Y., & Huang, X. (2023). The Metabolism of Reactive Oxygen Species and Their Effects on Lipid Biosynthesis of Microalgae. *International Journal of Molecular Sciences*, *24*(13), 11041. DOI: 10.3390/ijms241311041 PMID: 37446218

Wang, L., & Zhang, B. (2022). Cultivation of microalgae on agricultural wastewater for recycling energy, water, and fertilizer nutrients. In *Integrated Wastewater Management and Valorization Using Algal Cultures* (pp. 235–264). Elsevier. DOI: 10.1016/B978-0-323-85859-5.00006-3

Wang, L., Zhang, F., Shi, X., Zeng, C., Ahmad, I., Wang, G., Thapa, S., & Xu, X. (2023). Water resources system vulnerability in high mountain areas under climate change. *Journal of Cleaner Production*, *403*, 136789. DOI: 10.1016/j.jclepro.2023.136789

Wang, R., Wang, X., & Zhu, T. (2025). Research progress and application of carbon sequestration in industrial flue gas by microalgae: A review. *Journal of Environmental Sciences (China)*, *152*, 14–28. DOI: 10.1016/j.jes.2024.04.018 PMID: 39617540

Wang, S., Mukhambet, Y., Esakkimuthu, S., & Abomohra, A. E. L.-F. (2022). Integrated Microalgal Biorefinery – Routes, Energy, Economic and Environmental Perspectives. *Journal of Cleaner Production*, *348*, 131245. DOI: 10.1016/j.jclepro.2022.131245

Wani, N. R., Rather, R. A., Farooq, A., Padder, S. A., Baba, T. R., Sharma, S., Mubarak, N. M., Khan, A. H., Singh, P., & Ara, S. (2024). New insights in food security and environmental sustainability through waste food management. *Environmental Science and Pollution Research International*, *31*(12), 17835–17857. DOI: 10.1007/s11356-023-26462-y

Ward, A. J., Lewis, D. M., & Green, F. B. (2014). Anaerobic digestion of algae biomass: A review. *Algal Research*, *5*, 204–214. DOI: 10.1016/j.algal.2014.02.001

Webster, L. J., Villa-Gomez, D., Brown, R., Clarke, W., & Schenk, P. M. (2024). A synthetic biology approach for the treatment of pollutants with microalgae. *Frontiers in Bioengineering and Biotechnology*, *12*, 1379301. DOI: 10.3389/fbioe.2024.1379301 PMID: 38646010

Wei, H., Shi, Y., Ma, X., Pan, Y., Hu, H., Li, Y., Luo, M., Gerken, H., & Liu, J. (2017). A Type-I Diacylglycerol Acyltransferase Modulates Triacylglycerol Biosynthesis and Fatty Acid Composition in the Oleaginous Microalga, Nannochloropsis Oceanica. *Biotechnology for Biofuels*, *10*(1), 174. Advance online publication. DOI: 10.1186/s13068-017-0858-1 PMID: 28694845

Wetterwald, L., Leybros, A., Fleury, G., Delrue, F., Dimitriades-Lemaire, A., Chambonniere, P., & Hertz, A. (2023). Supercritical CO2 Extraction of Neutral Lipids from Dry and Wet Chlorella Vulgaris NIES 227 Microalgae for Biodiesel Production. *Journal of Environmental Chemical Engineering*, *11*(5), 110628. DOI: 10.1016/j.jece.2023.110628

Wicker, R. J. (2023). Bioprospecting for nordic photosynthetic consortia for wastewater treatment, carbon capture, and value creation.

Wijffels, R. H., & Barbosa, M. J. (2010). An outlook on microalgal biofuels. *Science*, *329*(5993), 796–799. DOI: 10.1126/science.1189003 PMID: 20705853

Wijffels, R. H., Kruse, O., & Hellingwerf, K. J. (2013). Potential of industrial biotechnology with cyanobacteria and eukaryotic microalgae. *Current Opinion in Biotechnology*, *24*(3), 405–413. DOI: 10.1016/j.copbio.2013.04.004 PMID: 23647970

Willis, L., Page, K. M., Broomhead, D. S., & Cox, E. J. (2010). Discrete free-boundary reaction-diffusion model of diatom pore occlusions. *Plant Ecology and Evolution*, *143*(3), 297–306. DOI: 10.5091/plecevo.2010.415

Wilson, M. H., Shea, A., Groppo, J., Crofcheck, C., Quiroz, D., Quinn, J. C., & Crocker, M. (2020). Algae-Based Beneficial Re-use of Carbon Emissions Using a Novel Photobioreactor: a Techno-Economic and Life Cycle Analysis. *BioEnergy Research 2020 14:1, 14*(1), 292–302. DOI: 10.1007/s12155-020-10178-9

Wong, P. Y., Mal, J., Sandak, A., Luo, L., Jian, J., & Pradhan, N. (2024). Advances in microbial self-healing concrete: A critical review of mechanisms, developments, and future directions. *The Science of the Total Environment, 947*, 174553. DOI: 10.1016/j.scitotenv.2024.174553 PMID: 38972424

Workman, M., Heap, R., Mackie, E., & Connon, I. (2023). Decision making for net zero policy design and climate action: considerations for improving translation at the research-policy interface: a UK Carbon Dioxide Removal case study. *Frontiers in Climate, 5*, 1288001. DOI: 10.3389/fclim.2023.1288001

Wu, G., Chong, J. W. R., Khoo, K. S., Tang, D. Y. Y., & Show, P. L. (2025). Upcycling food waste for microalgae cultivation toward lipid production in a closed-loop and system-integrated circular bioeconomy. *Biotechnology for Biofuels and Bioproducts, 18*(1), 74. DOI: 10.1186/s13068-025-02679-6 PMID: 40646635

Wu, G., Tham, P. E., Chew, K. W., Munawaroh, H. S. H., Tan, I. S., Wan, W. A. A. Q. I., Sriariyanun, M., & Show, P. L. (2023). Net zero emission in circular bioeconomy from microalgae biochar production: A renewed possibility. *Bioresource Technology, 388*, 129748. DOI: 10.1016/j.biortech.2023.129748 PMID: 37714493

Wu, P., Ding, P., Cao, Q. H., Zhang, C., Fu, B., Liu, H. B., Chen, C., & Liu, H. (2023). Amino acids as in-situ electron donors drive medium chain fatty acids production from sludge acidogenic fermentation liquid by electro-fermentation enhancement. *Chemical Engineering Journal, 476*, 146537. DOI: 10.1016/j.cej.2023.146537

Wykoff, D. D., Davies, J. P., Melis, A., & Grossman, A. R. (1998). The Regulation of Photosynthetic Electron Transport during Nutrient Deprivation in Chlamydomonas Reinhardtii. *Plant Physiology, 117*(1), 129–139. DOI: 10.1104/pp.117.1.129 PMID: 9576782

Wylie, L., Sutton-Grier, A. E., & Moore, A. (2016). Keys to successful blue carbon projects: Lessons learned from global case studies. *Marine Policy, 65*, 76–84. DOI: 10.1016/j.marpol.2015.12.020

Xiao, R., & Zheng, Y. (2016). Overview of Microalgal Extracellular Polymeric Substances (EPS) and Their Applications. *Biotechnology Advances, 34*(7), 1225–1244. DOI: 10.1016/j.biotechadv.2016.08.004 PMID: 27576096

Xiao, X., Li, W., Jin, M., Zhang, L., Qin, L., & Geng, W. (2023). Responses and Tolerance Mechanisms of Microalgae to Heavy Metal Stress: A Review. *Marine Environmental Research, 183*, 105805. DOI: 10.1016/j.marenvres.2022.105805 PMID: 36375224

Xia, W., & Shoemaker, C. (2021). GOPS: Efficient RBF surrogate global optimization algorithm with high dimensions and many parallel processors including application to multimodal water quality PDE model calibration. *Optimization and Engineering*, *22*(4), 2741–2777. DOI: 10.1007/s11081-020-09556-1

Xie, W., Aryanpur, V., Deane, P., & Daly, H. E. (2025). Negative emissions technologies in energy system models and mitigation scenarios - a systematic review. *Applied Energy*, *380*, 125064. DOI: 10.1016/j.apenergy.2024.125064

Xie, Y., Khoo, K. S., Chew, K. W., Devadas, V. V., Phang, S. J., Lim, H. R., Rajendran, S., & Show, P. L. (2022). Advancement of renewable energy technologies via artificial and microalgae photosynthesis. *Bioresource Technology*, *363*, 127830. DOI: 10.1016/j.biortech.2022.127830 PMID: 36029982

Xin, C., Addy, M. M., Zhao, J., Cheng, Y., Cheng, S., Mu, D., Liu, Y., Ding, R., Chen, P., & Ruan, R. (2016). Comprehensive Techno-Economic Analysis of Wastewater-Based Algal Biofuel Production: A Case Study. *Bioresource Technology*, *211*, 584–593. DOI: 10.1016/j.biortech.2016.03.102 PMID: 27039331

Xin, K., Cheng, J., Guo, R., Qian, L., Wu, Y., & Yang, W. (2024). Nuclear mutagenesis and adaptive evolution improved photoautotrophic growth of Euglena gracilis with flue-gas CO2 fixation. *Bioresource Technology*, *397*, 130497. DOI: 10.1016/j.biortech.2024.130497 PMID: 38408501

Xiong, J. Q., Kurade, M. B., & Jeon, B. H. (2017). Ecotoxicological Effects of Enrofloxacin and Its Removal by Monoculture of Microalgal Species and Their Consortium. *Environmental Pollution*, *226*, 486–493. DOI: 10.1016/j.envpol.2017.04.044 PMID: 28449968

Xiong, W., Cano, M., Wang, B., Douchi, D., & Yu, J. (2017). The plasticity of cyanobacterial carbon metabolism. *Current Opinion in Chemical Biology*, *41*, 12–19. DOI: 10.1016/j.cbpa.2017.09.004 PMID: 28968542

Xuan, J., He, L., Wen, W., & Feng, Y. (2023). Hydrogenase and nitrogenase: Key catalysts in biohydrogen production. *Molecules (Basel, Switzerland)*, *28*(3), 1392. DOI: 10.3390/molecules28031392

Xue, Q., Feng, X., Sun, Y., Si, Z., Zhang, L., Zhou, Z., Zhong, Y., Mei, K., Lin, G., & Zhao, H. (2025). A systematic review of concrete self-healing mechanism and analysis of new research directions. *Journal of Environmental Chemical Engineering, 13*(6), 119009. DOI: 10.1016/j.jece.2025.119009

Xu, H., Miao, X., & Wu, Q. (2006). High quality biodiesel production from a microalga *Chlorella protothecoides* by heterotrophic growth in fermenters. *Journal of Biotechnology, 126*(4), 499–507. DOI: 10.1016/j.jbiotec.2006.05.002 PMID: 16772097

Xu, L., Wim Brilman, D. W. F., Withag, J. A. M., Brem, G., & Kersten, S. (2011). Assessment of a Dry and a Wet Route for the Production of Biofuels from Microalgae: Energy Balance Analysis. *Bioresource Technology, 102*(8), 5113–5122. DOI: 10.1016/j.biortech.2011.01.066 PMID: 21334889

Xu, P., Shao, S., Qian, J., Li, J., Xu, R., Liu, J., & Zhou, W. (2024). Scale-up of microalgal systems for decarbonization and bioproducts: Challenges and opportunities. *Bioresource Technology, 398*, 130528. Advance online publication. DOI: 10.1016/j.biortech.2024.130528 PMID: 38437968

Xu, Z., Qi, H., Yao, D., Zhang, J., Zhu, Z., Wang, Y., & Cui, P. (2022). Modeling and comprehensive analysis of food waste gasification process for hydrogen production. *Energy Conversion and Management, 258*, 115509. DOI: 10.1016/j.enconman.2022.115509

Yaashikaa, P. R., Saravanan, A., Kumar, P. S., Thamarai, P., & Rangasamy, G. (2024). Role of microbial carbon capture cells in carbon sequestration and energy generation during wastewater treatment: A sustainable solution for cleaner environment. *International Journal of Hydrogen Energy, 52*, 799–820. DOI: 10.1016/j.ijhydene.2023.05.307

Yadav, C., Chu, K. H., Hassan, Z., Lee, J., & Jang, W. (2025). Nanocellulose sponges embedding metal oxide nanoparticles for adsorption and photodegradation of microplastics. *Chemosphere, 386*, 144614. DOI: 10.1016/j.chemosphere.2025.144614 PMID: 40779960

Yahaya, E., Yeo, W. S., Nandong, J., & Ngu, J. C. Y. (2025). CO2 fed microalgae cultivation in photobioreactor: Review on challenges and possible solutions. *Environmental Technology Reviews, 14*(1), 540–564. DOI: 10.1080/21622515.2025.2508945

Yamaoka, Y., Petroutsos, D., Je, S., Yamano, T., & Li-Beisson, Y. (2025). Light, CO2, and carbon storage in microalgae. *Current Opinion in Plant Biology, 84*, 102696. DOI: 10.1016/j.pbi.2025.102696 PMID: 39983365

Yang, R., Jin, X., Zhou, H., Ren, F., Zhang, X., Ma, Z., Yao, L., & Zhang, H. (2024). Regional Differences in Carbon Budgets and Inter-Regional Compensation Zoning: A Case Study of Chongqing, China. *Land 2024, Vol. 13, 13*(9). DOI: 10.3390/land13091495

Yang, C., Li, R., Zhang, B., Qiu, Q., Wang, B., Yang, H., Ding, Y., & Wang, C. (2019). Pyrolysis of Microalgae: A Critical Review. *Fuel Processing Technology, 186*, 53–72. DOI: 10.1016/j.fuproc.2018.12.012

Yang, H., & Xin, X. (2023). CO2 capture and lipid production performance of microalgae in the S-shaped photobioreactor under different culture modes. *Enzyme and Microbial Technology, 165*, 110194. DOI: 10.1016/j.enzmictec.2023.110194 PMID: 36682097

Yang, Q., Wu, J., & Li, G. (2025). Control of cell–cell communication and deciding when to grow. *Trends in Plant Science, 30*(6), 585–587. DOI: 10.1016/j.tplants.2025.01.003 PMID: 39893119

Yang, Y. F., Li, D. W., Balamurugan, S., Wang, X., Wei, D. Y., & Hong, Y. L. (2022). Chrysolaminarin Biosynthesis in the Diatom Is Enhanced by Overexpression of 1,6-β-Transglycosylase. *Algal Research, 66*, 102817. DOI: 10.1016/j.algal.2022.102817

Yang, Y., Tang, S., & Chen, J. P. (2024). Carbon capture and utilization by algae with high concentration CO(2) or bicarbonate as carbon source. *The Science of the Total Environment, 918*, 170325. DOI: 10.1016/j.scitotenv.2024.170325 PMID: 38278265

Yang, Z., Yan, H., Liu, H., Yang, L., Mi, G., & Wang, P. (2025). Enhancing Crop Nitrogen Efficiency: The Role of Mixed Nitrate and Ammonium Supply in Plant Growth and Development. *Biology (Basel), 14*(5), 546. DOI: 10.3390/biology14050546 PMID: 40427735

Yan, L., Xue, S., Cha, J., Wen, X., Wang, B., Shi, J., Li, P., Zhang, Y., & Xie, T. (2024). Effect of excess CO2 on semi-continuous microalgae systems: Carbon biofixation. *Environmental Technology & Innovation, 35*, 103645. DOI: 10.1016/j.eti.2024.103645

Yan, Z., Nakashima, K., Takano, C., & Kawasaki, S. (2025). Strategies for cost-optimized biocement production: A comprehensive review. *World Journal of Microbiology & Biotechnology, 41*(2), 67. DOI: 10.1007/s11274-025-04281-2 PMID: 39920409

Yaqoob, A. A., Ahmad, H., Parveen, T., Ahmad, A., Oves, M., Ismail, I. M., Qari, H. A., Umar, K., & Mohamad Ibrahim, M. N. (2020). Recent advances in metal decorated nanomaterials and their various biological applications: A review. *Frontiers in Chemistry*, *8*, 341. DOI: 10.3389/fchem.2020.00341 PMID: 32509720

Yaqoubnejad, P., Aminirad, H., & Taghavijeloudar, M. (2021). Optimization of Cultivation Condition of Newly Isolated Strain Chlorella Sorokiniana pa. 91 for CO2 Bio-Fixation and Nutrients Removal From Real Municipal Wastewater: Impact of Temperature and Light Intensity.

Yuan, W., Xu, E. G., Shabaka, S., Chen, P., & Yang, Y. (2024). The power of green: Harnessing phytoremediation to combat micro/nanoplastics. *Eco-Environment & Health*, *3*(3), 260–265. DOI: 10.1016/j.eehl.2024.04.001 PMID: 39234422

Yu, B. S., Pyo, S., Lee, J., & Han, K. (2024). Microalgae: A multifaceted catalyst for sustainable solutions in renewable energy, food security, and environmental management. *Microbial Cell Factories*, *23*(1), 308. DOI: 10.1186/s12934-024-02588-7 PMID: 39543605

Yukesh Kannah, R., Kavitha, S., Rajesh Banu, J., & Yeom, I. T. (2017). Synergetic effect of combined pretreatment for energy-efficient biogas production. *Bioresource Technology*, *232*, 235–246. DOI: 10.1016/j.biortech.2017.02.042 PMID: 28236758

Yusof, Z., Tong, Y. W., Selvarajoo, K., Parakh, S. K., & Foo, S. C. (2025). Overcoming challenges in microalgal bioprocessing through data-driven and computational approaches. *Current Opinion in Food Science*, *61*, 101253. DOI: 10.1016/j.cofs.2024.101253

Zachee, G., Kayiranga, A., Nizeyimana, J. C., Tian, S., Rugema, J., You, L., Huang, X., & Jian, Q. S. (2025). Removal of Antibiotics and Antibiotic Resistance Genes Using Microalgae-Based Wastewater Treatment System: A Bibliometric Review and Mechanism Analysis. *Journal of Water Process Engineering*, *72*, 107496. DOI: 10.1016/j.jwpe.2025.107496

Zammit, G., Zammit, M. G., & Buttigieg, K. G. (2023). Emerging technologies for the discovery of novel diversity in cyanobacteria and algae and the elucidation of their valuable metabolites. *Diversity*, *15*(11), 1142. DOI: 10.3390/d15111142

Zanoni, I., Amadori, S., Brigliadori, A., Costa, A. L., Ortelli, S., Giacò, P., Baldisseroto, C., Pancaldi, S., & Blosi, M. (2025). Hybrid materials for wastewater treatment: Synergistic coupling of Neochloris oleoabundans and TiO2 nanoparticles. *Nanoscale Advances*, *7*(12), 3803–3816. DOI: 10.1039/D5NA00236B PMID: 40395621

Zavala-Mendez, M., Vargas, A., & Carrillo-Reyes, J. (2022). Maximization of bio-hydrogen production from winery vinasses using on-line feedback control. *International Journal of Hydrogen Energy*, *47*(78), 33259–33271. DOI: 10.1016/j.ijhydene.2022.07.196

Zenda, T., Liu, S., Dong, A., & Duan, H. (2021). Revisiting sulphur—The once neglected nutrient: It's roles in plant growth, metabolism, stress tolerance and crop production. *Agriculture*, *11*(7), 626. DOI: 10.3390/agriculture11070626

Zeng, J., Wang, Z., & Chen, G. (2021). Biological characteristics of energy conversion in carbon fixation by microalgae. *Renewable & Sustainable Energy Reviews*, *152*, 111661. DOI: 10.1016/j.rser.2021.111661

Zhang, B. (2025). *Development and evaluation of a methanotroph-microalgae mixed-culture platform for integrated biogas utilization and valorisation* (Doctoral dissertation, Nanyang Technological University).

Zhang, X., An, L., Tian, J., Ji, B., Lu, J., & Liu, Y. J. B. T. (2023). Microalgal capture of carbon dioxide: A carbon sink or source?, *390*, 129824.

Zhang, B., Gong, M., & Dong, C. (2025). Microalgae photosynthetic carbon sequestration technology, financing subsidies and carbon trading mechanism: An evolutionary game mechanism modeling study from a biomechanical perspective. *Molecular & Cellular Biomechanics*, *22*(1), 1005–1005. DOI: 10.62617/mcb1005

Zhang, C. C., Zhou, C. Z., Burnap, R. L., & Peng, L. (2018). Carbon/nitrogen metabolic balance: Lessons from cyanobacteria. *Trends in Plant Science*, *23*(12), 1116–1130. DOI: 10.1016/j.tplants.2018.09.008 PMID: 30292707

Zhang, D., An, S., Yao, R., Fu, W., Han, Y., Du, M., Chen, Z., Lei, A., & Wang, J. (2022). Life cycle assessment of auto-tropically cultivated economic microalgae for final products such as food, total fatty acids, and bio-oil. *Frontiers in Marine Science*, *9*, 990635. DOI: 10.3389/fmars.2022.990635

Zhang, G. Z., Liu, Z. H., Zhang, J., Zhao, H., Xia, Y., Su, Y., Wang, H., & Wang, L. (2025). Algae-bacteria symbiosis for cracks active repair in cement mortar: A novel strategy to enhance microbial self-healing efficacy. *Construction & Building Materials*, *483*, 141815. DOI: 10.1016/j.conbuildmat.2025.141815

Zhang, J., Xue, D., Wang, C., Fang, D., Cao, L., & Gong, C. (2023). Genetic engineering for biohydrogen production from microalgae. *iScience*, *26*(8), 107255. DOI: 10.1016/j.isci.2023.107255

Zhang, M., Wang, Z., Zhang, X., Qian, X., & Shen, G. (2020). Biogas and quality fertilizer production from dry anaerobic digestion of rice straw with nitrogen addition. *Bioresource Technology Reports*, *11*, 100509. DOI: 10.1016/j.biteb.2020.100509

Zhang, S., & Liu, Z. (2021). Advances in the biological fixation of carbon dioxide by microalgae. *Journal of Chemical Technology and Biotechnology*, *96*(6), 1475–1495. DOI: 10.1002/jctb.6714

Zhang, W., Zhao, C., Cao, W., Sun, S., Hu, C., Liu, J., & Zhao, Y. (2020). Removal of pollutants from biogas slurry and CO2 capture in biogas by microalgae-based technology: A systematic review. *Environmental Science and Pollution Research International*, *27*(23), 28749–28767. DOI: 10.1007/s11356-020-09282-2 PMID: 32468373

Zhang, X., Lv, D., Li, B., Liu, Z., Yang, F., Tang, Q., Gao, Y., Zhao, H., Tan, Z., Du, P., Li, J., & Wang, J. (2025). Polydopamine-Functionalized Magnetic Algae Composite for Efficient Removal of Polystyrene Microplastics: Mechanistic Insights and Performance. *Colloids and Surfaces. A, Physicochemical and Engineering Aspects*, *137594*, 137594. Advance online publication. DOI: 10.1016/j.colsurfa.2025.137594

Zhang, Y. T., Jiang, J. Y., Shi, T. Q., Sun, X. M., Zhao, Q. Y., Huang, H., & Ren, L. J. (2019). Application of the CRISPR/Cas system for genome editing in microalgae. *Applied Microbiology and Biotechnology*, *103*(8), 3239–3248. DOI: 10.1007/s00253-019-09726-x PMID: 30877356

Zhao, S., Chen, H., Jia, C., Wang, Y., Xin, C., & Jiang, X. (2024). Comprehensive Evaluation and Scheme Optimization for Power Transmission and Transformation Projects with the Adoption of Carbon Reduction Technologies. *Energies 2024, Vol. 17*, *17*(3). DOI: 10.3390/en17030598

Zhao, B., & Su, Y. (2014). Process Effect of Microalgal-Carbon Dioxide Fixation and Biomass Production: A Review. *Renewable & Sustainable Energy Reviews*, *31*, 121–132. DOI: 10.1016/j.rser.2013.11.054

Zhao, B., & Su, Y. (2020). Macro assessment of microalgae-based CO2 sequestration: Environmental and energy effects. *Algal Research*, *51*, 102066. DOI: 10.1016/j.algal.2020.102066

Zhao, L., Xu, Y., Zhang, Y., Huang, Y., Huang, G., Hu, Z., Wu, C., & Tang, J. (2026). Performance and mechanisms of an algae-bacteria symbiotic system for treating low carbon-to-nitrogen ratio wastewater. *Biochemical Engineering Journal*, *226*, 109995. DOI: 10.1016/j.bej.2025.109995

Zhou, W., Wang, J., Chen, P., Ji, C., Kang, Q., Lu, B., Li, K., Liu, J., & Ruan, R. (2017). Bio-mitigation of carbon dioxide using microalgal systems: Advances and perspectives. *Renewable & Sustainable Energy Reviews*, *76*, 1163–1175. 10.1016/j.rser.2017.03.065. DOI: 10.1016/j.rser.2017.03.065

Zhuang, Y., Zhang, Y., Jin, H., Qi, D., & Chen, J. (2025). Unexpected ice-edge diatom blooms occur as thin phytoplankton layers in Barrow Canyon, Chukchi Sea. *Marine Environmental Research*, *206*, 107035. https://doi.org/https://doi.org/10.1016/j.marenvres.2025.107035. DOI: 10.1016/j.marenvres.2025.107035 PMID: 40024173

Zhu, J., Cheng, J., Zeng, J., Zhang, W., Liu, C., Effiong, K. S., & Hao, Q. (2026). A Review of Phytoplankton Sinking Rates: Mechanisms, Methodologies, and Biogeochemical Implications. *Biology (Basel)*, *15*(2), 130. DOI: 10.3390/biology15020130 PMID: 41594864

Zhu, J., Li, M., Yu, H., Zheng, Y., Yuan, L., Cao, Y., Liu, X., Sun, F., & Chen, C. (2024). Magnetic biochar enhanced microbial electrolysis cell with anaerobic digestion for complex organic matter degradation in landfill leachate. *The Science of the Total Environment*, *949*, 175013. DOI: 10.1016/j.scitotenv.2024.175013

Zhu, J., & Thompson, C. B. (2019). Metabolic regulation of cell growth and proliferation. *Nature Reviews. Molecular Cell Biology*, *20*(7), 436–450. DOI: 10.1038/s41580-019-0123-5 PMID: 30976106

Zhu, L. (2015). Biorefinery as a promising approach to promote microalgae industry: An innovative framework. *Renewable & Sustainable Energy Reviews*, *41*, 1376–1384. DOI: 10.1016/j.rser.2014.09.040

Zhu, T., Ning, P., Liu, Y., Liu, M., Yang, J., Wang, Z., & Li, M. (2025). Knowledge of microalgal Rubiscos helps to improve photosynthetic efficiency of crops. *Planta*, *261*(4), 78. DOI: 10.1007/s00425-025-04645-w PMID: 40042639

Zhu, X., Xiao, J., Wang, C., Zhu, L., & Wang, S. (2022). Global warming potential analysis of bio-jet fuel based on life cycle assessment. *Carbon Neutrality*, *1*(1), 25. DOI: 10.1007/s43979-022-00026-4

Zieliński, M., Dębowski, M., Kazimierowicz, J., & Świca, I. (2023). Microalgal carbon dioxide (CO2) capture and utilization from the European Union perspective. *Energies*, *16*(3), 1446. DOI: 10.3390/en16031446

Zimmerman, W. B., Tesar, V., Butler, S., & Bandulasena, H. C. (2008). Microbubble generation. *Recent Patents on Engineering*, *2*(1), 1–8. DOI: 10.2174/187221208783478598

Zobi, F. (2022, June). Diatom biosilica in targeted drug delivery and biosensing applications: Recent studies. In *Micro* (Vol. 2, No. 2, pp. 342-360). MDPI.

Zong, M., Wang, W., Ma, H., Cedrick, N., Sun, Y., Yan, X., Liu, H., Zhu, P., & Hua, M. (2025). Progress in Self-Repair Technology for Concrete Cracks via Biomineralization. *Materials (Basel), 18*(21), 5004. DOI: 10.3390/ma18215004 PMID: 41227962

Zuluaga-Astudillo, D., Ruge, J. C., & Caicedo-Hormaza, B. (2025). Micromechanical Characterization of Diatom Frustules of Multiple Origin. *Applied Sciences (Basel, Switzerland), 15*(2), 749. DOI: 10.3390/app15020749

Zuluaga-Astudillo, D., Ruge, J. C., Camacho-Tauta, J., Reyes-Ortiz, O., & Caicedo-Hormaza, B. (2022). Diatomaceous soils and advances in geotechnical engineering—Part I. *Applied Sciences (Basel, Switzerland), 13*(1), 549. DOI: 10.3390/app13010549

About the Contributors

Imran Ahmad is a Postdoctoral Researcher at the Institute for Water and Wastewater Technology (IWWT), Durban University of Technology (DUT), specializing in algae biotechnology and advanced wastewater treatment systems for environmental sustainability and circular bioeconomy applications. He holds an MPhil and PhD from Universiti Teknologi Malaysia, with focused expertise in biological wastewater treatment, microalgal bioremediation, nutrient recovery, and resource valorization. His research integrates fundamental science with engineering innovation, emphasising scalable, field-deployable solutions for sustainable water management and climate-responsive technologies. Dr Ahmad's technical strengths lie in the design, optimisation, and scale-up of photobioreactors, anaerobic treatment systems, and microalgae-based wastewater treatment technologies addressing critical challenges such as nutrient pollution, carbon emissions, and water reuse. His work demonstrates a strong balance between experimental research, process engineering, and real-world implementation. He previously served as a Bioprocess Engineer and Scientific Consultant at PhyEcoSyS, a technology-driven startup focused on microalgal carbon sequestration and wastewater bioremediation, where he translated laboratory research into industrially relevant environmental solutions. Dr Ahmad has contributed to several high-impact international initiatives, including the Indo-European Horizon 2020 PAVITR project and Liquid Trees, supporting the development of efficient wastewater treatment, algal cultivation, and bioenergy generation systems. Dr Ahmad also holds a utility innovation for a modified grease trap design optimised for restaurant wastewater, demonstrating his ability to translate research outputs into practical engineering tools with direct industry relevance. His scholarly contributions include publications in high-impact international journals and recognition through awards such as the My Membrane Award from the Malaysia Membrane Society (UTM) and acknowledgements from Japan's Algae Industry Incubation Consortium. In addition, he has edited and contributed to academic books published by leading publishers, including IWA Publishing, Springer, IGI

Global, and CRC Press, thereby strengthening global discourse in environmental engineering and sustainable technologies. With over eight years of academic experience, Dr Ahmad has mentored students, developed laboratory infrastructure, and fostered interdisciplinary collaborations, consistently advancing research that delivers measurable environmental and societal impact.

Faizal Bux, Director of the Institute for Water and Wastewater Technology (IWWT,) DUT has ~28 years of experience in the field water and wastewater treatment technology. His focus research areas include wastewater technology, algal biotechnology and bioenergy production. Prof Bux has also assumed the position of NRF SARChI (South African Research Chair Initiative) Chair from 2018. As a NRF B rated researcher and Academy of Science of South Africa (ASSAf) member, he has supervised 64 Masters and Doctoral students, hosted 23 PDF's and received numerous institutional awards. The Water Research Commission has been awarded Prof Bux the 'Knowledge Tree Award for Transformation and Redress' for his impact in growing PDI involvement in 2017. More recently, Prof Bux received the 2018 National Research Foundation (NRF) award, "Champion of Research Capacity Development and Transformation at South African Higher Education Institutions". He has acted/current editor for CLEAN Soil Air Water (John Wiley & sons, Germany), Environmental Science and Health Part A (Taylor Francis, USA), Biofuels Research Journal and Water Science and Technology and as a reviewer for >50 ISI Journals. He has contributed to 148 journal articles, 18 book chapters and >150 conference presentations and led 11 WRC projects. His citations are in excess of 5000 with an H Index of 33. Prof Bux currently holds the positions of the National coordinator and Chair of International Thematic Group (ITG): Water Resource and Pollution Research of the BRICS Networks University. In 2018 Prof Bux was nominated as Chair of the ITG BRICS-NU for 2 years with the option of a second term. He served 2 terms as a Member of IWA Management Committee specialist group Microbial Ecology and Water Engineering (MEWE) and on the Executive committee of IWA-SA as Treasurer. He was been elected as a member of the Academy of Science of South Africa in 2016. Prof Bux was admitted as a Fellow of the Water Institute of Southern Africa, Royal Society of Chemistry and more recently International Water Association. He has served on the scientific program committee/Advisory board/Chair/Invited speaker of various national and International conferences and holds memberships to many professional bodies. He has an extensive network of university collaborators globally and served as scientific advisor to NGOs both locally & internationally. He has been involved in bilateral projects with India, Brazil, Egypt, China, Algeria with continuing collaboration. He currently has bilateral projects with Japan, Oman, Poland (subject to approval) and global projects involving Oklahoma University (USA) and Aalborg University

(Denmark). He has played a key role in assisting the University to increase its research capacity and contribute to the Human Resource Capital Development of the country.

Nilamjyoti Kalita is an Assistant Professor & Head of the Department of Botany at Murkong Selek College, Assam, India. He earned his Ph.D. from Gauhati University, focusing on heavy metal remediation using cyanobacterial strains from oil-contaminated sites of Upper Assam. A university topper in his M.Sc. in Botany, he has strong expertise in molecular biology, algal taxonomy, and stress physiology. Dr. Kalita has published extensively in reputable Q1 international journals such as Journal of Hazardous Materials Advances, Food Biosciences, Journal of Polymers and the Environment, Heliyon, Plant Physiology and Biochemistry, Microbiological Research, etc., and has contributed to several Springer Nature-edited volumes. His research emphasizes the application of cyanobacteria and microalgae in bioremediation and sustainable technologies. He is a recipient of the Young Scientist Award (2026) from the Botanical Society of Assam, along with multiple academic honors, and serves as a reviewer for leading scientific journals.

***

Mubarak Ali. A, PG and Research Department of Microbiology, Sacred Heart College (Autonomous), Tirupattur, Tamil Nadu, India

Amaramma B Yaraladdi, completed her Master's degree in Life Sciences, Sri Sathya Sai University for Human Excellence, India.

Muhammad Nauman Khan holds a Ph.D. in Botany from Islamia College Peshawar, Pakistan, and is currently serving as an Assistant Professor at the University of Chakwal, Punjab, Pakistan. He is an accomplished researcher with more than 150 peer-reviewed international research articles and 40 book chapters published in reputed journals, achieving a cumulative impact factor of nearly 250. His academic activities include participation in international conferences through abstracts, oral presentations, and poster sessions. Dr. Khan is actively involved in scholarly publishing as a Subject Editor of Ethnobotany Research and Applications and an editorial board member of multiple journals. He has also served on organizing committees of major international conferences. Dr. Khan is a lifetime member of the Pakistan Botanical Society, the Weed Science Society of Pakistan, and the Khwarizmi Science Society.

Sangeetha Menon is an Assistant Professor in the Department of Life Sciences, School of Biological Sciences since June 2019 at Kristu Jayanti University,

Bengaluru, with 17 years of teaching and research experience. She holds an M.Phil, Ph.D. in Microbiology, and a Postgraduate Diploma in Bioinformatics. Her research interests span diverse areas including Medical Microbiology, Food and Dairy Microbiology, Forest Pathology, Biofertilizer technology, Environmental Microbiology, Microbial Genetics, Bioinformatics, Recombinant Technology, and Medicinal Plants. Dr. Menon is actively involved in guiding B.Sc and M.Sc projects and has a strong publication record with more than 45 research papers and book chapters in peer-reviewed journals and over 40 gene sequences submitted to NCBI. She has been awarded with several awards like Youth Icon Award, Research Excellence Awards and Best Women Assistant Professor Award. She has published 6 published patents and 1 granted patents and authored 8 books. She has completed several funded projects sponsored by ICFRE, KSCST and Bodhi Niketan Trust. Her contributions extend to serving on the editorial board of many reputed national and international peer reviewed and indexed journals. Dr. Menon has won best paper presentations award at several national and international conferences. She is actively involved in teaching undergraduate and postgraduate courses, mentoring students, and contributing to the department's research initiatives.

Ashish Kumar Nayak, Assistant Professor, Department of Civil Engineering, Prasad V. Potluri Siddhartha Institute of Technology, Vijayawada, Andhra Pradesh, 520007, India

P. Saranraj holds a Ph.D. in Microbiology from Annamalai University, Chidambaram, India, and is NET (ASRB–ICAR) qualified in Microbiology. He is currently serving as Assistant Professor and Head, Department of Microbiology, at Sacred Heart College (Autonomous), Tirupattur, Tamil Nadu, India. In addition, he is the Proprietor and Managing Director of JPS Scientific Publications, contributing actively to academic publishing and research dissemination. Dr. Saranraj has over 15 years of research experience and 10 years of teaching experience in the fields of Environmental Microbiology, Agricultural Microbiology, Industrial Microbiology, Bioremediation, Microbial Biotechnology, and Sustainable Ecosystem Management. His research work has focused on Microbial degradation of pollutants, Effluent treatment, Soil health improvement, Plant–microbe interactions, and Applied environmental sustainability. He has an extensive publication record, with more than 240 international research articles, over 90 books (international and national), and 70+ book chapters published with reputed academic publishers.

Ramya Raghavan, Assistant Professor, Department of Life Sciences, Sri Sathya Sai University for Human Excellence, India. Her expertise extends to the critical intersection of the exposome and Social Determinants of Health (SDOH).

Dr. Raghavan's work underscores how a holistic understanding of environmental exposures, from the molecular level to population-wide influences, is crucial for addressing health disparities and advancing a more equitable approach to disease prevention.

Index

A

B

C

D

E

F

G

H

I

L

M

N

P

R

S

T

U

W

www.ingramcontent.com/pod-product-compliance
Lightning Source LLC
LaVergne TN
LVHW081319300926
849161LV00008B/1812

* 9 7 9 8 3 3 7 3 7 6 3 0 1 *